BIOLOGY

Richard Fosbery
Adrian Schmit
Jenny Wakefield-Warren

Orders: please contact Bookpoint Ltd, 130 Milton Park, Abingdon, Oxon OX14 4SB. Telephone: +44 (0)1235 827720. Fax: +44 (0)1235 400454. Lines are open 9.00a.m.–5.00p.m., Monday to Saturday, with a 24-hour message answering service. Visit our website at www.hoddereducation.co.uk

First published in 2015 by

Hodder Education,

An Hachette UK Company

338 Euston Road

London NW1 3BH

Impression number 10 9 8 7 6 5 4 3 2 1

Year 2019 2018 2017 2016 2015

Cover photo © maw89 – Fotolia

Illustrations by Barking Dog Art

Typeset in 10.5/12 pt Bliss Light by Integra Software Services Pvt., Pondicherry, India

Printed in Italy

A catalogue record for this title is available from the British Library

ISBN 9781471809156

Contents

Get the most from this book

Welcome to the **OCR A Level Biology 1 Student's Book**. This book covers Year 1 of the OCR A Level Biology specification and all content for the OCR AS Level Biology specification.

The following features have been included to help you get the most from this book.

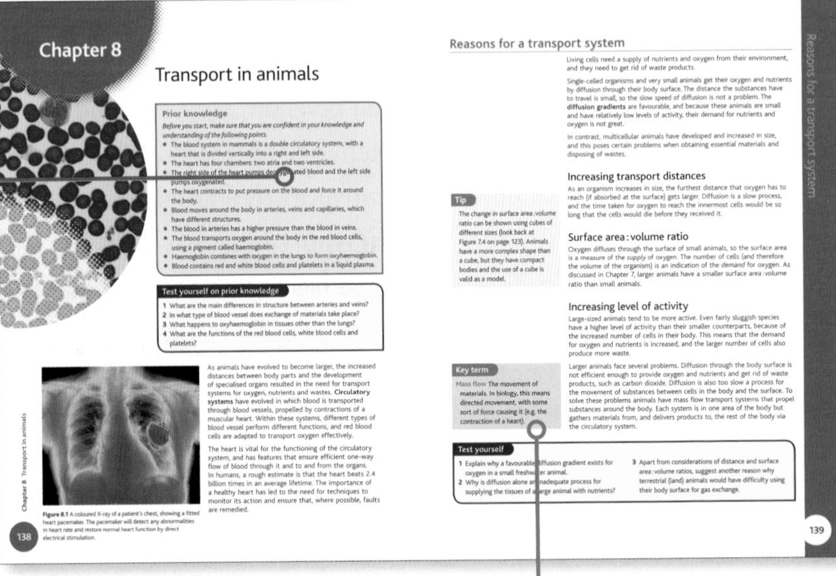

Prior knowledge

This is a short list of topics that you should be familiar with before starting a chapter. The questions will help to test your understanding.

Key terms and formulae

These are highlighted in the text and definitions are given in the margin to help you pick out and learn these important concepts.

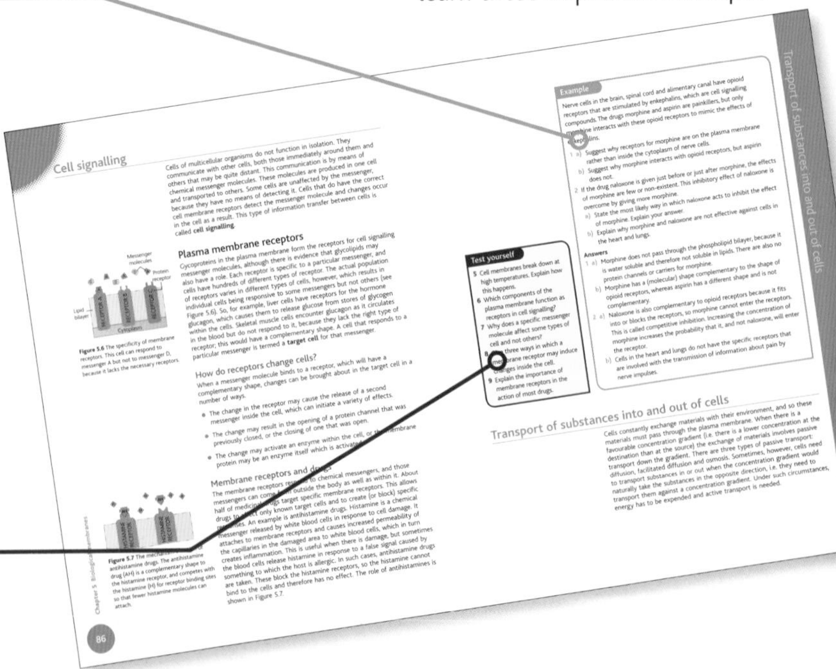

Examples

Examples of questions or calculations are included to illustrate chapters and feature full workings and answers.

Test yourself questions

These short questions, found throughout each chapter, are useful for checking your understanding as you progress.

Activities

These practical-based activities will help consolidate your learning and test your practical skills.

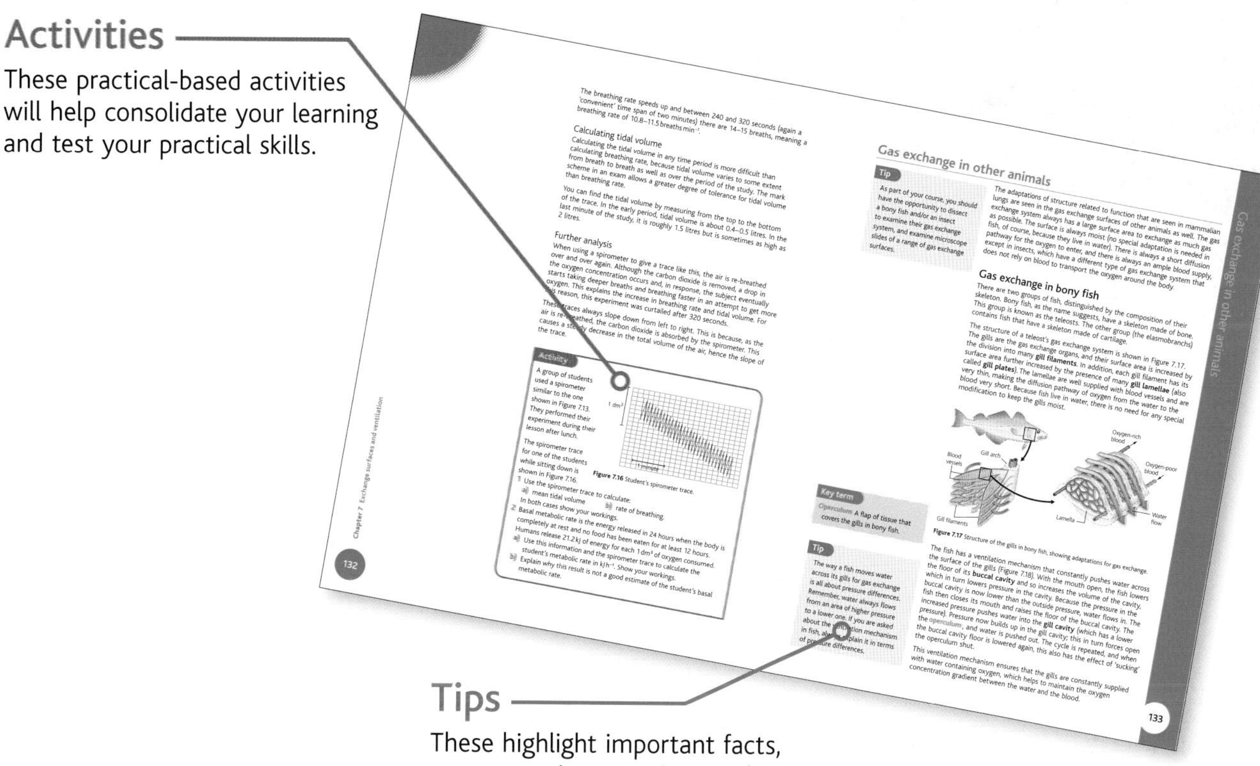

Tips

These highlight important facts, common misconceptions and signpost you towards other relevant chapters.

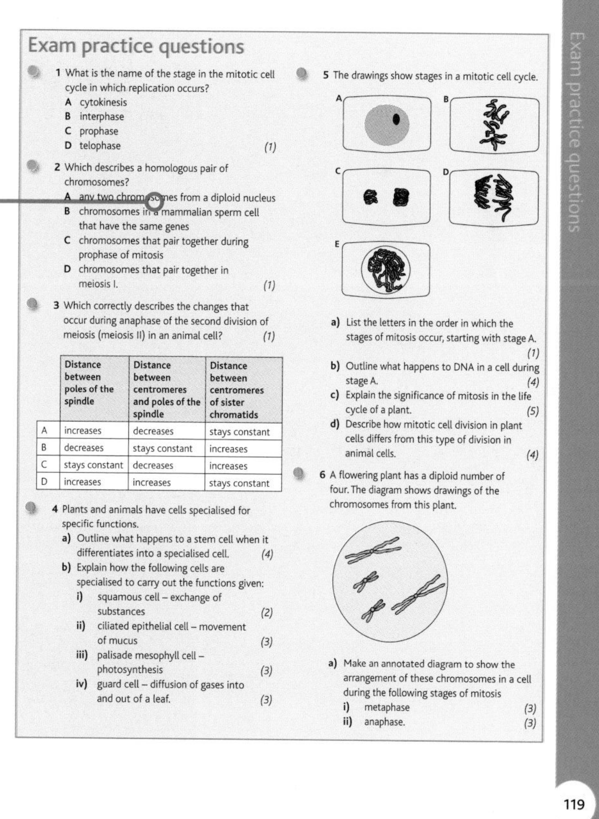

Exam practice questions

You will find Exam practice questions at the end of every chapter. These follow the style of the different types of questions you might see in your examination, including multiple-choice questions, and are colour coded to highlight the level of difficulty. Test your understanding even further, with Maths questions and Stretch and challenge questions.

- Green – Basic questions that everyone should be able to answer without difficulty.
- Orange – Questions that are a regular feature of exams and that all competent candidates should be able to handle.
- Purple – More demanding questions which the best candidates should be able to answer.
- Stretch and challenge – Questions for the most-able candidates to test their full understanding and sometimes their ability to use ideas in a novel situation.

Dedicated chapters for developing your **Maths** and **Practical skills** and **Preparing for your exam** can be found at the back of this book.

Acknowledgements

Special thanks to Dr Clare van der Willigen for her help during the writing of this book.

Chapter 1

Cell structure

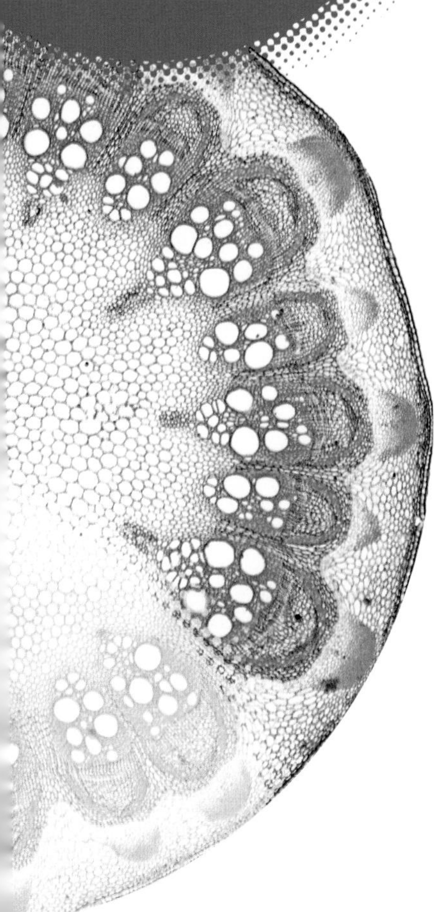

Prior knowledge

Before you start, make sure that you are confident in your knowledge and understanding of the following points.

- All living things are made up of cells.
- Animal and plant cells contain a nucleus, cytoplasm, a cell membrane, mitochondria and ribosomes.
- Plant cells (but not animal cells) also have a cell wall and a large central vacuole, and may contain chloroplasts.
- Cells (e.g. red blood cells) may develop a specialised structure that suits their function.
- Bacterial cells lack a nucleus, mitochondria and chloroplasts.
- The nucleus controls the cell's activities, because it contains genes.
- Mitochondria carry out aerobic respiration.
- Chloroplasts absorb light and use the energy for photosynthesis.
- Ribosomes make proteins.

Test yourself on prior knowledge

1 Chloroplasts are not present in all plant cells. Why not?
2 A sperm cell has a nucleus, a 'tail' and very little cytoplasm. Explain the value of these features in relation to the cell's function.
3 If a cell has many mitochondria, what does this suggest about the cell's function?
4 Suggest a reason why ribosomes were discovered much later than mitochondria.
5 How can bacterial cells function without a nucleus?

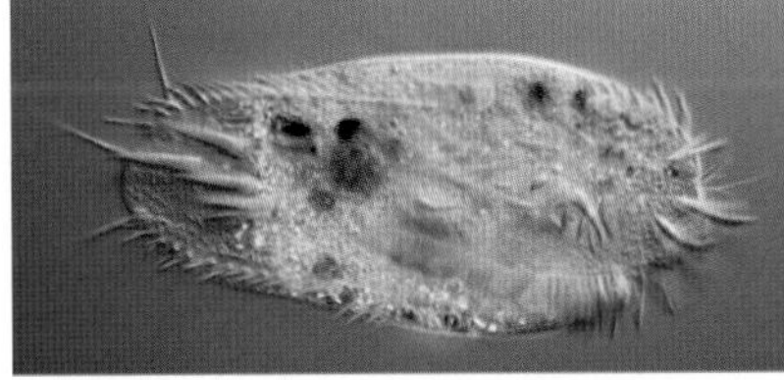

Figure 1.1 This *Stylonchia* species is a ciliated, unicellular, eukaryotic organism. *Stylonchia* species are carnivorous and feed on other single-celled organisms and bacteria. They have cilia, which beat the water and sweep prey into their oral cavities. There are 48 different mating types, which means that each individual can mate with 97% of the population. This photomicrograph was taken using a light microscope. The magnification is about ×330.

Key term

Organelle A specialised part of a cell that performs a particular function.

The cell is the basic unit of all living things. The features of cells found in single-celled organisms (the Protoctista), which live independently, and complex organisms like us are remarkably similar. All cells, for example, have mitochondria. An example of a single-celled organism is shown in Figure 1.1.

Cells are of two types: prokaryotic or eukaryotic. The only organisms with prokaryotic cells are bacteria and blue-green algae, but remember that bacteria outnumber all other forms of life – their numbers have been estimated as 5×10^{30} – so prokaryotic cells are the most common form of cell. Prokaryotic cells lack membrane-bound **organelles** (nucleus, mitochondria, etc.) and have features that are not seen in eukaryotic cells. Animals, plants and fungi are all eukaryotes, but each group has cells with distinctive features, even though they all have nuclei and have a number of common organelles.

Cells have within them a number of different organelles, each of which has a particular function related to the life of the cell. In multicellular organisms, cells may become specialised for particular functions, and their structure is modified as a result. Such cells are often grouped together in some way to form a tissue. This tissue may be part of an organ, which in animals may be a component of an organ system.

The study of cell structure has depended on the use of microscopes. Initially, microscopes used systems of lenses to magnify images formed by light, but in the twentieth century more powerful microscopes were developed that use electrons in place of light. Both light and electron microscopes continue to be used for slightly different purposes.

Studying cells by microscopy

Cells were first described and named by the Englishman Robert Hooke in the 1660s. The quality of lenses was improved over time, but the properties of light limit the magnifications possible. The invention of the electron microscope in 1931 allowed much more detailed study of the internal structure of cells, because of the increased magnifications that were then possible.

The two forms of microscope are shown in Figure 1.2.

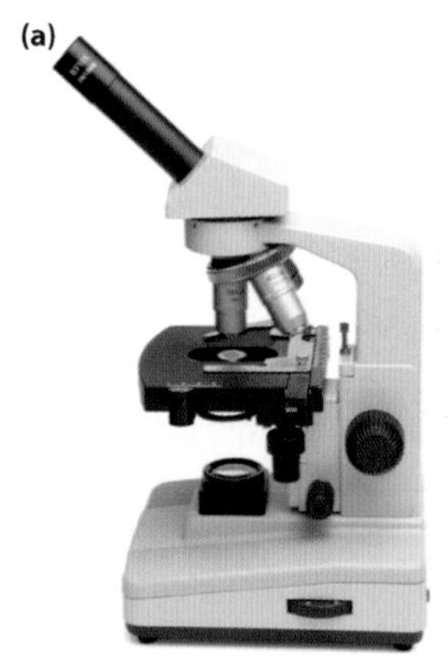

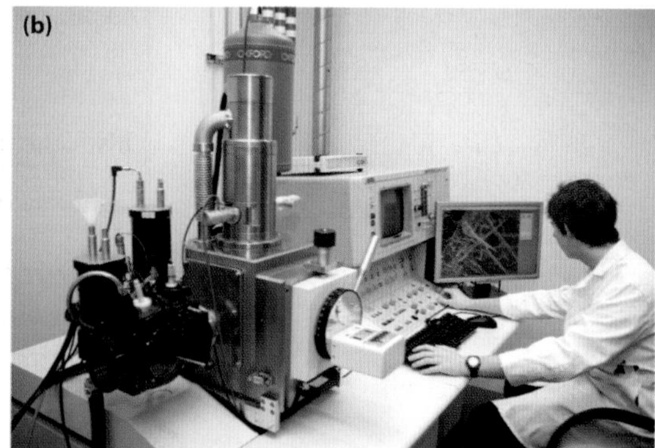

Figure 1.2 Modern (a) light microscope and (b) electron microscope.

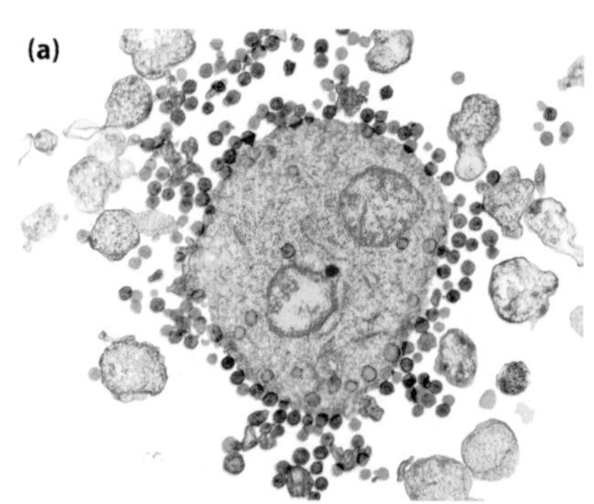

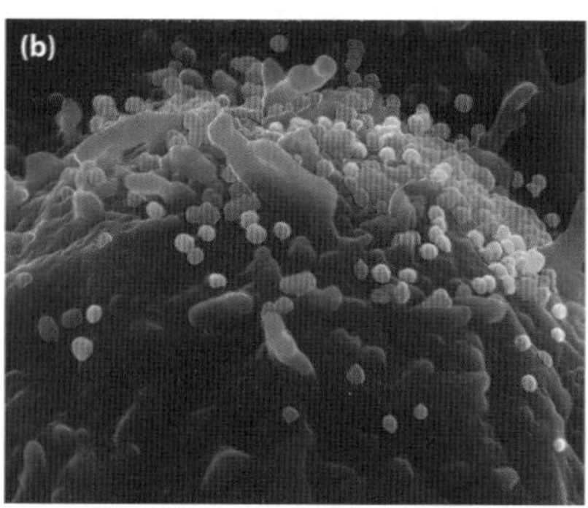

Figure 1.3 Budding HIV particles shown in (a) transmission electron micrograph (×11 000) and (b) scanning electron micrograph (×12 000). These images will have been black and shades of grey when produced, and colours have been added to make the different structures clearer.

Light microscopes are the type of microscope routinely used to examine specimens in science laboratories. With a light microscope, it is easy to see larger cellular structures (for example, cell walls, nuclei, chloroplasts and vacuoles) and at higher magnifications it is possible to see mitochondria. The internal structure of organelles cannot be seen, though. For this, and to see smaller organelles such as endoplasmic reticulum and ribosomes, the higher magnification and resolution of a transmission electron microscope is required.

Electron microscopes are of two types: **transmission electron microscopes**, in which electrons pass through the material, and **scanning electron microscopes**, in which the electrons are bounced off the surface of the material, so giving a three-dimensional view of the surface (see Figure 1.3b).

A relatively recent development in microscopy is the use of the **laser scanning confocal microscope**. In this technique, cells are stained with fluorescent dyes. A thick section of tissue (or even a small living organism) is scanned with a laser beam, which is reflected by the dyes. The laser beam is scanned across different depths of the section (see Figure 1.4). This has two advantages:

- The laser beam can be focused at a very specific depth, and this eliminates the blur seen in optical microscopes that is caused by out-of-focus tissue above the focal point. As a result, images are much clearer.

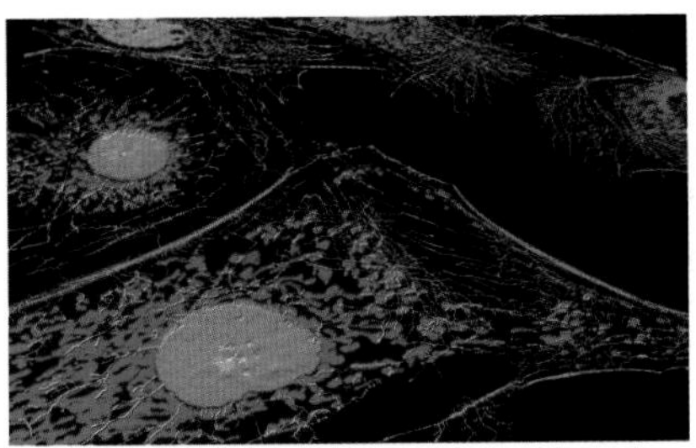

Figure 1.4 High-magnification laser scanning confocal microscopy image of lung cells (×1000). The technique allows the clear imaging of the cell cytoskeleton (green) and the cell organelles (red).

Tip

Although they both include the term 'scanning', the scanning electron microscope and the laser scanning confocal microscope use completely different technology. As the names suggest, one scans with electrons and the other with a laser beam. The laser scanning confocal microscope is not an electron microscope.

- Images can be taken at successive depths and fed into a computer, which can reconstruct a three-dimensional image of the tissue scanned.

Magnification and resolution

When you use microscopes, two properties of the instrument are key to understanding its performance: magnification and resolution.

Magnification is a measure of the ability of a lens or other optical instrument to magnify (enlarge) the size of something in an optical image.

Resolution refers to the ability of a microscope to distinguish two adjacent points as separate from each other. Instruments with a high resolution can distinguish two points as separate even when they are very close, and this increases the clarity of the image. Resolution is important in microscopy: magnifying a specimen a great deal is no use if the image of that specimen is not clear.

The magnification and resolution of different types of microscope are shown in Table 1.1 below. All the figures are approximate.

Table 1.1 Magnification and resolution of different types of microscope.

Instrument	Maximum magnification	Maximum resolution
Light microscope	×1 500	200 nm
Transmission electron microscope	×500 000	0.2 nm
Scanning electron microscope	×100 000	10 nm

The resolution shows how close (in nanometres) two points can be and still be distinguishable. The light microscope can distinguish between two objects separated by a distance of less than approximately half the wavelength of visible light (400–700 nm) used to illuminate the specimen – that is, 200 nm. The wavelength of light limits the resolution of a light microscope; the maximum resolution of 200 nm is rarely achieved.

Electrons have a much shorter wavelength than light, so a much higher resolution is possible with an electron microscope than with a light microscope.

Tip

A nanometre (nm) is 1×10^{-9} m or 0.000 000 001 m. A micrometre (μm) is 1×10^{-6} m or 0.000 001 m. See Chapter 18 for more about expressing very small numbers.

Although the magnification and resolution of a scanning electron microscope is much lower than that of a transmission electron microscope, this is unimportant because the two instruments have different uses. The scanning electron microscope is used to study external features, which are larger than the internal structure studied by transmission electron microscopy. In addition, transmission electron microscopes cannot study surface details, because they detect electrons that have passed through tissue.

Calculating the magnification of an image

When using a light microscope, it is easy to calculate the magnification of what you see: you simply multiply the magnification of the eyepiece lens (the one you look through) by the magnification of the objective lens (the one just above the slide).

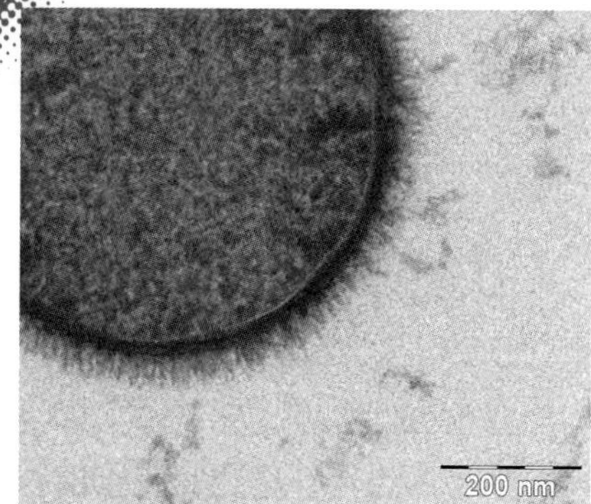

Figure 1.5 Photo of a *Bascillus subtilis*, showing a scale bar.

Tip

Every individual microscope is slightly different, so make a note of which microscope you used. You should also keep a note of the size of each eyepiece division for each objective lens, so you do not have to calibrate the lens each time.

$$\text{magnification of viewed image} = \text{magnification of eyepiece lens} \times \text{magnification of objective lens}$$

So, if the eyepiece is ×10 and the objective lens is ×20, then the overall magnification that you see is ×200.

What you have seen through the microscope, if it is significant, must be reported to others, either in a drawing or photo. It is important to indicate the magnification, not of what you were looking at, but of the image (the drawing or photo). Without this information, viewers have no idea of the size of the features in the image. The magnification of the image is usually either written down (e.g. ×400) or indicated by drawing a scale bar with a unit of measurement on it (see Figure 1.5).

Either method of showing magnification requires you to know the actual measurement of the structures seen. In a light microscope, you take measurements using an eyepiece graticule, calibrated with a micrometer slide. The micrometer slide has a scale engraved into it, calibrated in known divisions (e.g. 0.1 mm). The eyepiece graticule is a circular disc fitted into the eyepiece, which has a scale marked on it with equal divisions. What these divisions represent in terms of length varies as the objective lens is changed. The microscope's eyepiece graticule is calibrated for each objective lens, using the micrometer slide. The process is shown in Figure 1.6. You then know what each eyepiece division measures, in real terms, when using each objective lens, and so can use the graticule to measure any structure seen.

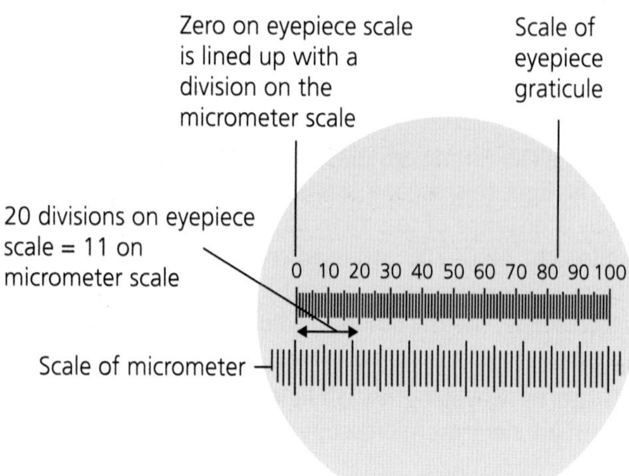

Figure 1.6 Calibrating using an eyepiece graticule and a micrometer slide.

Example

Calculate the real length of the eyepiece graticule divisions using Figure 1.6.

Answer

1 micrometer slide division = 0.1 mm

20 eyepiece divisions = 11 micrometer divisions
= 11 × 0.1 mm
= 1.1 mm

1 eyepiece division = $\frac{1.1}{20}$ mm
= 0.055 mm
= 55 µm (as 1 µm = 0.001 mm)

The eyepiece graticule can now be used to measure anything seen under the microscope. If you use the same objective lens, you also know that each division is equal to 55 µm.

Example

1 A photo of a tissue shows a structure that measures 20 mm (on the photo). The structure's real measurement is 200 µm. What is the magnification of the photo?

2 A microscope drawing shows a structure as being 54 mm long. The magnification is given as ×200. What is the real length of the structure?

Answers

1 magnification = $\frac{\text{length in picture}}{\text{real length}}$

Convert both measurements into a standard unit (mm in this case)

magnification = $\frac{20}{0.2}$ = 100

The magnification of the photo is ×100.

2 real length = $\frac{\text{length in the drawing}}{\text{magnification}} = \frac{54}{200}$
= 0.27 mm or 270 µm

Preparing sections for viewing under a microscope

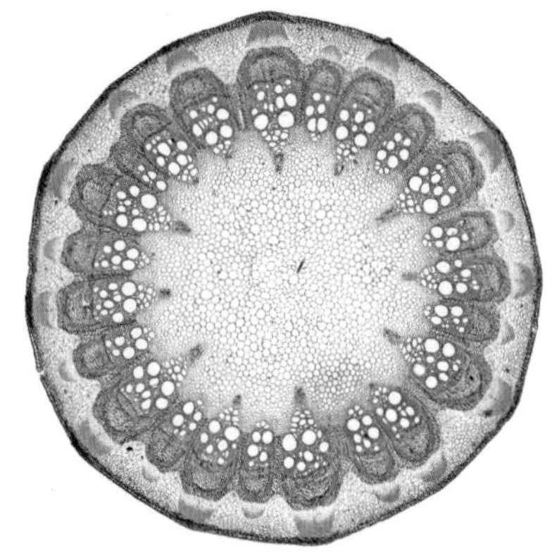

Figure 1.7 Cross section of a stem, stained with toluidine blue and phloroglucinol (red) to differentiate tissues.

A lot of living tissue is transparent (i.e. it lets light through) and does not block electrons. Such tissue does not show up well under light or electron microscopes, and so the tissue has to be stained so that you can see it.

To prepare biological material to view with the light microscope very thin sections of fresh specimens are cut with a sharp blade. Alternatively permanent sections can be made. The specimens are 'fixed' using a preservative such as formaldehyde, dehydrated using a series of ethanol solutions and then impregnated with paraffin wax or other resins to support the tissue. It is then sliced thinly with an instrument called a microtome. The paraffin is removed from the sections; they are stained and mounted using a resin and a cover slip. Specimens can also be frozen in liquid nitrogen or carbon dioxide and cut into very thin sections using an instrument called a cryostat.

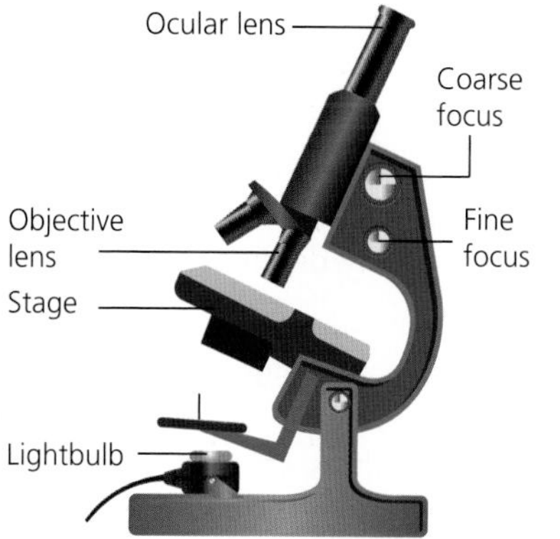

Figure 1.8 A light microscope.

Staining for light microscopy

Coloured dyes are used to stain tissues. These dyes absorb some colours of light and reflect others, making the structures that absorb the dye visible. Different tissues absorb different dyes, according to their chemical nature. Sections can therefore be stained with multiple dyes to show up different tissues within them (see Figure 1.7). This is called differential staining.

When you see a coloured photo taken under a light microscope, it is important to realise than many, if not all, of the colours seen are not natural. Some colours may be natural; chloroplasts, for instance, are naturally green and so do not need to be stained.

Staining for transmission electron microscopy

Transmission electron microscopes, which use electrons, need to use stains that absorb those particles. Electrons have no colour, and so the dyes make the parts look black or shades of grey. Heavy-metal compounds that absorb electrons are commonly used, for example phosphotungstic acid, osmium tetroxide and ruthenium tetroxide. If you see electron micrographs that are coloured (as are many of the electron micrographs in this book), this colouring is not part of the staining, but has been added afterwards using image-processing software.

Test yourself

1. Define the term *resolution*, as applied to microscopy.
2. Why are electron microscopes capable of a higher resolution than light microscopes?
3. Transmission and scanning electron microscopes are used for different types of task. Describe their different uses.
4. When calibrating an eyepiece graticule, why does each objective lens have to be calibrated separately?
5. The length of a feature on an electron micrograph is 30 mm. The magnification is given as ×200 000. What is the true length of the feature in micrometres (μm)?

Cellular organelles

All cells have within them specialised structures called organelles, which perform particular functions. The presence and numbers of these organelles can give a strong clue about the function of the cell that they are in.

Tip

Do not confuse the cytosol with the cytoplasm. The cytoplasm is the space between the nucleus and the plasma membrane. The cytosol is the fluid which surrounds the organelles.

Tip

Chapter 5 covers the topic of cell membranes.

In eukaryotic cells, many of the organelles are surrounded by one or more membranes. Prokaryotic cells have fewer organelles, and none of those are surrounded by a membrane.

This section deals with the structure and functions of organelles in eukaryotic cells, all of which are contained within a gel-like fluid known as the cytosol.

The nucleus

All eukaryotic cells contain a nucleus at some time in their lives, even though occasionally (e.g. in mammalian red blood cells) it is lost during the course of the cell's development. The nucleus contains the cell's DNA, which provides the template for making RNA. As a result, the nucleus indirectly controls all the activities of the cell. The nucleus is bound by a nuclear envelope (also known as the nuclear membrane), which consists of a double membrane; each membrane in the double membrane is similar in structure to the cell surface or plasma membrane (see Chapter 5). Nuclear pores are protein complexes with a channel through the centre.

The DNA in the nucleus exists in one of two forms: during cell division the threads of DNA coil up tightly and condense to form chromosomes; at all other times, the DNA is spread loosely through the nucleus, and it is then referred to as chromatin. Under the electron microscope, chromatin appears as a grainy material with no clearly defined structure (see Figure 1.9a).

A darkly staining region can often be seen within the nucleus in electron micrographs. This is the nucleolus, which is a special region of the nucleus responsible for the construction of ribosomes; ribosomes are discussed below. The ribosomes are small enough to leave through nuclear pores and reach the cytoplasm.

The appearance and structure of the nucleus is shown in Figure 1.9.

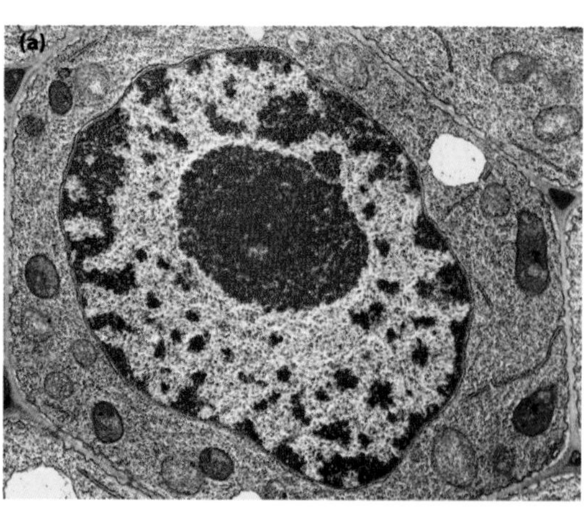

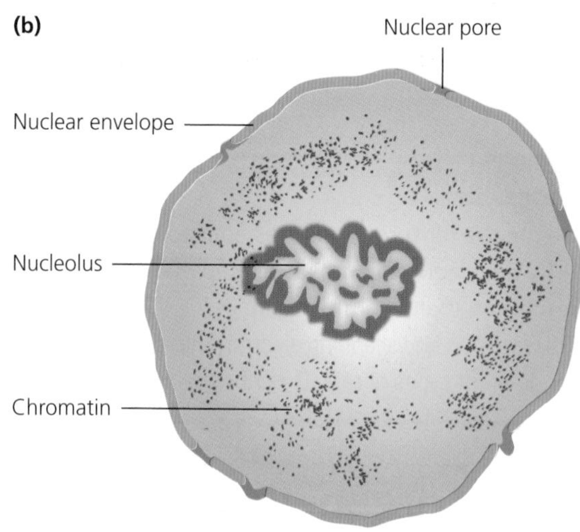

Figure 1.9 A nucleus in: (a) an electron micrograph (×25 000) and (b) a diagram, showing the structure of the nucleus.

The endoplasmic reticulum

The endoplasmic reticulum is the cell's transport system. The distances within a cell are small, but cells often need to transport materials to the outside. Chemicals often need to be 'processed' in parts of the cell and then exported to the outside. The transport, processing and export is all done by a system of folded internal membranes called the endomembrane system, consisting of the endoplasmic reticulum and the Golgi apparatus. The endoplasmic reticulum comes in two types: **rough endoplasmic reticulum** and **smooth endoplasmic reticulum**. These membranes form channels through the cytoplasm. The membranes join with the outer nuclear membrane to form a continuous membrane.

Rough endoplasmic reticulum

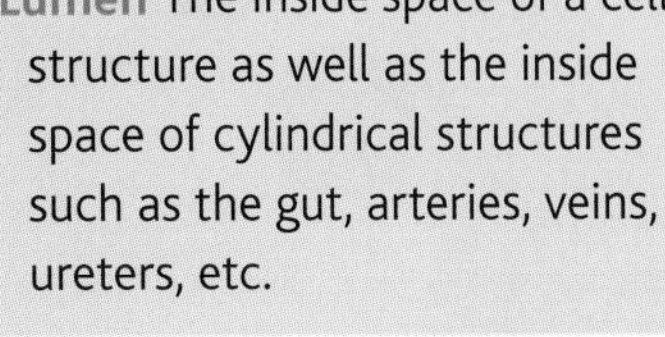

Key term

Lumen The inside space of a cell structure as well as the inside space of cylindrical structures such as the gut, arteries, veins, ureters, etc.

Rough endoplasmic reticulum (RER) is so called because ribosomes are embedded in the membrane. The ribosomes in the RER make proteins that are destined for transport and secretion (ribosomes that are free in the cytoplasm make proteins for the cell's own use). The new proteins enter the lumen of the RER and combine with carbohydrates. They are then transported to the Golgi apparatus (see below) for further processing.

Cells that produce a lot of proteins (e.g. enzyme and hormone-producing cells) have extensive RER. The appearance of the RER is shown in Figure 1.10.

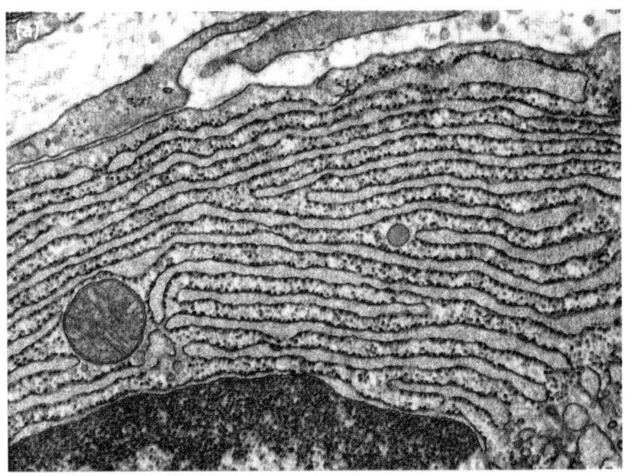

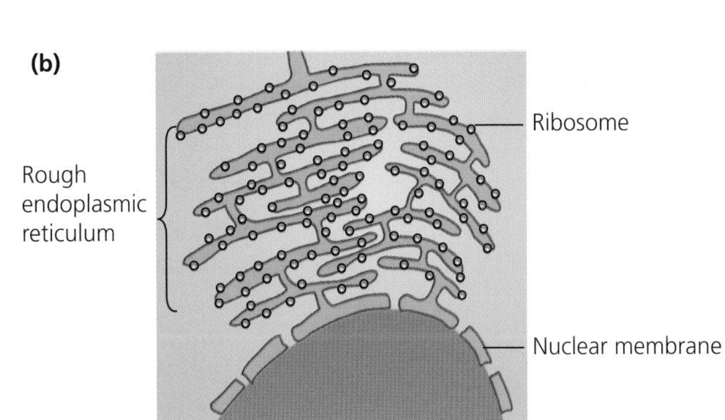

Figure 1.10 Rough endoplasmic reticulum in: (a) an electron micrograph (×18000) and (b) a diagram.

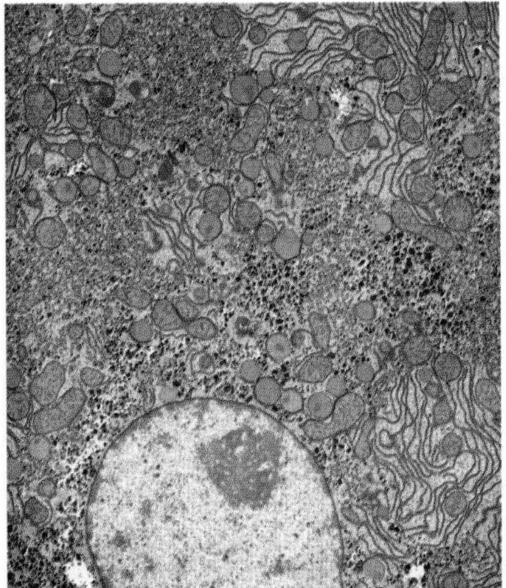

Figure 1.11 Electron micrograph of a cell with some smooth endoplasmic reticulum (×3300).

Smooth endoplasmic reticulum

The smooth endoplasmic reticulum (SER) is continuous with the RER, but is distinguished from it by a lack of ribosomes. The SER makes lipids, including the phospholipids for cell membranes; SER is therefore seen particularly in cells that produce a lot of lipids (e.g. liver cells). The SER also contains enzymes that are known to detoxify lipid-soluble drugs and some harmful products of metabolism.

The appearance of the SER is shown in Figure 1.11.

Golgi apparatus

The Golgi apparatus (also called the Golgi body or Golgi complex) is a series of membranous flattened sacs known as cisternae. The structure resembles smooth endoplasmic reticulum, but the cisternae are not interconnected. The Golgi apparatus is divided into two sections: the cis Golgi network and the trans Golgi network. The Golgi apparatus is shown in Figures 1.12 and 1.13.

> **Key term**
>
> **Secretion** The production and release of a substance from a gland or cell. The same word is also used for the substance secreted.

The function of the Golgi apparatus is to modify the proteins and lipids delivered to it from the RER and prepare them for **secretion**, then deliver them to the cell membrane, wrapped in a membrane (forming a vesicle). Vesicles pinched off the RER join the Golgi apparatus at the cis face, which is innermost. They are modified as necessary and passed to the trans network, where vesicles are formed. These vesicles move to the cell membrane and fuse with it, releasing the contents to the outside.

The Golgi apparatus is also responsible for the manufacture of lysosomes (which are described below) and the synthesis of cell wall materials in plants.

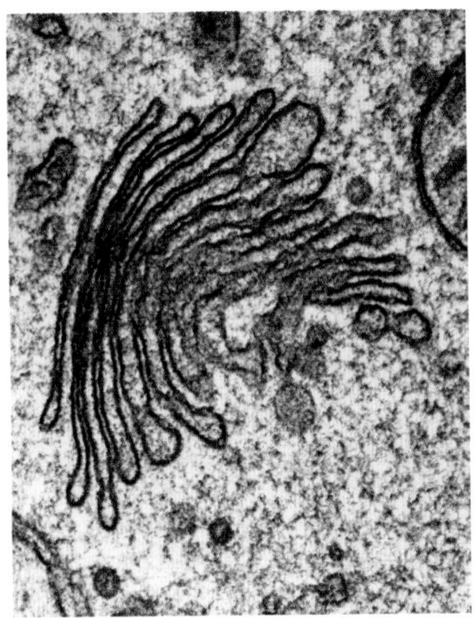

Figure 1.12 Electron micrograph of Golgi apparatus (×50 000).

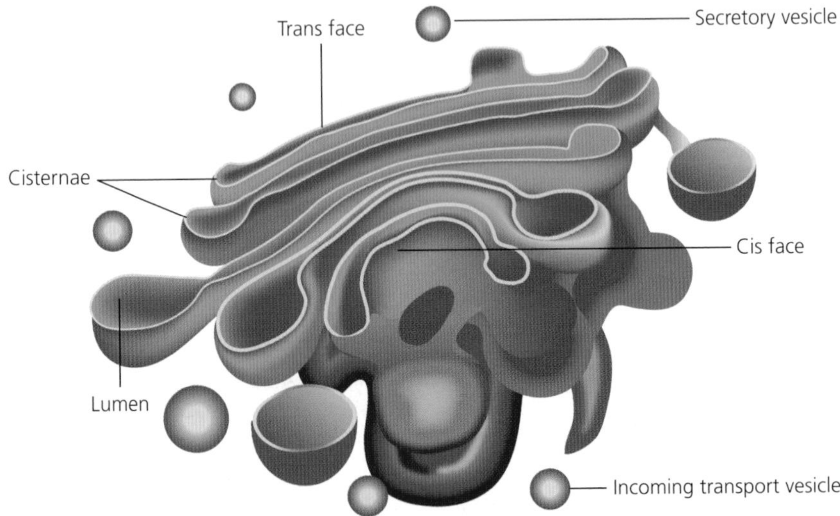

Figure 1.13 The Golgi apparatus, showing the structures involved in forming secretory vesicles containing proteins and lipids.

Ribosomes

We have already encountered ribosomes, which are made in the nucleolus and are found both free in the cytoplasm and attached to the membranes that form the RER. Ribosomes are usually present in large numbers. The function of the ribosomes is to manufacture proteins. Ribosomes consist of two types of molecule: ribosomal ribonucleic acid (rRNA) and protein. There are two components to a ribosome: a large subunit and a small subunit. These subunits are initially separate from each other but come together to form a complete ribosome by attaching to messenger RNA (mRNA) when protein synthesis is about to begin.

> **Tip**
>
> You will learn more about RNA and protein synthesis in Chapter 3.

The appearance and structure of ribosomes is shown in Figure 1.14.

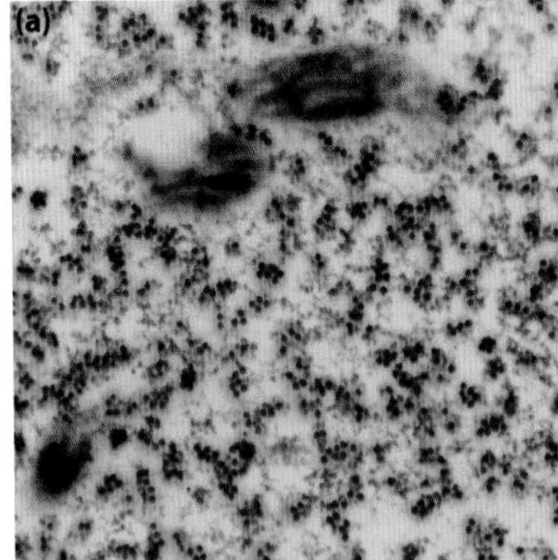

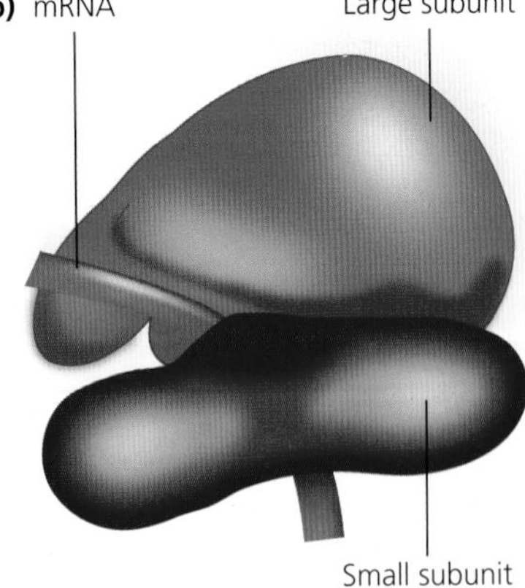

Figure 1.14 (a) Electron micrograph showing free ribosomes in the cytoplasm (×32000); (b) individual ribosome (large subunit and small subunit) attached to mRNA.

Tip

The S refers to the sedimentation coefficient in Svedberg units. The higher the number of Svedberg units, the faster the sedimentation and the larger the organelle.

Ribosomes are found in the cytosol of eukaryotic and prokaryotic cells, but the ones in eukaryotes are larger, and are referred to as 80S ribosomes (as opposed to the 70S ribosomes in prokaryotes).

Mitochondria

The mitochondria are membrane-bound organelles that carry out aerobic respiration. They are therefore most numerous in cells that are very active (e.g. muscle cells and nerve cells).

Tip

When you answer questions about the function of mitochondria, it is important to specify aerobic respiration, not just respiration. If no oxygen is available, anaerobic respiration does not occur in the mitochondria; it occurs only in the cytosol.

Mitochondria are organelles with a double membrane. The inner membrane is folded into a series of **cristae** (singular: crista). The appearance and structure of mitochondria is shown in Figure 1.15.

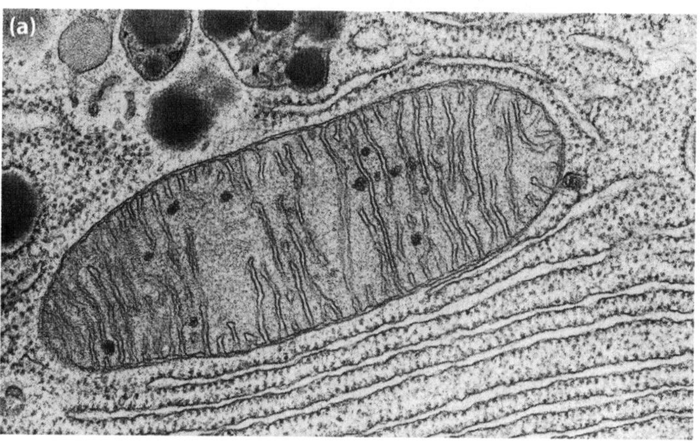

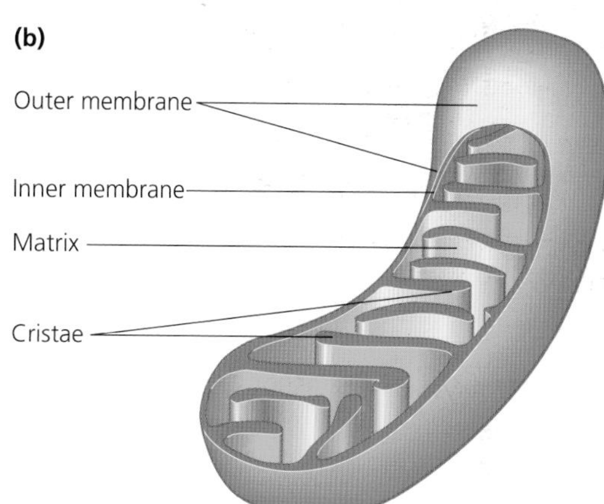

Figure 1.15 A mitochondrion: (a) electron micrograph (×1100) and (b) diagram.

Test yourself

6 State two functions of smooth endoplasmic reticulum.
7 Which organelles would you expect to be prominent in the following cells?
   a) Muscle cells
   b) Salivary gland cells
8 Why does the presence of ribosomes in mitochondria suggest that they may once have been independent prokaryotic organisms?
9 What is the name given to the folds in the inner membrane of a mitochondrion?

Mitochondria contain their own DNA and also possess 70S ribosomes. Mitochondria therefore resemble bacteria, which provides evidence for the idea that mitochondria may have evolved from prokaryotic organisms.

Chloroplasts

Chloroplasts are membrane-bound organelles found in some plant cells but never in animal cells. Their function is to absorb light energy for photosynthesis.

A chloroplast has a double outer membrane and a complex arrangement of internal membranes that form thylakoids, which are grouped into stacks called grana (singular: granum), joined by intergranal lamellae. Chlorophyll pigments, which absorb light, are found in the thylakoids. The space between the grana is filled with a thick fluid called the stroma. Some stages of photosynthesis take place in the grana, others in the stroma.

The internal structure of a chloroplast and its appearance under the electron microscope are shown in Figure 1.16. Like mitochondria, chloroplasts also contain their own DNA and 70S ribosomes.

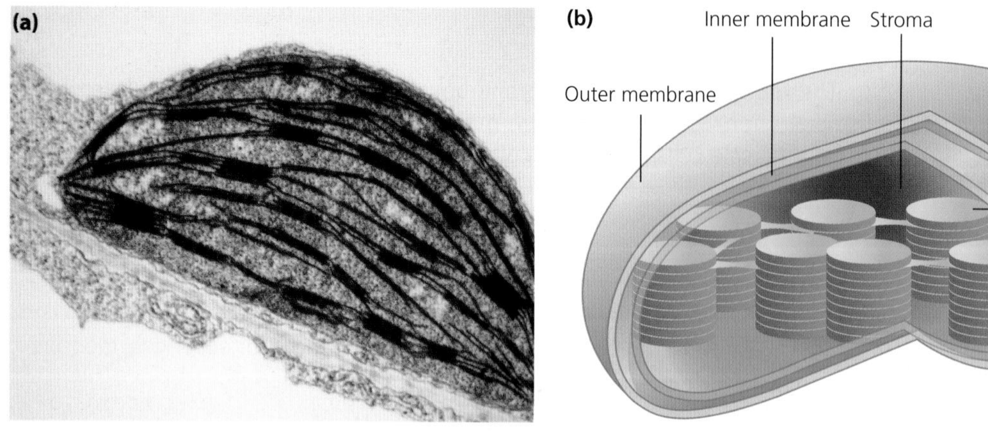

Figure 1.16 A chloroplast shown as: (a) an electron micrograph (×13750) and (b) a diagram.

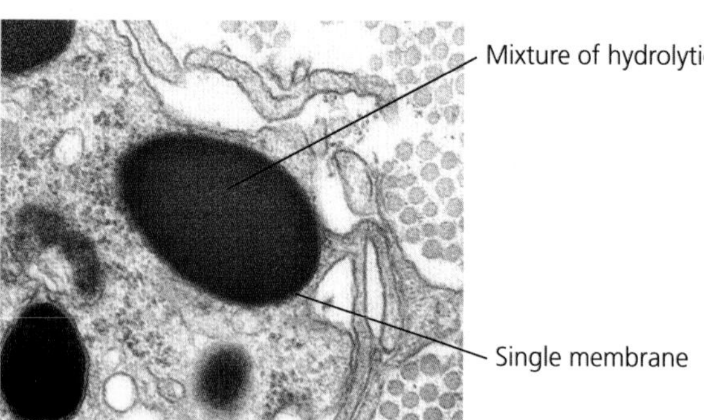

Figure 1.17 A lysosome (×13500).

Lysosomes

Lysosomes are concerned with the digestion of cellular material. To that end, they contain a variety of enzymes (about 40–50 in total) capable of breaking down proteins, nucleic acids, carbohydrates and lipids. The material that lysosomes take in and digest may come from outside the cell (e.g. bacteria) or from inside (e.g. damaged or old organelles). A single membrane separates the enzymes in lysosomes from the rest of the cytoplasm; however, even if the enzymes escaped into the cytoplasm they would do relatively little damage, because they work best in the acidic environment inside the lysosome; the cytoplasm is more or less neutral, and free lysosomal enzymes would be relatively ineffective.

Under the electron microscope (see Figure 1.17), lysosomes are darkly stained with no clearly distinguishable internal features.

Plasma membrane

The cell is bounded by a plasma membrane. The structure and functions of the plasma membrane are dealt with in detail in Chapter 5. The membrane acts like a barrier or gateway to the cell, controlling what goes in and out. This control results from the plasma membrane's differential permeability to different biological molecules, and also from the ability of protein carriers to pump substances into or out of the cell according to need.

The various other membranes in eukaryotic cells (around organelles, making up the endoplasmic reticulum, and surrounding the plant cell's central vacuole) are similar to the plasma membrane in that they are made of phospholipids and protein, although the specific proteins vary in the different membranes.

Centrioles and microtubules

The assembly of the spindle, a system of microtubules that separates the chromosomes during cell division, was thought to have been organised by the centrioles, but it is now known that the centrosomes do this. Only found in animal cells, centrioles are themselves constructed of microtubules arranged in nine triplets forming a cylinder. Centrioles occur in pairs near the nucleus (see Figure 1.18).

Key term

Microtubule Small tubular structure in the cytoplasm composed of tubulin, a globular protein that is arranged into spirals.

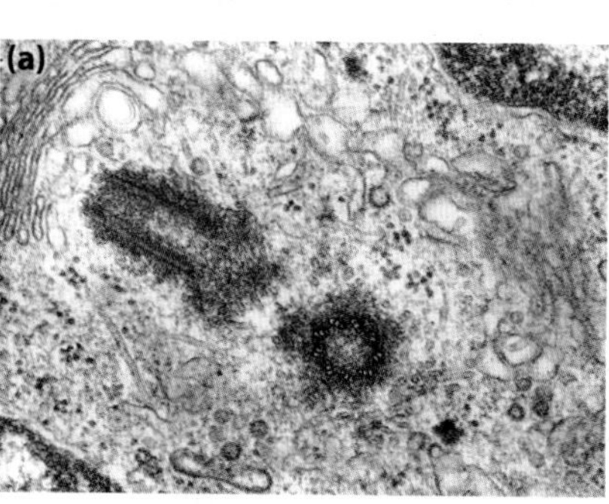

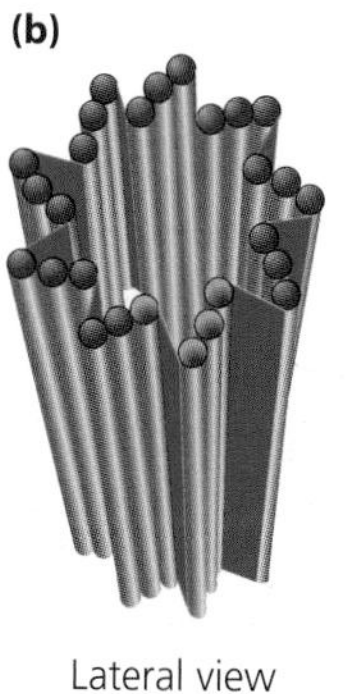

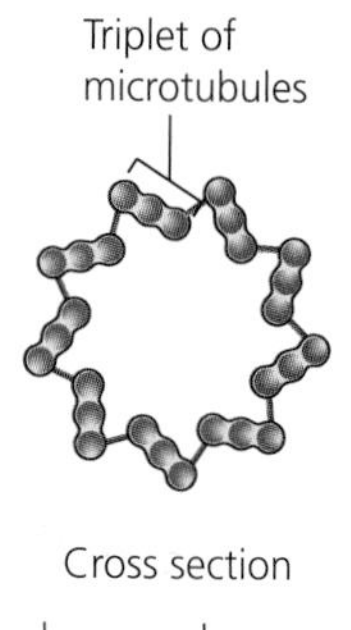

Figure 1.18 Centrioles: (a) electron micrograph (×30 000) and (b) diagram.

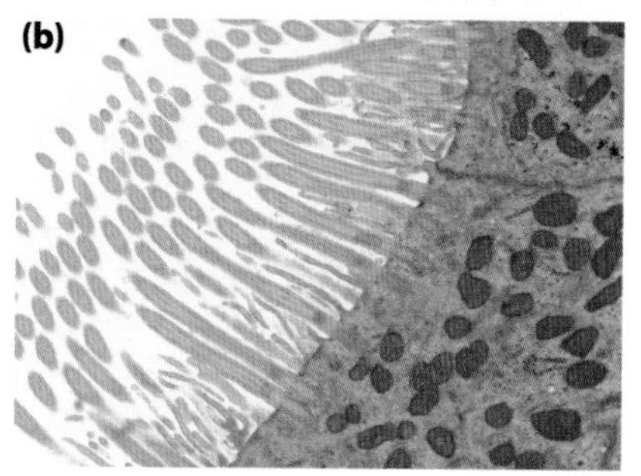

(b)

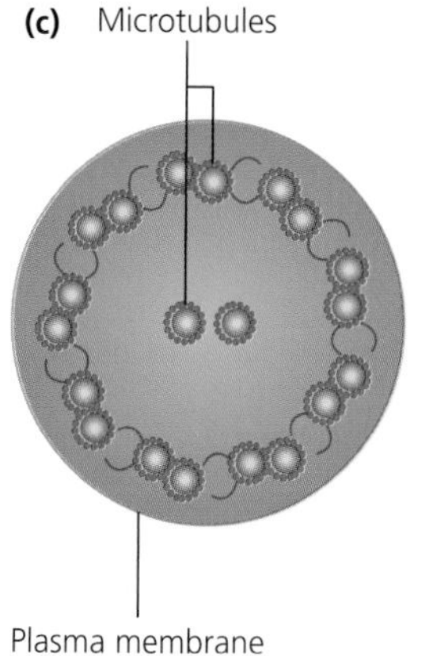

Figure 1.19 (a) Scanning electron micrograph of cilia in the gas exchange system (×4675); (b) transmission electron micrograph of cilia in section (×3850); (c) structure of a cilium or flagellum.

Flagella and cilia

Flagella (singular: flagellum) and cilia (singular: cilium) are structures composed of microtubules, found on the surface of cells, which move to create currents. The flagella and cilia may propel the cell along or (in the case of cilia only) move liquids across the surface of the cell.

Flagella are found in some single-celled organisms, where they move the cell around. A flagellum also forms the 'tail' of sperm cells. Flagella are usually found singly but sometimes occur in pairs.

Cilia are always present in large numbers, in both single-celled organisms and multicellular animals.

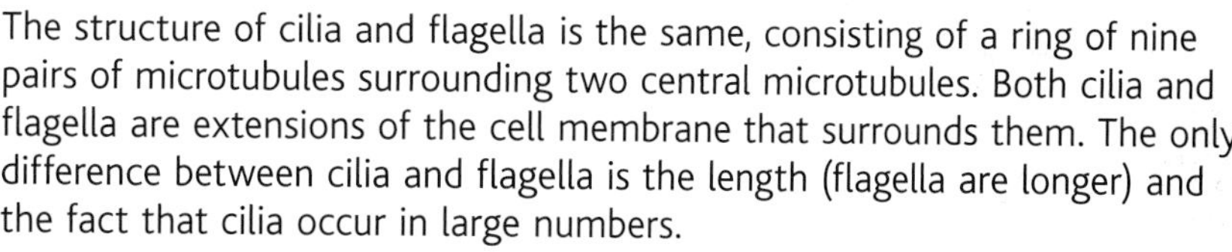
The structure of cilia and flagella is the same, consisting of a ring of nine pairs of microtubules surrounding two central microtubules. Both cilia and flagella are extensions of the cell membrane that surrounds them. The only difference between cilia and flagella is the length (flagella are longer) and the fact that cilia occur in large numbers.

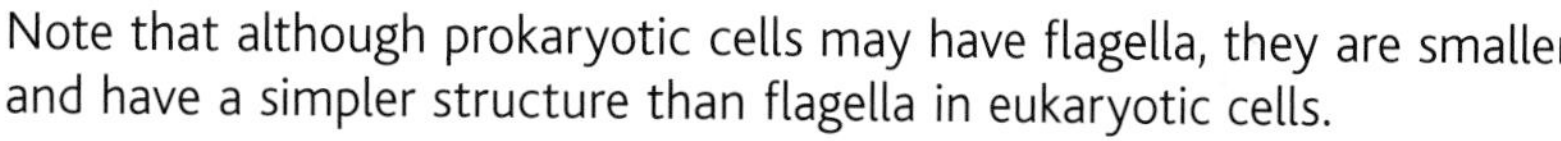
Note that although prokaryotic cells may have flagella, they are smaller and have a simpler structure than flagella in eukaryotic cells.

The structure of cilia and flagella is shown in Figure 1.19.

Cell wall

The cell wall is an extracellular layer that supports the cell and prevents it from bursting if water flows in.

In eukaryotes, cell walls are found around the cells of plants, many members of the Protoctista, and fungi. The cell walls of plants and algae are made of cellulose, but fungal cell walls consist of chitin. The appearance of a plant cell wall under the electron microscope is shown in Figure 1.20. In prokaryotes, the cell wall is made of murein, a polymer made of sugars and amino acids.

Tip

Cellulose is made of chains of glucose molecules (see Chapter 2). Chitin is similar to cellulose but contains nitrogen too.

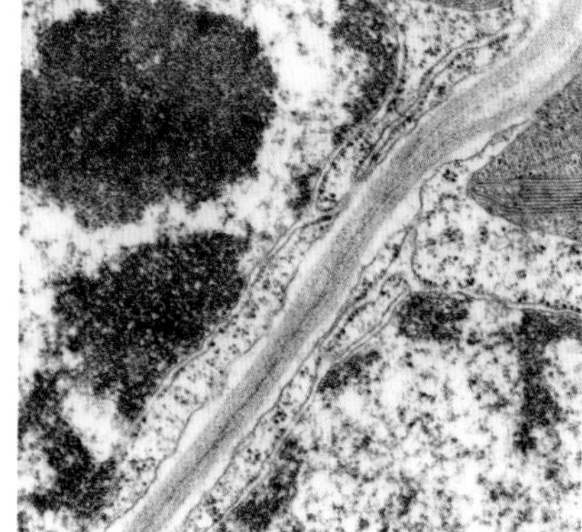

Figure 1.20 Electron micrograph of a plant cell wall.

Activity

Isolating cell components

So far we have described how microscopes are used to study cell structure. Another technique is cell fractionation. Cellular organelles can be isolated by breaking up the cells (homogenisation), which is done in an ice-cold solution that has the same solute concentration as the inside of the cells. The homogenate is spun at increasingly high speeds in a centrifuge; this process is called cell fractionation. Cell fragments, including organelles, move down the centrifuge tube depending on both their mass and the speed of the centrifuge: the heavier the organelle, the lower the centrifugal speed needed to compact the organelles into a pellet. The process is shown in Figure 1.21 below.

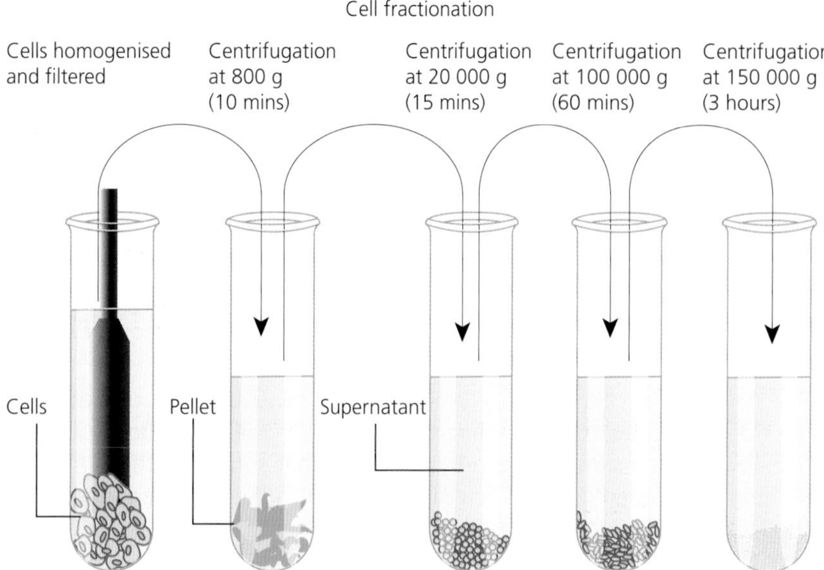

Figure 1.21 The stages in isolating cellular organelles.

The approximate sizes of different organelles are given in Table 1.2.

Table 1.2 Sizes of some organelles.

Organelle	Approximate size/μm
Nucleus (diameter)	7
Ribosome (diameter)	0.03
Mitochondrion (length × diameter)	3 × 1
Chloroplast (length × width)	5 × 3

1 Suggest why the homogenising is done in an ice-cold solution.
2 Suggest why it is important that the homogenising solution has the same solute concentration as the cells.
3 Suggest why the homogenate is filtered before use.
4 Stage 4 produces a pellet containing membrane fragments. If animal cells were used, suggest what would be found in the pellets at stages 2, 3 and 5.
5 Suggest why it may be difficult to isolate chloroplasts from plant cells using the method described above.

Organelles working together

Organelles allow division of labour in cells, with different organelles performing specialised functions. Many processes in the cell require the coordination of several different organelles, as, for example, in the production and secretion of proteins (Figure 1.22).

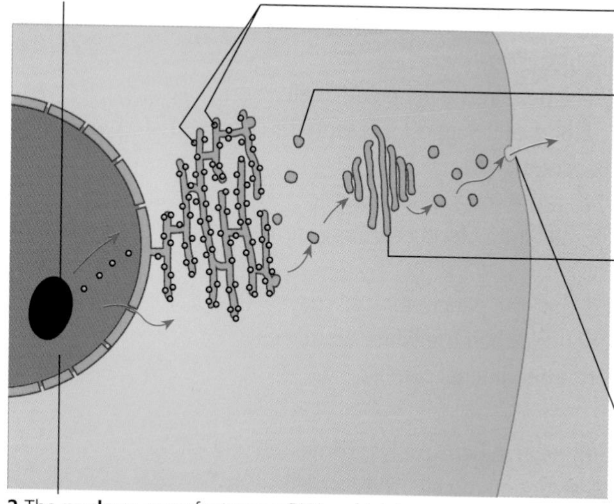
1 The **nucleolus** manufactures **ribosomes** for protein synthesis in the **RER**.
3 The **ribosomes** in the **RER** make proteins.
4 The **RER** processes the proteins which are then sent in vesicles to the **Golgi body**.
5 The **Golgi body** furthe processes the proteins and sends them in vesicles to the **plasma membrane**.
6 The vesicles fuse with the **plasma membrane** to secrete the finished protein product.
2 The **nucleus** manufactures mRNA, which is needed by **ribosomes** to make proteins.

Figure 1.22 Organelles involved in the production and secretion of proteins.

The cytoskeleton

Throughout the cytoplasm, a series of protein threads form the cytoskeleton. The threads are of two types: microtubules and microfilaments. The cytoskeleton has a number of different functions:

- Cellular movement: you have already seen that cilia and flagella contain microtubules that are responsible for moving them, so moving the cell (or moving liquids across the surface of the cell).
- Intracellular movement: the threads form a 'track' along which organelles can move from one part of the cell to another. Examples include the movement of vesicles, and the movement of chromosomes to either end of a cell during cell division.
- Strengthening and support: the cytoskeleton supports the organelles and also forms a sort of 'scaffolding' that strengthens the cell and helps to keep it in shape.

Key term

Intracellular Within a cell.

Microfilaments are solid strands, but the microtubules, as the name suggests, are tubular. The protein that makes up the two threads also differs: it is mostly actin in microfilaments, and mostly tubulin in microtubules.

Example

Figure 1.23 shows drawings of electron microscope sections of (a) a white blood cell and (b) a pancreatic cell. White blood cells form part of the immune system and ingest and destroy microorganisms. The pancreatic cell shown produces digestive enzymes.

1 Identify one organelle shown in the white blood cell but not in the pancreatic cell. Suggest a reason for its presence in the white blood cell.
2 Identify one organelle shown in the pancreatic cell but not in the white blood cell. Suggest a reason for its presence in the pancreatic cell.
3 The pancreatic cell has more rough endoplasmic reticulum (RER) than the white blood cell. Suggest a reason for this.
4 The drawings suggest that the pancreatic cell may contain more mitochondria than the white blood cell. Why might this information be inaccurate?

Answers

1 The organelle is the lysosome. Because lysosomes are involved in the destruction of cells, they would need to be present in white blood cells to destroy ingested microorganisms. Ingested microorganisms are destroyed by the proteolytic (protein digesting) enzymes that lysosomes contain.
2 The organelle is the Golgi apparatus. Its function is to package cell products ready for secretion. The pancreatic cells produce digestive enzymes. These enzymes would clearly need to be secreted from the cell, so the Golgi apparatus would be necessary.
3 The ribosomes attached to the RER produce proteins, which the RER then transports. The proteins for the cell's own needs are largely made in the ribosomes in the cytoplasm. Because the function of the pancreatic cell is to produce enzymes for secretion, the pancreatic cell would need a lot of RER to manufacture the enzymes.
4 You would see more mitochondria if you were looking at a larger area of the cell, but the scale bars on Figures 1.23(a) and (b) show that this is not the case – the magnification of both drawings is the same. The answer relates to an important biological principle of sampling. You are only looking at a small part of each cell. The parts seen may not be representative of the cell as a whole (i.e. the white blood cell may have dense concentrations of mitochondria in another part, or the area of the pancreatic cell seen may, by chance, have a cluster of mitochondria). The difference in the number of mitochondria is not huge.

(a)

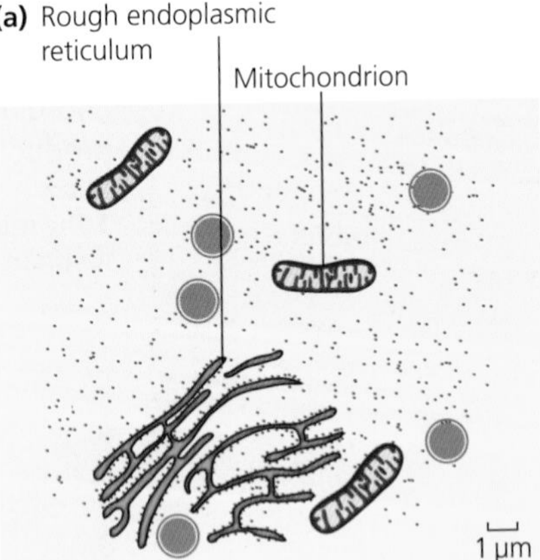

(b)

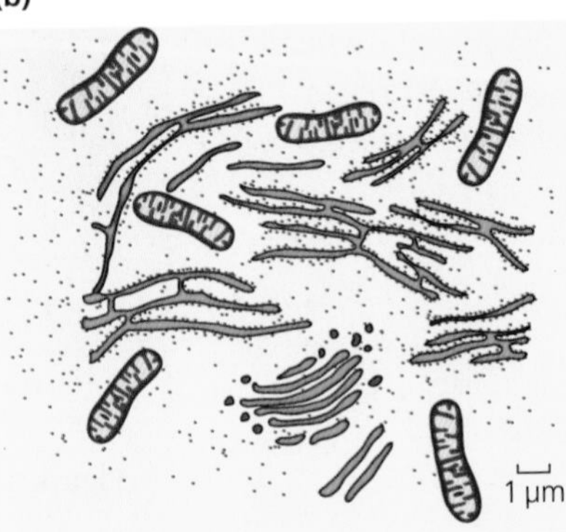

Figure 1.23 Drawings of electron microscope sections of: (a) a white blood cell and (b) a pancreatic cell.

Tip

The word 'suggest' in a question always means that you are expected to come up with a reasonable answer, even if that answer may not be correct in all respects.

Prokaryotic and eukaryotic cells

Cells are of one of two types: prokaryotic or eukaryotic. Prokaryotic cells are those of bacteria and blue-green bacteria. All other cells are eukaryotic. There are distinct differences in structure between prokaryotic and eukaryotic cells, as shown in Table 1.3.

Table 1.3 Features of eukaryotic and prokaryotic cells.

Feature	Eukaryotic cell	Prokaryotic cell
Nucleus	present	absent
DNA	linear and in the nucleus	circular and free in the cytoplasm
Cell wall	when present, made of cellulose (in plant cells) or chitin (fungal cells)	made of murein
Membrane-bound organelles	present	absent
Ribosomes	large (80S)	small (70S)
Capsule	absent	present
Pili	absent	sometimes present
Size	larger (2–200 μm)	smaller (less than 2 μm)

A prokaryotic cell is shown in Figure 1.24 but some features are not present in all prokaryotes.

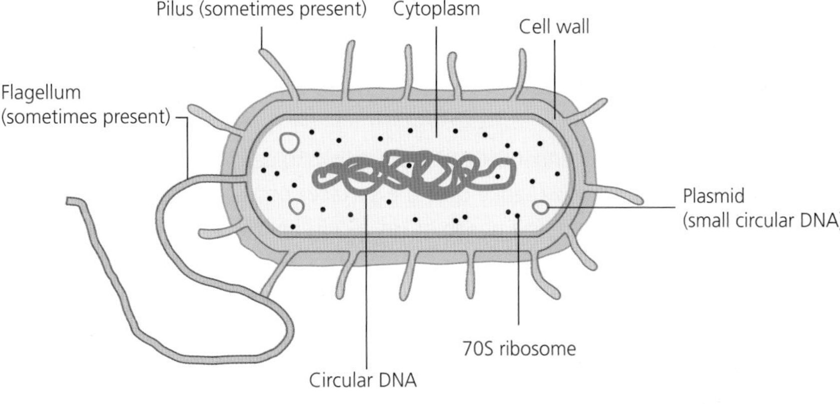

Figure 1.24 Prokaryotic cell.

Test yourself

10 Explain the relationship of the thylakoids and the grana in a chloroplast.

11 Suggest why lysosomes are very numerous in white blood cells.

12 Explain how the cell is protected from the protein-digesting enzymes that lysosomes contain.

13 What are the differences between cilia and flagella?

14 What are the differences between microfilaments and microtubules?

15 Give three features that can be found in prokaryotic cells but are never found in eukaryotic cells.

16 Name a feature found in all plant cells but never in animal cells.

Tip

If a question asks for structural differences between eukaryotic and prokaryotic cells, do not include size, because size is not classed as a structural feature.

Key term

Protoplast The part of a cell (in bacteria or plants) that is inside the cell wall.

Plant and animal cells

Within the eukaryotes, there are clear differences between plant and animal cells. Most of these differences involve the presence in plant cells of structures that are not found in animal cells – a cellulose cell wall (with plasmodesmata), chloroplasts and a large central vacuole containing cell sap. Plasmodesmata (singular: plasmodesma) are gaps in the cell wall through which cytoplasm connects the **protoplast** of one cell with an adjacent cell (see Figure 1.25). Of course, not all plant cells contain chloroplasts, only those that take part in photosynthesis.

The only organelles found in animal cells but not in plant cells are centrioles and flagella. Cilia are very rare in plant cells, occurring only in one primitive group called the cycads.

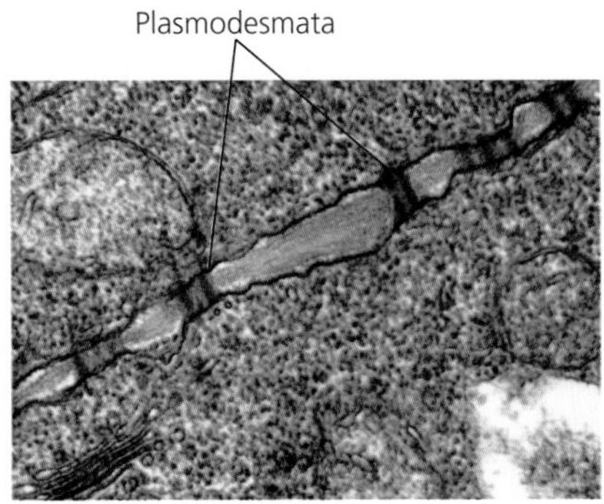

Figure 1.25 Cell wall with plasmodesmata (×16 000).

Exam practice questions

1 Which two organelles are both surrounded by two membranes?

A centriole and smooth endoplasmic reticulum
B chloroplast and mitochondrion
C Golgi apparatus and lysosome
D nucleus and rough endoplasmic reticulum *(1)*

2 The table below shows different combinations of cellular features.

	A	B	C	D
Relative number of secretory vesicles	few	many	few	many
Relative number of mitochondria	few	few	many	many
Extent of rough endoplasmic reticulum	large areas	small areas	small areas	large areas
Extent of smooth endoplasmic reticulum	small areas	large areas	small areas	large areas

Liver cells (hepatocytes) are metabolically active cells. They carry out many enzyme-controlled reactions to produce proteins and lipids that are exported into the blood. Which combination of cellular features is shown in the cytoplasm of hepatocytes? *(1)*

3 HeLa cells are cancer cells that are grown in cell culture and move readily over the surfaces of laboratory dishes. The photo of HeLa cells below was taken with a laser scanning confocal microscope. Microtubules are stained green and mitochondria red.

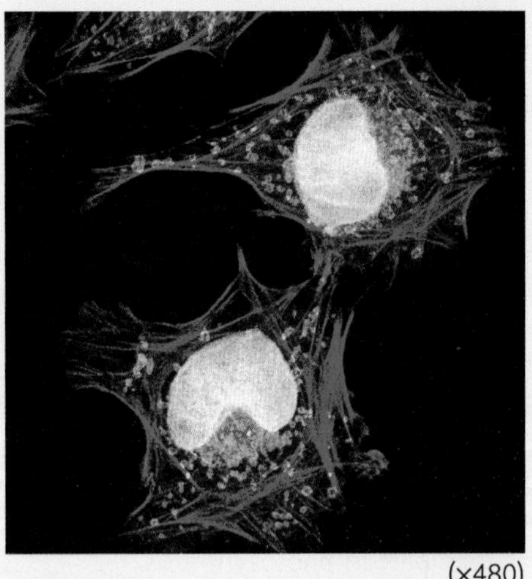

(×480)

a) i) State the cell structure that is stained yellow in these cells. *(1)*
ii) Describe the roles of microtubules in cells. *(4)*

b) Explain why further magnification of the photo above will not give more detail of the structure of mitochondria and microtubules. *(3)*

c) Use the photo above to explain the advantages of using a laser scanning confocal microscope to study cells. *(3)*

4 Stage micrometers are used to calibrate eyepiece graticules.
The diagram below shows a stage micrometer with divisions 0.1 mm apart, viewed through an eyepiece with a graticule.

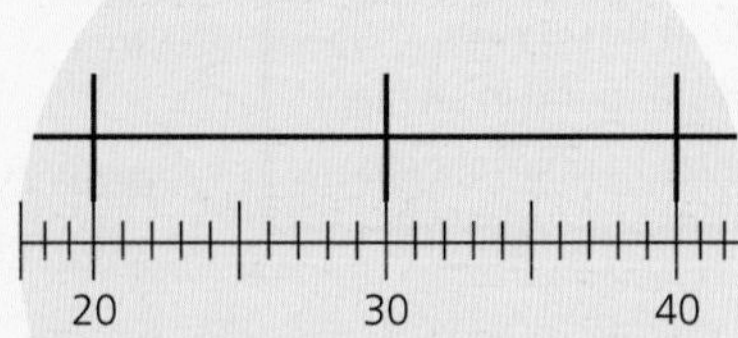

Using the eyepiece graticule at the same magnification, a student measured the length of a cell as 12 eyepiece units. What is the actual length of the cell?

A 10 μm
B 20 μm
C 100 μm
D 120 μm *(1)*

5 Some students investigated the storage of starch in the cells of banana fruits. They made a smear preparation of the banana flesh, irrigated it with iodine solution and looked at it with the high power of the microscope. One of the students made an annotated drawing of the cells.

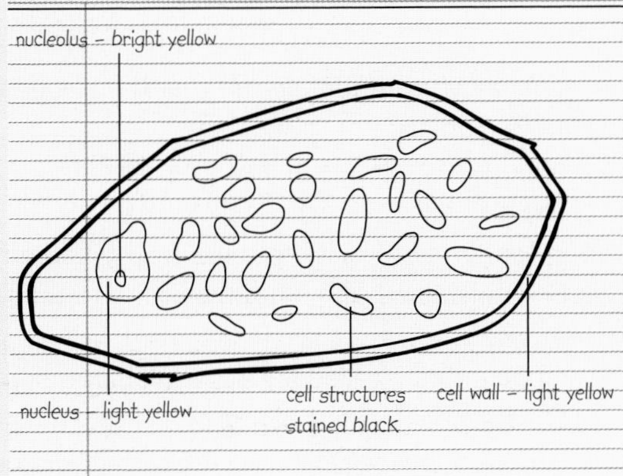

a) Explain why the students used iodine solution in their investigation. *(2)*

b) With reference to the student's drawing

i) Describe and explain the distribution of starch. *(2)*

ii) Explain the meaning of the term 'differential staining'. *(2)*

c) Explain in detail how the student could determine the actual width of the cell shown in the drawing. *(4)*

d) Explain what additional information about the cells could be obtained by using a scanning electron microscope. *(3)*

6 The photo below shows a single cell.

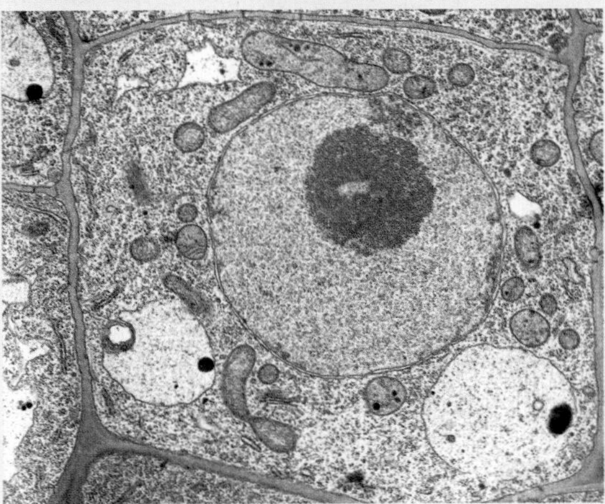

(×1400)

a) State the type of image shown in the photo above. Give reasons for your answer.

b) Identify the cell as far as you can. Give reasons for your identification.

c) Make a large labelled drawing of the cell. Annotate the drawing to show the functions of the structures you have labelled.

Chapter 2

Water, carbohydrates and lipids

Prior knowledge

Before you start, make sure that you are confident in your knowledge and understanding of the following points.

- Water is a small but important molecule in biological systems.
- Water is made up of the elements hydrogen and oxygen.
- Water has many important properties; it can move freely, its thermal properties allow life to be sustained, and it has the ability to act as a solvent for many substances.
- Carbon-containing compounds are called organic compounds or organic molecules. They are the essential building blocks of all living things.
- There are four major groups of biological molecules: carbohydrates, proteins, lipids and nucleic acids.

Test yourself on prior knowledge

1 What elements are contained in water and how are they bonded together?

2 Describe how the chemical bonding in water gives it two of its important properties. Name the properties.

3 Explain the feature of carbon that allows it to be used as an important building block in organic compounds.

4 Why are biological molecules called organic molecules?

5 Name the four groups of biological molecules and give an example of each.

Figure 2.1 There is huge variation in the organisms that live on Earth, but they are all made from the same molecules.

Constituents of cells

About 99% of living things are made up of only four elements: carbon, hydrogen, oxygen and nitrogen. These four elements are used, with a few other elements, to build all the molecules found in living cells.

The molecules formed are called biological molecules and consist of: water, which makes up a large proportion of living cells; carbon-based molecules that are combined with hydrogen and oxygen, such as carbohydrates and triglycerides; and carbon-based molecules that contain nitrogen in addition to the carbon, hydrogen and oxygen, such as proteins and nucleic acids.

Water and hydrogen bonding

Water is essential to life and so it is often the starting point when looking at biological molecules. The human body is approximately 80% water, and some plants and animals may be as much as 95% water.

Water consists of atoms of the elements hydrogen and oxygen, bonded together by **covalent bonds**. There are two hydrogen atoms and one oxygen atom.

Key term

Covalent bond A strong chemical bond formed by sharing one or more electrons between two atoms and so creating a molecule.

Key term

Polar molecule A molecule with regions of negative charge and positive charge.

Because the oxygen draws electrons from the hydrogen atoms, the water molecule forms a triangular shape with an unusual distribution of charges within it; this is known as a **polar molecule**. The oxygen has a slight negative charge (δ^-) and hydrogen a slight positive charge (δ^+), which results in an overall neutral electrical charge on the molecule. Water molecules are attracted to each other by weak bonds called hydrogen bonds, which form between the oxygen and the hydrogen atoms of adjacent water molecules (see Figure 2.2).

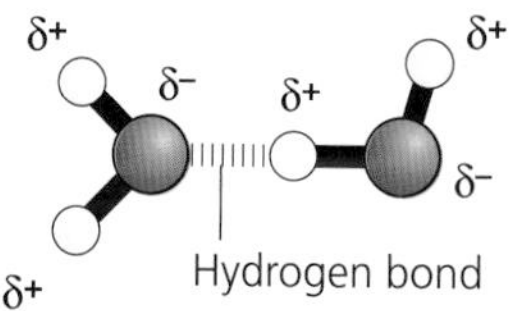

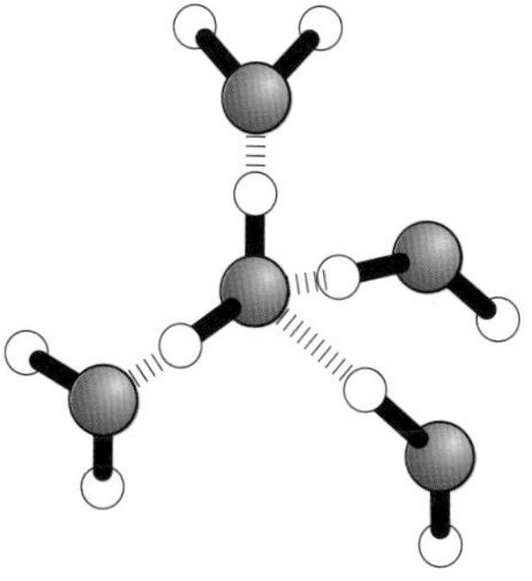

Figure 2.2 Hydrogen bonds in water.

Because the water molecules are attracted to each other they will flow together as a liquid and are attracted to charged particles and any charged surfaces. Other similar molecules do not behave in this way, because they have different properties. Hydrogen sulfide (H_2S), for example, is a gas at 0 °C, because it has a very low boiling point, whereas water is a liquid at 0 °C, which accounts for many of its unique properties.

Unique properties of water

Thermal properties

Hydrogen bonds between the water molecules constantly form and break, allowing the water molecules to flow.

Water is a liquid between 0 °C and 100 °C, which provides the perfect habitat for living things to survive. Water also acts as an efficient coolant, because it has a high specific heat capacity – it absorbs a great amount of heat relative to its mass.

Water molecules cannot easily escape unless heat is applied to weaken the bonds; at 100 °C water becomes a gas – called water vapour.

Individual water molecules lose kinetic energy as they cool and the hydrogen bonds do not break easily; the molecules form a more rigid structure – a solid (ice). At 0 °C, the solid water or ice is less dense than liquid water and so floats on its surface. The result is an insulating layer, which keeps the temperature of any water below it more constant. Living things can therefore survive in the water below the ice.

Water as a solvent

Key term

Solution A liquid (the solvent) with dissolved solids, liquids or gases (solutes).

The water molecules, known as the solvent, surround the solute molecules and keep them in **solution**, for example salt or sugar solutions as indicated in Figure 2.4. The dissolved solute reacts more easily than when the solute molecules are not dissolved, and takes part in metabolic processes. Because water is a polar molecule, any other polar molecules dissolve in water. Carbon-containing molecules with charged ionised groups such as carboxyl groups ($-COO^-$) or amino groups ($-NH_3^+$) easily dissolve by forming hydrogen bonds with the water hydroxyl groups (–OH).

Ionic substances such as sodium chloride (NaCl) release positively charged ions (cations) or negatively charged ions (anions), such as Na^+ and Cl^-, by forming clusters of water around the ions.

Tip

The properties of cohesion and adhesion of water are important in the transport of water in plants; see Chapter 9 to find out more about this.

The role of water in living things

The thermostable properties of water and water's ability to act as a solvent perform a vital role in living bodies, because they allow chemical reactions to take place within the cell and allow molecules to be dissolved in the cytosol. Water also makes an ideal transport medium because it dissolves polar molecules and ions, which can then be transported easily around living organisms, for example in the blood transport system of many animals.

Figure 2.3 A water droplet.

Ionic compounds like NaCl dissolve in water,

$NaCl \rightleftharpoons Na^+ + Cl^-$

with a group of orientated water molecules around each ion:

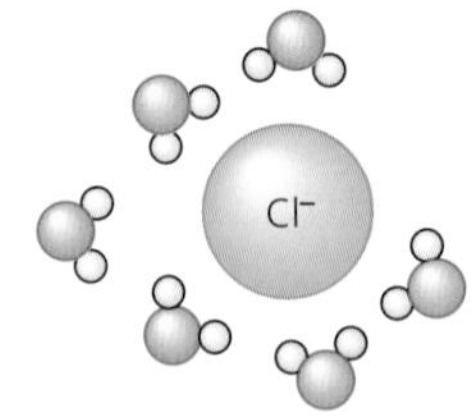

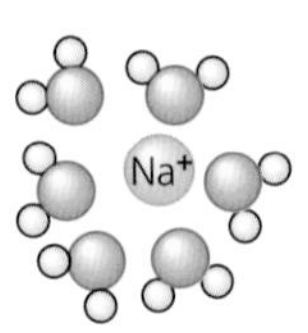

Sugars and alcohols dissolve due to hydrogen bonding between polar groups in their molecules (e.g. –OH) and the polar water molecules:

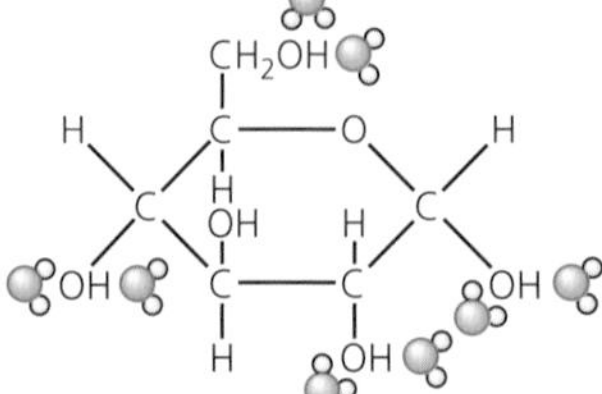

Figure 2.4 Water as a solvent.

Key terms

Cohesion A property of water in which water molecules are attracted to each other by hydrogen bonding, allowing the molecules to move together.

Tension A force that tends to stretch something.

Adhesion A property of water in which water molecules are attracted to surfaces such as the walls of the cells, vessels or tubes.

Tip

Hydrophobic molecules are important in membranes; see Chapter 5 to find out more.

The hydrogen bonds tend to pull water molecules towards each other, so they stick together. This is called cohesion. The hydrogen bonds also pull the molecules inwards at the surface, creating surface tension. You can easily see the effect of surface tension in a spherical water drop on a glass or on a waxy leaf. Some insects use this surface tension to 'skate' across the water surface. Cohesion allows water molecules to flow together in a column. Water also adheres to a surface and, as a result of both cohesion and adhesion, water can move up narrow tubes, for example a straw or a xylem vessel in the vascular tissue of a plant. Water also acts as a coolant for plants, as the water moves through the plant in the transpiration stream, and it provides a good habitat for many organisms in both fresh-water and salt-water habitats.

Non-polar substances cannot dissolve and are repelled by water – they are **hydrophobic** or water repelling, for example oils. This is an important property of the phospholipid bilayer of the cell membrane.

An abundance of carbon

In living things the most common element by mass is oxygen, followed by carbon.

Atoms share electrons with one another in order to achieve a full outer shell, which makes them stable. Carbon is a small atom but it makes an excellent building block, because the four pairs of electrons in the outer shell of the atom repel each other and so push away to shape the atom into a tetrahedron. These pairs of electrons can be shared by other atoms to form covalent bonds; this allows atoms to join together and form organic molecules. Carbon's four such bond sites (see Figure 2.5a) form covalent bonds that are stable and strong enough to make larger molecules.

a) Covalent bonds are formed by sharing of electrons, one from the carbon atom and one from the neighbouring atom it reacts with:

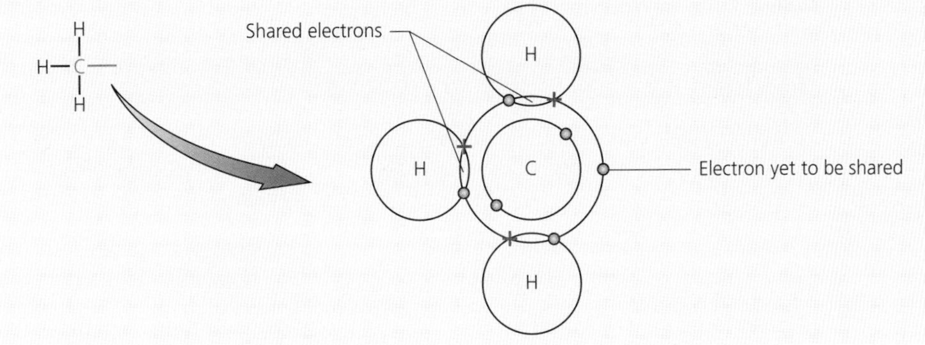

Shared electrons

Electron yet to be shared

b) Carbon atoms bond with other carbon atoms to form carbon 'skeletons':

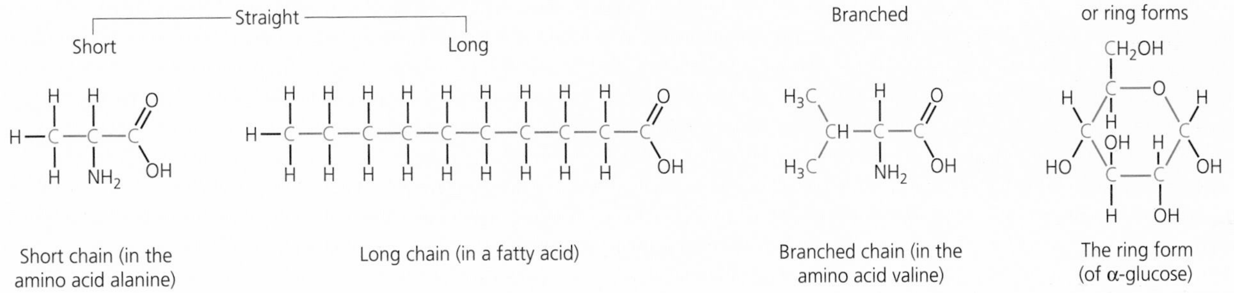

Straight

Short

Long

Branched

or ring forms

Short chain (in the amino acid alanine)

Long chain (in a fatty acid)

Branched chain (in the amino acid valine)

The ring form (of α-glucose)

Figure 2.5 (a) A covalent bond formed by sharing electrons between two adjacent atoms, one from the carbon and one from the adjacent hydrogen atom. In this case four hydrogen atoms have joined with the carbon, each by a covalent bond; (b) carbon atoms forming different carbon skeletons.

Tip

You need to understand the different forms that carbon-based molecules take but you do not need to learn them in detail.

Key terms

Monomer A single small molecule that may combine with other monomers to build up larger molecules called polymers.

Polymer A large molecule built up from many similar monomers joined together by covalent bonds to form a chain or a branched chain.

In the same way, carbon atoms may bond together and form chains, branched chains or ring shapes (see Figure 2.5b). Carbon elements also bond covalently with oxygen and hydrogen, and other atoms such as nitrogen and sulfur, and create many different large molecules. Sometimes carbon forms two bonds with another atom, called a double bond, for example in hydrocarbon chains where two carbons share two bonds. This is drawn as: C=C. In glucose, one of the oxygen atoms forms a double bond with an adjacent carbon, drawn as: C=O.

Building large molecules from small – monomers and polymers

A **monomer** is a single smaller molecule that may form covalent bonds with other similar smaller molecules to build a larger molecule. The bonds form as a result of a condensation reaction. A **dimer** is a molecule formed of two monomers joined together by a condensation reaction.

A **polymer** is a large molecule built up from many similar monomers joined together by covalent bonds to form a chain or a branched chain. Each bond forms as a result of a condensation reaction.

Table 2.1 lists the main polymers in carbohydrates, proteins and nucleic acids.

Table 2.1 The main polymers in biological molecules.

Biological molecules	Monomers	Polymers
Carbohydrates	monosaccharides, e.g. glucose, fructose, ribose and triose	polysaccharides, e.g. starch (amylose), glycogen and cellulose
Proteins	amino acids, e.g. glycine, valine and alanine	Polypeptides, e.g. amylase and lysozyme
Nucleic acids	nucleotides	Polynucleotides, e.g. DNA and RNA

Tip

To find out how condensation reactions are used in the formation of triglycerides see pages 33–4.

Each monomer joins to another by forming a specific covalent bond by a chemical reaction called a **condensation** reaction, in which a molecule of water (two hydrogen atoms and one oxygen atom) is released (see Figure 2.6). The same reaction is repeated each time another monomer is added to form the polymer. Condensation reactions are also known as dehydration reactions and are involved in the formation of many other macromolecules such as triglycerides.

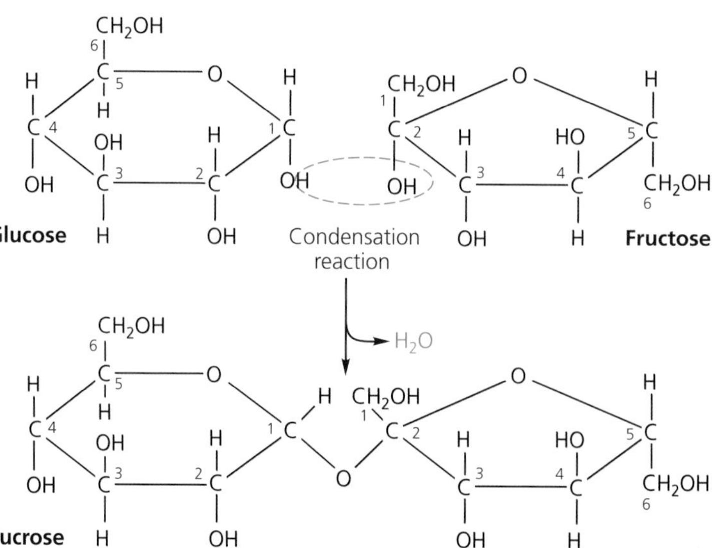

Figure 2.6 A condensation reaction between glucose and fructose, forming a glycosidic bond.

Key terms

Condensation A chemical reaction where two molecules are joined together with a covalent bond, forming a larger molecule and releasing one molecule of water.

Hydrolysis A chemical reaction where the covalent bond between two molecules is broken with the addition of a water molecule, separating the two molecules.

A **hydrolysis** reaction is the chemical reaction involved in splitting molecules joined by covalent bonds. In a hydrolysis reaction, a water molecule must be added to break the covalent bond, and the smaller molecules are then released (see Figure 2.7).

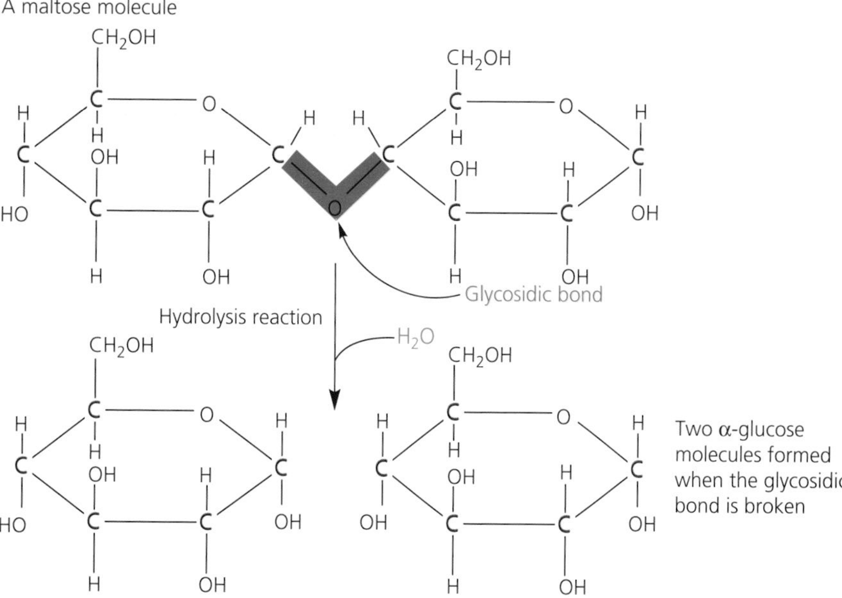

Figure 2.7 A hydrolysis reaction releasing two molecules of glucose from a molecule of maltose.

Test yourself

1 Describe a covalent bond and explain its importance in biological molecules.
2 Explain why the polar properties of water are important to living things.
3 Explain why properties of cohesion and adhesion are important to plants that live on land.
4 What are the main differences between hydrogen bonds and covalent bonds?

Figure 2.7 shows the hydrolysis of maltose releasing two glucose molecules. Hydrolysis also occurs in both sucrose and lactose, although in sucrose one glucose and one fructose molecule are released, and in lactose one glucose and one galactose molecule are released.

Polymers are often very large molecules, and they tend to be stabilised by many **hydrogen bonds**. These hydrogen bonds help the molecules to retain their final shape. The shape is crucial to the function of the molecules, and so these hydrogen bonds (and some other bonds) are important in maintaining the functioning of the polymers.

Carbohydrates

Key term

Cell respiration A process involving many enzyme-catalysed reactions that occur within cells and result in the release of energy, which is used to make adenosine triphosphate (ATP).

Carbohydrates are molecules forming a large proportion of the organic compounds in living cells. They act as an energy source; for example, glucose is used for energy release during cell respiration. Carbohydrates are also important as energy stores; for example, glycogen is stored in animal muscle and liver. Some carbohydrates are important structurally, for example cellulose in plant cell walls.

- All carbohydrates contain the elements carbon, hydrogen and oxygen in the specific ratio of two hydrogen atoms to each oxygen atom.
- Almost all carbohydrates have the general formula $(C_x(H_2O)_y)$.
- Carbohydrates include single simple sugars, for example glucose (see Figure 2.8), complex sugars such as sucrose and lactose, and polysaccharides such as starch, glycogen and cellulose.
- Some carbohydrates act as direct energy sources, e.g. glucose, and others as indirect energy sources, because they can be stored efficiently, e.g. glycogen.
- Other carbohydrates are structural molecules.

Three main groups of carbohydrates, each with their particular features and functions, are given in Table 2.2, although all share all the general properties and features of carbohydrates.

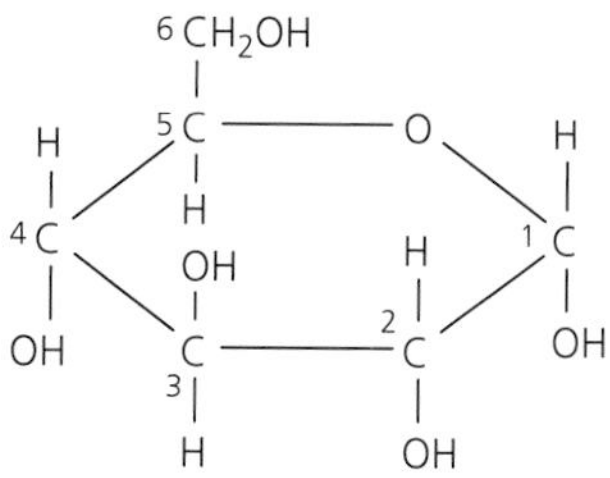

Figure 2.8 Structure of a common carbohydrate – glucose.

Table 2.2 Features of the main carbohydrates.

Features	Type of carbohydrates		
	Monosaccharides (simple sugars)	**Disaccharides (complex sugars)**	**Polysaccharides (complex carbohydrates)**
Type of molecule	single molecule	two molecules covalently joined	many molecules covalently joined to each other
Taste	sweet to taste	sweet to taste	not sweet
Solubility in water	soluble	soluble	insoluble
Number of glycosidic bonds	none	single glycosidic bond	many glycosidic bonds
Structure	exist as a single ring shape or as a straight chain	two rings joined	long chains, which may be branched and coiled, making them very compact
Roles	energy release, transported in blood, monomers for other carbohydrates	energy release, storage, and transport within plants	energy storage, structural component of cell walls
Examples	hexoses (6C) such as glucose, fructose and galactose; pentoses (5C) such as ribose and deoxyribose; trioses (3C) such as glyceraldehyde	sucrose maltose lactose	starch (amylose and amylopectin) glycogen cellulose

Monosaccharides

Monosaccharides are the simplest carbohydrates, often called simple sugars because they are small molecules. Glucose is particularly important because it is easily broken down in cell respiration to release energy, it is one of the molecules synthesised by green plants during photosynthesis, and is transported in the blood of many animals. All monosaccharides are used as monomers for other carbohydrates.

Glucose

Glucose exists in two forms: α-glucose and β-glucose. These forms of glucose are similar, but the –H and –OH groups on carbon number 1 (C1) are arranged differently. In α-glucose the –OH is below the carbon at position 1, and in β-glucose the –OH is above the C1. This structural difference leads to important differences in the way that α-glucose and β-glucose behave and how they affect the polymers formed from them. These two molecules, α-glucose and β-glucose, are called isomers.

Key term

Isomer Molecule containing the same number and types of atoms, but the atoms are arranged differently. Examples are α-glucose and β-glucose.

α-glucose (ring form)

β-glucose (ring form)

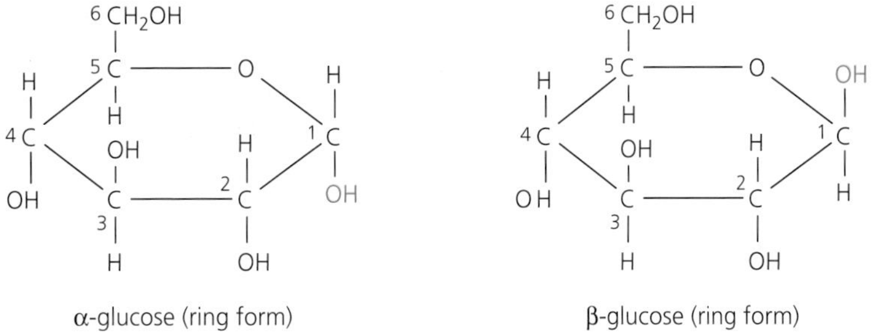

Figure 2.9 α- and β-glucose ring forms.

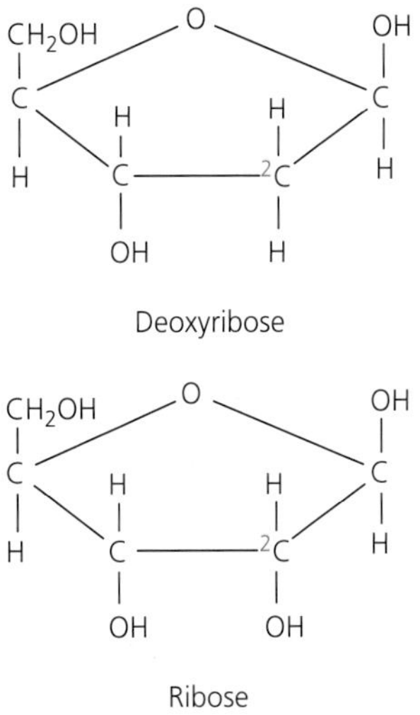

Glucose can be broken into smaller and simpler molecules by breaking the bonds between the atoms in a series of small steps to form carbon dioxide and water. Glucose is a relatively small molecule so diffuses easily, is water soluble (so is easily transported) and is the main respiratory substrate, resulting in energy being released. The main function of monosaccharides is therefore as an energy source.

Most organisms can only break down polymers of α-glucose, because they do not have the necessary enzymes to break down polymers of β-glucose.

Tip

Cellulose is a polymer of β-glucose (see pages 30–1). Animals do not have the cellulase enzymes needed to hydrolyse polymers of β-glucose.

Other monosaccharides

Pentose sugars with five carbon molecules, such as ribose and deoxyribose (see Figure 2.10), are important in forming nucleic acids. Triose sugars are three-carbon molecules, important intermediate molecules in both respiration and photosynthesis.

Deoxyribose

Ribose

Figure 2.10 A five-carbon ribose sugar and five-carbon deoxyribose sugar, showing the loss of one oxygen in deoxyribose at carbon number 2.

Disaccharides

Disaccharides are carbohydrates formed by two monosaccharides that join together using a covalent bond called a glycosidic bond. Each glycosidic bond is formed by the loss of two hydrogen atoms and one oxygen atom.

Key term

Glycosidic bond A covalent bond formed when two carbohydrate molecules are joined together by a condensation reaction.

When two α-glucose monomers are joined together, the bond formed is called a 1,4-glycosidic bond because the bond forms between C1 of one glucose and C4 on the other molecule. Maltose is the disaccharide formed when amylase breaks down starch by a hydrolysis reaction. When a glucose molecule and a fructose molecule combine, a molecule of sucrose is formed. The sugar found in our food is mainly sucrose. Digestive enzymes catalyse the hydrolysis of these sucrose molecules into glucose and fructose. Fructose is known as fruit sugar because it is a common sugar in fruits, although it is combined with a glucose molecule to form sucrose. An enzyme, called an isomerase, changes fructose into glucose. Lactose is milk sugar and is formed by a condensation reaction between glucose and galactose. All disaccharides can be easily hydrolysed into their monomers, and so provide another energy source.

Tip

Look at Figure 2.6, which shows the formation of a glycosidic bond between glucose and fructose. You need to understand the different forms that carbon-based molecules take but you do not need to learn them in detail.

Sucrose

Maltose

Lactose

Figure 2.11 Some disaccharide molecules: sucrose, maltose and lactose.

Test yourself

5 Use the diagram of glucose ring formation (Figure 2.8) to explain how the carbon molecules are numbered in the molecule.
6 What is an isomer? Give an example that illustrates your description.
7 Use Figure 2.9 to describe the differences between a molecule of α-glucose and a molecule of β-glucose.
8 Describe the role of an isomerase enzyme.
9 Use labelled diagrams to describe how a disaccharide molecule is formed.
10 Give three different examples of disaccharides and state the constituent monomers in each.

Activity

Investigating the Benedict's test for reducing sugars

A student carried out a Benedict's test by placing equal volumes of glucose solution and Benedict's solution in a boiling tube. The student heated the mixture for five minutes until the colour changed and then recorded the colour observed.

The student set up a control using distilled water instead of glucose, and again heated the mixture in the hot water bath for five minutes before recording the colour observed (see Table 2.3).

Refer to the outline method given above and the results in Figure 2.12 to answer the questions below.

1 Describe the different colours you would expect to see appearing in the tube as the temperature in the tube increased. What was the final colour of the glucose test?
2 What was the purpose of heating the distilled water with Benedict's solution?
3 Explain why it would not be possible to determine the exact volume or concentration of glucose using this particular test.

Table 2.3 Results of Benedict's test for reducing sugars.

Solution	Observations of the solution at the start	Observations of the solution after mixing and heating with Benedict's solution
Benedict's	bright blue	
Glucose	colourless and clear	orange
Distilled water	colourless and clear	pale bright blue

Figure 2.12 Test for reducing sugars. The results after heating a glucose solution (left) and distilled water with Benedict's solution.

Activity

Investigating the Benedict's test for non-reducing sugars

Another student tested a solution for non-reducing sugars using the Benedict's test. The student added 1 molar (1M) hydrochloric acid to the tube of solution and mixed the two by swirling the tube and heating the tube for five minutes in a water bath.

The student then added sodium hydrogen carbonate and left the tube until the fizzing stopped.

Then the student added Benedict's solution and heated the mixture in a hot water bath for five minutes. The student recorded any colour observations in a table (see Table 2.4).

Table 2.4 Results of Benedict's test for non-reducing sugars.

Solution	sucrose	sucrose pre-treated and pre-heated with HCl and sodium hydrogen carbonate
Observations at start	colourless and clear	colourless and clear
Observations immediately after adding Benedict's and mixing	bright blue and clear	bright blue and clear
Observations after heating	bright blue and clear	brick red cloudy precipitate

Make careful observations of the procedure and decide how it is different from the Benedict's test for reducing sugar. Compare the two sets of results given and use all these observations to answer the questions.

1 Describe and explain the colour changes observed in both tubes.
2 The student did not have a control tube. Suggest a suitable control that they could use.
3 What is the reason for using a control tube?
4 What would the results of a control test show you?
5 Why was the sucrose heated with hydrochloric acid at the start of the test? Describe the chemical reaction that takes place during this step.
6 Explain why sodium hydrogen carbonate is added to the tube before testing with Benedict's solution.

Tip

The Benedict's test is a qualitative test but it may be modified to form a semi-quantitative test by:

a) filtering and weighing the precipitate or
b) by diluting the reagents and using a colorimeter to determine quantitative differences.

Activity

Biological tests and interpreting them using colorimetry

When completing a biochemical test using colorimetry, a student first prepared colour standards in order to interpret the results. This involved using a range of different glucose concentrations, for example 0.1%, 0.5% and 2.0% glucose. To each tube of glucose concentration the student added an equal volume of dilute Benedict's solution and heated the tube in a water bath to at least 80 °C for nine minutes. Once cooled, the contents of each tube were filtered into separate colorimeter tubes and then placed in turn in the colorimeter.

A single colorimeter reading for each tube was recorded. The colorimeter was calibrated by using the maximum concentration of glucose to define absorbance of 0.00 arbitrary units.

The student muddled the tubes on the first trial, and so the teacher gave the following tip:

Place the tubes on a sheet of paper correctly numbered across the page, and set each tube on to its correct position on the page when not in the colorimeter, but do not write on the tubes or add any labels.

Once a reading for each glucose concentration was recorded, the student plotted a graph of absorbance against glucose concentration.

Now the student repeated the test using an unknown concentration of glucose mixed with the same volume of dilute Benedict's solution. The tube was heated in exactly the same way as before to the same temperature and for the same time.

The student again filtered the contents into a colorimeter tube and placed it in the colorimeter to record the absorbance, before using the colour standards graph to identify the value of the glucose concentration in the unknown solution.

Table 2.5 Results from the colorimeter

Glucose concentration/%	Absorbance/arbitrary units
0.0	0.92
0.1	0.90
0.5	0.72
1.0	0.38
1.5	0.17
2.0	0.00

1 Draw a graph of absorbance against glucose concentration.
2 What is the concentration of the unknown glucose solution?
3 Suggest why a more dilute Benedict's solution is used than is usual in the Benedict's test?
4 Explain the importance of filtering the glucose and Benedict's mixture once it has been heated and before it is placed in the colorimeter.
5 Why did the student muddle the tubes in the trial? (You need to read the teacher's tip above to understand the problem the student had with these tubes.)

Tip

Refer to Chapter 17 to learn how to use a colorimeter.

Polysaccharides

Polysaccharides are large polymers formed of many monosaccharides that combine by condensation reactions to create long chains. Each monomer added is attached to the chain by a glycosidic bond.

Starch

Starch is a polymer of α-glucose molecules; glycosidic bonds link each pair of α-glucose molecules. The chains coil up into a spring shape because of the shape of the monomer and the angle of the glycosidic bonds that join the C1 of one glucose to the C4 of the next, forming a 1,4-glycosidic bond.

Starch is actually two molecules: amylose, the long chain coiled into a spring, and amylopectin, which is a long branched chain with short side arms of glucose units attached to the main chain by 1,6-glycosidic bonds (see Figure 2.13). Although starch is insoluble, there are some differences in starch solubility which can be attributed to the different proportions of amylose and amylopectin in the starch.

The importance of starch is its function as a major carbohydrate-storage molecule in plants. In plant organs, such as potato tubers, starch is stored in amyloplasts (an **amyloplast** is a type of plastid) within the cells, and it is also stored in chloroplasts in the leaves.

Key term

Plastids Specialised membrane-bound organelles found in plant cells, for example chloroplasts and amyloplasts. Amyloplasts store starch grains.

The structure of starch makes it ideal as a storage molecule: it is compact and stable because of the coiling and folding. Individually, these bonds are weak, but when there are many of them the resulting molecule becomes much more stable. Starch is also insoluble, which means that it does not affect the water potential and osmotic properties of the cell. It can be easily broken down by enzymes into glucose when needed.

Tip

For more detail on water potential see Chapter 5.

Iodine solution is the test for starch, giving a blue-black colour when starch is present. The iodine ion fits into the helix of the amylose molecule and produces the typical blue-black colour of the positive starch test.

Amylose (a straight-chain polymer of α-glucose)

Amylopectin (a branched-chain polymer of α-glucose)

α-1,4-glycosidic linkages

α-1,6-glycosidic linkage

A starch molecule, showing both amylose and amylopectin.

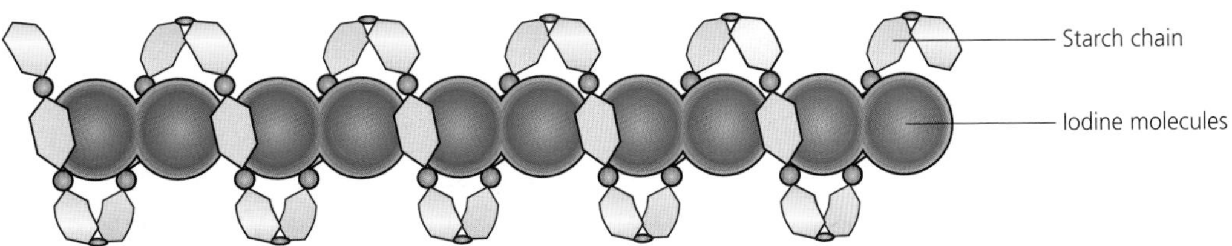

A starch–iodine complex showing the iodine ions within the amylose helix.

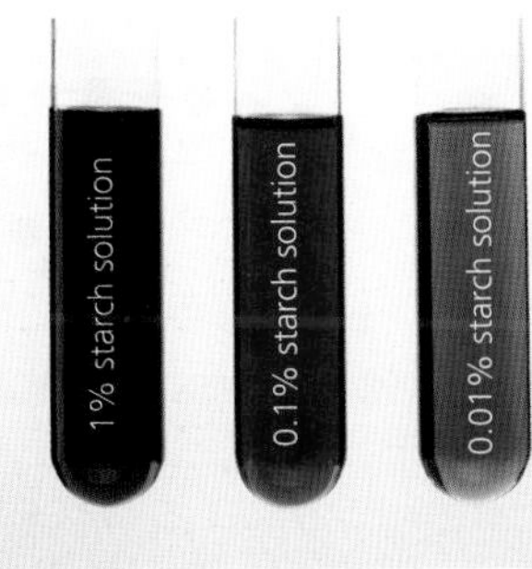

Figure 2.13 Starch.

Test for starch with iodine in potassium iodide solution; the blue-black colour comes from a starch–iodine complex (a) On a potato tuber cut surface (b) On starch solutions of a range of concentrations.

Activity

Investigating the iodine test for starch

During a practical session, a student tested for starch by mixing two drops of starch solution and two drops of iodine solution.

The student observed the colour of each solution before and after mixing the two types of solutions, and recorded the colours in a table (see Table 2.6).

The student also added two drops of distilled water to two drops of iodine.

The student then heated a mixture of starch and iodine, once the blue-black colour had appeared, in a test tube placed in a water bath.

Table 2.6 Results of iodine test for starch.

Solution	Observations of the solution at the start	Observations of the solution after mixing with iodine
Iodine	straw yellow	
Starch	colourless and almost clear	blue-black
Distilled water	colourless and clear	pale straw yellow

The colour returned to a straw yellow again. After the student cooled the mixture rapidly by holding the test tube under cold water, the blue-black colour reappeared.

Look at the procedures outlined here and at the results table then answer the questions below.

1 Explain why distilled water is described as being colourless and clear at the start.
2 Starch is described as 'almost clear'. From your understanding of the structure of starch, explain why a solution of starch would be almost clear.
3 Suggest the most suitable apparatus to use to add the drops of the two solutions. What precaution should be taken to ensure that the results are not contaminated in any way?
4 The result for starch and iodine solution shows a blue-black colour. Describe what has happened to form this colour.
5 Using the results of this investigation, and any further research, suggest why the blue-black colour first appeared, why it disappears when the mixture is heated, and why it reappears when the mixture is cooled. You need to look at the structure of the amylose molecule to help you answer this question.

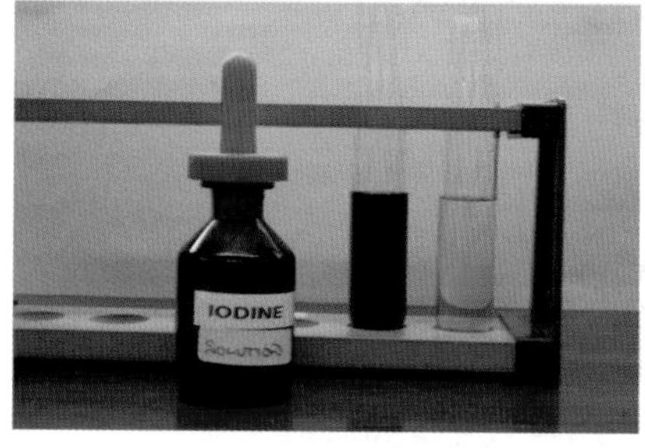

Figure 2.14 Starch test. Iodine solution is added to a starch solution (left) and distilled water.

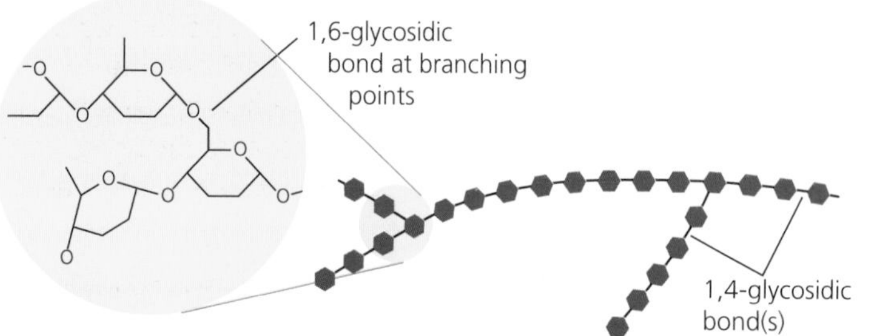

Figure 2.15 Branching chains of glycogen showing 1-4 and 1-6 glycosidic bonds.

Glycogen

Glycogen is another polymer made of α-glucose molecules. Glycogen is similar to amylopectin, because it forms long chains, but there are more branches and the chains tend to be shorter and so the molecules are very compact (Figure 2.15). Glycogen is stored as granules in animal cells such as liver and muscle cells. Like starch, glycogen is easily broken down into individual glucose molecules when the body needs energy.

Cellulose

Cellulose (Figures 2.16 and 2.17) is made of β-glucose molecules that are joined by glycosidic bonds, but the arrangement of –OH and –H groups on the C1 of the β-glucose means that adjacent glucose molecules only join if alternate molecules are inverted. The result is a straight glucose

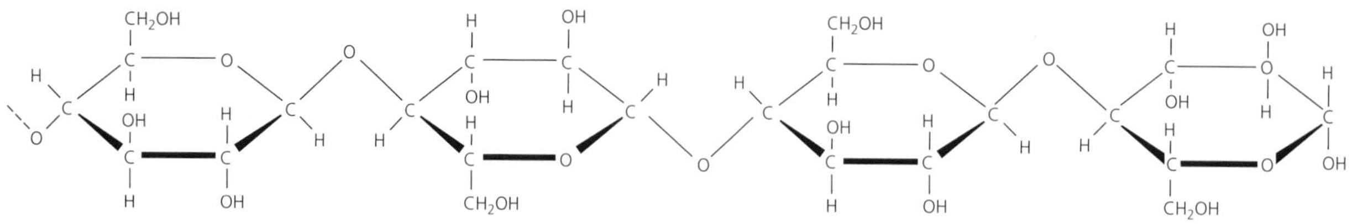

Figure 2.16 β-glucose molecules in a molecule of cellulose.

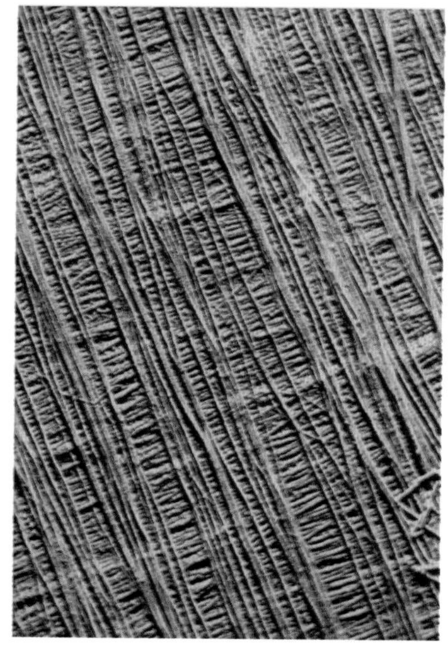

Figure 2.16 Cellulose.

chain with many –OH groups on the outside. This allows hydrogen bonds to form between the adjacent cellulose chains, creating many cross links. The cellulose chain is held together by the cross links to form microfibrils. Bundles of these microfibrils are also cross linked by hydrogen bonds to form fibres. The cellulose fibre is a very stable and strong structure forming part of the cell wall of plants. This differs from the structure of amylose, which is made of α-glucose units, discussed earlier in this chapter.

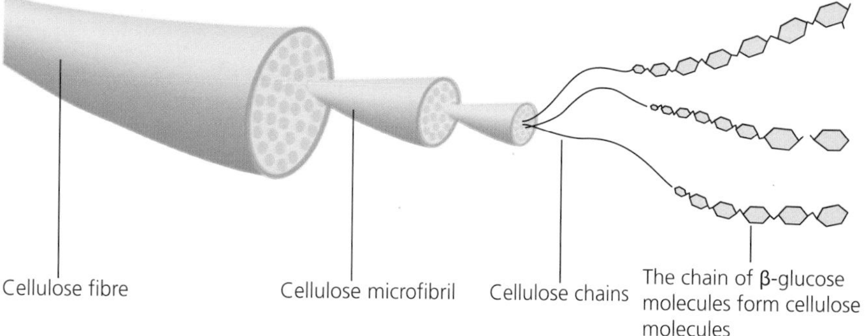

Figure 2.17 A cellulose molecule showing the cellulose chain, microfibrils and fibres.

Other carbohydrate polymers exist in other organisms, for example chitin in insect exoskeletons, and peptidoglycans in the cell walls of bacteria.

Relating structure to function

Tip

You will learn about protein synthesis in Chapter 3.

The structure of each carbohydrate molecule is related to its function. Simple sugars release energy easily, because they are small, simple molecules. Complex carbohydrates are more stable, compact and are mostly insoluble. These complex carbohydrates (starch and glycogen) can be broken down into individual glucose molecules when needed, but are less-ready supplies of energy.

Table 2.7 describes some carbohydrate molecules and their functions in organisms.

Table 2.7 Structure and function of different carbohydrate molecules.

Carbohydrate	Important features	Examples of molecules	Function in organisms
Monosaccharide	small, soluble	glucose (6C)	source of energy as easily hydrolysed; transported in the blood of animals
		ribose (5C)	part of all types of RNA molecules and ATP molecules, so it is an important part of protein synthesis
		deoxyribose (5C)	part of DNA for inherited information
Disaccharide	small, soluble	maltose	energy storage in some plants
		sucrose (made of glucose and fructose)	energy storage in some plants transported in plant phloem
		lactose (made of glucose and galactose)	the sugar found in milk; provides an important nutrient for young mammals
Polysaccharide	large complex molecules which are stable and insoluble	starch and glycogen	energy storage in plants (starch) and animals (glycogen)
		cellulose	structural and very strong; forms the main component of the cell wall in plants

Example

Inulin is the primary storage carbohydrate in a few species including Jerusalem artichoke, *Helianthus tuberosus*, and chicory, *Cichorium intybus*. These plants do not store much amylose, if any.

1 a) Copy and complete the table below to show how amylose differs from inulin.

Feature	Inulin	Amylose
Monomer	fructose	
Number of monomers	2,70	
Type of glycosidic bond	β 2–1	
Solubility in water	soluble	
Site of storage in a cell	vacuole	

b) Some forms of inulin have a terminal glucose. Explain what is meant by 'terminal glucose'.

c) Explain why polysaccharides such as amylose and glycogen are suitable as energy storage molecules.

Answers

1 a)

Feature	Inulin	Amylose
Monomer	fructose	α-glucose
Number of monomers	2–70	200–1000
Type of glycosidic bond	β 2,1	α 1,4
Solubility in water	soluble	insoluble
Site of storage in a cell	vacuole	amyloplasts/chloroplasts

b) 'Terminal glucose' refers to a molecule of glucose attached to the end of a polymer or a chain, in this case the end of a polymer made of fructose monomers.

c) Polysaccharides such as amylose and glycogen are suitable as energy storage molecules because they are insoluble. Because they are insoluble, they do not decrease the water potential of the cell (as would happen if the equivalent number of glucose molecules were stored instead of being converted to starch or glycogen). If glucose was stored, the cell would need a thicker cell wall to withstand the extra pressure potential.
Compared with many thousands of glucose molecules (which are also always surrounded by water molecules), polysaccharides such as starch and glycogen are compact; they do not take up as much space.
Glucose needs to be readily available, because it is easily respired and so provides energy quickly, and soluble so that it can be transported easily to where it is required. Glucose is easily removed from the ends of the polysaccharide molecules, α 1,4 glycosidic bonds are easy to break by hydrolysis, and glycogen is branched so there are many places from which glucose can be removed.

Test yourself

11 Describe the main differences between starch and glycogen.

12 Explain the features of starch that make it an ideal storage molecule.

13 How are the cellulose fibres arranged in a plant cell wall? Find out how they are held together and describe their importance in the cell wall.

14 What can you find out about chitin and peptidoglycans and their role in insects, fungi and bacteria?

Lipids

Lipids are macromolecules. They are not polymers, because they are not formed of many similar monomer units. Lipids include triglycerides (fats and oils), phospholipids, waxes, and steroids such as cholesterol.

All lipids are made of the elements carbon, hydrogen and oxygen, but the proportion of oxygen is low compared with that in carbohydrates. All lipids dissolve in organic solvents such as alcohol, but not in water, and so are described as **hydrophobic**.

The role of lipids in living organisms

- Lipids are important molecules in living cells because they have a high energy yield and so act as an important energy source, which can be respired.
- They can also act as an energy store: in plants lipids are often stored as lipid droplets, and in animals as fat in adipose tissue.
- Lipids act as an insulating layer, forming thermal insulation under the skin of mammals and an electrical insulator around some of the nerve cells in vertebrates and some invertebrates. They also form the waxy cuticle around leaves and stems of plants.
- All biological membranes are made from lipids, and the group of hormones called steroids are lipids.

Triglycerides include fats and oils, which are both formed from a condensation reaction between two groups of molecules: fatty acids and glycerol. The bond formed is called an ester bond or ester link.

The glycerol molecule is an alcohol and is the same in all fats and oils, but the fatty acid molecules are different depending on the molecule of fat or oil formed.

Glycerol (Figure 2.18) is a small molecule of just three carbon atoms and three oxygen atoms arranged as part of three –OH groups, which include three of the hydrogen atoms. The remaining hydrogen atoms are H groups arranged around the carbon.

Fatty acids are large molecules consisting of a hydrocarbon chain (see Figure 2.19) with a carboxyl group at one end. This is the end that reacts with the –OH group of the glycerol to form the ester bond. A methyl group (–CH_3) at the other end makes the whole chain hydrophobic. Because there are three –OH groups on the glycerol, three fatty acids bond to the glycerol by three condensation reactions, so that three molecules of water are lost and three ester bonds formed for each triglyceride. Figure 2.20a shows a typical fatty acid.

There are many different fatty acids, which can vary considerably in the length of the hydrocarbon chain, although the carboxyl end is always the same. This hydrocarbon chain gives the triglyceride its hydrophobic properties, because the chain has the charges equally distributed and so they do not form hydrogen bonds with water. The number of carbon atoms in the hydrocarbon chain varies according to the type of fatty acid, but the chain is usually between 14 and 22 carbon atoms long.

```
   H  H  H  H  H  H  H  H  H  H  H  H  H  H  H    O
   |  |  |  |  |  |  |  |  |  |  |  |  |  |  |   //
H— C— C— C— C— C— C— C— C— C— C— C— C— C— C— C— C
   |  |  |  |  |  |  |  |  |  |  |  |  |  |  |   \
   H  H  H  H  H  H  H  H  H  H  H  H  H  H  H    O— H
```

Figure 2.19 An unsaturated fatty acid.

Key terms

Fatty acid A molecule with a hydrocarbon (fatty) chain and a carboxylic acid (which contains a carboxyl (COOH) group).

Glycerol A three-carbon alcohol molecule that forms the basic structure to which the fatty acids are joined in a triglyceride.

Ester bond The bond formed when an organic acid such as a fatty acid joins to an alcohol such as glycerol by a condensation reaction. (A triglyceride with three fatty acids joined to the glycerol has three ester bonds.)

Hydrocarbon chain A chain of carbon atoms bonded together, with hydrogen atoms bonded on each side of each carbon.

```
    H
    |
H — C — OH
    |
H — C — OH
    |
H — C — OH
    |
    H
```

Figure 2.18 A glycerol molecule.

Tip

A triglyceride is a large molecule because it has a long hydrocarbon tail – but it is not a polymer, because it is made of four smaller molecules that are not similar to each other.

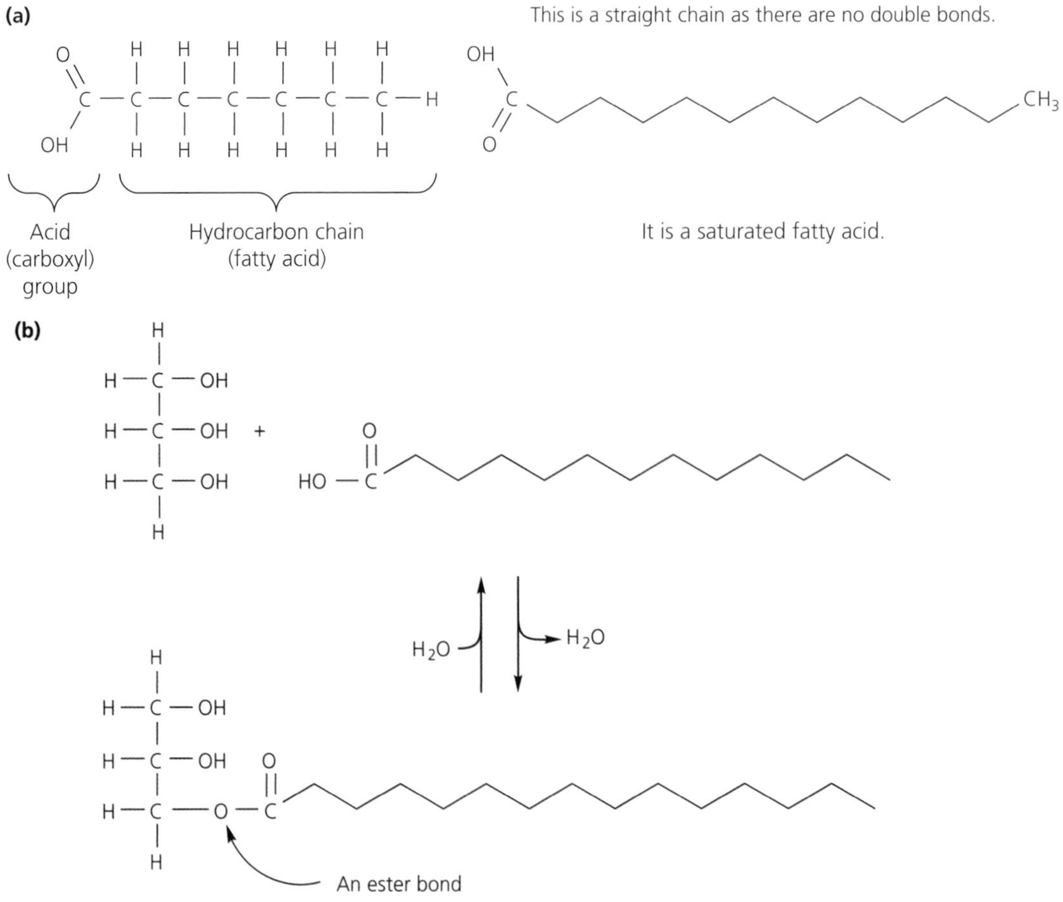

(a)

This is a straight chain as there are no double bonds.

Acid (carboxyl) group

Hydrocarbon chain (fatty acid)

It is a saturated fatty acid.

(b)

An ester bond

Figure 2.20 (a) A typical fatty acid; (b) the formation of an ester bond in a monoglyceride.

Triglycerides are efficient energy stores: they are energy rich and are not water soluble, so they do not affect the water potential of the cell, which is vital with storage molecules. When oxidised during respiration, triglycerides release hydrogen molecules which combine with the oxygen to form metabolic water. Many desert animals rely on metabolic water to survive. Fat is also an excellent insulator, because it is a poor conductor of heat and it aids buoyancy, so is useful to aquatic animals such as whales. Fat is also used for protection around many of the vital organs.

Activity

Testing for fats – the emulsion test

During a practical lesson, an emulsion test was carried out using oil as the source of fat. Ethanol was added and mixed well to dissolve the oil. The mixture was poured into another tube containing water to make the emulsion. A photograph of the results was taken and is shown in Figure 2.21.

1 Give a detailed description of the appearance of the mixture in the tube after adding the oil and ethanol mix to the water.
2 Suggest a suitable control for this test.
3 Find out about the translucency test for fats and compare this with the emulsion test. Decide which test gives the most accurate results.

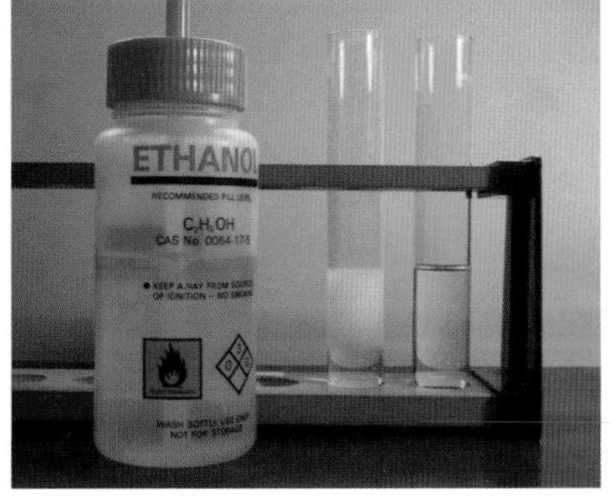

Figure 2.21 The results of an emulsion test.

Key terms

Saturated fatty acid A fatty acid with a long hydrocarbon chain where none of the carbon atoms are linked with double bonds. Each carbon atom in the chain is linked to two hydrogen atoms.

Unsaturated fatty acid A fatty acid with a long hydrocarbon chain where there are double bonds between some of the carbon atoms. Where there are double bonds, only one hydrogen atom is bonded to each of the carbon atoms on either side of the double bond.

Saturated or unsaturated fatty acids?

When all the carbon atoms in a fatty acid are joined by single bonds, the molecule is described as **saturated**, and when combined with glycerol forms a saturated fat. Typically, saturated fatty acids are animal fatty acids such as stearic acid and butyric acid.

An **unsaturated** compound is one where the hydrocarbon chain has at least one double bond; this forms a mono-unsaturated fatty acid. The double bond has two bonds between two of the carbon atoms in the chain, C=C, which means that there are fewer positions for the hydrogen atoms to bond to the carbon, so there are fewer hydrogens in the chain. Where there are many double bonds, a poly-unsaturated fat is formed. Typically, poly-unsaturated fatty acids are found in plants, for example oleic acid.

The double bond in an unsaturated fatty acid changes the shape of the hydrocarbon chain so that it kinks at the bond. The hydrocarbon chains do not lie straight together but instead push apart, making unsaturated fatty acids less solid and more fluid.

Figure 2.22 shows a saturated and an unsaturated fatty acid.

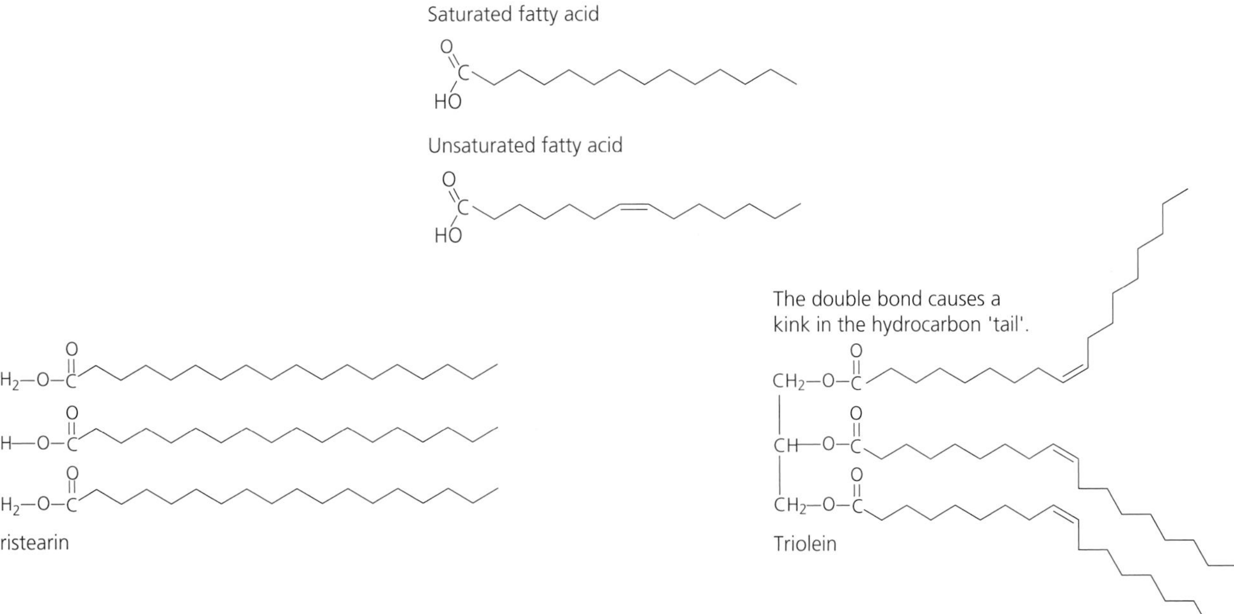

Figure 2.22 A saturated fatty acid and an unsaturated fatty acid.

Phospholipids

Phospholipids (see Figure 2.23) are present in all biological membranes.

Tip

Cell membranes are covered in Chapter 5.

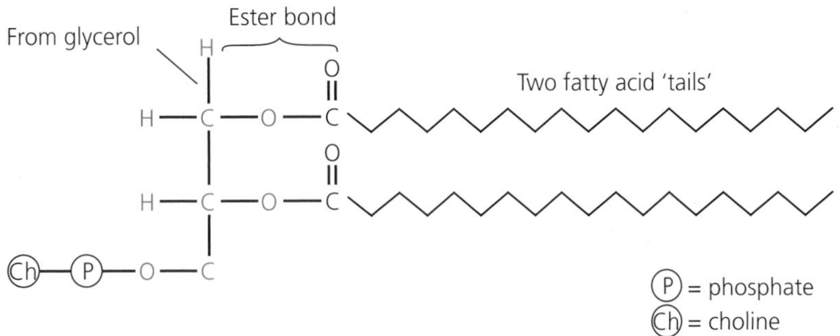

Figure 2.23 A phospholipid molecule.

Phospholipids are similar to triglycerides, with a glycerol molecule and fatty acids combined by ester bonds. However, phospholipids have only two fatty acid chains attached. The third –OH group on the glycerol molecule is occupied by a phosphate group, which is also bonded by a condensation reaction. In most phospholipids the phosphate group is attached to a nitrogen-containing water-soluble group such as choline. The result is a phosphate 'head' – the original glycerol combined with the phosphate group to form a hydrophilic end to the molecule. The two fatty acid chains form the double fatty acid 'tail'; these fatty acid chains are hydrophobic and form a hydrophobic core facing away from the water-based cytoplasm and external fluid. The result is a membrane barrier to the cell that prevents certain substances entering and leaving the cell. The same type of membrane surrounds the organelles. The fact that phospholipid molecules are part hydrophobic and part hydrophilic gives the cell membrane many of its properties.

Differences between phospholipids and triglycerides

Table 2.8 summarises some of the differences between phospholipids and triglycerides.

Table 2.8 Properties of triglyceride and phospholipid molecules related to their functions in living organisms.

	Triglyceride	Phospholipid
Structure	one glycerol molecule and three fatty acids	one glycerol molecule, two fatty acids and a phosphate group
Function	an energy source and an energy store; acts as an insulation layer, e.g. under the skin or around nerves and a protective layer around organs; a waxy layer on plant leaves	the molecule is part hydrophobic and part hydrophilic, giving it specific properties, for example it acts as a barrier to many ions and molecules; basis of cell membranes when carbohydrate chains are attached; forms glycolipids used in cell signalling

Cholesterol

Cholesterol (see Figure 2.24) is a molecule similar to phospholipids, with hydrophobic and hydrophilic regions. It is an important part of cell membranes in all eukaryotic cells. In vertebrates, it is made in the liver and is transported all over the body in the blood. Apart from its importance in cell membranes it is also used to make steroid-based hormones such as testosterone, oestrogen and progesterone.

Tip

Testosterone, the oestrogens and progesterone are examples of steroid hormones that are made from cholesterol.

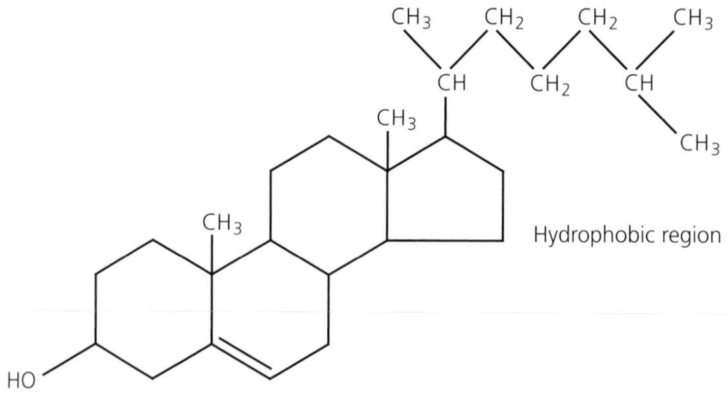

Figure 2.24 A cholesterol molecule.

Tip

The presence or absence of cholesterol in a membrane is one of the differences between eukaryotic cells and prokaryotic cells. See Chapter 5 on cell membranes.

Test yourself

15 Describe the structure of a saturated fatty acid and state one important use of this type of molecule in cells.
16 What is a double bond? Describe the effect this may have on the structure of a fatty acid.
17 What are the main structural differences between a triglyceride and a phospholipid?
18 What type of bond is used to join the molecules of fatty acid to the glycerol molecule, and how are these bonds formed?
19 Describe the differences in the properties of a phospholipid molecule, a cholesterol molecule and a triglyceride.
20 Suggest one important structural aspect that may account for some of the differences between triglycerides, phospholipids and cholesterol molecules.

Exam practice questions

1 Which one of the following involves condensation reactions?

A breaking down triglycerides to fatty acids and glycerol
B dissolving glucose in water
C forming hydrogen bonds between cellulose molecules
D synthesising glycogen *(1)*

2 Solutions of four carbohydrates were tested with Benedict's solution. Which row shows the expected results when solutions of these carbohydrates are boiled with Benedict's solution? *(1)*

	fructose	glucose	sucrose	starch
A	blue	blue	red	red
B	blue	red	red	blue
C	red	blue	blue	red
D	red	red	blue	blue

3 a) The diagram below shows a molecule of water.

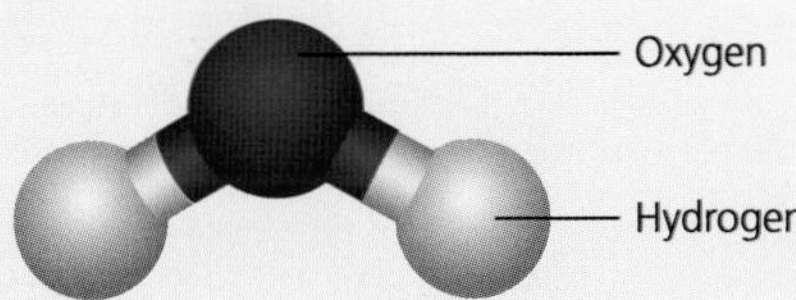

Copy the diagram above and show how the molecule of water interacts with three other water molecules. *(4)*

b) List three roles of water in living organisms. *(3)*

4 Which best describes a suitable test for the presence of lipids in a biological sample? *(1)*

A Add ethanol to the sample and shake until it has all dissolved.
B Add ethanol to the sample, shake and then pour the solution into water.
C Add water to the sample and shake until it has all dissolved.
D Add water to the sample, shake and then pour the solution into ethanol.

5 Triglyceride molecules are produced by condensation of glycerol and fatty acids.

a) i) Name the bond that forms between glycerol and a fatty acid. *(1)*

ii) State how many molecules of water are formed when a molecule of a triglyceride is synthesised from glycerol and fatty acids. *(1)*

iii) State three roles of triglycerides in mammals. *(3)*

b) Linoleic acid is an unsaturated fatty acid with 18 carbon atoms. It has two double bonds (C=C) immediately after the sixth and ninth carbon atoms from the methyl ($-CH_3$) end of the hydrocarbon chain.

i) Make a diagram to show the structure of a molecule of linoleic acid. *(4)*

ii) Explain why linoleic acid is described as an unsaturated fatty acid. *(3)*

c) Phospholipids and cholesterol are components of cell membranes.

i) Explain how the structure of phospholipids makes them suitable for cell membranes. *(3)*

ii) Describe three ways in which the structure of cholesterol differs from the structure of phospholipids. *(3)*

6 During germination of seeds, maltose is formed by the breakdown of starch. Maltose is further broken down to form glucose.

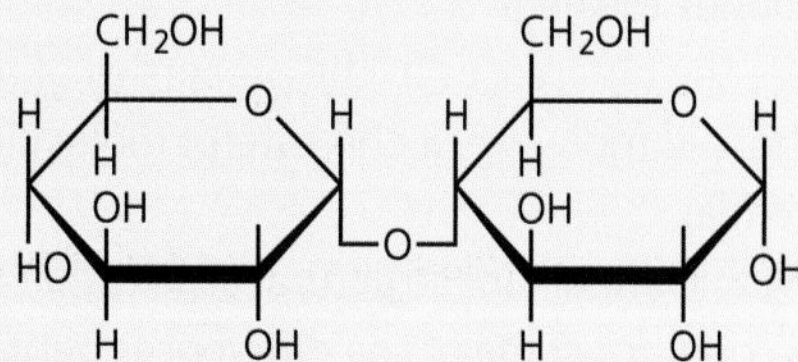

a) i) Copy the diagram of maltose and use it to show how maltose is broken down into molecules of glucose. *(2)*

ii) Name the type of reaction that occurs when starch is broken down to glucose during the germination of seeds. *(1)*

b) Explain why maltose and glucose are soluble in water, but starch is not. *(2)*

c) The table below shows the composition of the seeds of four species of flowering plant.

Substance in seeds	Percentage of dry mass of seeds			
	Sunflower	Hemp	Maize	Wheat
Starch	0	0	50–70	60–75
Sugars	2	2	1–4	0
Lipids	52	33	5	2
Proteins	21	25	10	14
Fibre	10	32	7	12

Discuss the composition of the seeds from the four species with respect to the functions of the substances shown in the table. *(5)*

7 a) The diagram below shows a molecule of β-glucose.

Describe how α-glucose differs from β-glucose. *(1)*

b) The diagram below shows the end of a growing molecule of cellulose.

Copy and complete the diagram to show how a molecule of β-glucose is added to the end of the growing molecule of cellulose. *(4)*

c) Suggest how cellulose is a suitable material for the cell walls of plants. *(5)*

Stretch and challenge

8 The fruit of the cotton plant, *Gossypium hirsutum*, contains seeds surrounded by cellulose fibres known as lint. As the fruits ripen, sugars may become attached to the lint. Some sugars come from the plant itself and some fall on the lint after egestion by aphids that feed on cotton plants. The sugars are a mixture of reducing sugars and non-reducing sugars. Lint covered in sugars is sticky.
During processing, the lint is separated from the seeds and made into cotton fibres. The lint is more likely to stick to the processing equipment if it is covered in sugars.
Explain how you could use the Benedict's test to find out how much of each sugar is attached to lint from different samples of cotton.

Chapter 3

Proteins and nucleic acids

Prior knowledge

Before you start, make sure that you are confident in your knowledge and understanding of the following points.

- Proteins have a complex structure.
- Proteins are made up of smaller 'building blocks' or monomers called amino acids.
- Proteins consist of the elements carbon, hydrogen and oxygen together with nitrogen, and some contain sulfur.
- There are many different types of proteins, which all have a different structure.
- Proteins are important for the growth of new cells and tissues, for enzymes and for some structural molecules.
- Nucleic acids contain the elements carbon, hydrogen, oxygen, nitrogen and phosphorus, but never sulfur.
- Nucleic acids have important roles in storing and retrieving genetic information and in controlling cell functions.
- The nucleic acids are DNA and RNA.
- The nucleic acids DNA and RNA are similar in structure but have some differences.

Test yourself on prior knowledge

1 What elements are in both proteins and in nucleic acids?

2 a) Name the element that is found in protein but not nucleic acids.

b) Name the element that is in nucleic acids but not proteins.

3 Name the monomer of proteins and the monomer of a nucleic acid.

4 Describe the role that proteins play in living things. How is this different from the role of nucleic acids?

5 What are the similar features of the molecules DNA and RNA? Describe the main ways in which RNA is different in structure from DNA.

Early research

The middle of the twentieth century was an exciting time to be a molecular biologist. In 1944, Oswald Avery, Colin MacLeod and Maclyn McCarty, working in New York, suggested that DNA was the genetic material in living things, however for a few years after this many scientists still thought that proteins, being more complex and versatile, were the molecules of heredity.

Figure 3.1 Rosalind Franklin (1920–58) was an X-ray crystallographer. Her X-ray images of DNA played an important part in the discovery of the structure of DNA.

In 1948, Linus Pauling predicted the process of protein translation from an RNA molecule (discussed later in this chapter) and also how genes might replicate. In February 1953 he proposed the model of a triple helix. Shortly after, in April 1953, James Watson and Francis Crick, working at the Cavendish Laboratory in Cambridge, published a short paper in the scientific journal *Nature* proposing a structure of DNA of a double helix with the two backbones on the outside. They suggested this, not from scientific research, but by piecing together all the information they had and by taking some inspired leaps of imagination to fill in the gaps. Watson and Crick had gleaned small amounts of information about Pauling's work from Watson's sister, who was dating Pauling's son; they also saw images of X-rays diffracted through DNA molecules on to an X-ray plate that Watson had obtained from the laboratory of two other scientists, Maurice Wilkins and Rosalind Franklin. Wilkins shared Watson and Crick's Nobel Prize in 1962, which acknowledged the important part his research findings played in the discovery. Rosalind Franklin did not share the Nobel Prize as she died in 1958, four years before it was awarded. The prize is not awarded posthumously so her important contribution was not acknowledged.

Proteins

Proteins are made up of carbon, hydrogen and oxygen, just like carbohydrates and fats, but the element nitrogen is also present, and some proteins contain sulfur as well. Proteins form a large part of all the organic molecules found in living organisms: roughly 60% of the molecules in cells are protein excluding water.

Proteins have many different functions. They are the basic materials for cell growth, cell repair and replacement of materials. Some proteins have a structural role, such as in cytoplasm, muscle, collagen and elastin in skin, collagen in bone and keratin in hair, or as protein carrier molecules in cell membranes. Others form antibodies, enzymes or many human hormones, and as a result are vital to metabolic processes.

Basic building block of a protein – an amino acid

Amino acids are the small molecules from which all the different types of protein are formed. They are the monomers that, when joined together, form proteins. The sequence of amino acids that are joined together determines the type of protein formed and hence its function.

Amino acids all have the same structure: an amino group ($-NH_2$) at one end of the molecule and a carboxyl group ($-COOH$), also called an organic acid group, at the other end. These two groups are attached centrally with a carbon atom that carries a hydrogen atom on one side and an R group on the other. It is the R group that is different in each of the 20 amino acids that make up proteins. Once amino acids are joined to one another by peptide bonds they are often referred to as amino acid residues.

> **Key term**
>
> **R group** A side chain that acts as a functional group as it helps to determine the internal bonds and so the shape of the polypeptide and hence its final function. Also called a residual group.

The smallest amino acid is glycine, which has a single hydrogen atom as its R group. Other amino acids may have a very large R group: it could be an acidic group as found in acidic polar amino acids (e.g. aspartic acid with $-CH_2-COOH$ as the R group); a basic group found in basic polar amino acids (e.g. arginine with $(-CH_2)_3-HN-CNH-NH_3$); or a polar amino acid

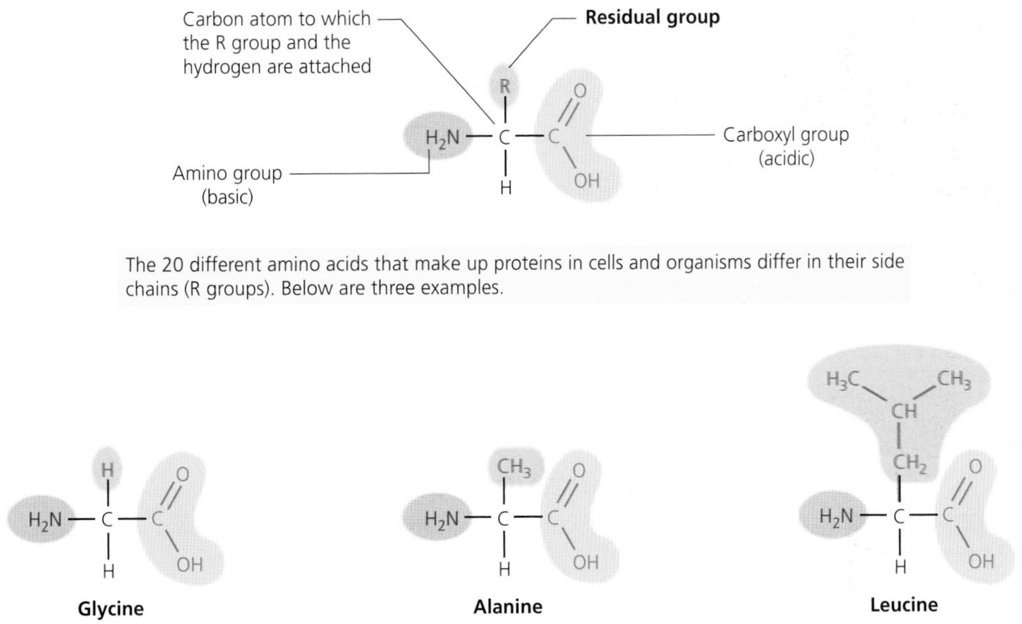

Carbon atom to which the R group and the hydrogen are attached

Residual group

Carboxyl group (acidic)

Amino group (basic)

The 20 different amino acids that make up proteins in cells and organisms differ in their side chains (R groups). Below are three examples.

Glycine

Alanine

Leucine

Figure 3.2 The structure of a basic amino acid and three examples of different amino acids.

(e.g. serine with $-CH_2OH$). All polar or charged R groups are hydrophilic groups, creating a hydrophilic end to the molecule. Non-polar amino acids are hydrophobic (e.g. valine with its $-CH-(CH_3)_2$ R group).

Forming peptide bonds – condensation and hydrolysis

Proteins are very large molecules. They are called **polymers** because they are made up of many similar subunits or molecules – the amino acids.

Two amino acids are joined by a condensation reaction, with one molecule of water lost. The resulting covalent bond is called a **peptide bond**, and the molecule formed is a dipeptide. This bond can be broken by a hydrolysis reaction using one molecule of water.

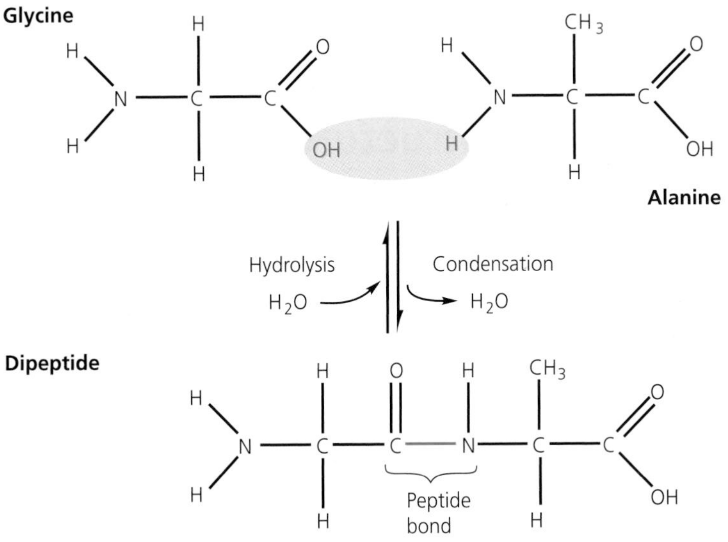

Glycine

Alanine

Hydrolysis H_2O

Condensation H_2O

Dipeptide

Peptide bond

Figure 3.3 A peptide bond between two amino acids, forming a dipeptide.

Tip

See Table 14.5 on page 264, which gives information about the 20 amino acids that make up proteins.

Key term

Peptide bond The bond that forms when two amino acids are joined by a condensation reaction with the loss of one water molecule. The bond may be broken by a hydrolysis reaction.

Test yourself

1 State the different roles that proteins play in a cell, and for each describe the role's importance to the cell.
2 Why are amino acids known as 'building blocks'?
3 What is the name of the covalent bond formed when amino acids join together? Describe how this bond forms.
4 What process is needed to break this bond?
5 What is the significance of the R group in an amino acid?
6 State the smallest R group in an amino acid, and name the amino acid. Name an example of an amino acid with a large R group.

Polypetides

Key term

Polypeptide A polypeptide is a polymer of amino acids, each joined to the others by a covalent bond called a peptide bond.

When a chain of amino acids is formed, each new amino acid is joined by a peptide bond, and the resulting chain is called a peptide. Longer chains of amino acids are called **polypeptides**.

Polypeptides differ in the number, type and order of the amino acids joined in the chain. Some polypeptides contain a smaller variety of amino acids, and others may have most or all of the 20 different amino acids.

Tip

Each peptide or polypeptide is formed by the process of protein synthesis involving ribosomes on the endoplasmic reticulum and messenger RNA (mRNA). Refer to the information on protein synthesis on page 53.

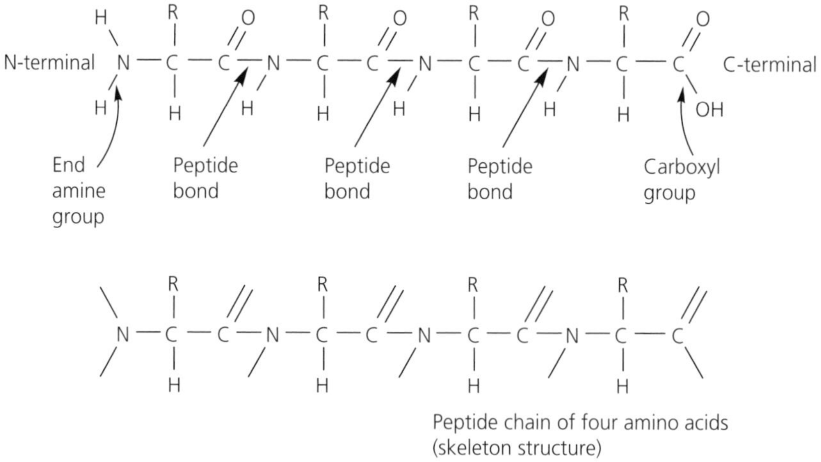

Figure 3.4 A peptide of four amino acid residues showing the position of the C-terminal and the N-terminal of the molecule.

Levels of protein structure

The shape of a protein molecule is determined by the amino acids and their arrangement in the chain. The shape of the protein molecule is vital in determining the properties of that protein and its function in the living organism.

Proteins are of different sizes. An example of a small protein is the human insulin protein, which has 51 amino acids. Other proteins may have over 600 amino acids.

All proteins are organised at different levels – primary, secondary, tertiary and quaternary – although only some have a quaternary structure.

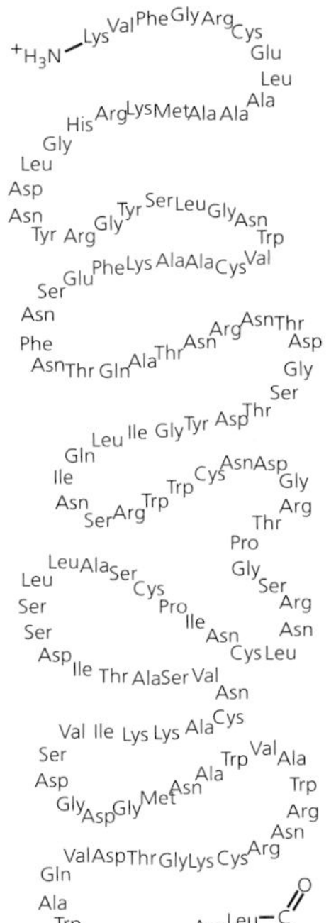

Figure 3.5 Primary structure of a protein.

Primary structure

The primary structure of a protein is the sequence, type and number of the amino acids in the amino acid chain, as well as the position of the disulfide bonds if present. This primary structure varies enormously in different proteins, which is why so many different types of protein are possible from the 20 amino acids. Each protein is made up of many amino acids, each joined to the chain by a condensation reaction forming a peptide bond. Each chain has its own unique structure as a result of the amino acids present. The structure and amino acids present determine its function. If one amino acid in the chain is changed, the function can be disrupted or even stopped completely.

Secondary structure

Once the primary structure is formed, the chain takes a particular shape by folding or coiling as a result of bonds that form between certain amino acids in the chain. The order of amino acids determines where the bonds form and so what shape occurs.

There are two main forms of secondary folding: the **alpha helix** and the **beta pleated sheet**. The alpha helix is held permanently in place by hydrogen bonds between amino acids in one part of the chain and those a little further along the chain. The beta pleated sheet folds in a concertina-like way, with hydrogen bonds connecting to an adjacent pleated sheet.

Hydrogen bonds are weak bonds, but the number of them present in such protein molecules helps maintain the stability of the molecule overall.

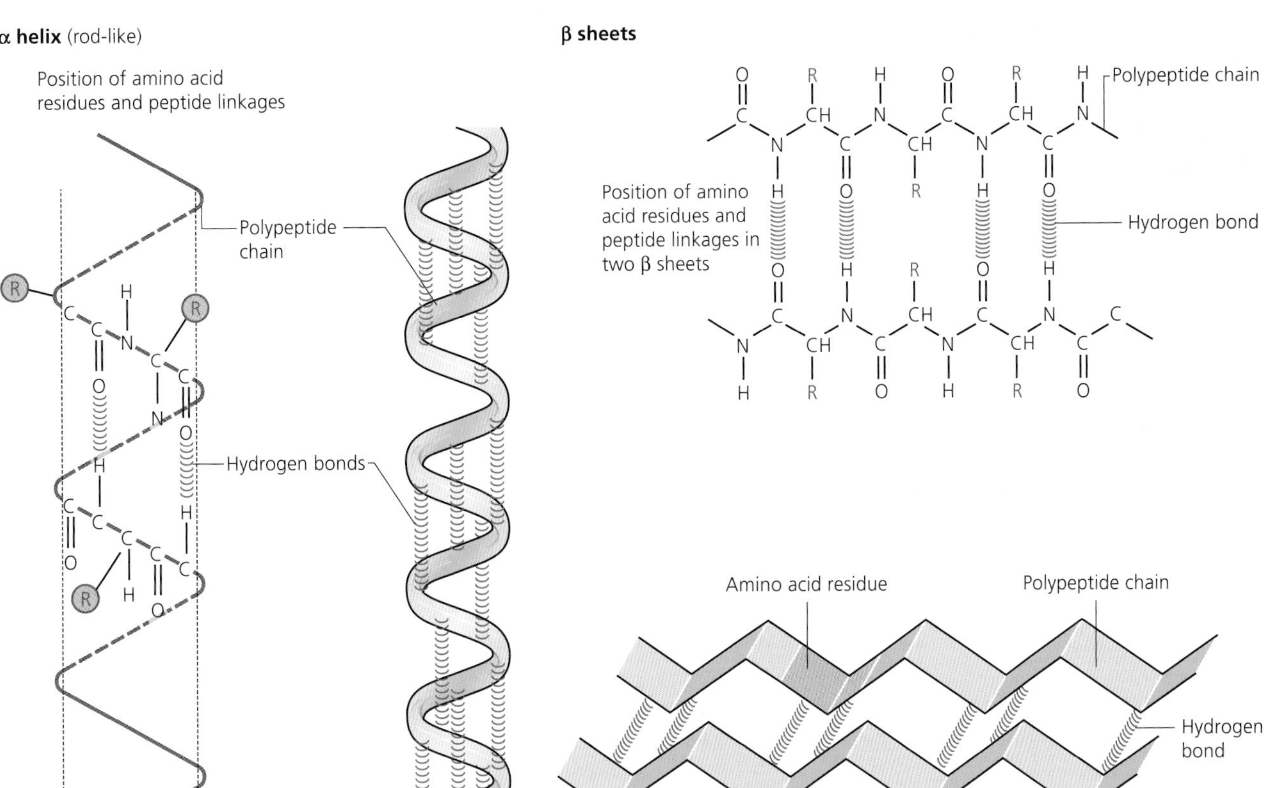

Figure 3.6 Secondary structure of a protein showing an alpha helix and a beta pleated sheet.

Tertiary structure

The tertiary structure is the three-dimensional shape of the protein molecule that occurs when the secondary structure becomes further coiled or twisted into a more complex shape. In each, the final shape depends on how the structure twists or coils. These complex shapes are held together by different types of bond that form because of the position of certain amino acids in the chain. Again, the sequence of the amino acids in the chain determines the exact shape of the molecule.

The two main types of tertiary structure are globular and fibrous.

When the protein folds and coils to form a three-dimensional shape, it is called a **globular protein**. This 3D shape makes globular proteins important molecules in metabolic processes. All enzymes, for example, are globular proteins. The globular shape allows enzymes to have a specifically shaped area called the active site, which is complementary to the specific substrate that the enzyme catalyses. Globular protein molecules are usually soluble and, in addition to enzymes, form protein-based hormones, plasma proteins, antibodies and the components of the cytoskeleton.

If the protein twists into a long fibrous structure, the protein is called a **fibrous protein**. These fibrous proteins are important in structural roles such as that of keratin in hair and nails, and collagen in skin, bone and cartilage. Fibrin, the blood-clotting protein, is a fibrous protein.

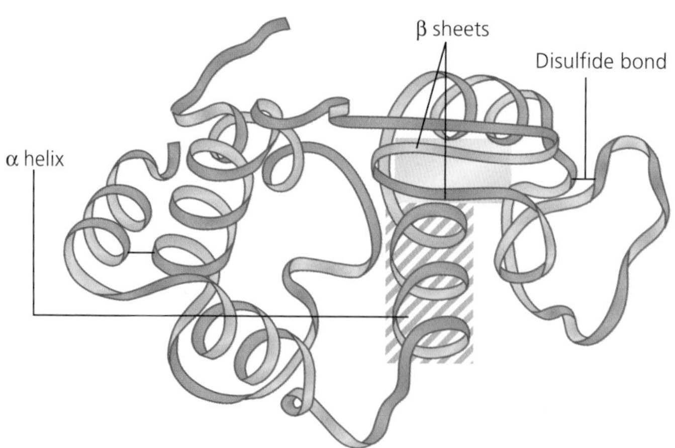

Figure 3.7 Tertiary structure of a protein showing the disulfide bonds forming the cross links to stabilise the molecule.

In both globular and fibrous proteins, a number of different types of bond stabilise the molecule and hold it in place. Some of these are weak bonds such as hydrogen bonds. Individual **hydrogen bonds** may be broken easily, but because they are present in large numbers they help to stabilise molecules. **Ionic bonds** and **hydrophobic interactions** also form between different R groups. Other bonds are covalent bonds – strong bonds that are not easily broken – such as the **disulfide bonds**.

Tip

The tertiary three-dimensional shape is vital to the function of the protein molecule.

Key terms

Globular protein Proteins with a 3D (three-dimensional) shape that forms after twisting and folding of the secondary structure.

Fibrous protein Proteins with a long, rope-like shape that forms after twisting of the secondary structure.

Hydrogen bond The attraction between the slight positive charge (δ^+) of a group (e.g. –NH) on one amino acid and the slight negative charge (δ^-) on another (e.g. –CO).

Ionic bond The attraction between a positively charged R group of one amino acid and a negatively charged R group of another.

Hydrophobic interaction The association between the hydrophobic R groups of amino acids (e.g. $-CH_3$ on alanine), where water is excluded.

Disulfide bond Covalent bond between sulfur atoms of the R groups of two cysteine amino acids.

Example

Trypsin is an enzyme that is secreted by the pancreas.

Figure 3.8 shows a ribbon model of the tertiary structure of trypsin.

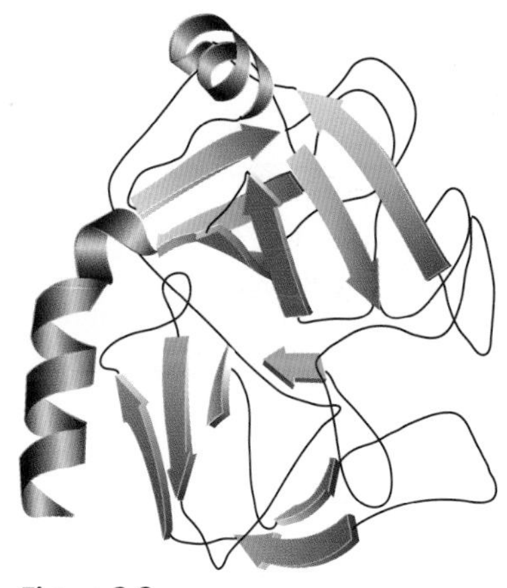

Figure 3.8

1 The primary structure of trypsin is not visible in the model. Describe how the primary structure would be depicted.
2 Figure 3.8 shows that trypsin has two forms of secondary structure.
   a) Name the two forms of secondary structure visible in the diagram of trypsin.
   b) State the type of bond that stabilises the secondary structure of proteins.
3 Explain how the tertiary structure of a protein such as trypsin is stabilised.

Answers

1 The primary structure would show all the amino acids in the correct sequence with peptide bonds joining each amino acid.
2 a) alpha helix and beta pleated sheet
   b) hydrogen bonds
3 The tertiary structure is stabilised by disulfide bonds, covalent bonds between sulfur-containing R groups, e.g. cysteine, and non-covalent bonds between R groups, e.g. hydrogen bonds, ionic bonds and hydrophobic interactions.

Key terms

Prosthetic group A non-protein part of a protein molecule that is permanently attached to the molecule and is vital for the normal functioning of the molecule.

Conjugated protein A globular protein with a prosthetic group, such as haem in haemoglobin.

Tip

The haem group in haemoglobin is responsible for carrying oxygen. Haem is a prosthetic group – a porphyrin ring with Fe (iron) in the centre. The Fe ion is either Fe^{2+} or Fe^{3+}.

Quaternary structure

Some proteins are made up of more than one polypeptide chain. In this case, two or more chains are held together and function as a whole. The protein will not function unless all the subunits are together. In some cases the polypeptide subunits are identical, while in other cases the subunits contain different types of polypeptide. In some an inorganic molecule or ion is also required for the protein to function.

Globular proteins – haemoglobin and enzymes

Haemoglobin is an example of a globular protein molecule with a quaternary structure and a haem group containing an inorganic ion – the metal iron, which gives the typical red colour. Adult human haemoglobin has four polypeptide chains – two α chains and two β chains – each associated with a haem group. The four chains are held together by a number of different bonds to make a stable globular molecule with a specific shape designed to carry out the molecule's function of oxygen transport. The haem part of the molecule is a **prosthetic group**, which makes haemoglobin a **conjugated protein**. You can see a diagram of haemoglobin on page 156 in Chapter 8.

Other globular protein molecules include enzymes (for example amylase and catalase, which rely on their three-dimensional shape in order to carry out their function of catalysing chemical reactions) antibodies and certain protein-based hormones such as insulin.

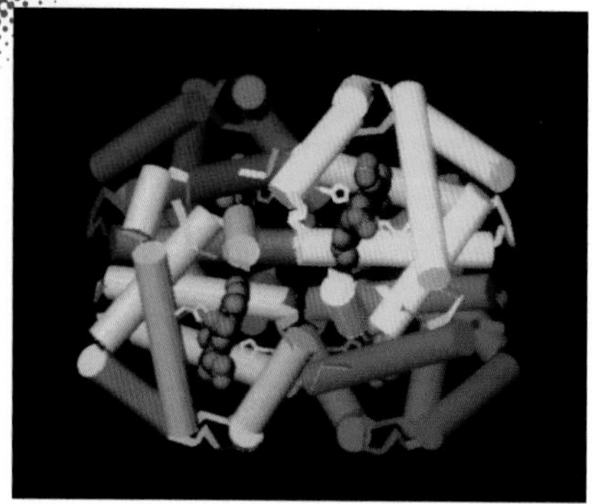

Figure 3.9 A haemoglobin molecule showing the four chains and iron attached.

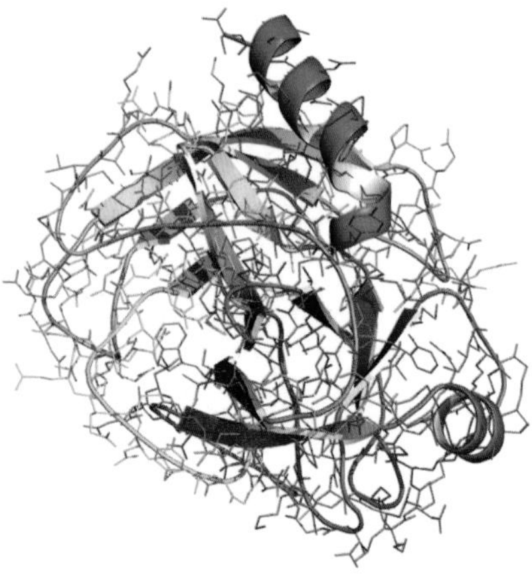

Figure 3.10 An enzyme molecule.

Tip

Refer to Chapter 4 to find out more about enzymes.

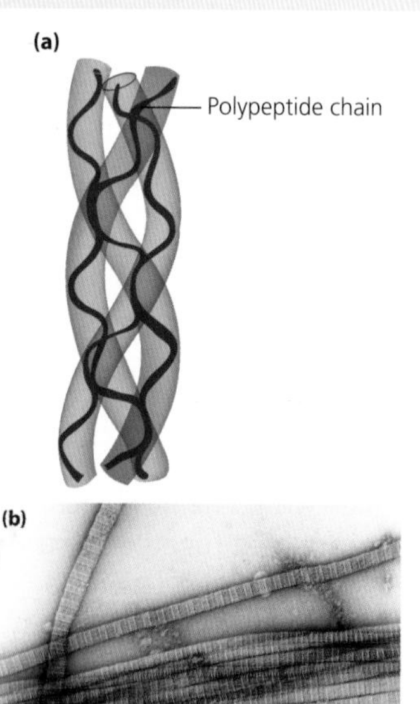

Figure 3.11 Collagen molecule showing its quaternary structure.

Insulin is a hormone produced by the beta cells of the islets of Langerhans in the pancreas. The amino acid chain was sequenced in 1955 by Fred Sanger and was the first protein to have its sequence determined. In humans it is composed of two polypeptide chains with 21 amino acid residues in chain A and 30 amino acid residues in chain B. The chains are held together by three disulfide bridges. Insulin is an important hormone in controlling glucose levels in the blood.

Fibrous protein – collagen

Collagen is an example of a fibrous protein with a quaternary structure; it has three polypeptide chains twisted around each other like a plait or rope. Each chain is made up of three repeating amino acids. Strength is provided by many hydrogen bonds and covalent bonds between the chains. The chains in turn form a collagen fibril that links with other fibrils to form a collagen fibre. Collagen is important, because it provides support and strength in many structures in the body such as the heart and arteries, tendons and bone, and cartilage and skin.

Table 3.1 compares the features of haemoglobin and collagen.

Keratin and elastin are examples of other fibrous proteins. Elastin has the ability to stretch and return to its original shape; it is found in connective tissue, skin, tendons and bone.

Table 3.1 Comparison between haemoglobin and collagen.

Feature	Haemoglobin	Collagen
Type of protein structure	globular protein	fibrous protein
Number of polypeptides	4	3
3D structure	folded into a compact ball-like shape	twisted into long fibres
Helical structure	folded into a right-handed alpha-helical structure	wound into a left-hand helical structure
Solubility in water	soluble	insoluble
Types of amino acid	most of the 20 different amino acids are present in the haemoglobin molecule	made up of just a few different types of amino acid; three amino acids, one of which is glycine form most of the collagen molecule
Prosthetic group	contains the prosthetic group haem	no prosthetic group
Role	transport of oxygen	provides strength in many areas of the body such as artery walls, tendons, cartilage and bone, and also provides the elasticity of the skin

Activity

Biuret test for proteins

In order to find the colour for the positive protein test, a student used the biuret test on a sample of protein suspension. Equal volumes of protein suspension and sodium hydroxide were shaken to mix the two solutions. A small amount of copper sulfate was then added drop by drop, shaking between each addition. The student then recorded the colour changes observed (see Table 3.2). The teacher gave the following hint: keep a tube of the original blue colour of the copper sulfate alongside the biuret test solution, to compare the colours.

Figure 3.12 Results of biuret test on protein suspension and control solution.

Table 3.2 Student's results table.

Solution	Before adding the reagents	After adding both reagents
Protein suspension	cloudy suspension	mauve colour
Control solution	cloudy suspension	copper sulfate blue

1 Suggest a suitable liquid that may be used in the control tube.
2 Explain why the protein is referred to as a *suspension* rather than a *solution*.
3 Explain why this test is a *qualitative test* and not a *quantitative test*.
4 What changes must be made to make this a quantitative test?

Importance of inorganic ions in biological processes

Key term

Cofactor A non-protein chemical compound that is needed for the protein's biological activity.

Biological processes involve enzymes and their substrates and also a number of important inorganic ions. Inorganic ions in animals and plants are essential for vital cellular activity. They contribute to the osmotic pressure of body fluids, may act as cofactors and may provide other important functions.

Test yourself

7 Describe the importance of the primary structure of a protein in determining the shape and form of the final shape of that protein.
8 There are two main types of tertiary structure. Describe the differences between the two types and give an example of each.
9 Describe the quaternary structure of a protein.
10 Describe a named example of a protein that has a quaternary structure (other than haemoglobin).
11 Haemoglobin has a prosthetic group. What is a prosthetic group?
12 What is the prosthetic group in haemoglobin? Describe the role it plays in the body.

Tip

See Chapter 5 to find out about osmosis.

Tip

You will learn about the roles of these ions throughout your A Level course.

Table 3.3

	Inorganic ion	Chemical symbol	Role of ion
Cations	Calcium ions	Ca^{2+}	• part of bone and enamel structure as calcium phosphate • a cofactor in blood clotting • an ion involved in nerve transmission across the synapse and in muscle contraction
	Sodium ions	Na^{+}	• an electrolyte • essential function in nerve transmission • essential in water reabsorption in loop of Henle and collecting duct
	Potassium ions	K^{+}	• an electrolyte • essential in nerve transmission • essential in water reabsorption in loop of Henle and collecting duct • used in plant guard cells as part of the stomatal opening mechanism
	Hydrogen ions	H^{+}	• in hydrogen bonds, which are common bonds in many biochemical molecules • involved in ATP formation • involved in the control of blood pH and in the transport of carbon dioxide
	Ammonium ions	NH_4^{+}	• an intermediate ion in the deamination of proteins
Anions	Nitrate	NO_3^{-}	• nitrogen source for green plants to manufacture proteins; it is the form taken up from the soil
	Hydrogen carbonate	HCO_3^{-}	• involved in carbon dioxide transport in the blood, with H^{+}
	Chloride	Cl^{-}	• the shift of chloride ions into and out of red blood cells maintains pH balance during carbon dioxide transport
	Phosphate	PO_4^{3-}	• as phospholipids, phosphates form part of cell membranes • forms calcium phosphate, an important constituent of bone for giving strength • constituent of ATP and nucleic acids
	Hydroxide	OH^{-}	• one of the important ions in bonding between biochemical molecules

Biosensors

Chromatography and biosensors offer methods for obtaining quantitative results. A biosensor is a very precise and accurate analytical device. It converts a biological response into an electrical signal using a catalyst, usually a highly specific and stable enzyme.

Tip

Refer to Chapter 17 to learn how to perform chromatography.

A common example of a biosensor is the blood glucose biosensor. This uses the enzyme glucose oxidase to break down blood glucose. The enzyme, which includes FAD, oxidises the glucose and then uses two electrons to reduce the FAD to $FADH_2$. This is in turn oxidised by the transducer (the electrode in the device) which creates a current. This current is a measure of the glucose concentration. A drop of blood is obtained from the patient using a sterile lancet or pricker, which is squeezed onto a test strip. This strip is inserted into the biosensor meter which displays the blood glucose reading as a digital figure.

$$ATP + H_2O \rightarrow ADP + P_i + 30.5\,kJ\,mol^{-1}$$

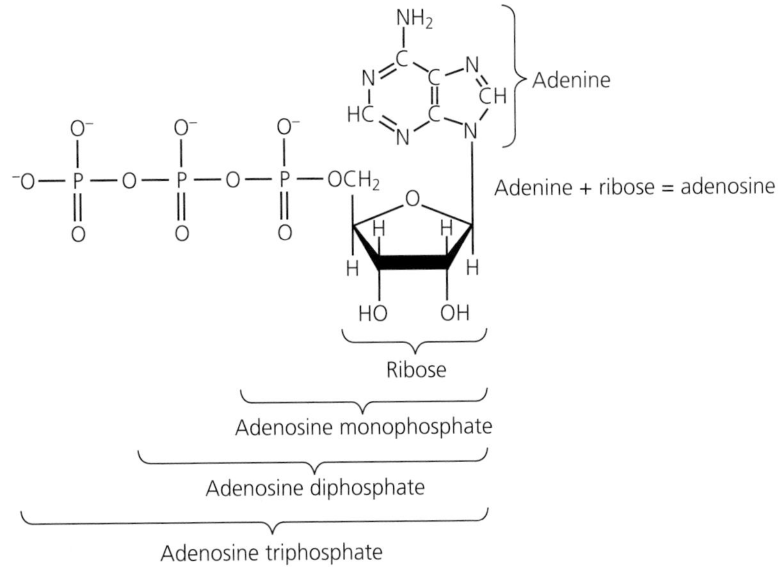

Figure 3.13 Structure of ATP, ADP and AMP

Structure of ADP and ATP

Adenosine triphosphate or ATP is a compound that transfers energy within cells. It is the universal energy currency. It is composed of the nitrogenous base adenine covalently bonded to the pentose sugar ribose (forming adenosine) and three phosphate groups forming a short chain. During energy transfer the final phosphate group is removed by hydrolysis to release energy and phosphate, leaving the compound ADP (adenosine diphosphate).

Nucleic acids

There are two types of nucleic acid: deoxyribonucleic acid (**DNA**) and ribonucleic acid (**RNA**). Nucleic acids are vital molecules because they carry the genetic code in all living things and are important in controlling cellular activity and protein synthesis.

Nucleic acids are made up of carbon, hydrogen, oxygen, nitrogen and phosphate.

DNA is a double-stranded **polynucleotide**, which means it is made up of many nucleotide molecules joined to each other with covalent bonds, which form by condensation reactions.

Key terms

DNA Deoxyribonucleic acid is a double-stranded polymer of nucleotide molecules that carries the information for protein synthesis. It contains the pentose sugar deoxyribose.

RNA Ribonucleic acid is a single-stranded polymer of nucleotide molecules containing the pentose sugar ribose. There are three types of RNA, which all have important roles in protein synthesis.

Polynucleotide A polymer of nucleotide monomers covalently bonded together.

Organic nitrogenous base A nitrogen-containing organic compound that is a constituent of nucleotides. There are five of these in nucleic acids.

Structure of a nucleotide as a monomer

Each nucleotide is a monomer and is the basic building block of the nucleic acid molecules.

Unlike the monomers of other biological molecules, nucleotides are made up of three biological molecules that are bonded together with covalent bonds formed by condensation reactions.

The subunits are:

- a pentose (5C) sugar molecule – either ribose or deoxyribose – both of which contain five carbon atoms
- an **organic nitrogenous base**: adenine (A), cytosine (C), guanine (G), thymine (T) or uracil (U)
- a phosphate group.

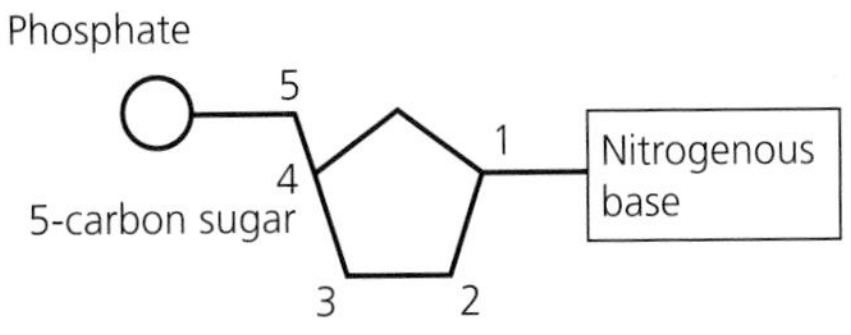

Figure 3.14 A nucleotide.

DNA contains the bases adenine, cytosine, guanine and thymine. RNA contains the bases adenine, cytosine, guanine and uracil.

Note that RNA has the same bases, except that uracil replaces thymine.

Organic bases

Organic nitrogenous bases are of two types: *purines* or *pyrimidines*. Purines are larger than pyrimidines. Purines consist of two carbon–nitrogen rings, whereas pyrimidines consist of a single carbon–nitrogen ring.

Pyrimidines:

Thymine

Cytosine

Uracil

Purines:

Adenine

Guanine

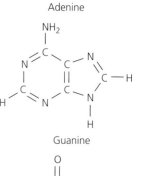

Figure 3.15 Molecules of purines and pyrimidines.

> **Tip**
>
> Quick memory aid: think of a nucleotide as made of three components looking a little like a house (the sugar) with a satellite dish (phosphate group) attached on one side and a garage (the nitrogenous base) attached on the other side.

Structure of DNA – a polynucleotide

Nucleotides are joined together by condensation reactions to form a chain, the polynucleotide. The phosphate group of one nucleotide is joined to the sugar molecule of the next nucleotide in the chain, and this is repeated for each nucleotide added. This forms one of the two sugar–phosphate backbones of the DNA molecule. At this stage, the chain is a single chain with the organic bases attached and projecting out from the chain. Only nucleotides with the same pentose sugar are attached to each other to form the chain.

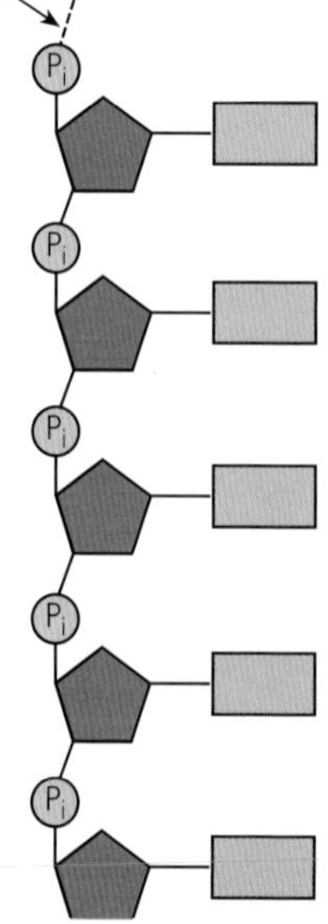

Nucleotides are added at this end of the growing polynucleotide
Nucleotides become chemically combined together, phosphate to pentose sugar, by covalent bonds, with a sequence of bases attached to the sugar residues. Up to 5 million nucleotides condense together in this way, forming a polynucleotide (nucleic acid).

Figure 3.16 Part of a nucleotide chain showing the nucleotide bases projecting outwards.

Tip

The double-stranded molecule looks like a ladder, with the two sugar–phosphate backbones forming the ladder supports and the organic bases forming the rungs between, because they project from the backbone.

Tip

RNA is also a nucleic acid but has important differences in both structure and function from DNA.

DNA – the nucleic acid

The nucleic acid forms when two polynucleotide chains join together, by hydrogen bonds between the nitrogenous bases, to form a double-stranded molecule.

When the nucleic acid contains the sugar deoxyribose and the base thymine, the molecule is known as DNA.

The bases always pair up in a specific way: a pyrimidine always joins with a purine, because only this pairing allows hydrogen bonds to form between the bases. This specific pairing is called **complementary base pairing**. In DNA, adenine always pairs with thymine, and guanine with cytosine. This pairing occurs because adenine forms two hydrogen bonds with thymine, and guanine forms three hydrogen bonds with cytosine, and so they can only pair in this combination. The importance of the base pairing is that the molecule becomes very stable and is always the same width along the chain, because the bases hold the backbones the same distance apart.

The double strand of DNA is formed when two parallel polynucleotide strands are joined together by the hydrogen bonds formed between nitrogenous base pairs that project out from the sugar–phosphate backbone.

The strands are described as **antiparallel**, because they run in opposite directions to each other.

Figure 3.17 The double strand of DNA showing the antiparallel chains.

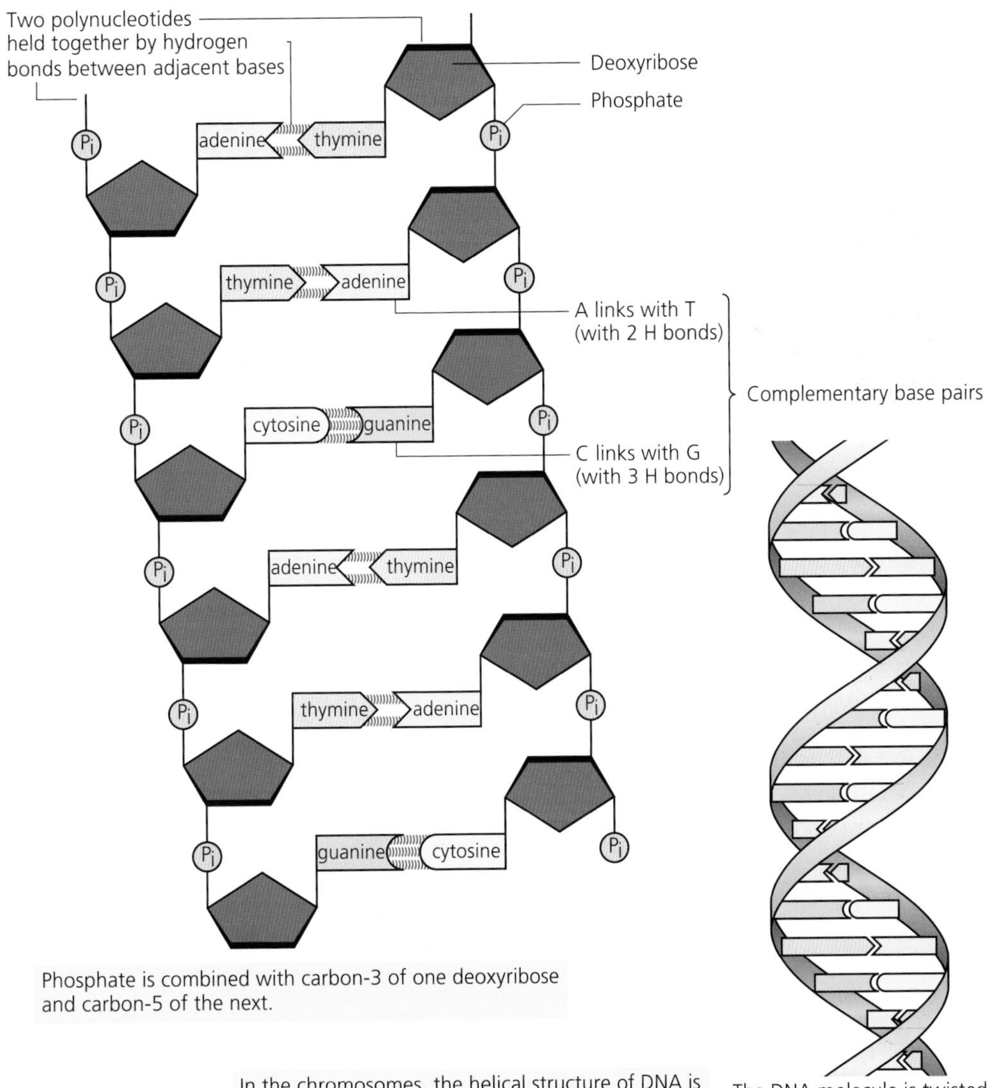

Key term

Semi-conservative replication
Each new DNA molecule is formed of one conserved strand from the original molecule, which acts as the template, and one strand of new nucleotides.

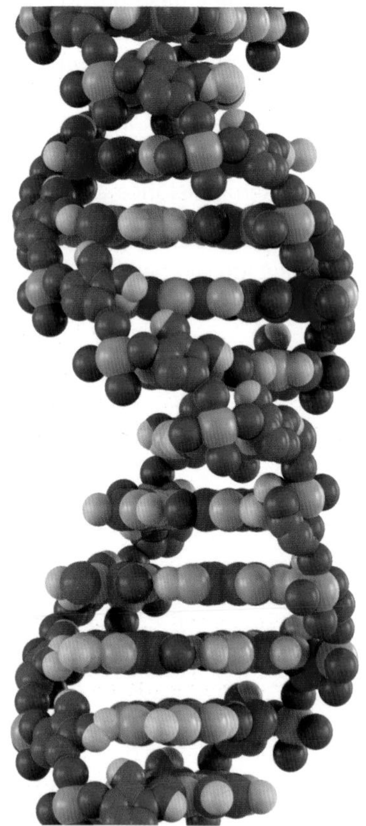

Figure 3.18 A double helix.

Tip

See Chapter 6 for details of cell division and the cell cycle, and for a definition of a chromatid.

Tip

The role of the enzyme polymerase is to attach to one end of the nucleotide chain and covalently bond other nucleotides to the chain to form the polynucleotide. The enzymes are specific, and so there are two types: DNA polymerase and RNA polymerase.

Once formed, the DNA molecule is twisted into a helical shape with both backbones twisting, and so it is called a double helix.

Tip

A double helix cannot be compared to a spiral staircase, because the spiral staircase involves one backbone remaining upright and the other twisting around it to form level rungs, whereas in DNA both backbones twist. A double helix is therefore similar to a ladder that has been twisted around both uprights.

DNA semi-conservative replication

The DNA replicate must be an exact copy to form two sister chromatids, and so the process must be accurate.

To replicate, the DNA double helix unwinds and then unzips down the centre of the two strands, with the hydrogen bonds between the bases breaking leaving two separate strands. The enzyme helicase is involved in this process and also in holding the two separate strands apart.

The two separate strands act as templates for new double DNA molecules to be formed. New DNA nucleotides are attached to the exposed bases; they line up with and are attached to the original separated strands. Exact copies are made because of complementary base pairing. Adenine always pairs with thymine, and cytosine always with guanine. Hydrogen bonds form between the base pairs, and then the enzyme DNA polymerase catalyses the condensation reaction to covalently bond each nucleotide to the adjacent one along the sugar–phosphate backbone.

Two exact DNA copies have now been made. Each copy carries one strand of the original DNA that acted as the template (it is conserved) and one strand of new nucleotides. This is called **semi-conservative replication**.

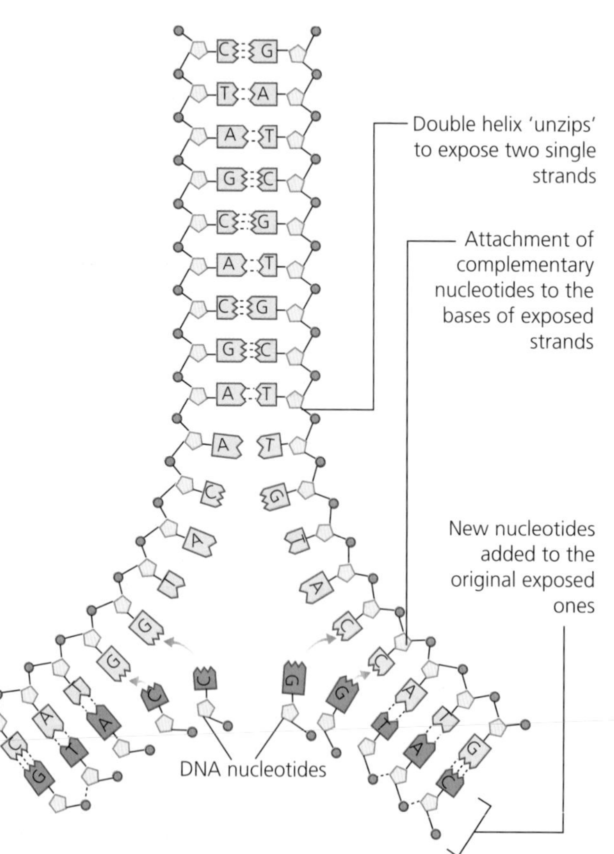
Double helix 'unzips' to expose two single strands
Attachment of complementary nucleotides to the bases of exposed strands
New nucleotides added to the original exposed ones
DNA nucleotides

Figure 3.19 DNA semi-conservative replication.

Key terms

Triplet code The three nitrogenous bases read along the DNA strand that code for amino acids.

Transcription The process of copying the code exactly to form a template of mRNA.

The gene – the genetic code

The importance of DNA is its ability to carry, as a code, the information to make proteins. The code for DNA is a **triplet code** in which three bases code for an amino acid.

There are 20 different amino acids and 64 different combinations of triplet codes; therefore some triplets code for the same amino acids. There may be up to six different codes for the same amino acid, although these usually only differ by one base; for example, the amino acid serine is coded for by the triplets TCT, TCC, TCA, TCG, AGT and AGC (in which A is adenine, T is thymine, C is cytosine and G is guanine). The advantage of having different codes for the same amino acid is that if a mutation causes a base change, the triplet may still code for the same amino acid and therefore not change the protein produced.

Some amino acids have only one triplet code; the amino acids methionine and tryptophan are examples of this. Three codes (AGT, GAT and AAT) act as full stops signalling the end of the message.

DNA is an extremely long molecule that codes for numerous proteins. Each short length of DNA known as a gene is in fact a sequence of bases that code for the amino acids making up a single protein. Because no two proteins are the same length, the length of the gene also varies; some genes may be as short as 50–100 nucleotides, but most are thousands of nucleotides long.

Protein synthesis

Transcription

Within each gene one of the DNA strands is known as the 'coding' strand and the other is known as the 'template' strand. When the cell needs to produce a protein from this gene the enzyme RNA polymerase binds to the DNA and unzips the two strands. The RNA polymerase then moves along the template strand in the 3′ to 5′ direction. As it moves along it pairs free RNA nucleotides with complementary nucleotides on the DNA template strand. These RNA nucleotides are then covalently linked with phosphodiester bonds to create a polynucleotide chain, growing in the 5′ to 3′ direction, known as mRNA. Note that, because the new mRNA has been formed by complementary pairing with the DNA template strand, it will have the same sequence of nucleotides (with uracil instead of thymine) as the DNA coding strand.

Translation

The mRNA, once formed, leaves the nucleus via the nuclear pore for the ribosomes, where it acts as the template for proteins to be synthesised. Each triplet is called a **codon**.

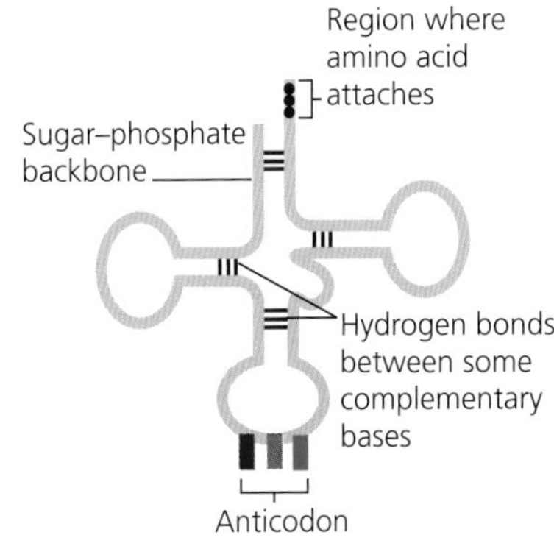

Figure 3.20 A molecule of tRNA.

There are different short transfer RNA (tRNA) molecules for each amino acid, and so they pick up the correct amino acid and carry it to the correct position on the mRNA template. At the opposite end of the tRNA is an **anticodon**, which is a triplet complementary to the mRNA codon. This ensures that the amino acids are correctly sequenced along the mRNA.

The ribosome reads the code three bases (one codon) at a time. Each amino acid is in turn attached by a condensation reaction to form a peptide bond with the next amino acid. The ribosomes move along the chain and hold the tRNA and amino acid in place by temporary hydrogen bonds that break once the peptide bond is formed, leaving

Key term

Translation The process by which information in the code from the sequence of bases in messenger RNA controls the sequence of amino acids that a ribosome assembles.

the tRNA to move off to collect another amino acid and the ribosome to move along the chain. This process continues until the chain of amino acids is complete and a primary protein structure is formed. The process is called **translation**, because the code is translated into the amino acid sequence.

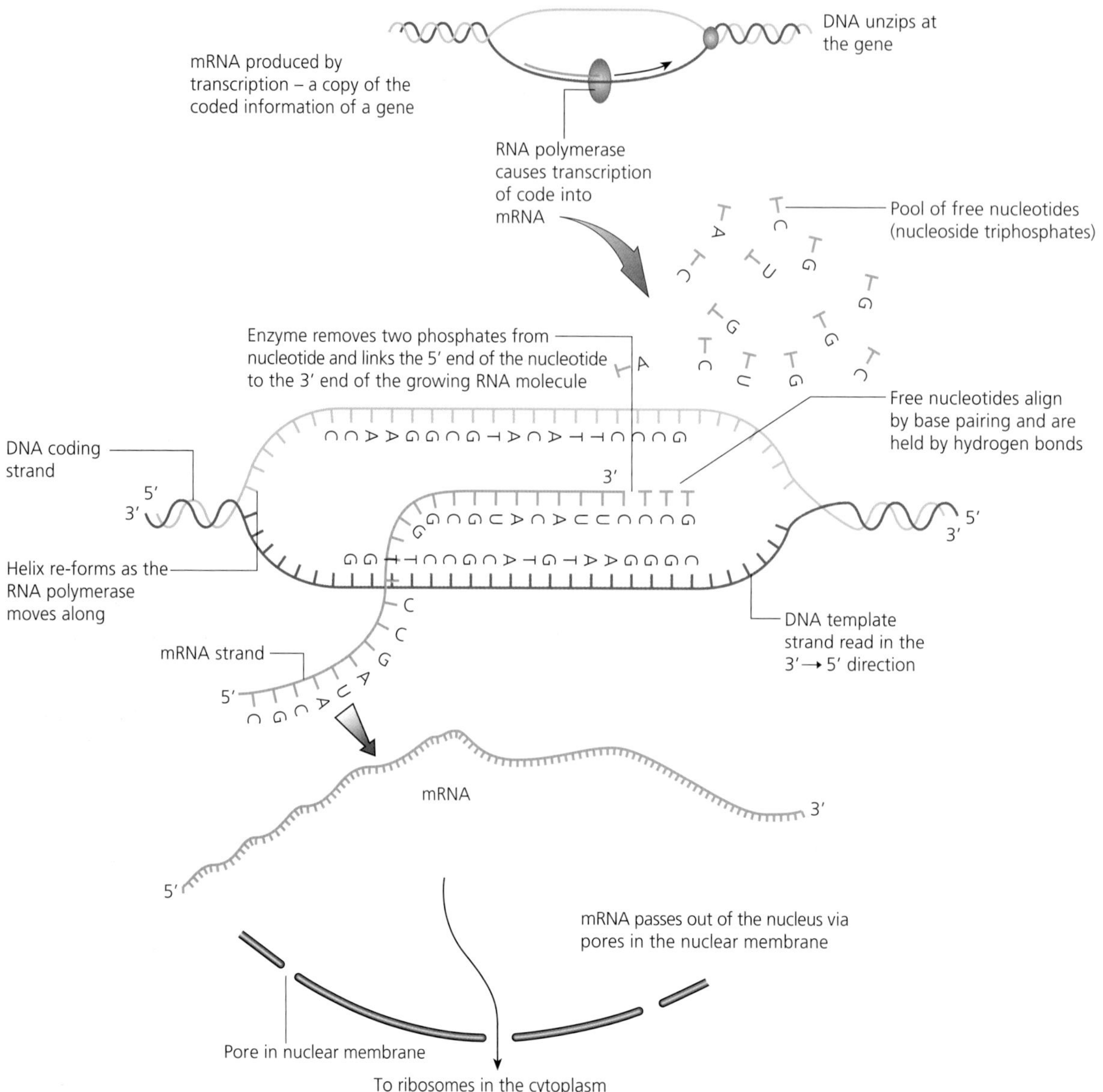

Figure 3.21 Protein synthesis.

RNA – a polynucleotide

RNA is a single-stranded polynucleotide, which means it is made up of a number of RNA nucleotide molecules joined to each other with covalent bonds that form by condensation reactions. RNA forms relatively short lengths of up to a few thousand nucleotides.

Some essential differences between DNA and RNA are outlined in Table 3.4.

There are three types of RNA: messenger RNA (mRNA), transfer RNA (tRNA) and ribosomal RNA (rRNA).

Messenger RNA

When the DNA has 'unzipped', mRNA is made by attaching RNA bases to the short section of exposed DNA. RNA nucleotides attach to the exposed DNA bases with hydrogen bonds, building a new mRNA molecule. The complementary base pairing means that the correct bases are always attached. The mRNA therefore copies the information from the DNA code – this is **transcription**.

Transfer RNA

tRNA molecules are short chains of RNA that fold on to themselves to form some parts that are doubled, creating a clover-shaped molecule. These tRNA molecules carry amino acids to the mRNA and the ribosomes, where an enzyme forms peptide bonds between each pair of amino acids to form a polynucleotide chain.

Ribosomal RNA

rRNA molecules are short chains of RNA which, when attached to the ribosomal protein molecules, form ribosomes. These are the site of protein synthesis.

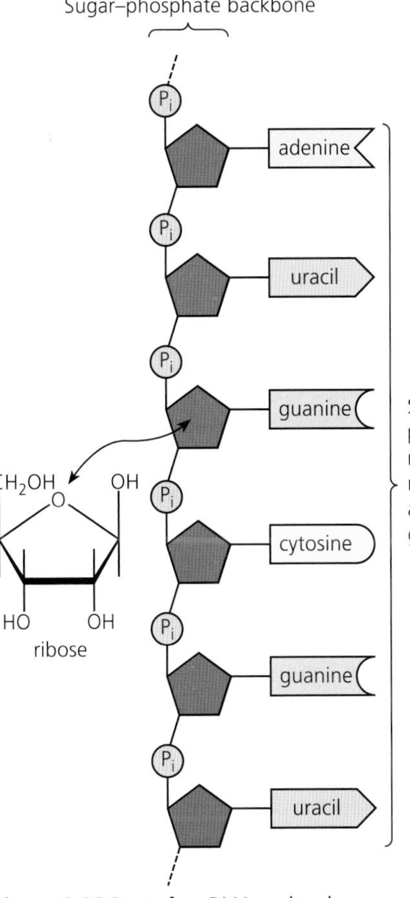

Figure 3.22 Part of an RNA molecule.

Table 3.4 Differences between DNA and RNA.

Feature	DNA	RNA
Basic shape	double strand	single strand
Nitrogenous bases	adenine, thymine, cytosine and guanine	adenine, uracil, cytosine and guanine
Type of pentose sugar	deoxyribose	ribose
Helical nature of the molecule	a double helix of two antiparallel polynucleotide chains	one polynucleotide chain twists into a helix and coils back on itself, so some sections of the chain join to other parts of the chain for tRNA and rRNA, mRNA is linear
Types of molecule	one type of DNA with numerous variations due to the arrangement of the bases	three types of RNA – mRNA, tRNA and rRNA – each having a different role in protein synthesis
Role	long term storage of genetic information; sequence of bases codes for assembly of amino acids to make proteins	protein synthesis

Example

In the 1950s, Erwin Chargaff, working in Paris, determined the relative quantities of the four bases in DNA in different organisms. His results provided important evidence for the model of DNA proposed by James Watson and Francis Crick in 1953. Some of Chargaff's data is shown in Table 3.5.

Table 3.5 Extracts from Chargaff's data.

Organism	Adenine %	Thymine %	Guanine %	Cytosine %
Escherichia coli (bacterium)	24.7	23.6	26.0	25.7
A yeast	31.3	32.9	18.7	17.1
Wheat	27.3	27.1	22.7	22.8
Octopus	33.2	31.6	17.6	17.6
Sea urchin	32.8	32.1	17.7	17.3
Chicken	28.0	28.4	22.0	21.6
Human thymus gland	30.9	29.4	19.9	19.8
Human liver	30.3	30.3	19.5	19.9
Human sperm	30.7	31.2	19.3	18.8

1 Comment on Chargaff's data, explaining how it proved useful to Watson and Crick.

2 Table 3.6 shows Chargaff's data for the DNA extracted from a virus.

Table 3.6 Chargaff's data for the DNA extracted from a virus.

Organism	Adenine %	Thymine %	Guanine %	Cytosine %
Virus	24.0	31.2	23.3	21.5

a) Explain how the results for the virus differ from the results for all the organisms listed in Table 3.5.

b) Suggest why the results for the virus are different from those for all the organisms in Table 3.5.

Answers

1 Data from the table shows that for humans, chickens, yeast, wheat, octopus and sea urchins, the percentage of adenine equals the percentage of thymine, the percentage of cytosine equals the percentage of guanine, and the percentage of purines (adenine plus cytosine) equals the percentage of pyrimidines (thymine plus guanine), allowing for a small amount of experimental error. This suggests that there is pairing between the bases. Thymine in one polynucleotide pairs with adenine on the other polynucleotide; similarly, cytosine pairs with guanine.
Purine plus purine is too big, and pyrimidine plus pyrimidine is too small; the base pairing of purine plus pyrimidine fits into the width of DNA (2 nm). It also suggests that bases face inwards rather than outwards, and the hydrogen bonding between the bases stabilises the double-helical structure of DNA.
The data shows that the percentage of adenine plus thymine does not necessarily equal the percentage of cytosine plus guanine, and that the DNA from different species has a different adenine–thymine:cytosine–guanine ratio. These ratios are the same in different tissues from a particular species. This suggests that DNA could be the genetic material (and not protein as was widely thought to be the case).

2 a) The percentages of adenine and thymine, as well as cytosine and guanine, are not the same, but three percentages for adenine, guanine and cytosine are similar.

b) Virus DNA is single-stranded DNA, not a double-stranded double helix as for the organisms in Table 3.5.

Comparing the roles of DNA and RNA

Table 3.7 compares the roles of DNA and RNA.

Table 3.7 The roles of DNA and RNA.

Feature of the role	DNA	RNA
Information code	The base sequence is a code used to build proteins – it is an information store.	RNA does not store information, but copies the DNA code by hydrogen bonding RNA nucleotides to the DNA bases and forming a chain of mRNA.
Replication	Complementary base pairing means code can be copied exactly to form new DNA strands. Each new DNA molecule has one of the old polynucleotide chains and one new chain.	Complementary base pairing allows a new mRNA molecule to be built by copying from the DNA code – this is transcription.
Size of molecule and information stored	Large amounts of information can be stored because the strand is extremely long.	Large amounts of information can be copied as mRNA, but small amounts of information are copied each time and so the molecules are much shorter, usually no more than a few thousand nucleotides long.
Stability of the molecule	A stable molecule, because of the covalent bonds between and within the nucleotides, and the hydrogen bonds between all the bases in the molecule. It is further stabilised by twisting into a double helix.	Less stable. It can easily be formed from RNA nucleotides attaching to the DNA it is copying, and is easily broken down once used and then reformed into another mRNA as needed. tRNA and rRNA are more stable than mRNA.
Copying information	Hydrogen bonds between bases mean the code can be unzipped to copy or read information.	Hydrogen bonds between the complementary bases of DNA and mRNA mean the exact copy is formed and then can easily be broken to allow mRNA to move out of the nucleus through the nuclear pore. tRNA and rRNA are not involved in copying information.

DNA precipitation

Isolating DNA from cells is an important technique in molecular biology used as the starting point for a number of other investigations. The 'Marmar preparation' is one method used to precipitate and isolate DNA. The cells are first disrupted by breaking the cell and nuclear membranes using a concentrated detergent solution. Filtering the resulting suspension removes cell debris and membrane fragments, leaving the soluble proteins and the DNA. A protease enzyme is then used to remove the protein leaving only the DNA, which can be precipitated using ice cold ethanol. The resulting white precipitate can then be used for analysis or in other investigations.

Test yourself

13 Name the monomers of nucleic acids.
14 How does the structure of these monomers differ from the monomers of polysaccharides and proteins?
15 Name the five organic nitrogenous bases in nucleic acids.
16 Describe the double-helix structure of DNA.
17 Compare and contrast the structure and function of DNA and RNA.
18 What are the different types of RNA and what different roles do they play in protein synthesis?
19 Make a table to compare the chemical elements that make up carbohydrates, lipids, proteins and nucleic acids.

Exam practice questions

1 Which of the following is a correct match between the macromolecule and the bonds between its monomers?

A cellulose, hydrogen bonds
B nucleic acid, phosphodiester bonds
C phospholipid, peptide bonds
D protein, disulfide bonds (1)

2 Which applies to all proteins that have quaternary structure?

A composed of a single polypeptide with α helices
B composed of four polypeptides with a complex 3D shape
C composed of two or more polypeptides
D composed of more than two polypeptides (1)

3 The diagram below shows a short length of a molecule of DNA.

a) i) Name K–N. (4)

ii) The two polynucleotides in each molecule of DNA are antiparallel. Explain why they are described as *antiparallel*. (3)

b) State four ways in which a short length of mRNA differs from the DNA shown in the diagram. (4)

c) State the roles of mRNA, tRNA and rRNA in cells. (3)

4 The diagram shows two amino acids, glycine and alanine.

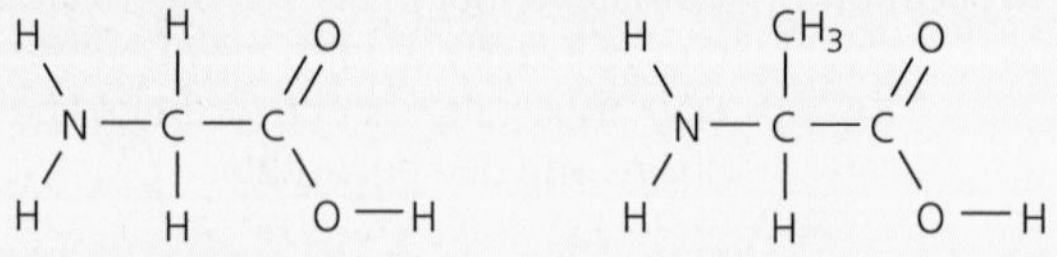

a) i) Copy and complete the diagram to show how the two amino acids are joined to form a dipeptide. (4)

ii) Name the type of reaction that you have drawn. (1)

b) State three ways in which amino acids differ from α-glucose. (3)

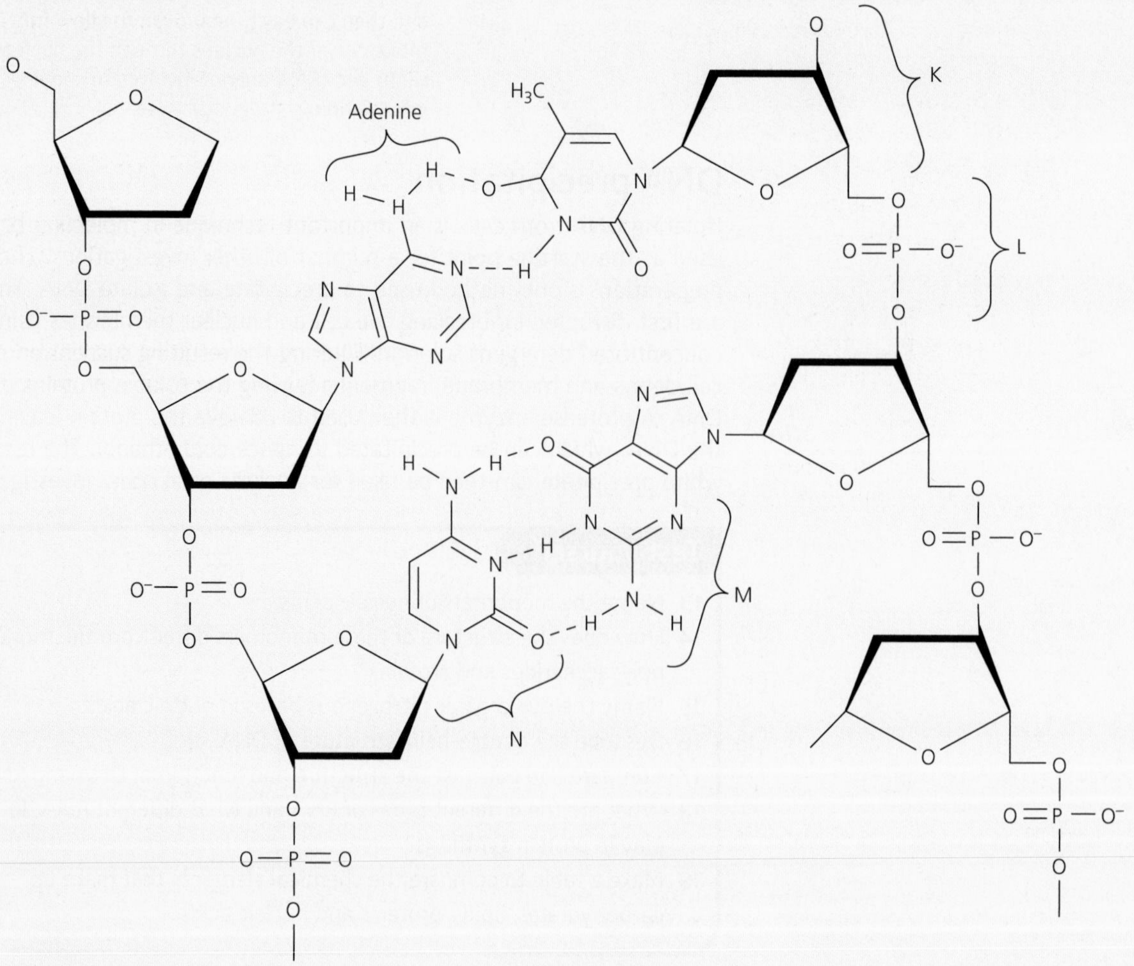

c) Outline how a tripeptide is broken down into its constituent amino acids. *(2)*

d) Collagen and elastin are two structural proteins found in the walls of arteries. Outline the contributions of these two proteins to the function of arteries. *(3)*

5 Lysozyme is a protein that catalyses the breakdown of components of bacterial cell walls. The gene for human lysozyme has been inserted into organisms such as yeast and rice to produce large quantities of the protein for use in research and as a food preservative. The table shows a short sequence of the DNA in the gene that codes for the first seven amino acids of the protein.

1	AAG	TTC	lysine
2	GTT	CAA	valine
3	TTT	AAA	phenylalanine
4	GAG	CTC	glutamic acid
5	AGA	TCT	arginine
6	TGC	ACG	cysteine
7	GAA	CTT	glutamic acid

a) Use the table above to explain the following features of the genetic code.

i) a triplet non-overlapping code *(4)*

ii) a degenerate code *(2)*

b) Human lysozyme has been produced by cells of yeast and rice. Explain how this shows another important feature of the genetic code. *(2)*

c) Suggest why the triplet at position 6 is important in determining the activity of lysozyme in killing bacteria. *(3)*

6 Amino acids differ from one another in the structure of their R groups (side chains). The diagram shows the R groups of four amino acids.

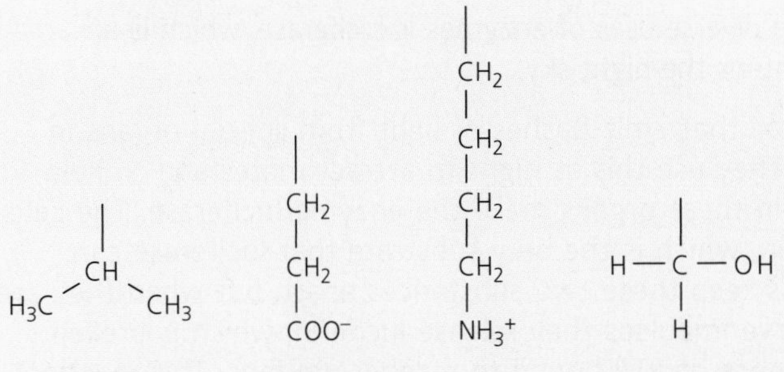

Valine: $-CH(CH_3)_2$ (CH bonded to H_3C and CH_3)

Glutamic acid: $-CH_2-CH_2-COO^-$

Lysine: $-CH_2-CH_2-CH_2-CH_2-NH_3^+$

Serine: $H-C-OH$ with H below

Valine Glutamic acid Lysine Serine

The R groups of amino acids can be polar or non-polar. Some of the polar R groups are basic and others are acidic. Which shows the correct identification of the R groups? *(1)*

	A	B	C	D
Non-polar	glutamic acid	valine	serine	lysine
Polar and basic	valine	lysine	glutamic acid	serine
Polar and acidic	serine	glutamic acid	lysine	valine
Polar and neutral	lysine	serine	valine	glutamic acid

Stretch and challenge

7 Plantaricin-423 is a polypeptide isolated from a strain of the bacterium *Lactobacillus plantarum*. This polypeptide is composed of 37 amino acids, but their sequence is not known.
The first steps in determining the sequence of amino acids in a protein involve hydrolysing the protein completely and using chromatography.

Suggest the reason for these two steps and outline how they are carried out.

8 The biuret test can be used as a simple qualitative test to find out if protein is present in a sample of biological material. The biuret test can also be used quantitatively to estimate the concentration of protein in different solutions.

Explain how the biuret test can be used to determine the concentration of protein in different colourless fluids.

Chapter 4

Enzymes

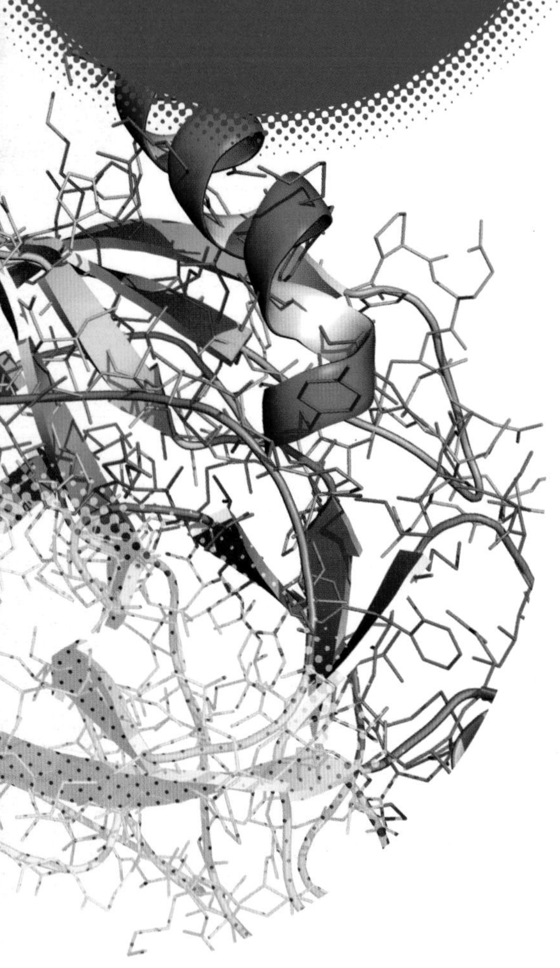

Prior knowledge

Before you start, make sure that you are confident in your knowledge and understanding of the following points.

- All catalysts speed up chemical reactions and are not used up during the reaction.
- In a catalysed reaction, substrate molecules are converted to product molecules.
- Enzymes are biological catalysts made by cells. Almost all chemical reactions that occur within living organisms are catalysed by enzymes.
- Enzymes are globular proteins, each type with a specific three-dimensional shape.
- The active site is the part of an enzyme molecule that interacts with the substrate.
- Enzymes are specific to a particular reaction because they can only interact with one type of substrate. This is because only one type of substrate molecule fits into the active site.
- The rate of any enzyme-controlled reaction is affected by temperature, pH, enzyme concentration and substrate concentration.

Test yourself on prior knowledge

1. Define the term *catalyst* and give an example of a chemical catalyst and a biological catalyst.
2. State how biological catalysts differ from chemical catalysts.
3. Describe a globular protein.
4. What is the importance of the active site in an enzyme?
5. Enzymes are described as being specific. Explain what is meant by this.
6. List four factors that influence the activity of enzymes.

Figure 4.1 The firefly, *Photinus lucicrescens*, with its glowing abdomen.

Life is completely dependent on enzymes. Enzymes form the biological molecules that make up living tissues and they break down the biological molecules that provide organisms with energy and building materials.

An example of the diverse uses of enzymes is luciferase, which is an enzyme that brightens the night sky.

Fireflies are beetles that emit flashes of light from special organs in their abdomens. They use this at night to attract mates and to help catch prey. Cells in these organs make the enzyme luciferase. The cells also make luciferin, which is the only substrate that luciferase can catalyse. The cells keep these two substances apart, but when they are stimulated by nerve impulses they release luciferin, which is broken down in the presence of ATP bound to magnesium ions. This reaction releases light.

The luciferin–luciferase reaction has been developed into a very sensitive test for ATP. Each time a molecule is detected, a flash of light is given off.

This is used to detect the presence of ATP released by organisms such as bacteria and fungi. The test can be used for assessing the cleanliness of kitchens and food-processing plants, the suitability of water for drinking, and the effectiveness of waste water treatment plants.

The roles of enzymes

Key terms

Catalyst A substance that increases the rate of a chemical reaction but does not become altered or changed during the reaction, so can be reused over and over again.

Enzyme A protein molecule made by cells that acts as a catalyst and increases the rate of a chemical reaction.

Enzymes are biological catalysts that speed up reactions, without undergoing permanent change. These reactions need high temperatures, high pressures, extremes of pH and high concentrations of reactants – all factors that would kill organisms. So, without enzymes these chemical reactions would occur so slowly, if at all, that life would be impossible.

Enzymes catalyse all the chemical reactions that occur in organisms. A **catalyst** is a chemical that speeds up a chemical reaction without changing the nature of the reaction or being changed itself; this means that it can be used over and over again.

Enzymes are very specific catalysts that are made by cells and catalyse the metabolic reactions of organisms. Enzymes are globular proteins with complex tertiary structures. Some are single polypeptides and some have two or more polypeptides so have quaternary structure (see page 45 in Chapter 3).

Enzymes are effective in small quantities and remain unchanged by the reaction, so can be used again many times. Reactions that enzymes catalyse are reversible. Enzymes do not determine the direction of a reaction, but its speed.

Enzyme activity is affected by a number of factors, including temperature, pH, enzyme concentration and substrate concentration. Their activity is often dependent on the presence of other compounds that are either attached loosely or bound permanently.

Sites of activity of enzymes

All enzymes are proteins, so are made by the process of protein synthesis inside cells (see pages 53–54 in Chapter 3). Most enzymes remain inside cells, but many cells are specialised to synthesise enzymes that are secreted to work outside cells in the environment or inside a body cavity, such as in the gut or the blood.

Intracellular enzymes

Key term

Intracellular enzymes Enzymes that catalyse reactions within cells.

Many different reactions occur simultaneously inside cells at low concentrations, low temperatures and at a pH near 7.0. These reactions are catalysed by **intracellular enzymes** that function exclusively within cells. The enzyme catalase is an intracellular enzyme. It acts on hydrogen peroxide, which is the waste product of metabolic processes in cells; it is a powerful oxidising agent and therefore extremely toxic, so it needs to be removed or destroyed very quickly. The enzyme catalase acts on hydrogen peroxide to break it down into water and oxygen within seconds, whereas normally hydrogen peroxide left uncatalysed would take months to degrade. Some cells use hydrogen peroxide to kill pathogens, cells infected with viruses and cancer cells. Many intracellular enzymes are free in solution, for example in the cytosol, nucleoplasm, mitochondrial matrix and the stroma of chloroplasts. Many are fixed in place, for example on

Key term

Extracellular enzymes Enzymes that catalyse reactions outside cells.

either side of the cell surface membrane and in the inner membranes of mitochondria and chloroplasts. Some enzymes of the digestive system are fixed to the cell membrane, for example sucrase, attached to the cell membranes of cells lining the gut.

Figure 4.2 It takes about 12 to 48 hours to digest a meal like this. Without the extracellular enzymes we make in our digestive system, it would take far longer. Life would not be possible without enzymes to catalyse reactions.

Extracellular enzymes

Extracellular enzymes catalyse reactions that occur outside cells, such as those involved in the breakdown of food in the gut. These enzymes are needed to digest the food we eat, because large molecules cannot be absorbed until they are broken down into simple ones.

Many digestive enzymes are secreted from the cells that make them to the gut lumen, where they act on the food; for example, amylase, which hydrolyses starch into maltose, and trypsin, which breaks down protein molecules into peptides and amino acids.

Some organisms rely entirely on extracellular digestion in which the enzymes are secreted outside their body to digest food. Fungal hyphae secrete all the enzymes they need to digest their food directly onto the food, where it is digested into simple molecules and then absorbed through the hyphal walls.

Tip

Extracellular enzymes work outside cells – some work in the environment, others work inside body cavities such as the lumen of the gut, through which food travels.

Tip

The enzymes given as examples in the rest of this chapter catalyse breakdown reactions. However, enzymes that catalyse reactions that build up biological molecules, such as starch for energy storage, collagen for support and DNA for information storage, are just as important.

Tip

The name of an enzyme and or group of enzymes often includes the name of the substrate or the type of reaction that the enzyme(s) catalyse followed by 'ase'. For example, carbohydrases catalyse the hydrolysis of carbohydrates. They are hydrol**ase** enzymes, because they catalyse hydrolysis reactions.

Figure 4.3 This mould fungus is growing over the surface of the orange and into its interior to gain energy, by using enzymes for extracellular digestion.

Mechanism of enzyme action

The active site and enzyme specificity

The specificity of an enzyme is determined by the shape of its active site. This particular area of the enzyme molecule has a specific shape that is complementary to the shape of the substrate molecule. This allows the substrate molecule to fit exactly into the active site, with what is known as a complementary fit. Other molecules without this specific shape do not fit into the enzyme's active site, and so the enzyme cannot catalyse them.

Tip

The active site is the part of an enzyme that has a complementary shape to the substrate or a particular part of the substrate. The substrate does not have an active site and is not complementary to the whole enzyme, only to the active site.

Active sites are described as grooves or clefts (see Figure 4.4). They are composed of a small number of amino acids within a polypeptide. The features of the active site and the type of substrate that it accepts are determined by the R groups of these amino acids. Some can have non-polar R groups, so provide a hydrophobic interior to the active site that accepts non-polar substrates. Others are polar R groups that form temporary ionic bonds with substrate molecules.

Key term

Specificity The relationship between an enzyme and the only type of molecule that fits into its active site.

There are different degrees of specificity. Some enzymes are specific only to one reaction. Others are less specific and catalyse a number of reactions of the same type or act on a particular linkage; for example, a protease acts on peptide bonds.

Subtilisin is the enzyme in the cleaning fluid for contact lenses. It is a general protease that breaks peptide bonds between any pair of amino acid residues in a polypeptide. Trypsin and chymotrypsin, which are mammalian proteases secreted into the small intestine, are much more specific and only break peptide bonds between certain amino acids (see Table 4.1 and Figure 4.5).

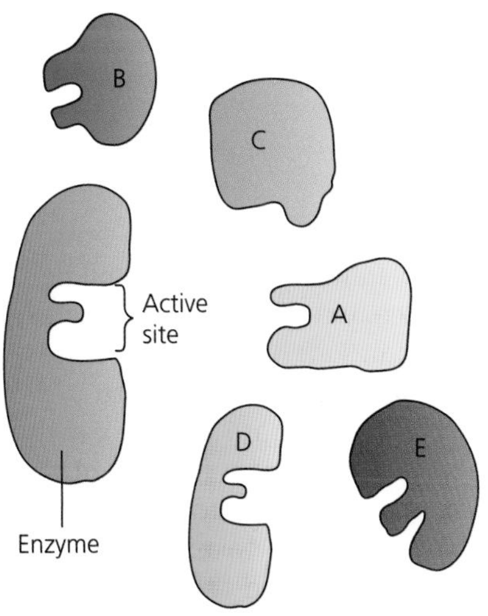

Figure 4.4 The active site of this enzyme has a shape that is complementary to the shape of its substrate, molecule A. The other molecules have shapes that are not complementary.

Table 4.1 The specificity of three proteases.

Enzyme	Specificity of protease activity (site of peptide bonds broken)
trypsin	next to arginine or lysine
chymotrypsin	next to phenylalanine, tyrosine and tryptophan
subtilisin	between any pair of amino acids

Tip

Notice that in Figure 4.4 substrate A has a shape **complementary** to the active site. It is a common error to say that the substrate has the same shape as the active site. D has the same shape, but that will not fit. The substrate has to have a complementary shape.

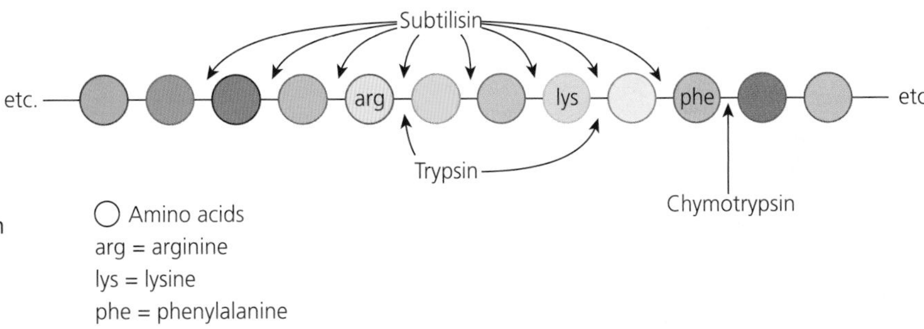

Figure 4.5 Degrees of specificity: subtilisin catalyses the breakdown of all peptide bonds within a polypeptide; trypsin and chymotrypsin are more specific.

Key terms

Enzyme–substrate complex The complex that forms when a substrate fits into an active site.

Lock and key hypothesis The idea that an enzyme's active site is complementary in shape to the substrate in the same way as a key (substrate) and a lock (enzyme).

Enzyme–product complex The complex that forms in the active site after the reaction is complete but before the product or products leave.

During an enzyme-catalysed reaction, substrate molecules fit into the active site of the enzyme; usually a much larger molecule. An enzyme functions by combining with the substrate molecule involved in the reaction to form an enzyme–substrate complex. There are two ideas about how enzymes interact with their substrates: the lock and key hypothesis and the induced-fit hypothesis.

Lock and key hypothesis

The idea of the substrate fitting exactly into the active site is called the lock and key hypothesis. This is the idea that the substrate acts as a key fitting exactly into a lock which is the active site of the enzyme. Interactions between the R groups within the active site and the substrate stabilise the enzyme–substrate complex. The substrate is altered and forms a product. The complex is now called an enzyme–product complex. However, the enzyme itself remains unchanged, and so the enzyme releases the **product** and leaves itself as a free enzyme molecule ready to accept another substrate molecule.

Key term

Induced-fit hypothesis The theory that the active site of an enzyme changes shape during the binding of a substrate molecule, and this puts strain on the substrate molecule contributing to the reaction.

The lock and key hypothesis was proposed in 1890. Research work since then has revealed that the interaction between substrate and the enzyme active site is not as simple as first thought, and suggests that the active site changes shape to accommodate the substrate.

Tip

Collisions between the substrate and enzyme are random, but once the correct substrate meets the active site of the enzyme it is held in place until the reaction has been completed. This complex is called the enzyme–substrate complex.

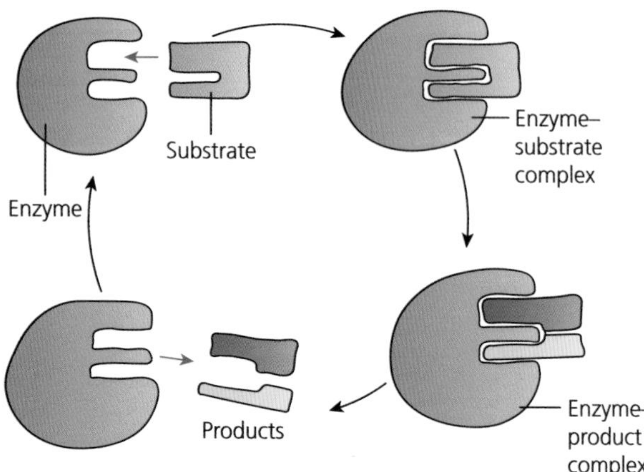

Figure 4.6 The lock and key mechanism for enzyme action. The shape of the substrate is an exact fit for the active site. Once the reaction is over, the active site is free to take another substrate molecule.

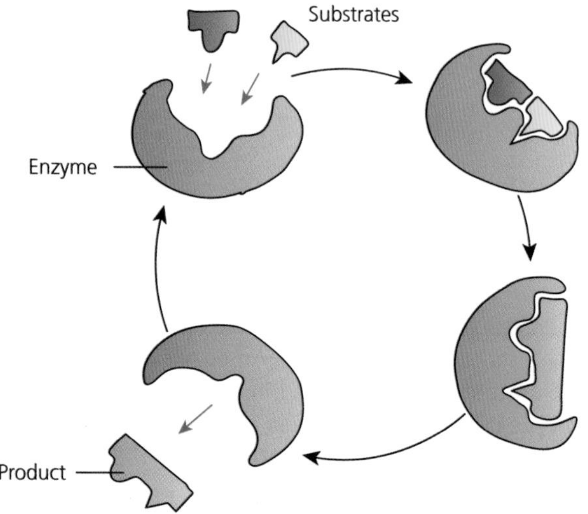

Figure 4.7 Induced fit. The active site changes shape as the substrate molecule binds in place.

Induced-fit hypothesis

The lock and key theory does not fully explain how enzymes and substrate molecules are able to collide successfully and form the enzyme–substrate complex. Because the collisions are random, it would be quite difficult for enzymes and substrate molecules to collide and form the enzyme–substrate complex if the active site were a fixed shape. It is also important to note that proteins are not the fixed structures that the lock and key theory suggests. To take account of these points, a modified theory called the **induced-fit hypothesis** was proposed in 1959. This suggests that an active site does not have a complementary shape to the substrate until it has moulded around the substrate so the two fit together closely. Amino acids with specific R groups within the active site are brought closer to the substrate, hold it in place and put the substrate under strain, so making the reaction proceed to form the product(s).

All enzymes speed up a reaction that would take place extremely slowly without a catalyst.

Activation energy

Reactions that occur in organisms involve compounds that are very stable. An energy barrier prevents a chemical reaction happening or only allows it to happen very slowly. Hydrogen peroxide, for example, decomposes to water and oxygen over a period of six months, because it takes this time to accumulate enough kinetic energy to overcome the energy barrier. However, add catalase and there is a very vigorous reaction that is over in seconds.

Key term

Activation energy The energy that is needed to overcome the energy barrier to allow a reaction to occur. Enzymes provide an alternative pathway with a lower activation energy for the reaction.

The energy required to allow the reaction to overcome the barrier is called the **activation energy**. This energy barrier may be overcome by increasing the temperature and/or pressure or by adding a catalyst. Enzymes do not lower the overall energy change of the reaction; instead they provide an alternative pathway with a lower activation energy. This allows reactions to occur without the need for extreme conditions that would kill cells. Once the enzyme's active site and the substrate molecule collide forming an enzyme–substrate complex, the activation energy is lowered because a different pathway for the reaction is followed. This is possibly due to the tensions created within the complex by the

Test yourself

1 Define the term *enzyme*.
2 Sketch a graph to explain what is meant by *activation energy*.
3 How does subtilisin differ from trypsin in its mode of action?
4 Explain the importance of enzymes in organisms.
5 Explain how the lock and key hypothesis shows how enzymes function.
6 Explain how the induced-fit hypothesis differs from the lock and key hypothesis.

charges within the amino acids at the active site and the links between the amino acids and the substrate. The reacting molecules effectively become an intermediate molecule that quickly forms products with a lower energy level than that of the substrate.

You can think of the energy barrier as a hump at the top of a hill, and the activation energy as the energy required to push a ball over this hump so it can roll down. In a similar way, enzymes provide the activation energy for substrate molecules to react and form the product. Without the enzyme the activation energy required to allow this action is large, but with the enzyme the energy needed is very much reduced, as shown in Figure 4.8.

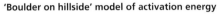

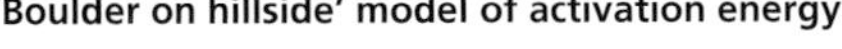

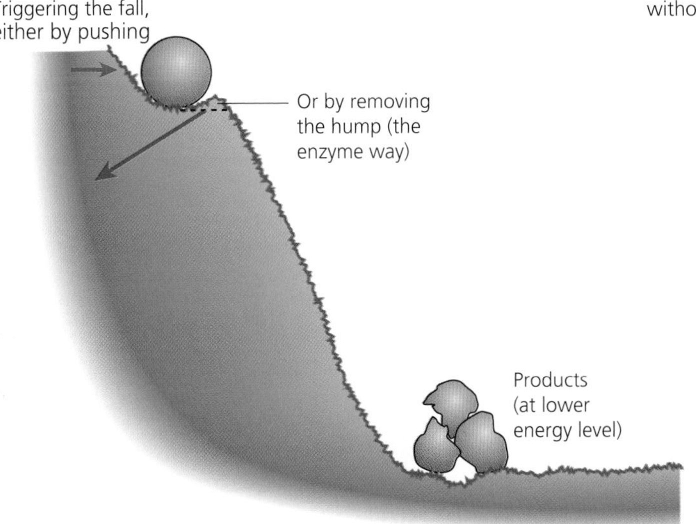

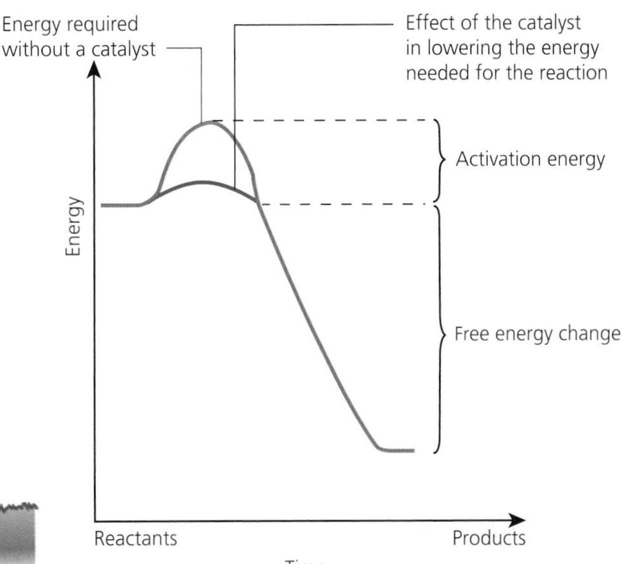

Figure 4.8 A model of activation energy.

Investigating enzyme activity

The activity of enzymes is determined by finding the rate at which a reaction proceeds. When enzyme molecules are added to substrate molecules, they form enzyme–substrate complexes. A reaction occurs and the product molecules leave the active sites. Over time the number of substrate molecules decreases and the chance of a substrate molecule entering an active site decreases. This means that the number of product molecules formed decreases.

You can follow the course of an enzyme-catalysed reaction by seeing how long it takes for the substrate to disappear or the product to appear. For example, a solution of milk protein is cloudy. If a protease such as trypsin is added, the cloudiness disappears as the protein is hydrolysed. The reaction can be followed with this procedure.

1 Add 10 cm^3 of a solution of milk powder to test tube 1.

2 Add 1 cm^3 of a protease solution to test tube 2.

3 Put both test tubes in a water bath at 25 °C for five minutes to equilibrate.

Tip

Keeping the enzyme solution and substrate solution separately until they are both at the desired temperature for the reaction is called equilibration.

4 After five minutes, pour the contents of test tube 2 into test tube 1, return test tube 1 to the water bath and start a timer.

5 Watch carefully and time how long it takes for the cloudiness to disappear. Use a test tube of water to judge when the end point has been reached.

The time taken for the reaction to occur is **not** the rate of reaction. If there is a fast rate of reaction then the reaction will be completed in a short time, and if there is a slow rate of reaction it will take much longer to be complete. In circumstances like this you can calculate the rate as the reciprocal of time: $\frac{1}{t}$ in which t = the time taken to reach an end point in seconds. The unit for rate in this case is seconds^{-1} (written as s^{-1}).

If the time taken for the cloudiness to disappear is 500 s, then the rate is 0.002 s^{-1}, which can be written in standard form as $2.0 \times 10^{-3} s^{-1}$.

Often it is not possible to see a change happening, such as the disappearance of cloudiness or the appearance of a colour. If amylase is added to starch, for example, it is better to use iodine solution to follow the disappearance of starch, because it is hydrolysed (see page 29 in Chapter 2). To do this you take samples from the reaction mixture at intervals of time and test them with iodine solution and record the colour. You should continue taking samples until two consecutive results give a yellow colour indicating that all the starch has been hydrolysed.

Activity

Following the course of an enzyme-catalysed reaction

Some students investigated the reaction in which starch is hydrolysed by amylase. They obtained quantitative results with a colorimeter. First they had to measure the optical density of solutions of starch of different concentrations when mixed with a dilute iodine solution. They followed this procedure.

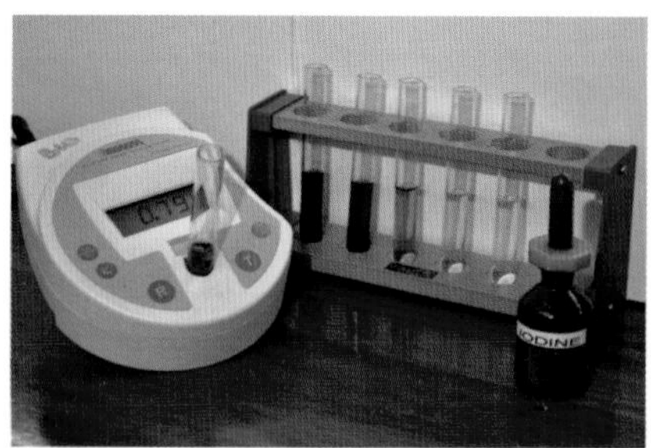

Figure 4.9 A colorimeter.

Method

1 Make a 10 g dm^{-3} solution of starch and use it to make up these solutions: 1.0, 2.0, 3.0, 4.0, 5.0, 6.0, 7.0, 8.0 and 9.0 g dm^{-3}

2 Add equal volumes of each solution to test tubes and the same volume of iodine solution to each test tube.

3 Place each test tube in a colorimeter and measure the absorbance of the solution, as in Figure 4.9.

Table 4.2 gives some colorimeter readings for known concentrations of starch.

Table 4.2 Colorimeter readings for known concentrations of starch.

Concentration of starch/mg dm^{-3}	0	1000	2000	3000	4000	5000	6000	7000	8000	9000	10000
Absorbance	0.00	0.30	0.52	0.70	0.85	0.96	1.07	1.16	1.24	1.28	1.30

Questions

1 Describe how to make the dilutions of starch from a 10 g dm^{-3} solution.

2 Plot a calibration graph using the figures in Table 4.2.

The students then made a solution of 10 000 mg dm^{-3} starch. They added 1 cm^3 amylase solution to 10 cm^3 of the starch solution and took samples at two-minute intervals for 16 minutes. The colorimeter readings for their samples are in Table 4.3.

Table 4.3 Colorimeter readings.

Time/min	0	2	4	6	8	10	12	14	16
Colorimeter reading: absorbance	1.32	0.60	0.30	0.10	0.09	0.08	0.06	0.04	0.01

Use the calibration graph to determine the concentration of starch at each sampling time.

3 Copy the table of results and add a third row to show the concentrations of starch for each sample.
4 Draw a graph to show the decrease in starch concentration over time.
5 Describe the trend shown in your graph.
6 Explain why the rate of reaction is not constant throughout the 16 minutes.
7 Suggest a suitable control for this experiment and justify why it would be suitable.
8 State two limitations of this method that may affect the quality of the results obtained.
For each limitation
- explain how it influenced the quality of the results
- describe how you would modify the procedure to overcome the limitation.

Tip

Look at Chapter 17 for more information about colorimetry.

Determining initial rates of reaction

If you follow the course of an enzyme-catalysed reaction, you can either take samples to:

- follow the disappearance of a substrate, or
- follow the appearance of a product.

Tip

Taking a sample as soon as the enzyme is added to the substrate is called taking a reading at **time zero**.

You can see from the graph you have drawn of the disappearance of starch in the activity above that the rate is fastest at the beginning and decreases with time as the concentration of starch decreases. Eventually no substrate is left, so the reaction stops. The probability of successful collisions between enzyme molecules and substrate molecules is highest at the beginning of the reaction at time zero, when the enzyme is added to the substrate. To compare the activity of different enzymes or the same enzyme under different conditions, it makes sense to determine the rate of reaction when there is the maximum substrate concentration and the rate is not limited by a decrease in substrate as happens if you determine the rate by timing until all the substrate is gone.

To follow the appearance of a product you can use the enzyme catalase, which is common in many animal and plant tissues and in yeast. Catalase catalyses the decomposition of hydrogen peroxide.

$$2H_2O_2 \rightarrow 2H_2O + O_2$$

Usually catalase is extracted from plant material that contains the enzyme such as potato, lettuce or celery. The plant material is cut up, liquidised in a blender and filtered. The filtrate is the extract and contains catalase. The diagram shows how the apparatus for following the reaction may be set up. You could also use a gas syringe to collect the oxygen produced.

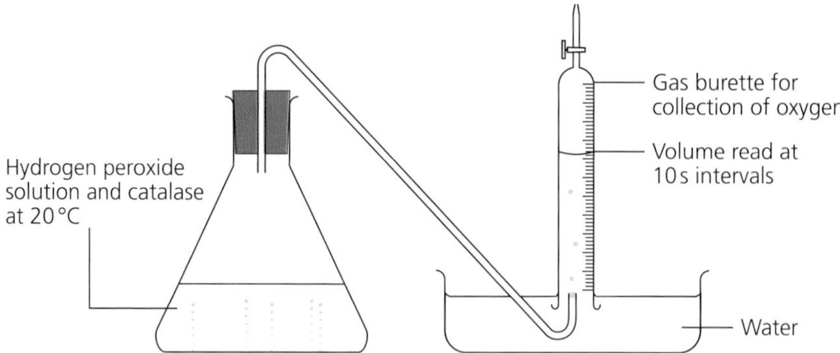

Figure 4.10 How to follow the reaction in which hydrogen peroxide is decomposed to oxygen and water.

Table 4.4 shows the results obtained, which are then plotted on a graph (Figure 4.11).

Table 4.4 Volume of oxygen over time as hydrogen peroxide is decomposed to oxygen and water.

Time/s	0	10	20	30	40	50	60	70	80	90	100
Volume of oxygen collected/cm^3	0.0	2.6	3.9	4.8	5.4	5.8	6.0	6.1	6.2	6.2	6.2

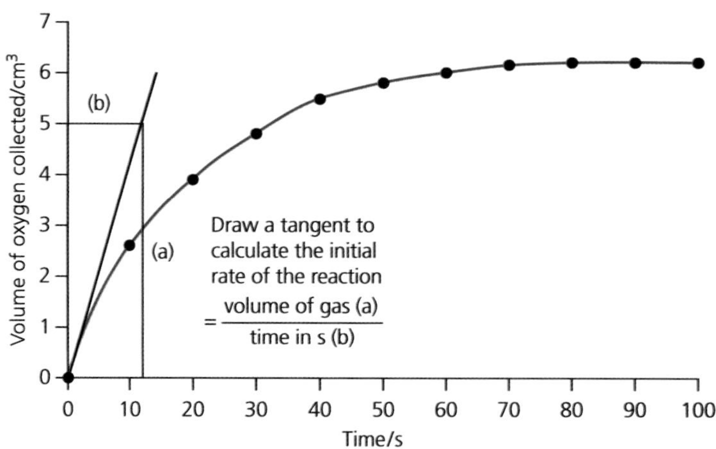

Figure 4.11 Graph of the volume of oxygen produced when an extract of catalase is added to hydrogen peroxide.

The initial rate is determined by either:

- calculating the rate from the first sample, or
- taking a tangent to the curve drawn on the graph.

In using the second method, the points that you use to calculate the rate should be separated by half the length of the tangent; the rate in this example is $0.3\,cm^3\,s^{-1}$. In all the examples in the next section, the rates of reaction have been determined by taking the initial rate.

Factors affecting enzyme activity

Enzyme molecules are proteins. Their stability is affected by high temperatures and extremes of pH. Under these conditions the weak bonds that hold the tertiary structure of enzymes together break and the important shape of the enzyme is lost.

Tip

Successful collisions are those in which an enzyme–substrate complex is formed. Remember that the kinetic energy at the moment of collision is not relevant, because the enzyme is not actually **reacting** with the substrate.

The activity of enzymes is also influenced by external factors. You have seen how in a reaction enzymes and substrate molecules must collide. Three factors that can increase the probability of molecules colliding are:

- temperature
- substrate concentration
- enzyme concentration.

pH also affects the activity of enzymes.

The effect of temperature

As with other molecules in a solution, enzyme and substrate molecules move continually and randomly as a result of their kinetic energy. This results in the two types of molecules colliding; some of the collisions are successful, and so enzyme–substrate complexes form.

An increase in temperature increases the kinetic energy of the molecules; the molecules move faster and the chances of collisions also increases. There will be **more** successful collisions as a result and so more product will be formed. This causes an increase in the reaction rate.

Key term

Optimum temperature The temperature at which the rate of a process is at its maximum. Above or below this temperature the rate of reaction is slower. The optimum temperature for enzymes varies according to the habitat to which an organism is adapted.

Enzymes catalyse the reaction faster as the temperature increases, but only up to a point, because all enzymes have a temperature at which they work best. The temperature at which a particular enzyme works best is called its **optimum temperature**. At higher temperatures than this the enzyme loses its stability and begins to be disrupted. Optimum temperatures for enzymes vary widely. The optimum for humans and many mammals is about the same as their body temperature, between 35 °C and 40 °C. The optimum temperature for many plant enzymes is about 20 °C, and some bacterial enzymes have optimum temperatures in excess of 80 °C.

Figure 4.12 shows the effect of temperature on a mammalian enzyme.

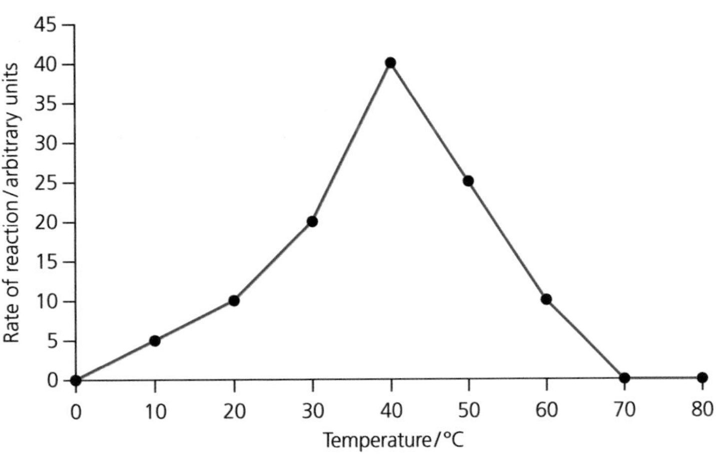

Figure 4.12 The effect of temperature on an enzyme-catalysed reaction.

The rate of reaction increases up to a maximum rate of 30 arbitrary units at 40 °C. At temperatures above this maximum, the rate decreases steeply, and there is no activity at 70 °C. The graph has rates plotted at 10 °C intervals; straight lines are drawn between the plotted points because we do not know what happens between these temperatures. The optimum temperature is anywhere between 30 °C and 50 °C. The only way to find out is to repeat the investigation, taking readings at smaller intervals of temperature within that range.

Key term

Denaturation A change in the tertiary structure of proteins such as enzymes, which means it no longer functions. Thermal denaturation is usually irreversible; denaturation by changes in pH is often reversible.

Tip

Enzymes are protein molecules and not living organisms, so do not say that an enzyme is 'killed' at high temperatures.

As the temperature increases, the movement of the molecules increases and causes the molecules to vibrate. The increased kinetic energy and vibration put a strain on the enzyme molecule, and the weaker hydrogen and ionic bonds that stabilise the enzyme's tertiary structure begin to break. This breaking causes the tertiary structure of the enzyme to change, and the specific shape of the active site is disrupted and lost. When the shape is no longer held in its correct form, the complementary shape of the active site is also lost and so the enzyme can no longer form enzyme–substrate complexes. The enzyme is now **denatured** and the reaction rate slows down as more and more enzymes become affected. When an enzyme is denatured like this, it is a permanent change and cannot be reversed. Covalent bonds, such as peptide bonds and disulfide bonds, are not broken by increases in temperature.

The temperature at which an enzyme is denatured varies, but most enzymes found in living organisms become denatured at temperatures higher than 60 °C. Some bacteria living in extreme environments have enzymes that are able to withstand very high temperatures such as 80 °C or more. These enzymes are **thermostable**. The proportion of enzymes that are denatured in a solution depends not only on the temperatures that they are exposed to but also how long they are left at those temperatures before they are tested to see if they are still active. Left for a short time at 60 °C, many enzymes may retain their activity; kept at 60 °C for 30 minutes or longer, very few remain active.

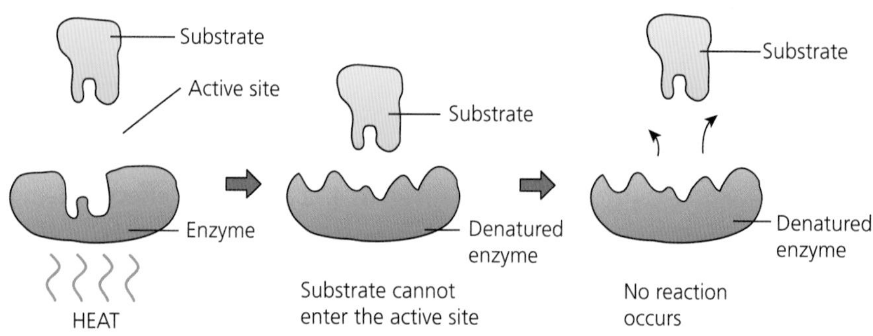

Figure 4.13 At high temperatures the enzyme is denatured and the active site changes shape so it is no longer complementary to the substrate.

Key term

Temperature coefficient The ratio between the activities of a process, such as an enzyme-catalysed reaction, at two different temperatures. If the two temperatures are 10 °C apart then the temperature coefficient is written as Q_{10}.

Temperature coefficient

The temperature coefficient for any process is the ratio between the rates of that process at two different temperatures. If you look carefully at Figure 4.12 you will see that the rate of reaction increases by a factor of 2 for every 10 °C increase in temperatures below the maximum rate at 40 °C. For most enzyme-catalysed reactions the rate of the reaction doubles for every 10 °C rise in temperature. This is described as the **temperature coefficient** (Q_{10} = 2) and is given as an equation:

$$\text{temperature coefficient} = \frac{\text{rate of reaction at } (x + 10)\,°\text{C}}{\text{rate of reaction at } x\ °\text{C}}$$

where x = any chosen temperature

For example, the rate at 20 °C is 10 arbitrary units and at 10 °C the rate is 5 arbitrary units.

Substituting into the equation above, x is 10 °C and 20 °C is (x + 10 °C), so

$$Q_{10} = \frac{10}{5} = 2$$

To calculate values for Q_{10} from rate of reaction graphs, use intercepts at two temperatures 10 °C apart and read off the rate of reaction. Substitute the figures taken from the graph into the equation above. There is no unit for Q_{10} because it is a ratio. If there is no scale given on the vertical axis for rate, you can still calculate the Q_{10} by measuring the distances on the vertical axis with a ruler and using these as being equivalent to the rates.

Activity

The effect of temperature on enzymes

The activity of enzymes at different temperatures may be investigated by placing reaction mixtures in water baths set to a suitable range of temperatures. The substrate and enzyme solutions are equilibrated at the temperatures chosen and then mixed together and kept at these temperatures (0–80 °C in this example). The results of such an investigation on hydrolysis of milk protein by a protease are shown in Table 4.5.

Table 4.5 Rate of reaction and time for cloudiness to disappear after protease added to milk protein.

Temperature /°C	Time for cloudiness to disappear/s	Rate of reaction $\times 10^{-3}$/s^{-1}
0	no reaction	0.0
10	1400	
15	960	1.04
20	650	
25	480	
30	360	
35	240	
40	185	
50	440	
55	850	
60	no reaction	0.0
70	no reaction	0.0

1 Explain how the students equilibrated the test tubes.
2 Copy the table and complete it by calculating all the missing rates of reaction.
3 Plot a graph to show the effect of temperature on the rate of reaction.
4 Calculate the temperature coefficient (Q_{10}) for the following temperature ranges:
10–20 °C; 20–30 °C; 30–40 °C
5 Use the graph to describe the effect of temperature on the activity of the protease.
6 Explain the results shown in the graph.
7 State **two** limitations of this method that may affect the quality of the results obtained.
For each limitation:
- Explain how the limitation influenced the quality of the results.
- Describe how you would modify the procedure to overcome the limitation.

8 Suggest what results you would expect if a thermostable protease was used instead.

The effect of pH on enzyme activity

pH is a measure of the concentration of hydrogen ions (H^+) in solution. The pH scale is a logarithmic scale, which means that a change in 1 pH unit, for example from pH 7 to pH 6, is equal to a change in hydrogen ion concentration of ×10. When there are more hydrogen ions present, the solution becomes increasingly more acidic and the pH decreases. As with temperature, enzymes function over a particular range of pH, with an optimum pH within that range. Above and below the optimum pH the enzyme performs less well, and beyond the extremes of the range the enzyme does not function at all, because it is denatured.

The optimum pH is specific for each enzyme; for some, the pH range in which they operate is narrow; for others it is wider. Gastric protease (pepsin) is an enzyme in the stomach that has an optimum of pH 2, while pancreatic protease has an optimum pH of pH 8.5. Most intracellular enzymes have an optimum pH of about 6.5, which is the pH of cytosol. Lysosomal enzymes have a much lower optimum pH, which means they rarely function if released into the cytosol.

The shape of an active site is determined by interactions, such as hydrogen bonding and ionic bonding, between R groups of the amino acids in the active site. As the hydrogen ion concentration changes, some of these bonds break and the active sites become less effective. At certain values of pH the enzymes lose their shape and are no longer active; they are denatured at these values of pH.

In addition, the charges on R groups are important in the formation of the enzyme–substrate complexes: in the induced-fit model these charges allow the substrate to enter and become linked within the active site, allowing the enzyme–substrate complex to form. Any changes in the charge therefore affect the ability of the active site to link successfully with the substrate molecules. A pH close to the optimum will denature the enzyme but unlike with the effect of temperature this is likely to be reversed. However at the extremes of the pH range the enzyme will be permanently denatured.

Reaction mixtures may be prepared using buffer solutions to maintain a constant pH. Buffer solutions are available to maintain different values of pH.

Figure 4.14 shows the activity of three enzymes at different ranges of pH. The enzymes investigated in this experiment were:

- pepsin
- catalase
- alkaline phosphatase.

The rates of reaction were determined and are shown in Figure 4.14.

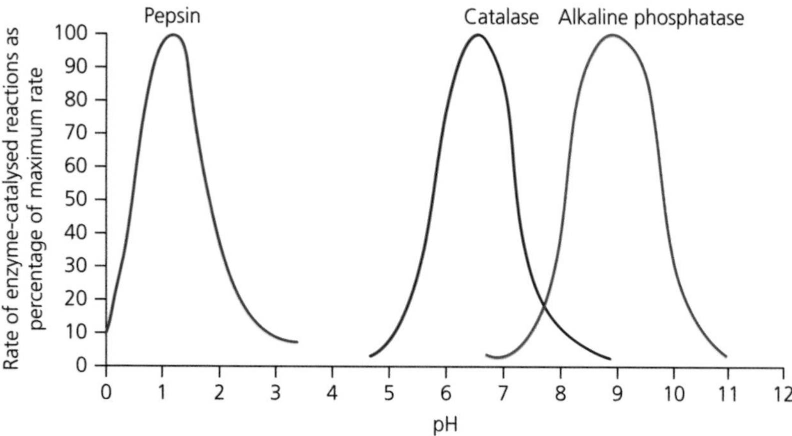

Figure 4.14 The effect of pH on three enzymes. Many enzymes like these function most efficiently over a narrow range of pH.

Look carefully at the graph and see how:

- Each enzyme is active over a range of pH.
- As pH increases, the activity of each enzyme increases until it reaches a maximum.
- Maximum activity occurs at the optimum pH for each enzyme.
- At values of pH above the optimum pH, the activity decreases.
- At a certain pH there is no activity.

The effect of substrate concentration

Substrate concentration determines the probability of collisions occurring between substrate molecules and enzyme molecules.

At a low substrate concentration there are few substrate molecules in the solution so fewer will collide and form an enzyme–substrate complex with the enzyme, so the rate will be slower.

To investigate the effect of different substrate concentrations on the activity of catalase, the procedure on pages 67–8 can be used. Different concentrations of hydrogen peroxide are prepared by diluting a 20 volume solution. ('20 volume' means that when 1 cm^3 of hydrogen peroxide is decomposed, 20 cm^3 of oxygen is produced. A 20 volume solution is 1.67 mol dm^{-3}.) Table 4.6 shows the results of an investigation like this.

Table 4.6 Effect of substrate concentration on the activity of catalase, measured as the volume of oxygen produced at different time intervals.

Concentration of H_2O_2/mol dm^{-3}	Volume of oxygen/cm^3						
	time/s						
	0	10	20	30	60	90	100
0	0.0	0.0	0.0	0.0	0.0	0.0	0.0
0.2	0.0	0.7	1.4	1.9	3.2	4.4	4.6
0.4	0.0	1.3	2.1	2.8	4.0	5.2	5.5
0.6	0.0	1.8	2.6	3.2	4.6	5.7	6.0
0.8	0.0	2.1	3.2	3.9	5.2	5.9	6.0
1.0	0.0	2.4	3.8	4.7	5.7	6.0	6.0

The results in Table 4.6 are plotted on the graph in Figure 4.15.

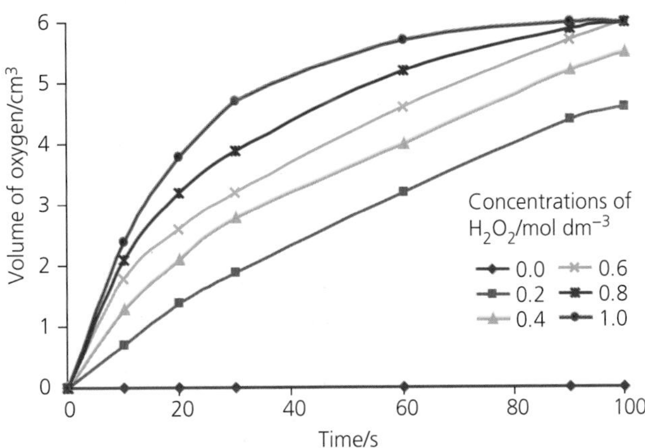

Figure 4.15 The progress of the reaction with different concentrations of hydrogen peroxide.

Tangents to the curves in Figure 4.15 are used to determine the initial rates. These initial rates are shown in Table 4.7 and then plotted on a graph of rate of reaction against substrate concentration in Figure 4.16.

Table 4.7 The effect of concentration of hydrogen peroxide on the rate of its decomposition by catalase (processed data).

Concentration of H_2O_2/ mol dm^{-3}	Initial rate of reaction × 10/cm^3 s^{-1}
0.2	0.07
0.4	0.13
0.6	0.18
0.8	0.21
1.0	0.21

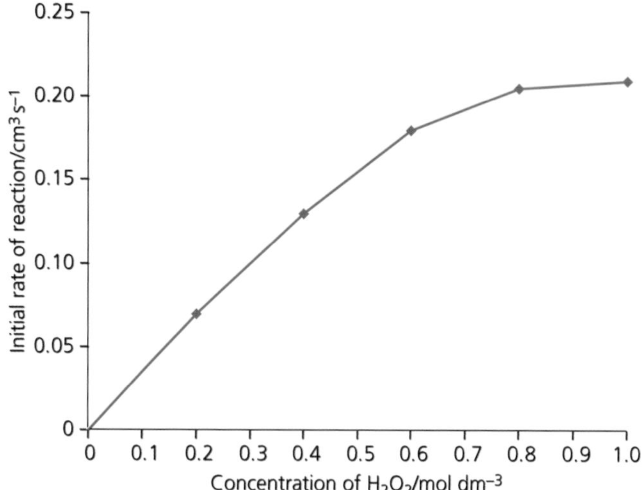

Figure 4.16 How increasing the concentration of substrate influences the initial rate of the decomposition of hydrogen peroxide by catalase. Note that the *x*-axis is substrate concentration and not time. Do not confuse this graph with the time-course graph in Figure 4.15.

You can see from the Figure 4.16 that as the substrate concentration increases to substrate concentration of 0.8 mol dm^{-3} the rate of reaction increases.

At a concentration of 0 there was no reaction, so the line must start at the origin. The rate increases as more substrate molecules are available and there are more successful collisions with the active sites of enzymes. In fact for most of the low concentrations of substrate, the rate is directly proportional to the substrate concentration. Up until the concentration of 0.8 mol dm^{-3} of hydrogen peroxide the enzyme activity is limited by the concentration of substrate, because if the concentration is increased the rate increases. At concentrations greater than 0.8 mol dm^{-3} the rate remains constant at 0.2 cm^3 s^{-1}. The rate is **not** limited by the substrate concentration, because increasing the concentration has no effect. The rate must be limited by something else.

At higher substrate concentrations, the number of enzyme molecules present limits the reaction. The enzyme molecules are unavailable, because they are all involved in enzyme–substrate complexes. Effectively the enzyme active sites are all occupied. Once this happens, the reaction can only proceed at the same rate as the enzyme turnover rate, i.e. how quickly the enzymes can react, create the product and release it so they are free to meet another substrate molecule. At this stage, the initial rate of the reaction for other reaction mixtures is therefore the same, unless the enzyme concentration is increased.

Tip

Starting lines at the origin: even if you do not have the result for 0 mol dm^{-3}, it is common sense that no oxygen would be produced, so the line starts at the origin, 0,0.

Test yourself

7 The time for an enzyme-catalysed reaction to reach its end point was 83 s. How would you calculate a rate of reaction from this result?

8 The rate of a reaction was calculated as 0.000 45 s^{-1}. How would you write this so it is easier to plot on a graph?

9 In an investigation, 600 mg of protein was broken down by a protease in 85 minutes. What is the rate in mg h^{-1}? A student thought this was a very slow rate. Suggest the likely conditions that might have given such a low rate.

10 If an instruction tells you to equilibrate an enzyme solution and a substrate solution, what would you do?

11 **a)** Sketch a graph to show the time course for a reaction in which the substrate disappears.

b) Sketch a separate graph to show the time course for a reaction in which product appears.

12 Sketch a graph for an enzyme that is not active below 10 °C, has an optimum temperature of 90 °C and is inactive at 120 °C.

The effect of enzyme concentration

If the enzyme concentration is increased, more active sites will be available for the substrate molecules to fit into, and more enzyme–substrate molecules will form. As a result the reaction rate will rise and continue to rise until other factors become limiting and therefore important; these factors included substrate concentration, pH and temperature. At this point the reaction has reached its maximum rate for this substrate concentration, temperature and pH until one of the factors is changed.

The results of an investigation on the effect of protease concentration on the rate of the reaction are shown in Table 4.8.

Table 4.8 Effect of protease concentration on rate of reaction.

Concentration of protease/%	Time taken for reaction mixture to go colourless/s				Rate of reaction × 10^{-3}/s^{-1}
	Replicates			Mean	
0.10	555	570	560	421.3	2.4
0.25	166	174	170	127.6	7.8
0.50	92	104	98	73.6	13.6
0.75	73	79	75	56.9	17.6
1.00	49	41	47	34.5	29.0

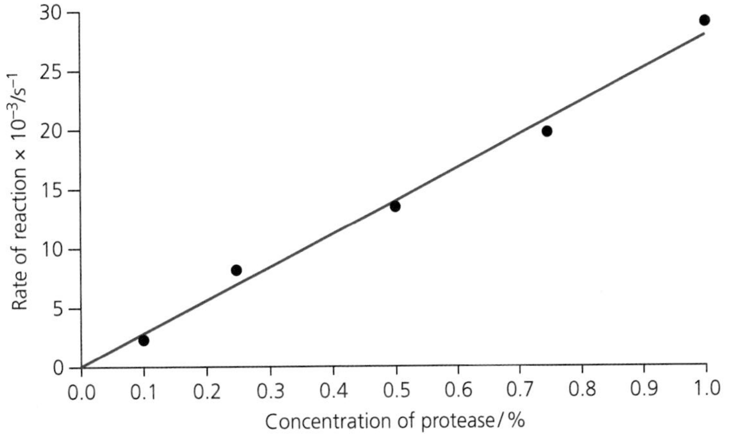

Figure 4.17 The effect of enzyme concentration on the rate of an enzyme-catalysed reaction.

Interactions between enzymes and other substances

Key terms

Cofactor A substance not made from amino acids that is required by an enzyme to function. They may be a permanent part of the molecule (prosthetic group) or a temporary part.

Inorganic cofactors Inorganic ions such as zinc, iron and nickel that an enzyme needs to function.

There are substances other than substrates that interact with enzymes. These are cofactors and inhibitors.

Some enzymes are inactive until they combine with a substance that changes their tertiary structure so that the active site is able to bind with substrate molecules. Other enzymes are active until they are stopped because of an interaction with an inhibitor molecule.

Cofactors, coenzymes and prosthetic groups

Some enzymes can only function if another non-protein substance is present with them that helps the enzyme to function correctly. These are called **cofactors**. Some enzymes require **inorganic** ions in order to function correctly. These ions help to stabilise the enzyme structure or take part in

Key term

Coenzyme A small organic non-protein cofactor. Coenzymes are involved in enzyme-catalysed reactions by donating or accepting hydrogen ions or chemical groups such as phosphate groups between different enzyme-catalysed reactions.

the reaction at the active site. Amylases, which digest starch to maltose, require the presence of calcium ions and chloride ions to function correctly.

Larger organic cofactors are called **coenzymes**. Some of these are bound permanently to the enzyme, often in or near the active site. Others only bind temporarily to the active site during the reaction. Coenzymes are used to link different enzyme-catalysed reactions in a metabolic sequence, such as respiration and photosynthesis.

Many of the B group of vitamins are used to make important coenzymes in metabolic reactions:

- Pantothenic acid is a component of coenzyme A.
- Nicotinic acid is used in the synthesis of the coenzymes NAD and NADP.
- The coenzyme FAD is based on vitamin B_1 – riboflavin.

NAD and FAD are alternately reduced and oxidised in the reactions in which they take part, which transfers energy in the form of hydrogen ions in respiration. NADP fulfils a similar role in chloroplasts during photosynthesis. Other coenzymes, such as ATP and coenzyme A, transfer chemical groups. ATP transfers phosphate groups between respiration and energy-consuming processes in cells. Coenzyme A transfers the acetyl group ($-CH_3CO$) from glucose and from fatty acids during respiration.

Tip

At this stage, you do not need to know the names of the coenzymes that play an essential role in respiration. It is enough to know that vitamins of the B group are required to make these coenzymes.

Any cofactor that becomes a permanent part of an enzyme is a prosthetic group. These prosthetic groups contribute to the three-dimensional shape of the enzyme and so are vital to the enzyme's ability to function. For example, the enzyme carbonic anhydrase, found in red blood cells, is important in the conversion of carbon dioxide and water to carbonic acid. Its prosthetic group is a zinc ion that forms an important part of its active site. FAD is the prosthetic group of the mitochondrial enzyme succinate dehydrogenase.

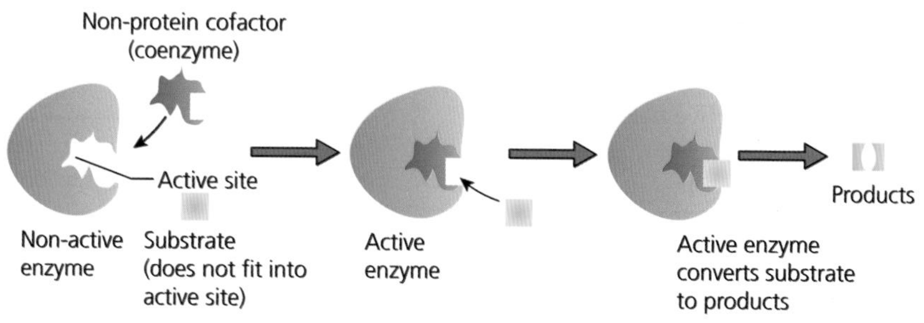

Figure 4.18 Diagram showing an enzyme with a cofactor (coenzyme).

Enzyme inhibitors

Key term

Non-reversible inhibition The inhibition that occurs when an inhibitor combines permanently with an enzyme and completely inactivates it.

An inhibitor is any substance that slows down or stops an enzyme-catalysed reaction by affecting the enzyme in some way.

Some inhibitors form covalent bonds with enzymes and inhibit them permanently; these are **non-reversible inhibitors** or irreversible inhibitors. If this inhibition occurs in a cell or an organism, the enzyme is completely inactivated. The cell must produce more of the enzyme or enzymes that are inhibited. This can only happen by activating the gene or genes so that they are transcribed and translated.

For example, heavy metal ions such as lead and mercury form covalent bonds with sulfur-containing R groups on proteins. Non-reversible inhibitors such as lead and mercury are serious poisons. The enzyme ferrochelatase, which is involved in making haem for haemoglobin, is very sensitive to inhibition by lead.

Other substances bind temporarily with enzymes, and their inhibitory effect can be reversed by a change in the environment of the enzyme; these inhibitors are **reversible inhibitors**.

The two groups of reversible enzyme inhibitors are

- competitive inhibitors
- non-competitive inhibitors.

The two groups are dependent on where the inhibitor binds. Competitive inhibitors bind to the active site. Non-competitive inhibitors bind to any region of an enzyme *other than* the active site; these sites are known as allosteric sites.

> **Key term**
>
> **Reversible inhibition** The inhibition that occurs when an inhibitor combines temporarily with an enzyme. The inhibition is reversed and the enzyme becomes active again when the inhibitor is no longer attached to the enzyme. Reversible inhibitors can be competitive or non-competitive.

Competitive inhibitors

Competitive inhibitors are molecules that have the same shape as the substrate molecules, or have one part of their molecule with the same shape. They are described as competitive because they compete with substrate molecules to fit into active sites. If they collide with the enzyme, they fit into the active site of the enzyme in the same way as the substrate would do, effectively blocking the active site from forming a complex with a substrate molecule. When an enzyme–inhibitor complex is formed the enzyme is not able to catalyse any substrate molecules, and so the reaction overall slows down as fewer enzyme–substrate complexes form.

If the number of inhibitor molecules is low compared with the number of substrate molecules, the reaction rate is not slowed much. However, if more inhibitor molecules are present than substrate molecules then more will successfully compete for the enzyme active site and occupy it, and so fewer enzyme–substrate complexes can form. Increasing the substrate concentration allows the reaction rate to increase, providing the number of enzyme and inhibitor molecules remains unchanged.

> **Key terms**
>
> **Competitive inhibition** The inhibition that occurs when an inhibitor with the same shape as the substrate combines with the active site, blocking access for the substrate. This type of inhibition is reversed by increasing the concentration of substrate.
>
> **Non-competitive inhibition** The inhibition that occurs when an inhibitor combines with an allosteric site on an enzyme. The tertiary structure changes so that the active site is no longer able to accept the substrate. This type of inhibition is not reversed by increasing the concentration of substrate.

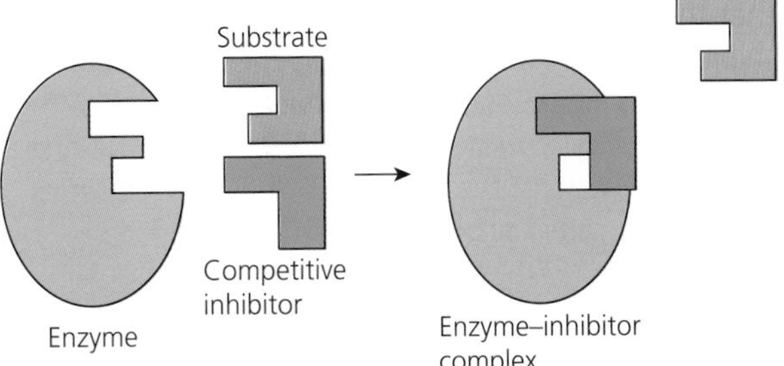

Figure 4.19 Competitive inhibition.

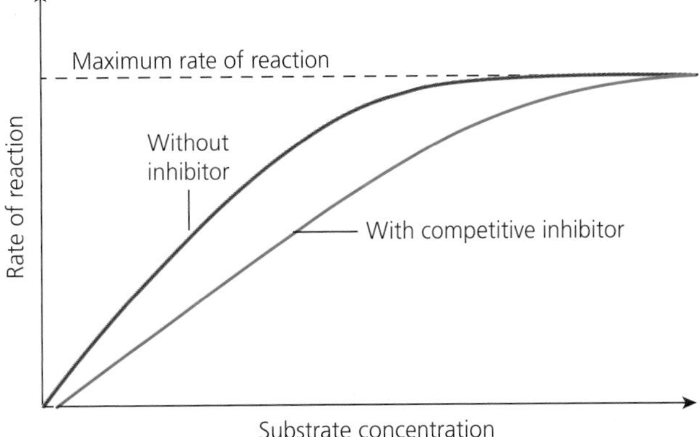

Figure 4.20 The effect of a competitive inhibitor on the rate of an enzyme-controlled reaction.

Tip

An inhibitor slows down the rate of an enzyme-catalysed reaction by interacting with the enzyme, *not* the substrate.

Non-competitive inhibitors

Non-competitive inhibitors do not compete with the substrate for the active site of the enzyme. Instead these inhibitors attach to the enzyme at a different part of the molecule, called the allosteric site, and the result of the attachment causes the shape of the active site to change. The substrate can therefore no longer fit into the active site and an enzyme–substrate complex cannot form, so the reaction rate slows down.

Because the inhibitor and the substrate are not competing, any increase in the concentration of substrate molecules does not speed up the reaction; the important aspect is how many inhibitor molecules are present. Once all the enzyme allosteric sites are occupied by inhibitor molecules, the reaction stops and adding additional substrate molecules cannot have any effect.

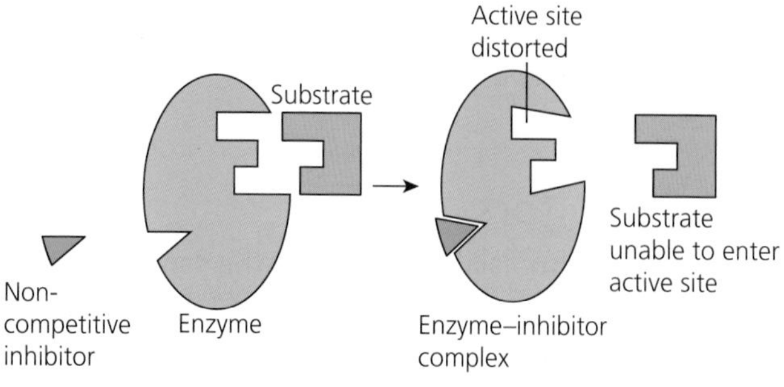

Figure 4.21 Non-competitive inhibition.

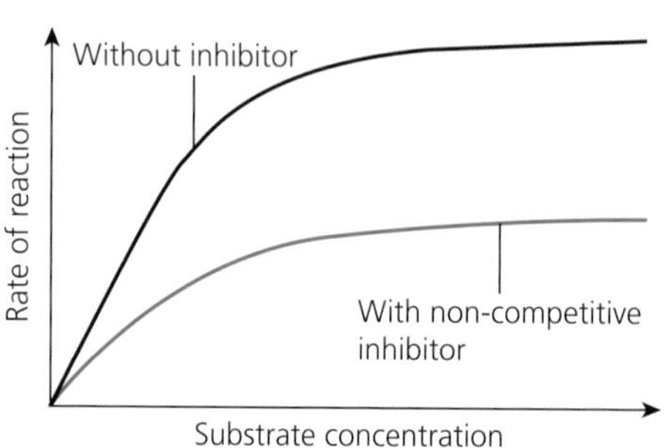

Figure 4.22 The effect of a non-competitive inhibitor on the rate of an enzyme-catalysed reaction.

Controlling metabolic processes

The activity of enzymes is controlled by inhibitors. For example, the enzyme β-galactosidase catalyses the hydrolysis of the glycosidic bond in lactose, as shown in the equation:

$$\text{lactose} + \text{water} \xrightarrow{\beta\text{-galactosidase}} \text{glucose} + \text{galactose}$$

Galactose is one of the products of this reaction and acts as a competitive inhibitor. This slows down the activity of the enzyme when there are plenty of product molecules available.

Most intracellular enzymes catalyse reactions as part of a sequence, as you can see in Figure 4.23.

Since enzymes increase the rate of metabolic reactions so significantly, often by as much as 10 million times, it is vital that there is control on metabolic pathways to make sure that there is not an excess of some products at the expense of others. Inhibitors play an important role in the control of metabolic pathways like the one in Figure 4.23. The product molecules of the last reaction in the series catalysed by enzyme 4 act as inhibitors of the first enzyme in the series – when this happens, the product of the enzyme series controls its own production. This is called **product inhibition**.

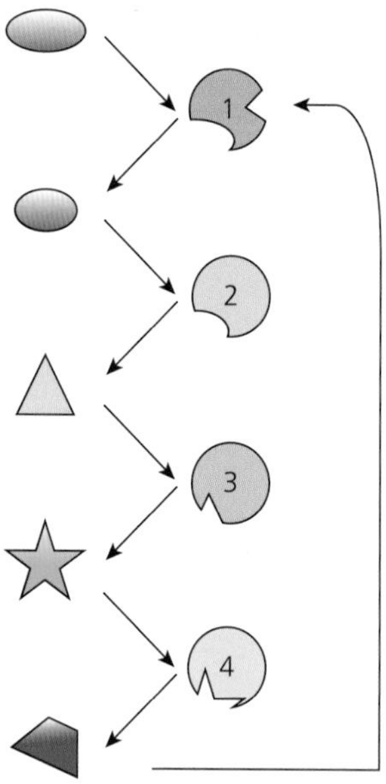

Figure 4.23 A metabolic pathway with enzymes working in series. The product of the reaction catalysed by enzyme 4 is a non-competitive (allosteric) inhibitor of enzyme 1.

Inhibitors as poisons

Some inhibitors are poisons. For example, cyanide is a non-reversible inhibitor of the enzyme cytochrome oxidase, which is the mitochondrial enzyme that catalyses the last reaction of aerobic respiration. This inhibition will almost certainly be fatal, because there is no way to reverse the inhibition other than by making new enzymes, and that takes time.

Tip

A metabolic poison is a chemical that stops a metabolic process in a living cell or in the body. Many metabolic poisons act as inhibitors of one enzyme in a metabolic pathway and result in the reaction being stopped.

Cyanide is a **metabolic poison**, because it prevents a metabolic reaction from taking place.

Malonate, another metabolic poison, is a competitive inhibitor of a different mitochondrial enzyme, succinate dehydrogenase.

Researchers use metabolic poisons to better understand how enzymes function in metabolic processes such as respiration.

Inhibitors as medicinal drugs

Some inhibitors act in a beneficial way, because they inhibit enzymes that may have harmful consequences in some individuals. You may recognise at least two of these inhibitors.

Table 4.9 Inhibitory action and benefit of some medicinal drugs.

Medicinal drug	Inhibitory action	Benefit of the inhibitor
Penicillin	non-reversible inhibition of transpeptidase enzyme responsible for forming cross links in bacterial cell walls	bacteria are destroyed
Aspirin (acetylsalicylic acid)	non-reversible inhibition of the COX enzyme involved in producing prostaglandins for stimulating inflammation and pain	reduces inflammation and gives pain relief
Eflornithine	non-reversible inhibitor of ornithine decarboxylase, an enzyme essential for cell growth	used for treatment of African trypanosomiasis (sleeping sickness)
Statins	competitive inhibition of HMG-CoA reductase, an enzyme involved in cholesterol synthesis	reduces the concentration of cholesterol in the blood

Test yourself

13 Give an example of an inorganic cofactor and discuss the importance of cofactors in enzyme-catalysed reactions.

14 Describe the mode of action of non-reversible enzyme inhibitors.

15 Describe the action of a competitive enzyme inhibitor and compare this with the action of a non-competitive inhibitor.

16 Name a non-reversible enzyme inhibitor and a reversible competitive enzyme inhibitor that is used as a medicinal drug.

17 Describe the benefits of using enzyme inhibitors in medicine.

Exam practice questions

1 Results from enzyme investigations can be qualitative or quantitative.
Which of the following is a qualitative result?

A decrease in the concentration of substrate
B final colour of the reaction mixture
C increase in the mass of product
D time taken for all substrate to disappear *(1)*

2 Amylase is an enzyme that requires a cofactor.
What is the cofactor? *(1)*

A Cl^- **B** Fe^{2+} **C** Na^+ **D** Zn^{2+}

3 The diagram shows an enzyme-catalysed reaction.

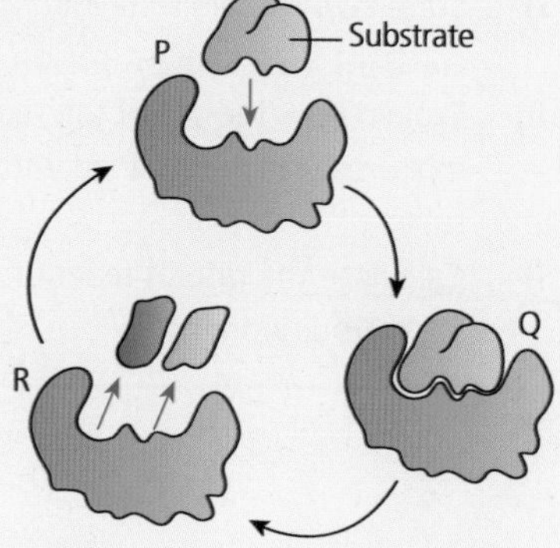

a) Use the diagram to explain the mode of action of the enzyme. *(5)*

Intracellular enzymes are involved in the synthesis of the structural compounds cellulose and collagen. Both are secreted through the cell surface membrane to form important extracellular components.

b) State the smaller molecules from which cellulose and collagen are synthesised. *(2)*

c) Hydroxylases are enzymes that synthesise collagen. These enzymes require vitamin C and iron as cofactors.

 i) Explain the role of cofactors in enzyme-catalysed reactions like those catalysed by hydroxylases. *(2)*

 ii) Name another protein that requires iron in order to function. *(1)*

4 Cells in the pancreas secrete trypsinogen, an inactive form of trypsin. When it reaches the conditions in the small intestine, trypsinogen is converted to trypsin.

a) i) State the role of trypsin in the small intestine. *(3)*

 ii) Explain why trypsin is secreted from cells in the pancreas in an inactive form. *(3)*

Students investigated the effect of pH on the activity of trypsin. Their results are shown in the table.

Tube	pH	Time taken for substrate to disappear/s	Rate of reaction $\times 10^{-3}/s^{-1}$
A	4	no reaction	0.00
B	6	360	2.78
C	7	248	
D	8	126	7.94
E	9	533	1.88
F	10	no reaction	0.00

b) i) State **two** factors that should be kept constant in this investigation. *(2)*

 ii) Explain how the pH of the reaction mixtures was maintained within each tube. *(2)*

c) i) Calculate the rate of reaction for tube C. Show your working. *(2)*

 ii) Use the results in the table to describe the effect of pH on the rate of the reaction catalysed by trypsin. *(4)*

 iii) Explain why there is no reaction at pH 4 (tube A) in contrast to the activity at pH 8 (tube D). *(4)*

5 The enzyme HMG-CoA reductase reduces a coenzyme in one of the steps in the metabolic pathway that produces cholesterol. The enzyme is only active in the liver.

Fluvastatin is a drug that is taken to reduce the synthesis of cholesterol.

Fluvastatin selectively and competitively inhibits HMG-CoA reductase.

a) Explain how fluvastatin acts to inhibit the action of the enzyme. *(3)*

b) Suggest why it is important that fluvastatin is a selective inhibitor of the enzyme. *(2)*

c) The chemical glyoxylate is a non-competitive inhibitor of HMG-CoA. Explain how the action of glyoxylate differs from the action of fluvastatin. *(4)*

d) Coenzyme A is synthesised from pantothenic acid, one of the vitamins in the B group. Explain the role of coenzymes in the functioning of enzymes. *(3)*

e) The activity of some enzymes within cells is controlled by product inhibition. Explain what this means. *(2)*

6 Catalase can be extracted from many plant tissues. Some students were set the task of comparing the activity of catalase from a number of different vegetables. To start, they made extracts of catalase from each vegetable and used the extracts to determine the initial rate of reaction using a 10 volume hydrogen peroxide solution.

Describe, in detail, the procedure that the students should follow when investigating the activity of catalase from lettuce leaves.

7 Catalase is one of the fastest-acting enzymes. Plan an investigation to find out the effect of changing the concentration of catalase on the rate of production of oxygen using a 10 volume hydrogen peroxide solution.

Chapter 5

Biological membranes

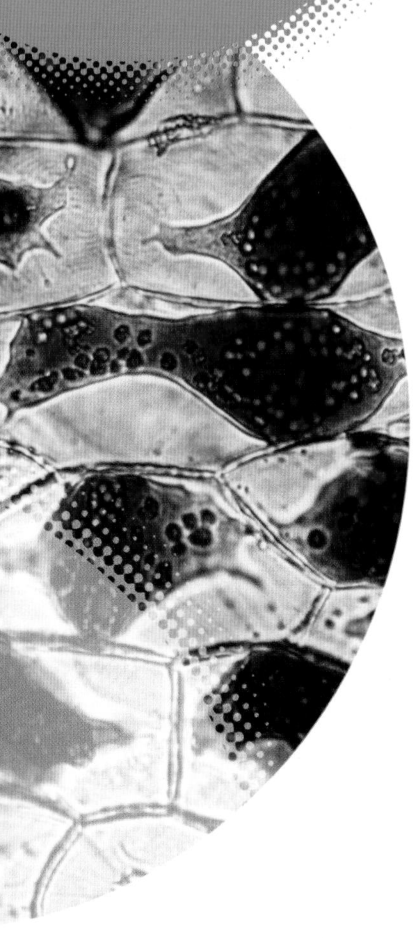

Prior knowledge

Before you start, make sure that you are confident in your knowledge and understanding of the following points.

- Substances diffuse into and out of the cell through the cell membrane.
- Diffusion is the net movement of a substance from a region of high concentration to a region of low concentration, and involves random movement of molecules.
- The rate of diffusion increases when there is a shorter distance to travel, a greater concentration gradient and/or a greater surface area.
- The cell membrane is partially permeable – it lets some substances through but prevents the movement of others.

Test yourself on prior knowledge

1 What is a concentration gradient?
2 What are the two special features of osmosis that distinguish it from other instances of diffusion?
3 What happens to a non-woody plant if its cells become flaccid?
4 Explain why plant cells do not burst when placed in water, whereas animal cells do.
5 Explain the term *net movement*.

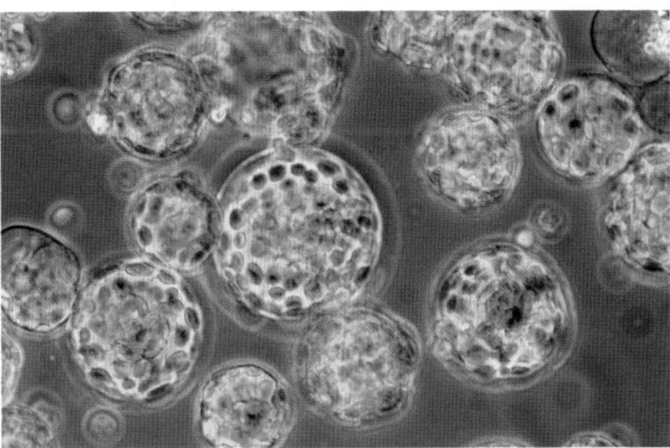

Figure 5.1 A protoplast is a plant cell that has had its cell wall removed chemically so that it is only enclosed by the plasma membrane. This is a mixture of protoplasts of tobacco, *Nicotiana tabacum* (green) and carrot, *Daucus carota* (red). Protoplasts are useful in plant science research because they are easier to manipulate for genetic experiments and other studies. (×230)

The plasma membrane is the gateway to the cell. It is a barrier that is **partially permeable**. Some substances travel freely through it; others cannot penetrate. The membrane has mechanisms that can adjust its permeability so that certain substances are allowed through in some circumstances but not in others.

There are four mechanisms by which molecules can move through the membrane. **Diffusion**, **facilitated diffusion** and **osmosis** are all passive processes in which substances move down a concentration gradient. When the cell needs to move a chemical against a concentration gradient, energy has to be used; this is called **active transport**.

Osmosis is the net movement of water molecules from a more dilute solution (which has more water molecules) to a more concentrated solution (which has fewer water molecules) across a partially permeable membrane.

When a plant cell contains as much water as possible, it is fully turgid. If an animal cell takes in too much water, it bursts. This is called cell lysis. This occurs if cells are in hypotonic solutions (lower solute concentration outside the cell than inside).

A plant cell can lose so much water that its membrane starts to pull away from the cell wall; it is said to be plasmolysed. If an animal cell loses water, it shrivels. It is said to be crenated. This happens if cells are in hypertonic solutions with a higher solute concentration outside the cell than inside the cell. An isotonic solution is a solution with the same solute concentration inside the cell as outside, so there is no net water loss or gain.

Substances can also enter or leave the cell by means of **endocytosis** and **exocytosis** respectively. In these processes, the substances do not actually pass through the membrane.

Tip

The membrane that surrounds the cytoplasm may be called the *plasma membrane*, or the cell surface membrane. The term cell membrane can be ambiguous as most eukaryotic cells have many internal membranes as well as their plasma membrane that separate the cytoplasm from the outside.

Structure and role of the cell membranes

In addition to the plasma membrane, which surrounds the cell, there are many internal membranes around organelles and vacuoles and forming the endoplasmic reticulum.

There are some variations in structure of membranes, mostly related to the proteins they contain. The chloroplasts and mitochondria are each surrounded by two membranes that have an unusual membrane structure. The outer membranes of these organelles have pores that allow small molecules to pass through easily, and the inner membrane contains many proteins that perform important functions in photosynthesis and respiration.

Membranes control the movement of substances through them, but they are also involved in **cell signalling**, containing proteins that act as receptors for chemicals from other cells that can cause specific changes in cellular activity. More details of cell signalling are given later in this chapter.

Fluid mosaic model of the plasma membrane

The structure of the plasma membrane is shown in Figure 5.2. This is the **fluid mosaic model**, which was first proposed by Singer and Nicolson in 1972. The membrane is about 7 nm thick and is mainly composed of lipid and protein, and in surface view is a pattern of protein and lipid molecules. The phospholipids move around in the lipid layer and the proteins can also move, so the structure is fluid. Figure 5.2 shows the fluid mosaic model.

The lipid part of the membrane consists of a double layer, a **bilayer**, the vast majority of which is made up of **phospholipids**, although **glycolipids** are also scattered through it. Phospholipids naturally form a bilayer when they are in contact with water or an aqueous medium such as exists in living organisms. This is because the phospholipid molecule has a **hydrophilic polar** head section and a **non-polar hydrophobic** tail. The

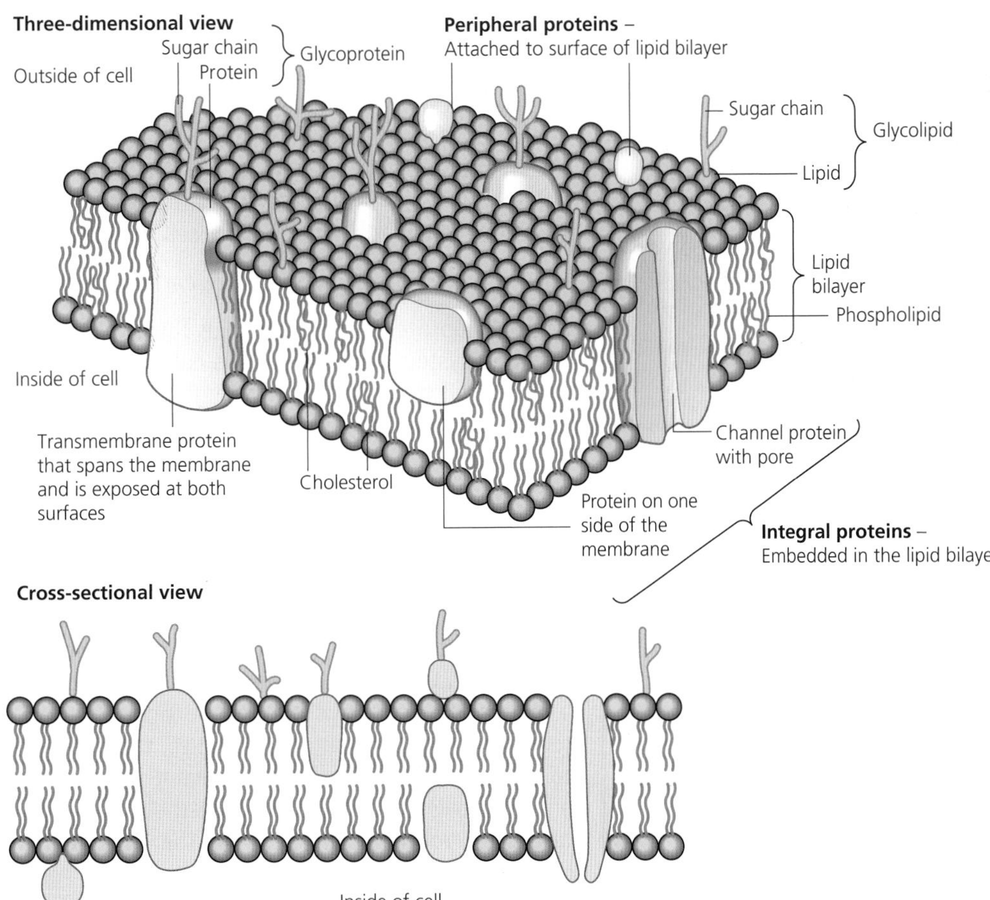

Figure 5.2 Structure of a eukaryotic plasma membrane.

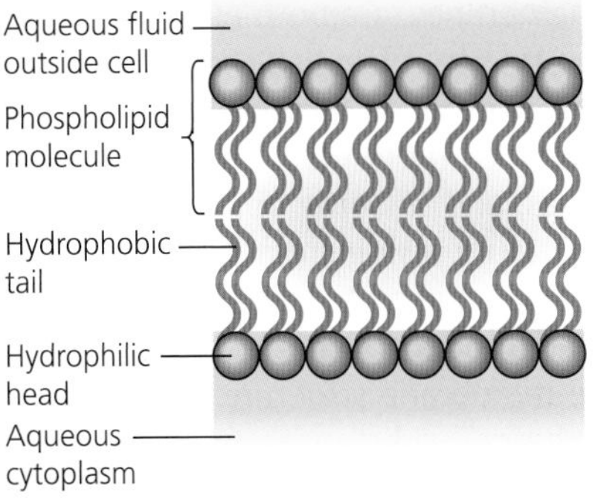

Figure 5.3 Structure of the phospholipid bilayer.

Tip

Look back at Chapter 2 to remind yourself about polar molecules

term 'hydrophilic' indicates an attraction to water (which is also a polar molecule), whereas 'hydrophobic' refers to a repulsion from water.

The most stable structure for the lipid to adopt is a bilayer, because the heads of the lipids are attracted to water and the tails repelled; the heads are then in the aqueous medium and the tails are as far away from it as possible (see Figure 5.3). The hydrophobic tails also attract each other, further stabilising the structure.

Proteins of various sorts are found in the membrane, and they have a variety of functions.

Channel and carrier proteins

Certain proteins span the width of the membrane. These are known as **transmembrane proteins**, and their function is to transport water-soluble charged particles through the membrane, because such substances cannot get through the lipid bilayer. The intrinsic proteins are of two sorts: **carrier proteins** and **channel proteins**. The channel proteins form a sort of passageway through which water and polar substances can pass by diffusion down a concentration gradient. The carrier proteins are rather different. They can change shape and, in doing so, move substances from one side of the membrane to the other. This may be down a concentration gradient, or by facilitated diffusion, or against a concentration gradient, which requires energy and is known as active transport. You will learn more about these processes later in the chapter.

Key term

Antigen A molecule – for example, a polysaccharide or protein – that is foreign to the body and that stimulates an immune response and the production of antibodies. The term is shortened from '**anti**body **gen**erator'.

Test yourself

1 What is the collective name for all the membranes in a cell?
2 Which scientists first suggested the fluid mosaic model of cell membranes?
3 Why do phospholipids naturally form a bilayer when in an aqueous environment?
4 What is the difference between a channel protein and a carrier protein?

Glycoproteins and glycolipids

Glycoproteins are proteins that have a short carbohydrate chain attached. These carbohydrate chains protrude from the plasma membrane. These chains can form hydrogen bonds with water and, in doing so, they stabilise the membrane structure. They are also important in cell signalling (see below), acting as receptors for certain molecules (e.g. hormones) and triggering specific changes in the cell when that molecule binds.

Glycolipids are lipids with a short carbohydrate chain attached.

Both glycoproteins and glycolipids form the surface **antigens** by which the immune system can identify the cell as belonging to the body (or identify it as foreign if transplanted into the body of another individual). We shall look at this in more detail later.

Cholesterol

Cholesterol has gained rather a bad reputation because of the role of excess cholesterol in heart disease, but it plays an important role in the plasma membrane. There is a lot of cholesterol in the plasma membrane, and its main function is to maintain a suitable level of fluidity in the membrane. Like phospholipid molecules, the cholesterol molecule has both a hydrophilic and a hydrophobic portion. This allows cholesterol to bind to the phospholipids and, in doing so, this prevents the membrane being too fluid. On the other hand, adjacent fatty acids could come together and crystallise, which would make the membrane more rigid. The presence of cholesterol molecules between the fatty acid chains prevents this happening and so keeps the membrane fluid.

Factors affecting membrane structure and permeability

Proteins and lipids, which make up cell membranes, are both affected by temperature, and so it is no surprise to find that the behaviour of membranes is affected by changes in temperature. As temperature rises, the lipid component of the membrane becomes more fluid. This reduces its effectiveness as a barrier to polar molecules, and so some polar molecules get through. In addition, raising the temperature also increases the speed of any diffusion that can take place. A critical change occurs at around 40 °C, however, because at this temperature many proteins start to denature. When proteins denature, this disrupts the membrane structure and the membrane no longer acts as an effective barrier; substances can pass freely through what is left of it. This change, unlike the increased fluidity of the lipids at lower temperatures, is irreversible.

Most of the cell membrane is made from lipids, so organic solvents, which dissolve lipids, can severely disrupt the membrane structure, leading to increased permeability. For example, you may have noticed that your skin feels strange if it has been exposed to ethanol or the acetone in nail varnish remover.

Activity

The effect of temperature on membranes in beetroot cells

Students investigated the effect of temperature on the permeability of cell membranes using beetroot cells. Beetroot cells have pigmented cell sap that is held inside by the tonoplast (the membrane surrounding the vacuole) and the plasma membrane (see Figure 5.4). If membrane permeability is increased, the sap leaks out into the surrounding solution. The amount that comes out can be measured using a **colorimeter**. This device detects light passing through the solution. It can be set to measure absorbance (how much light does not get through) or transmission (the amount of light that does get through). In this experiment, the absorbance setting was used; therefore, the higher the absorbance, the greater the permeability of the membrane.

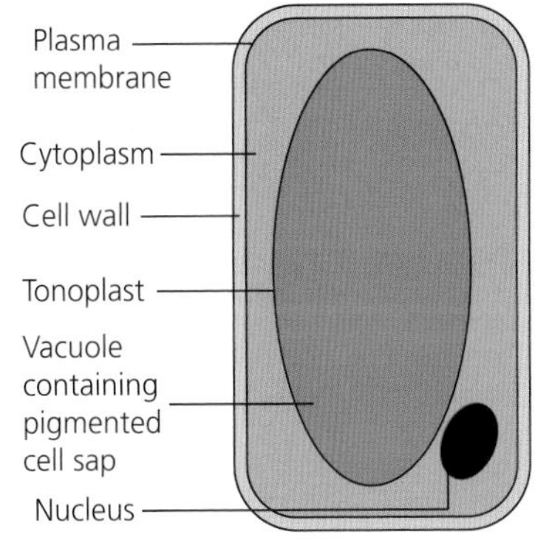

Figure 5.4 Beetroot root cell.

Method

1 A cork borer was used to cut pieces of uniform diameter from the beetroot.
2 The beetroot cylinders were cut into 10 mm lengths.
3 The beetroot pieces were rinsed with distilled water until the rinsing water showed no pink colour.
4 Individual cylinders were placed in tubes containing 5 cm³ of distilled water, maintained at temperatures of 4 °C, 20 °C, 30 °C, 40 °C, 50 °C and 60 °C.
5 For each temperature, five pieces were placed in separate tubes (i.e. five repeats).
6 The tubes were left for 20 minutes.
7 The solution surrounding each piece of beetroot was placed in a colorimeter, set to measure absorbance.

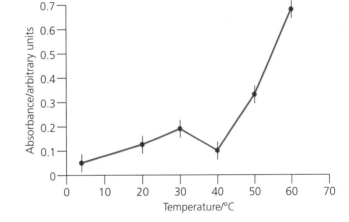

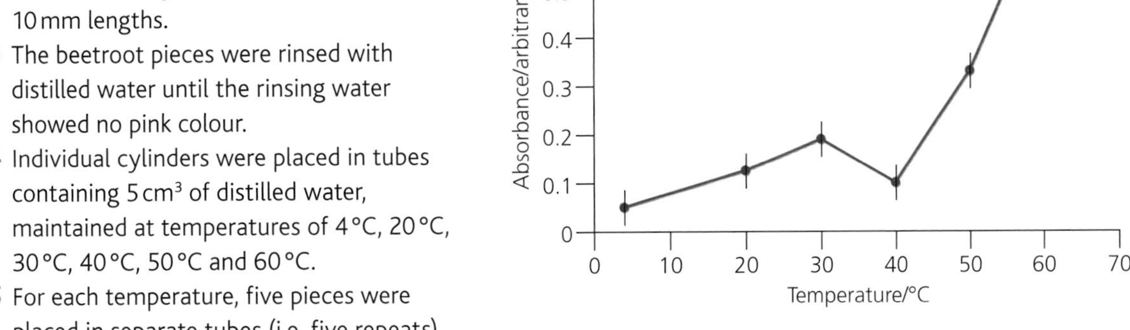

Figure 5.5 The effect of temperature on membrane permeability.

The results are shown in Figure 5.5. The error bars show the range of results obtained for each temperature.

Questions

1 List the factors that have been controlled in the experimental method.
2 Due to the number of discs required, it was not possible to use the same beetroot for all temperatures. Suggest why this is unlikely to matter.
3 Explain the reason for Step 3.
4 The students suggested that the result at 40 °C was anomalous. Suggest why they may be wrong in this assumption.
5 Explain the shape of the graph
   a) between 4 °C and 30 °C
   b) between 40 °C and 60 °C.
6 Suggest, with reasons, what might happen to the absorbance readings at temperatures higher than 60 °C.

Cell signalling

Cells of multicellular organisms do not function in isolation. They communicate with other cells, both those immediately around them and others that may be quite distant. This communication is by means of chemical messenger molecules. These molecules are produced in one cell and transported to others. Some cells are unaffected by the messenger, because they have no means of detecting it. Cells that do have the correct cell membrane receptors detect the messenger molecule and changes occur in the cell as a result. This type of information transfer between cells is called **cell signalling**.

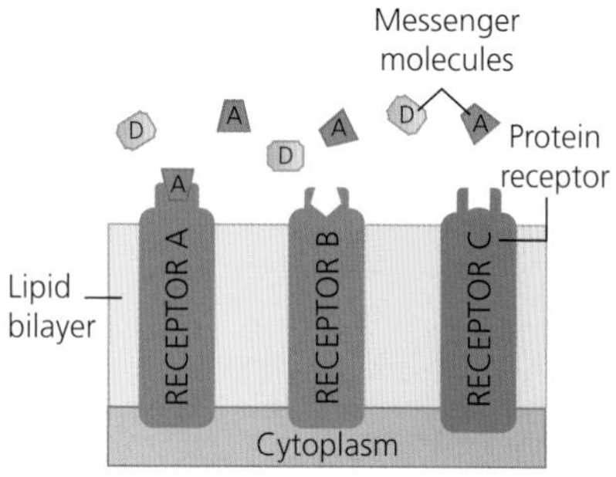

Figure 5.6 The specificity of membrane receptors. This cell can respond to messenger A but not to messenger D, because it lacks the necessary receptors.

Plasma membrane receptors

Gycoproteins in the plasma membrane form the receptors for cell signalling messenger molecules, although there is evidence that glycolipids may also have a role. Each receptor is specific to a particular messenger, and cells have hundreds of different types of receptor. The actual population of receptors varies in different types of cells, however, which results in individual cells being responsive to some messengers but not others (see Figure 5.6). So, for example, liver cells have receptors for the hormone glucagon, which causes them to release glucose from stores of glycogen within the cells. Skeletal muscle cells encounter glucagon as it circulates in the blood but do not respond to it, because they lack the right type of receptor; this would have a complementary shape. A cell that responds to a particular messenger is termed a **target cell** for that messenger.

How do receptors change cells?

When a messenger molecule binds to a receptor, which will have a complementary shape, changes can be brought about in the target cell in a number of ways.

- The change in the receptor may cause the release of a second messenger inside the cell, which can initiate a variety of effects.
- The change may result in the opening of a protein channel that was previously closed, or the closing of one that was open.
- The change may activate an enzyme within the cell, or the membrane protein may be an enzyme itself which is activated.

Membrane receptors and drugs

The membrane receptors respond to chemical messengers, and those messengers can come from outside the body as well as within it. About half of medicinal drugs target specific membrane receptors. This allows drugs to affect only known target cells and to create (or block) specific responses. An example is antihistamine drugs. Histamine is a chemical messenger released by white blood cells in response to cell damage. It attaches to membrane receptors and causes increased permeability of the capillaries in the damaged area to white blood cells, which in turn creates inflammation. This is useful when there is damage, but sometimes the blood cells release histamine in response to a false signal caused by something to which the host is allergic. In such cases, antihistamine drugs are taken. These block the histamine receptors, so the histamine cannot bind to the cells and therefore has no effect. The role of antihistamines is shown in Figure 5.7.

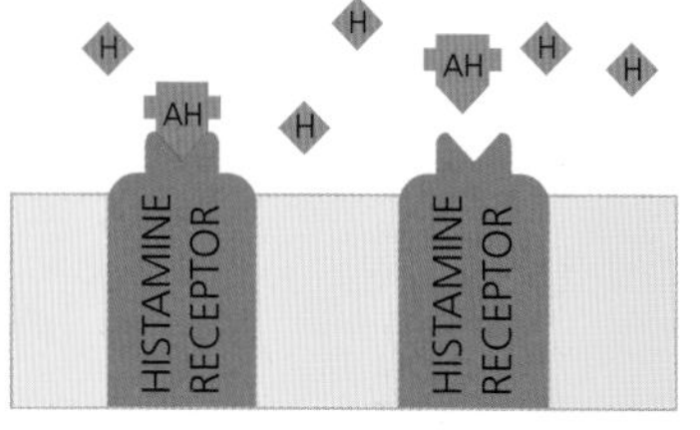

Figure 5.7 The mechanism of action of antihistamine drugs. The antihistamine drug (AH) is a complementary shape to the histamine receptor, and competes with the histamine (H) for receptor binding sites so that fewer histamine molecules can attach.

Example

Nerve cells in the brain, spinal cord and alimentary canal have opioid receptors that are stimulated by enkephalins, which are cell signalling compounds. The drugs morphine and aspirin are painkillers, but only morphine interacts with these opioid receptors to mimic the effects of enkephalins.

1 a) Suggest why receptors for morphine are on the plasma membrane rather than inside the cytoplasm of nerve cells.
   b) Suggest why morphine interacts with opioid receptors, but aspirin does not.

2 If the drug naloxone is given just before or just after morphine, the effects of morphine are few or non-existent. This inhibitory effect of naloxone is overcome by giving more morphine.
   a) State the most likely way in which naloxone acts to inhibit the effect of morphine. Explain your answer.
   b) Explain why morphine and naloxone are not effective against cells in the heart and lungs.

Answers

1 a) Morphine does not pass through the phospholipid bilayer, because it is water soluble and therefore not soluble in lipids. There are also no protein channels or carriers for morphine.
   b) Morphine has a (molecular) shape complementary to the shape of opioid receptors, whereas aspirin has a different shape and is not complementary.

2 a) Naloxone is also complementary to opioid receptors because it fits into or blocks the receptors, so morphine cannot enter the receptors. This is called competitive inhibition. Increasing the concentration of morphine increases the probability that it, and not naloxone, will enter the receptor.
   b) Cells in the heart and lungs do not have the specific receptors that are involved with the transmission of information about pain by nerve impulses.

Test yourself

5 Cell membranes break down at high temperatures. Explain how this happens.
6 Which components of the plasma membrane function as receptors in cell signalling?
7 Why does a specific messenger molecule affect some types of cell and not others?
8 Give three ways in which a membrane receptor may induce changes inside the cell.
9 Explain the importance of membrane receptors in the action of most drugs.

Transport of substances into and out of cells

Cells constantly exchange materials with their environment, and so these materials must pass through the plasma membrane. When there is a favourable concentration gradient (i.e. there is a lower concentration at the destination than at the source) the exchange of materials involves passive transport down the gradient. There are three types of passive transport: diffusion, facilitated diffusion and osmosis. Sometimes, however, cells need to transport substances in or out when the concentration gradient would naturally take the substances in the opposite direction, i.e. they need to transport them against a concentration gradient. Under such circumstances, energy has to be expended and active transport is needed.

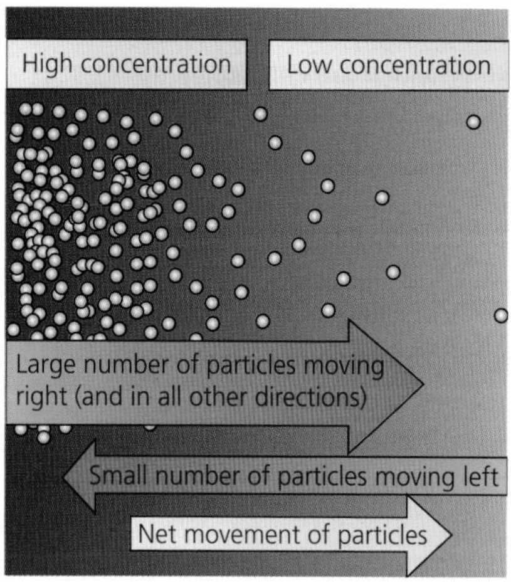

Figure 5.8 The mechanism of diffusion.

Diffusion

Diffusion is the net movement of atoms or molecules from an area of higher concentration to an area of lower concentration.

Diffusion is a passive process, and no energy is required. Particles are constantly moving, although the state of a substance may restrict the extent of this movement. In solids, the movement is confined to vibration and the particles stay in a fixed position. As a result, very little diffusion takes place in solids. In liquids and even more so in gases, the particles are free to move from place to place and so diffusion can occur readily. This movement is random – there is no way the particles can 'know' where they are going – but naturally results in the particles spreading out from an area where they are concentrated. The mechanism is shown in Figure 5.8.

In cells, however, the plasma membrane forms a partially permeable barrier. Even with a concentration gradient in their favour, some particles cannot pass through the lipid bilayer. They need help, and this comes in the form of facilitated diffusion.

Tip

When discussing and writing about diffusion and osmosis, always refer to **net** movement rather than just movement. Particles are in constant motion, and some move from left to right and others from right to left. 'Net movement' implies movement in both directions, whereas 'movement' suggests just one direction.

Facilitated diffusion

Facilitated diffusion assists larger molecules, polar molecules and ions to pass through the membrane. They do so using proteins in the membrane – either channel proteins or carrier proteins. The two routes have different mechanisms, but in both cases the process is passive and no metabolic energy is required.

Ions and small polar molecules are transported through the membrane by channel proteins, which function like pores in the membrane. Different channel proteins transport different substances, and in many cases the channels can be opened and closed to regulate flow. The mechanism is shown in Figure 5.9.

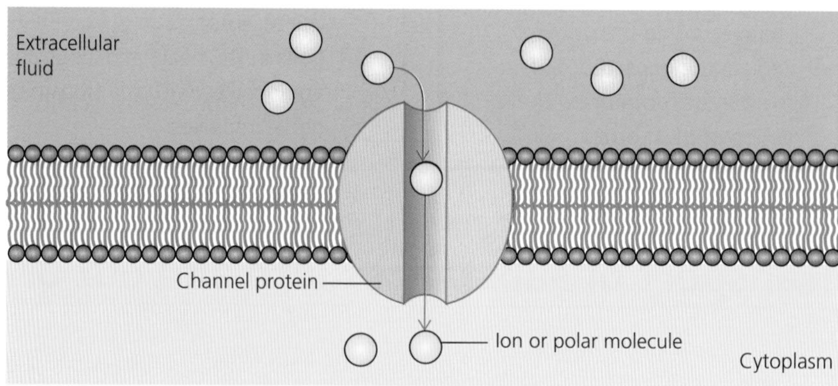

Figure 5.9 Facilitated diffusion through a channel protein.

Larger molecules use carrier proteins that are specific to the substance being transported. The molecule attaches to the protein, which then changes shape; this change of shape transfers the molecule to the other side of the membrane. The shape change requires no metabolic energy from the cell and so the process is passive. The mechanism is shown in Figure 5.10.

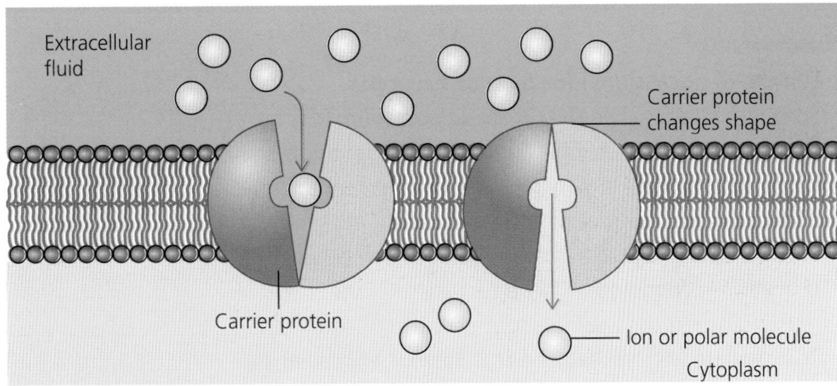

Figure 5.10 Facilitated diffusion by carrier proteins.

Active transport

Particles cannot pass passively up a concentration gradient. The cell has to provide energy to move them, which gives the process of **active transport** its name.

> **Tip**
>
> Look back at Chapter 3 to remind yourself about ATP.

Active transport is carried out by carrier proteins in the membrane, like facilitated diffusion is, but not by channel proteins. **Adenosine triphosphate (ATP)** provides the energy. The mechanism of active transport is shown in Figure 5.11.

As in facilitated diffusion, each carrier is specific and only transports a certain substance or sometimes a restricted range of substances. Some protein carriers are involved in two-way active transport. They pump one substance into the cell and another out at the same time.

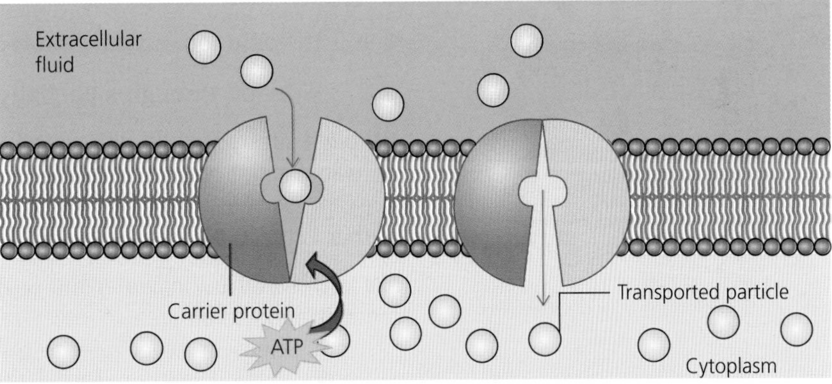

Figure 5.11 Active transport by carrier proteins.

> **Tip**
>
> Respiratory inhibitors such as cyanide stop active transport; as a result, this prevents the formation of the ATP needed for respiration. If a question describes an experiment in which a respiratory inhibitor is used, active transport is likely to be involved somewhere in the answer required.

Activity

Uptake of potassium ions by root hair cells

Some students studied the uptake of potassium ions by root hair cells under aerobic and anaerobic conditions. Root hair cells were grown in culture solution containing different concentrations of potassium ions. One set of cultures was kept in aerobic conditions, the other in anaerobic conditions. After a certain time, the cells were removed and homogenised. The filtered homogenate was then placed in an instrument called a flame photometer, which burns the ions and measures the intensity of the colour. Potassium burns with a lilac flame, and the intensity of that colour indicates how much potassium is present in the cells.

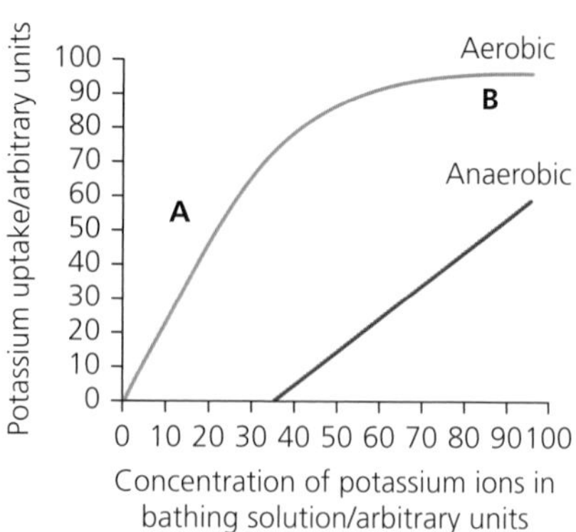

Figure 5.12

The results are shown in Figure 5.12.

1 Suggest **two** pieces of evidence from the data that support the hypothesis that uptake in section A is by active transport.
2 Anaerobic conditions kill the cells. Explain how potassium can be absorbed in anaerobic conditions.
3 Suggest why the aerobic curve levels at B.
4 The data shows potassium **uptake**, not overall potassium content. Using the data, suggest what the approximate potassium content of the cells was (in arbitrary units) at the start of the experiment. Give a reason for your answer.

Osmosis

Osmosis is the net movement of water molecules from a more dilute solution to a more concentrated solution across a partially permeable membrane. Two factors define osmosis:

- It is the diffusion of **water molecules** only.
- It is diffusion **through a partially permeable membrane**.

Although substances do diffuse through membranes, the general term 'diffusion' is not restricted to that sort of movement. Diffusion of substances other than water through the membrane is never called osmosis, neither is the movement of water through, for instance, the cytoplasm.

Like with all diffusion, water molecules move from an area of higher concentration to an area of lower concentration, but here there is a problem. Scientists never refer to the 'concentration' of a solvent such as water; the term is restricted to solutes **in** a solvent. It is incorrect, therefore, to refer to 'water concentration'. Instead, the concentration of water in a solution has a special term – water potential.

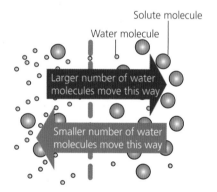

Figure 5.13 Mechanism of osmosis.

Key term

Water potential A measure of the relative tendency of water to move from one area to another. Water moves from an area of higher water potential to an area of lower water potential. Therefore, the lower the water potential, the greater the tendency for water to move to the area.

Tip

Remember that osmosis is a two-way process, so the correct term to use is **net movement of water**. It is incorrect to say that osmosis stops, because it never does while water molecules are present on either side of a membrane. The correct phrase to use is **reach an equilibrium**, in other words, the movement of water in each direction is equal.

Tip

Turgor, flaccid and plasmolysis are terms that only apply to plant cells – do not use them when discussing animal cells. Crenation, on the other hand, is only used when describing animal cells.

Tip

In exam answers, if you use the terms hypertonic, isotonic and hypotonic, you must always define them in terms of water potential.

Key term

Plasmolysis The shrinking of cytoplasm away from the cell wall of a plant cell when water is lost due to osmosis, resulting in space between the cell wall and cell membrane.

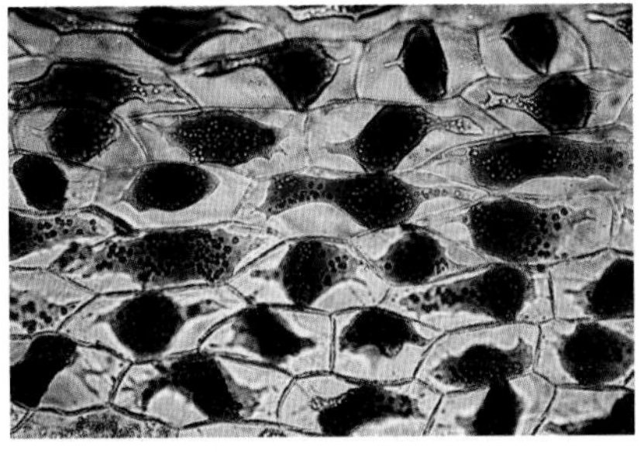

Figure 5.15 Plasmolysis in a plant cell. Notice that the cytoplasm has pulled away from the cell wall. (×235)

Osmosis is a very useful process for living things, but it also has potential hazards: if the solution bathing cells has a significantly different water potential to that of the cells, damage can result. This is shown in Figure 5.14.

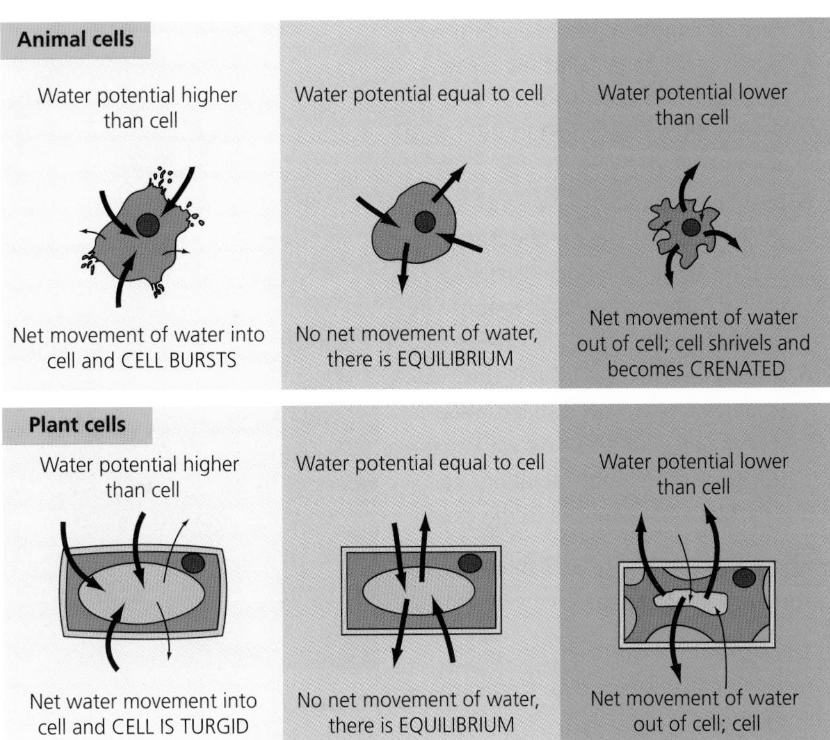

Figure 5.14 Osmosis in animal and plant cells.

In animal cells, the bursting of the cells in a **hypotonic** solution is fatal, and **crenation** in a **hypertonic** solution usually is, too. Hypotonic refers to higher water potential, hypertonic is lower water potential, and isotonic is the same water potential.

For plant cells, being **turgid** is a good thing – it is essential, in fact, for the provision of support in non-woody plants. The cells become rigid and this prevents the plant from wilting. Plasmolysis, however, can result in cell death, because the cytoplasm pulls away from the cell wall. The appearance of a plasmolysed cell under the microscope is shown in Figure 5.15. At the point when the cell is not turgid but has not yet plasmolysed, it is referred to as **flaccid**.

Activity

Plasmolysis in onion cells

Plasmolysis occurs in plant cells when the cell membrane pulls away from the cell wall. This may happen by osmosis or it may happen if the cell loses water by evaporation. Cells can recover if they receive sufficient water. The epidermis from the fleshy scale leaves in red onion bulbs makes good material to observe plasmolysis.

A student followed these instructions to observe plasmolysis in onion epidermal cells.

Method

1. Cut a red onion into quarters and pull off part of a fleshy scale leaf. Peel off a single layer of red cells from the concave surface of the leaf. Cut the piece so it is small enough to fit beneath a coverslip.
2. Place the piece of epidermal tissue on a slide. Cover it with two drops of distilled water. Add a cover slip.

3 Look at the epidermal cells through a microscope, starting with the low power lens. Take a photograph of the cells.
4 Peel off another piece of epidermis from the onion leaf. Put the piece on a slide and add two drops of 1.0 mol dm^{-3} potassium nitrate solution.
5 Examine the epidermal cells through the microscope. Take a photograph of the cells and compare the appearance of these cells to those in distilled water.
6 After a few minutes irrigate the epidermal cells with distilled water as shown in Figure 5.16. This will remove the potassium nitrate solution. Describe what happens to the cells.

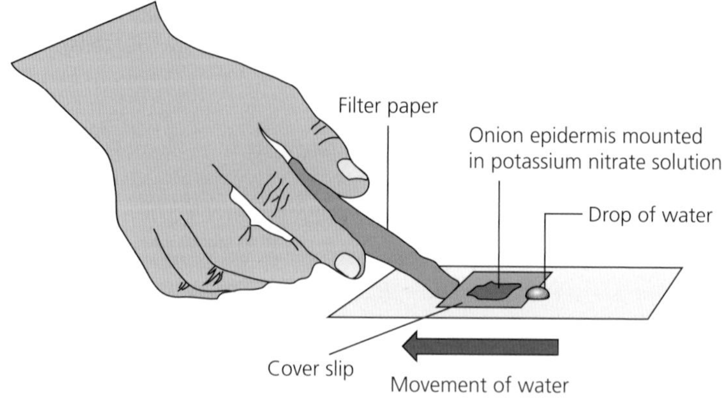

Figure 5.16 Method of changing the solution around the onion epidermis.

Figure 5.17 and Figure 5.18 show the student's photographs.

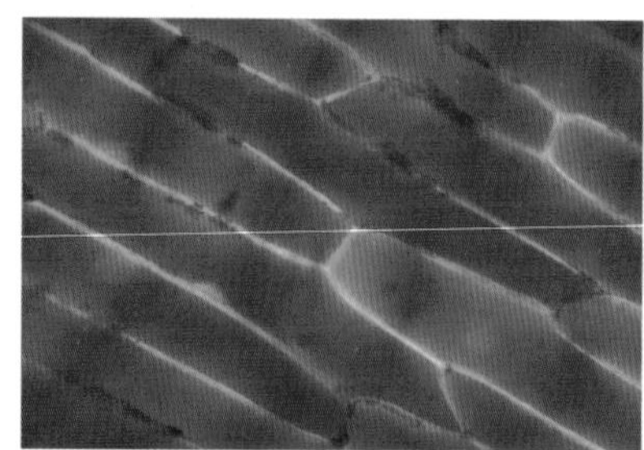

Figure 5.17 Epidermal cells immersed in distilled water.

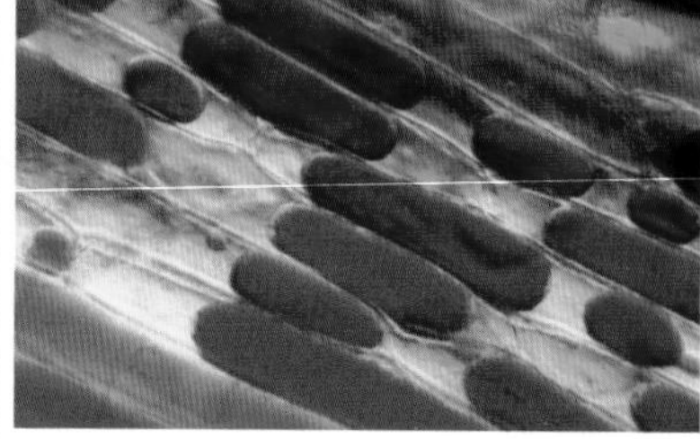

Figure 5.18 Epidermal cells immersed in 1.0 mol dm^{-3} potassium nitrate solution.

The student then prepared more pieces of epidermal tissue and immersed them in solutions of potassium nitrate with the following water potentials: –500 kPa, –1000 kPa, –1200 kPa, –1500 kPa, –1800 kPa and –2000 kPa.

The student counted the number of cells that were plasmolysed in three fields of view in each solution.

Table 5.1 shows the results that the student recorded in each field of view.

Table 5.1 The effect of solutions of different water potential on the percentage plasmolysis of cells of onion epidermis.

Water potential of potassium nitrate solution/kPa	Field of view					
	1		2		3	
	Number plasmolysed	Total counted	Number plasmolysed	Total counted	Number plasmolysed	Total counted
0	0	20	0	18	0	23
–500	0	17	0	15	0	21
–1000	4	18	5	21	5	19
–1200	10	17	12	21	11	22
–1500	16	19	18	20	18	22
–1800	20	21	19	19	22	22
–2000	21	21	16	16	18	18

Questions

1 Use the term **water potential** to explain the student's results shown in Figure 5.17 and Figure 5.18.
2 When the cells shown in Figure 5.18 were irrigated with water they gradually returned to the condition shown in Figure 5.17. Explain how this happened.
3 Calculate the mean percentage of cells that were plasmolysed in each of the seven solutions of potassium nitrate as shown in Table 5.1. Make a table to show the mean percentage of cells in each solution.
4 Plot a graph of percentage plasmolysis against water potential.
5 Describe the effect of water potential on the onion epidermal cells as shown in your graph.
6 Explain why the student was told to make counts of cells in three fields of view in solution and not just one.
7 Explain why you plotted percentage plasmolysis on your graph rather than total number of plasmolysed cells.

Endocytosis and exocytosis

Key term

Invagination Folding in to form a pocket.

Endocytosis is the process by which cells absorb molecules by engulfing them. There are two forms of endocytosis: **phagocytosis** and **pinocytosis**. These are shown in Figure 5.19.

Pinocytosis is the method by which small particles and fluid are taken in. The plasma membrane **invaginates** inwards and then the membranes fuse around the molecules to form a small vesicle.

In phagocytosis, protrusions called **pseudopodia** (singular: pseudopodium or pseudopod) extend from the cell and wrap themselves around a larger particle (e.g. a microorganism). Once again, the membrane fuses to seal the microorganism into a vesicle.

In both pinocytosis and phagocytosis, the particles end up inside the cell but are not actually in the cytoplasm – they are in a vesicle separated from the cytoplasm by a membrane. Exocytosis occurs at the end of phagocytosis to excrete the undigested particles.

Test yourself

10 Explain why facilitated diffusion is sometimes necessary in cells, in addition to simple diffusion.
11 What is the difference between a channel protein and a carrier protein?
12 Which molecule provides the energy for active transport?
13 What does the term *osmotic equilibrium* mean?
14 Why can an animal cell never be plasmolysed?
15 Describe the benefit to plants of having turgid cells.
16 Explain the difference between exocytosis and phagocytosis.

Pinocytosis

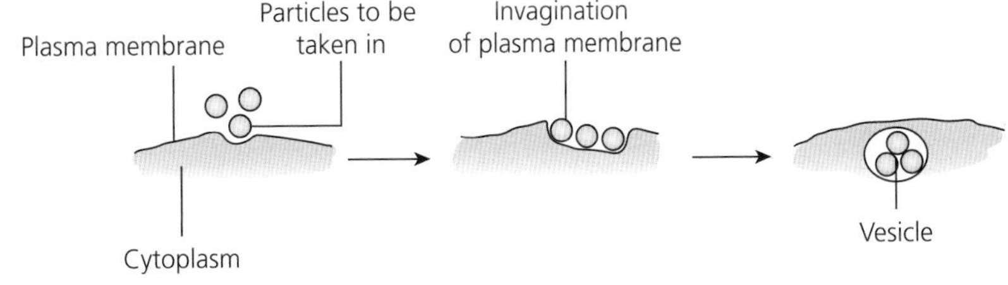

Phagocytosis

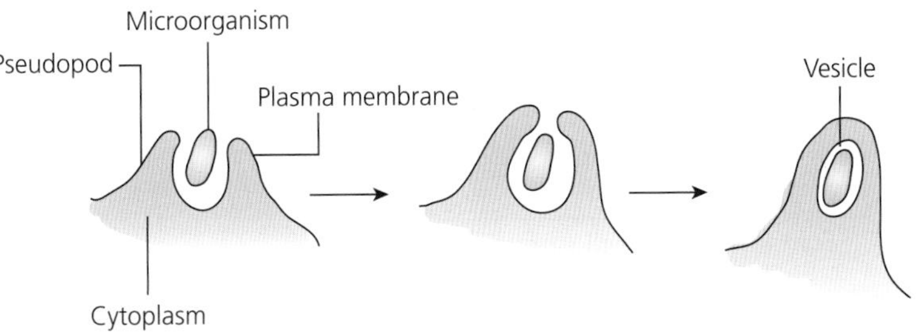

Figure 5.19 Pinocytosis and phagocytosis.

Substances are secreted and egested from a cell by the process of exocytosis. This is rather like a reversal of the process of pinocytosis, and is shown in Figure 5.20.

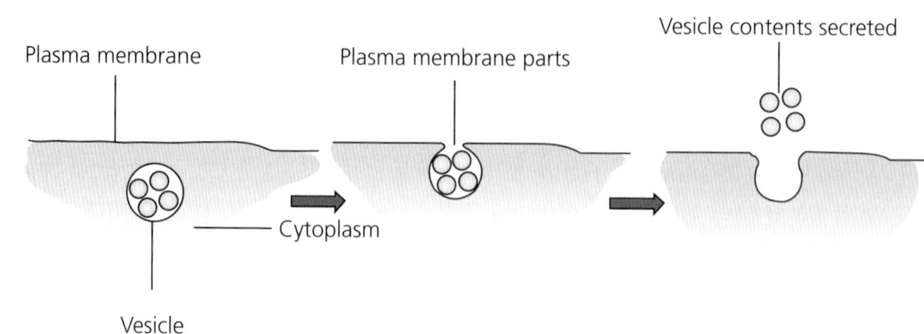

Figure 5.20 Exocytosis.

Both endocytosis and exocytosis are active processes that require ATP as an immediate source of energy.

Exam practice questions

1 Which is the definition of diffusion?

A net movement of molecules and ions from a region of high concentration to a region of low concentration

B net movement of molecules and ions from a region of low concentration to a region of high concentration

C net movement of molecules and ions from one region to another against a concentration gradient

D net movement of molecules and ions from one region to another across a cell surface membrane *(1)*

2 Which of these components forms the bilayer in all biological membranes?

A cholesterol **B** glycolipids
C phospholipids **D** proteins *(1)*

3 The diagram shows two receptors, **H** and **J**, on the surface of a cell. Which shows the cell signalling compounds that bind to these two receptors? *(1)*

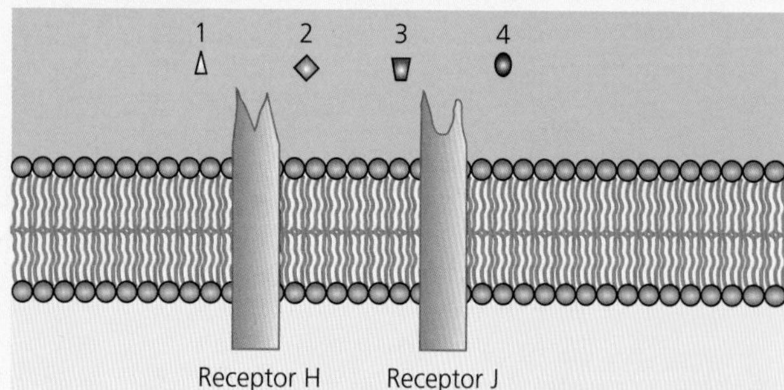

	Cell surface receptors	
	H	J
A	1	2
B	1	4
C	2	3
D	3	4

4 The diagram shows a cross section of a small region of a cell surface membrane.

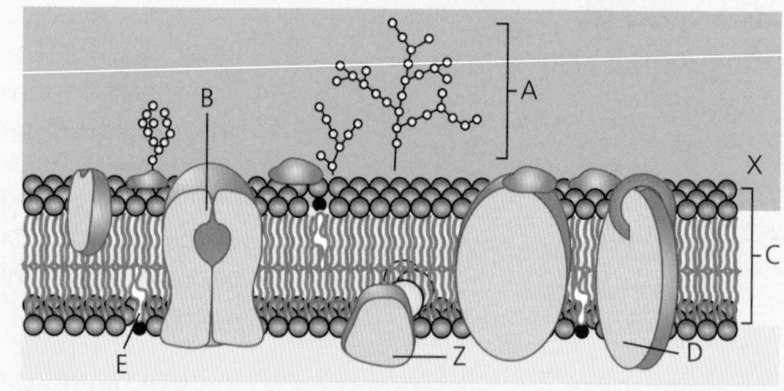

a) State which is the most likely width of the membrane shown in the diagram.

70 µm 17.0 µm 7.0 µm 70 nm 17.0 nm 7.0 nm 70.0 pm 17.0 pm 7.0 pm *(1)*

b) State the roles of the components of the membrane labelled **A–E**. *(5)*

c) The diagram represents the fluid mosaic structure of cell membranes. Explain why:

i) membranes are described as fluid mosaic structures *(3)*

ii) the side labelled **X** is the extracellular surface of the membrane. *(2)*

d) i) Use the diagram to explain how membranes can act as sites of chemical reactions. *(2)*

ii) State **four other** functions of membranes within cells. *(4)*

5 Blood plasma has a concentration equivalent to that of a 9 g dm^{-3} sodium chloride (NaCl) solution. A student investigated the effect of sodium chloride solution on red blood cells from a mammal by transferring small samples of red blood cells into a range of sodium chloride solutions between 0 and 15 g dm^{-3}.

The student followed this procedure:

1. Use a graduated pipette to transfer $10\,cm^3$ of each solution to labelled test tubes.
2. Put $0.5\,cm^3$ of blood into each solution.
3. Mix the blood with each solution by inverting each tube three times.
4. Use a syringe to remove samples from each tube and use a cell counter to determine the number of red blood cells in each sample.
5. Calculate the number of red blood cells present in each sample as a percentage of the number in the $9\,g\,dm^{-3}$ solution.
6. Plot the results on a graph.

a) State two variables that are controlled in this investigation. *(2)*

b) The student's graph is shown below.

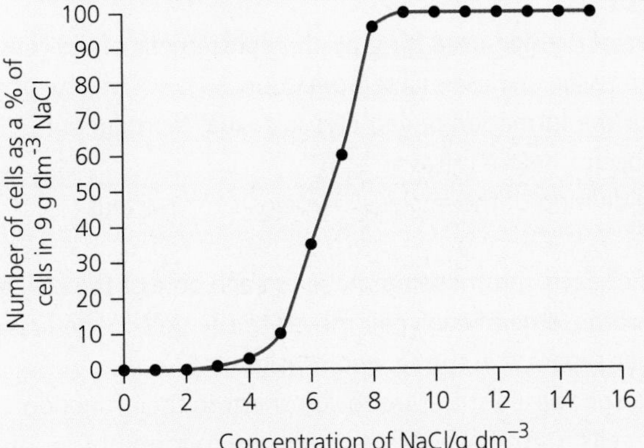

Describe the student's results. *(5)*

c) State an assumption that the student has made in calculating the number of cells remaining as a percentage of the number in the $9\,g\,dm^{-3}$ NaCl solution. *(1)*

d) The student took small drops of blood from each solution and observed the red blood cells. The student made the following records in a laboratory notebook for three solutions.

NaCl solution/g dm⁻³	Percentage of cells	Appearance of cells	Appearance of contents of test tube
2	0	no cells visible	clear red solution
9	100	many cells visible	red suspension
15	100	many cells visible	red suspension

Explain, using the term **water potential**, the results in the student's notebook. *(9)*

e) A further sample of blood was added to a $20\,g\,dm^{-3}$ solution of sodium chloride. Predict the likely results. *(3)*

6 Aquaporins are membrane proteins that allow the movement of water across membranes in plant and animal cells. The diagram below shows an aquaporin.

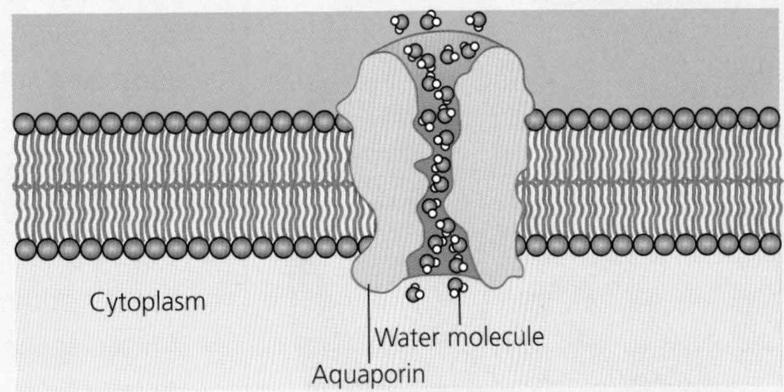

a) Explain why aquaporins are necessary for the movement of water across cell membranes. *(2)*

b) State **four** factors that influence the rate at which water molecules can pass across a cell membrane into a plant cell. *(4)*

c) The volume of plant cells increases over a period of time after they are placed into distilled water.

Explain why

i) cell volume increases *(4)*

ii) after a while the cell volume does not increase any further. *(3)*

Stretch and challenge

7 Glucose is absorbed across the cell surface membrane of yeast cells by facilitated diffusion. The rate of uptake of glucose by yeast can be investigated by using a solution of glucose and a suspension of yeast cells. It is possible to determine the quantity of glucose absorbed by yeast cells in a fixed length of time.

Plan an investigation to find out if the rate of glucose uptake by yeast cells is influenced by the concentration of glucose in the surrounding solution. Make a prediction, devise a method and explain how you would obtain and analyse the results.

Chapter 6

Cell division, diversity and organisation

Prior knowledge

Before you start, make sure that you are confident in your knowledge and understanding of the following points.

- Most body cells contain chromosomes in matched pairs.
- Gametes have half the number of chromosomes of body cells.
- Most body cells in an individual have the same number of chromosomes, but this varies between species.
- Before a cell divides, its nucleus must divide. Nuclei divide in two ways: mitosis and meiosis.
- Mitosis is the type of division used for growth, replacement of old cells, repair of damaged tissue and asexual reproduction.
- Meiosis is used for the formation of gametes in sexual reproduction.
- DNA replication occurs before mitosis.
- Mitosis is a single division of the nucleus; in meiosis the nucleus divides twice.
- In mitosis, copies of each chromosome move to each pole of the cell, but in meiosis one chromosome of each pair moves to the pole in the first division. The copies separate in the second division.
- Cells often have a specialised structure to suit them to their function.
- Groups of similar cells form tissue, tissues associate to form organs, and different organs work together to form organ systems.

Test yourself on prior knowledge

1. Explain why body cells have matched pairs of chromosomes.
2. The pairs of chromosomes in a body cell are 'matched', but not identical. Explain.
3. How many chromosomes are found in a human body cell?
4. Why is a separate form of nuclear division (meiosis) necessary to produce gametes?
5. Red blood cells transport oxygen. In what ways are their structure specialised for that function?

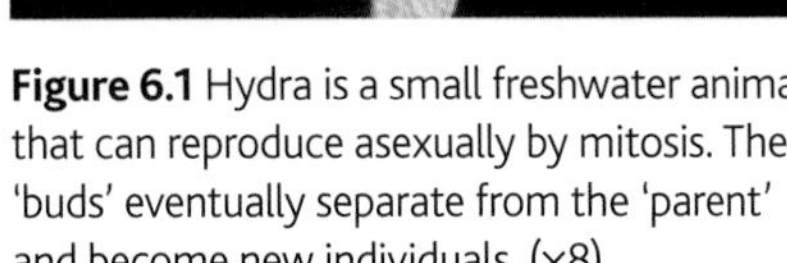

Figure 6.1 Hydra is a small freshwater animal that can reproduce asexually by mitosis. The 'buds' eventually separate from the 'parent' and become new individuals. (×8)

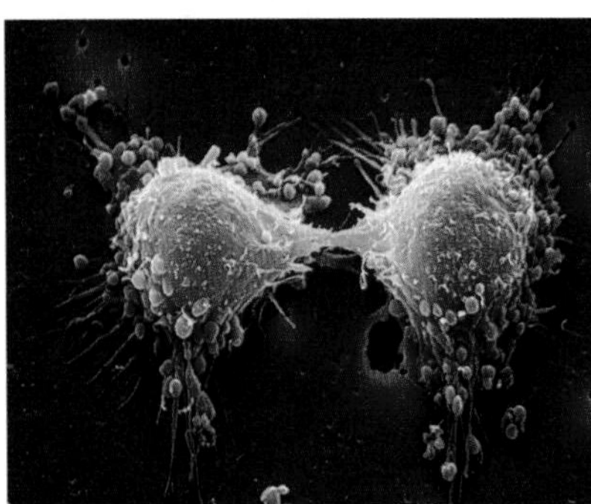

Figure 6.2 Coloured scanning electron micrograph of two prostate cancer cells in the final stage of cell division (cytokinesis). (×500)

Cell division is fundamental to the continuation of life. Many single-celled organisms reproduce asexually by mitosis (see Figure 6.1). More complex organisms use cell division by mitosis to grow and cell division by meiosis to reproduce. The control of cell division is complex and can fail. For example, when mitosis runs uncontrollably, the rapid formation of tissue growths called tumours produces the condition we call cancer (Figure 6.2).

You have learned how cells have a structure that enables them to carry out living processes. However, all the cells in a multicellular organism originated from a single cell, either by asexual reproduction or from a **zygote** formed

Key term

Differentiation The process by which a cell develops to become more distinct in form and function.

Tip

Look again at Chapter 3 to remind yourself about the structure of DNA and how DNA replicates.

Tip

This chapter is about cell division in eukaryotes.

Tip

The terms mitosis and meiosis are so similar that you need to make sure you spell them correctly.

by the fertilisation of a female **gamete** (sex cell) by a male gamete. In order for multicellular organisms to develop, the original cell needs to duplicate itself many times, and also to **differentiate** into different forms, each suited to their particular purpose. The key components in this cell division are the chromosomes.

Each new cell must receive a full set of chromosomes just like the ones in its parent cell, and so the chromosomes must duplicate before cell division can take place.

The process by which cells divide to form new cells for growth, replacement or repair of tissues is called **mitosis**. However, the formation of gametes requires a special type of nuclear division, called **meiosis**. When one gamete fertilises another, the nuclei fuse; in order for the new organism to have the same number of chromosomes as its parents, each gamete must only contain one chromosome from each homologous pair. For example, in humans, where there are 46 chromosomes, the egg and sperm cells contain only 23.

As well as specialising into different types of cell, the cells must arrange themselves in an ordered structure in order to form a whole organism. Cells that have specialised in a particular way come together to form **tissues**, and tissues may be grouped with others to form **organs**. Finally, organs may be linked structurally or functionally to form **organ systems**.

The cell cycle

When a cell has divided, it takes a period of time for the new cells to mature before dividing again. The cycle of division, growth and maturity and then another division is called the **cell cycle**. The length of the cell cycle varies for different types of cell, with more specialised cells generally having a longer cell cycle. The length of the cell cycle of a variety of cells is shown in Table 6.1.

Table 6.1 Duration of cell cycle for some different human cell types.

Cell type	Duration of cell cycle
Embryo cells	8–60 minutes
Yeast cells	1.5–3 hours
Intestinal epithelial cells	about 12 hours
Bone marrow cells	about 18 hours
Stomach epithelial cells	about 24 hours
Hepatocytes (liver cells)	about 1 year

Certain highly specialised cells (e.g. muscle and nerve cells) lose the power to divide altogether. Mammalian red blood cells are also incapable of dividing, because they lose their nucleus during the course of development.

The cell cycle and the events associated with the various stages are shown in Figure 6.3.

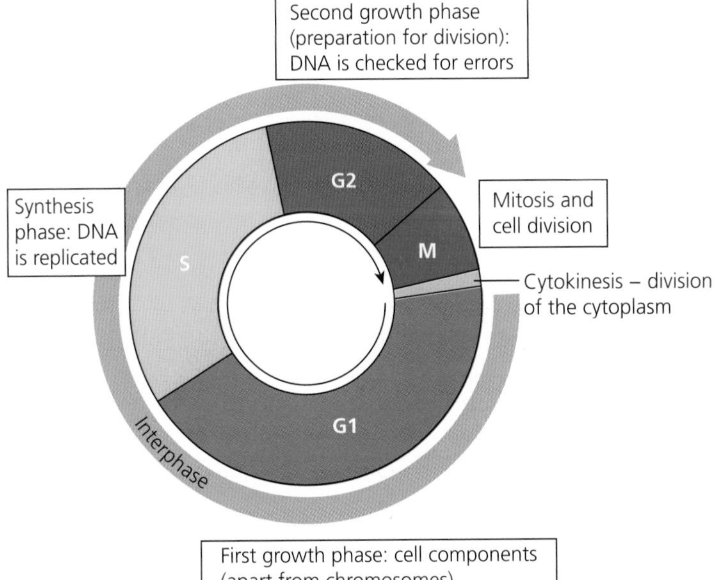

Figure 6.3 Events of the cell cycle.

Mitosis and **cytokinesis** occupy only a small percentage of the cell cycle (approximately 5%, but this varies for different cell types). Following division, the cell enters a first growth phase (G1) followed by a synthesis phase (S phase) in which the DNA is duplicated. There is then a second growth phase (G2), at the end of which the DNA starts to coil tightly, to become visible as chromosomes at the beginning of mitosis.

Key term

Cytokinesis The division of the cytoplasm after nuclear division.

Tip

Remind yourself about semi-conservative replication by looking at Chapter 3.

During this phase the genetic information in replicated DNA is checked for possible errors by a suite of enzymes with specific proof-reading and repair functions. If any are detected, the cell may destroy itself to prevent passing on mutations. There are four points in the cell cycle, known as **checkpoints**, where these checks take place.

1 During the G1 phase, the chromosomes are checked for damage. If damage is detected, the cell does not proceed into the S phase until the DNA has been repaired.

2 During the S phase a check is made that all the chromosomes have replicated. If they have not, the cell cycle is stopped.

3 During the G2 phase, another check is made for DNA damage that may have occurred during replication. Once again, the cycle may be delayed to repair the DNA.

4 The final checkpoint occurs during metaphase. This check identifies whether the chromosomes have correctly attached to the spindle fibres before anaphase proceeds.

The parts of the cell cycle when mitosis is not occurring are collectively referred to as **interphase**.

Key terms

Checkpoint A point during the cell cycle when the genetic information in replicated DNA is checked for possible errors.

Interphase The part of the cell cycle when the cell is not undergoing cell division.

Activity

DNA levels in dividing cells

The mass of DNA in a cell can be measured using a technique called *flow cytometry*. A fluorescent dye that binds to nucleic acids is added to cells and then the level of fluorescence is measured – the more fluorescence, the more DNA is present. The cells are treated with the enzyme RNAase, which removes RNA, before carrying out the flow cytometry. The readings are calibrated by reference to the nucleated red blood cells of chickens, because the mass of DNA in those cells is known.

The mass of DNA in dividing cells was measured at different times over a 24-hour period, and the results are shown in the graph (Figure 6.4).

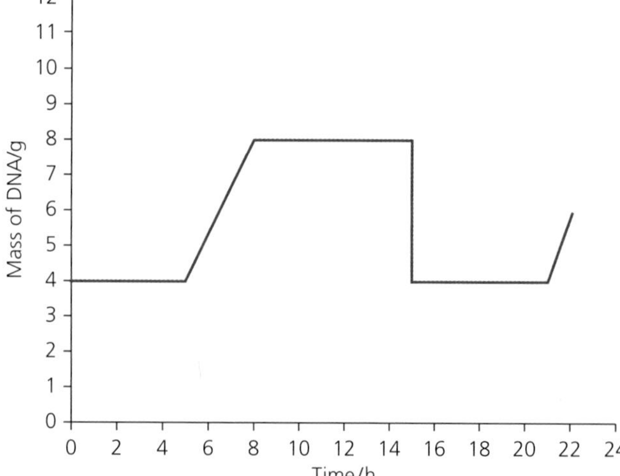

Figure 6.4 Mass of DNA in dividing cells over time.

1 Explain the changes in mass of DNA shown in the graph.
2 Use the data in the graph to explain how you know that the cell division occurring is mitosis and not meiosis.
3 It is important that all (or nearly all) of the cells observed are actively dividing. Explain why.
4 Why is it necessary to remove the RNA before carrying out flow cytometry?
5 Explain why 'The readings are calibrated by reference to the nucleated red blood cells of chickens, because the mass of DNA in those cells is known.'
6 What is the length of the cell cycle in these cells?

Mitosis

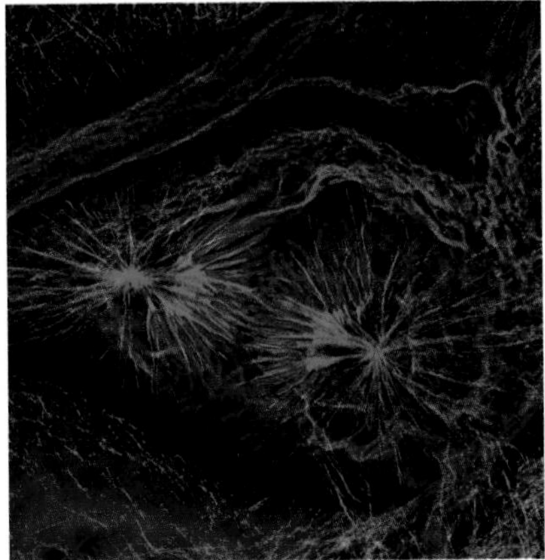

Figure 6.5 Fluorescent micrograph of an amphibian cell during anaphase. You can clearly see the spindle apparatus (green) and chromosomes (blue). (×900)

When new cells are required for growth, repair of tissues or asexual reproduction, mitosis is the type of nuclear division that provides them. The new cells require a full set of chromosomes, and so those in the original cell, or **mother cell**, must duplicate in order to give the two sets that are needed. Scientists divide the process of mitosis into stages for convenience, but it is important to understand that it is a continuous process with no 'pauses', so that one stage blends smoothly into the next. The stages of mitosis are shown in Figure 6.6.

There are four stages in a mitotic division: **prophase**, **metaphase**, **anaphase** and **telophase**. Interphase is not a stage of cell division, it is the part of the cell cycle during which cell division is **not** occurring. The events of the different stages in animal cells are described below.

Prophase

The replication of DNA occurs in the S (synthesis) phase of interphase, when the DNA is uncoiled and not visible, so that by the time prophase starts and the chromosomes begin to appear, they are already divided into 'sister' **chromatids**, held together at a single point known as the **centromere**. The chromosomes become visible during prophase, because the DNA undergoes supercoiling. Each chromosome at this stage has two molecules of DNA as a result of replication. This is in preparation for division.

By the end of prophase, the chromosomes have become visible, the nucleolus has disappeared and the nuclear membrane has broken down.

Key term

Supercoiling The twisting of the DNA around its own axis, winding the helix more tightly.

Metaphase

At the start of metaphase in animal cells, the centrosomes (each consisting of a pair of centrioles, duplicated, like the DNA during interphase) move to opposite ends of the cell called poles. The centrosomes send out microtubules to form a **spindle**, which forms a framework that later guides the chromatids to the opposite poles. The centromere of each chromosome then attaches to the spindle at the centre of the structure, called the **equator**. This attachment involves a protein structure on each chromatid, called a **kinetochore**.

Anaphase

The centromeres divide and the newly-separated chromatids (now called chromosomes) are pulled by their centromeres to opposite poles of the cell. The spindle fibres shorten at both ends so pulling the chromatids apart.

Tip

There can be confusion between chromosomes and chromatids. When a chromosome consists of two sister chromatids, each part is called a chromatid and the two chromatids together are the chromosome. In anaphase, when the chromatids separate, each newly-separated chromatid becomes a chromosome. By telophase you should be talking about chromosomes rather than chromatids.

Telophase

In telophase, new nuclear membranes form around each group of chromosomes, which now uncoil again. New nucleoli form in each nucleus.

Cytokinesis

After telophase, the cytoplasm then divides in the process called cytokinesis. The cytoplasm near the equator tucks inwards, splitting the cytoplasm (see Figure 6.6). The organelles are shared between the two cells, having increased in number during interphase. The Golgi apparatus produces vesicles that form the new section of plasma membrane.

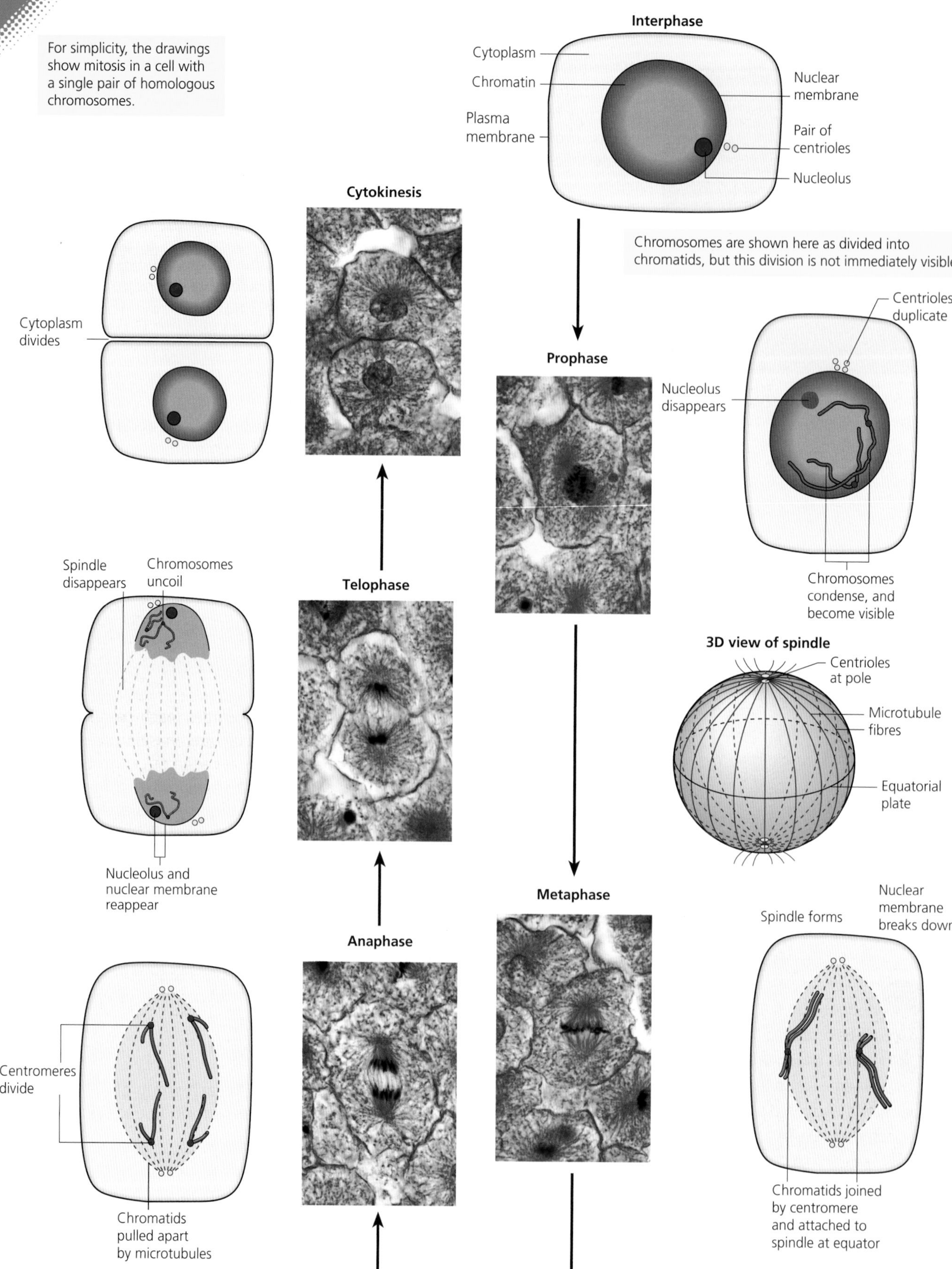
For simplicity, the drawings show mitosis in a cell with a single pair of homologous chromosomes.
Interphase
Cytoplasm
Chromatin
Plasma membrane
Nuclear membrane
Pair of centrioles
Nucleolus
Chromosomes are shown here as divided into chromatids, but this division is not immediately visible.
Prophase
Centrioles duplicate
Nucleolus disappears
Chromosomes condense, and become visible
3D view of spindle
Centrioles at pole
Microtubule fibres
Equatorial plate
Metaphase
Spindle forms
Nuclear membrane breaks down
Chromatids joined by centromere and attached to spindle at equator
Anaphase
Centromeres divide
Chromatids pulled apart by microtubules
Telophase
Spindle disappears
Chromosomes uncoil
Nucleolus and nuclear membrane reappear
Cytokinesis
Cytoplasm divides

Figure 6.6 Mitosis in an animal cell.

Example

Body cells go through a regular cycle of cell division by mitosis. The length of this cycle varies in different cells. The cell cycle of bone marrow cells takes 18 hours; the different phases of the cycle are shown in Figure 6.7.

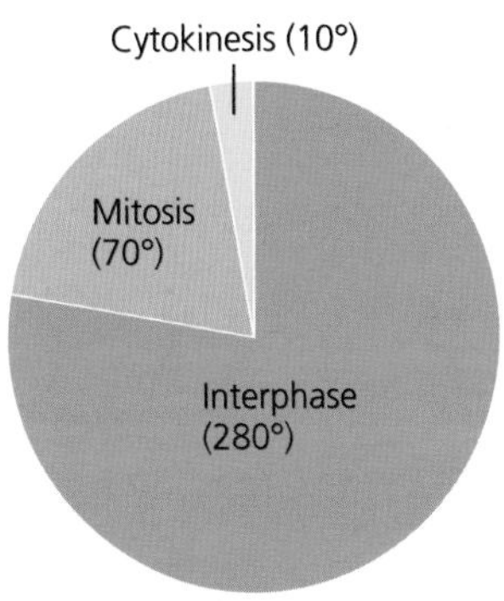

Figure 6.7 Phases of the cell cycle in bone marrow.

In cancer, the cell cycle runs out of control, rapidly producing cells that do not differentiate properly. A mutation of a gene known as the p53 gene has been linked with an increased risk of some cancers. The normal p53 gene produces p53 protein, which has several roles in the cell cycle:

- It can activate DNA repair proteins when DNA has sustained damage.
- It delays the cell cycle if DNA damage is detected.
- It activates another protein, called p21, which acts as a 'stop' signal for cell division.

The mutated p53 gene does not produce p53 protein.

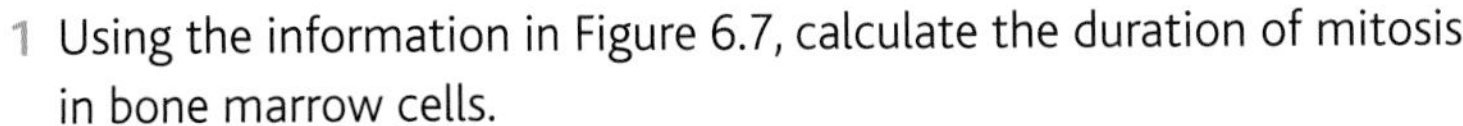

1 Using the information in Figure 6.7, calculate the duration of mitosis in bone marrow cells.
2 At which parts of the cell cycle would the p53 protein activate DNA repair proteins?
3 Suggest two reasons why a mutation in the p53 gene can result in cancer.

Answers

1 The complete cell cycle (360°) is 18 hours long. On the pie chart, mitosis is shown as 70°. Therefore mitosis occupies $\frac{7}{36}\left(\frac{70}{360}\right)$ of the cell cycle.
Mitosis in bone marrow cells takes 3.5 hours.
2 The DNA is checked for damage and repaired if necessary at the G1 and G2 stages of interphase.
3 There would be no p53 protein. DNA repair proteins would not be activated, so damaged DNA would not be repaired and the cell cycle would continue, passing the damaged DNA on to the new cells.
No p21 protein, which acts as a 'stop signal' for cell division, would be produced. Cell division would continue, without the p21 protein to stop it when appropriate.

Tip

It may not be enough just to say 'the G1 and G2 stages' – it is better to show that you know that these occur in interphase. Note that the word 'parts' (i.e. plural) in the question indicates that a single stage is not enough.

Tip

It is important that you give a full answer – i.e. that you mention the absence of the p53 and p21 proteins, not just the effects of this absence.

Mitosis in plants

Mitosis in plants is very similar to the process described in animals but plants do not have centrioles, although they still form a spindle. Also, during cytokinesis the vesicles from the Golgi apparatus form the new cell membrane and the cell wall.

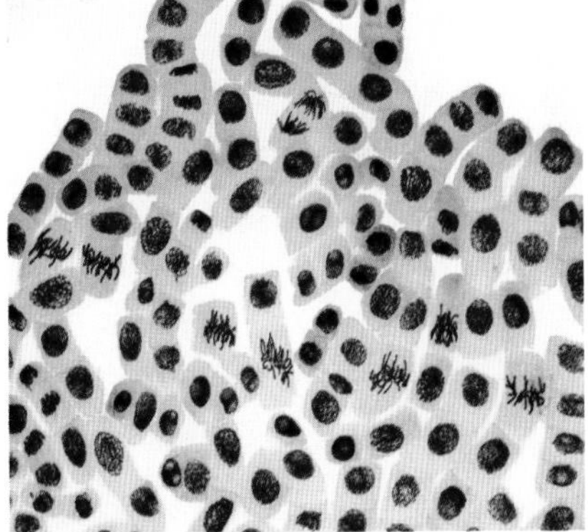

Figure 6.8 Photomicrograph of onion (*Allium* species) showing cellular mitosis. Four photomicrographs have been added together to form this image. (×250)

Mitosis and asexual reproduction

Asexual reproduction, where a single organism gives rise to one or more new organisms, in eukaryotes always involves mitosis to produce the new cells that form the new organism.

Yeast produces cells that are genetically identical to the original: all the organisms produced by asexual reproduction are also genetically identical, both to the parent and to each other, because exact copies of the DNA are made during replication (see Figure 6.9).

Key term

Asexual reproduction Reproduction which involves the production of offspring from a single individual.

Asexual reproduction can be advantageous, because it is much quicker than sexual reproduction, but the lack of genetic variation means that the population has less resistance to disease and environmental change. If one organism is susceptible, all will be – so the whole population may be wiped out. The only genetic variation that occurs in a population produced by asexual reproduction is as a result of mutation.

Asexual reproduction is far more common in plants than in animals, because plants cannot move from place to place to find a mate. The ability to reproduce asexually is very useful in such circumstances.

Budding in yeast

Yeast, which is a unicellular fungus, reproduces asexually by a process called **budding**. In budding, the nuclear division is by mitosis, but the division of the cytoplasm occurs in a different way from that described above.

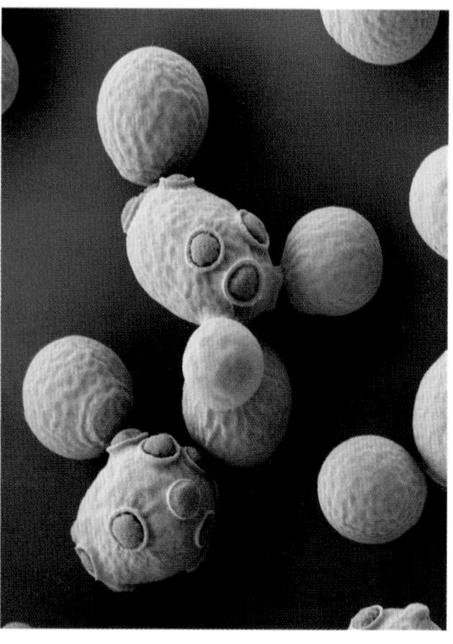

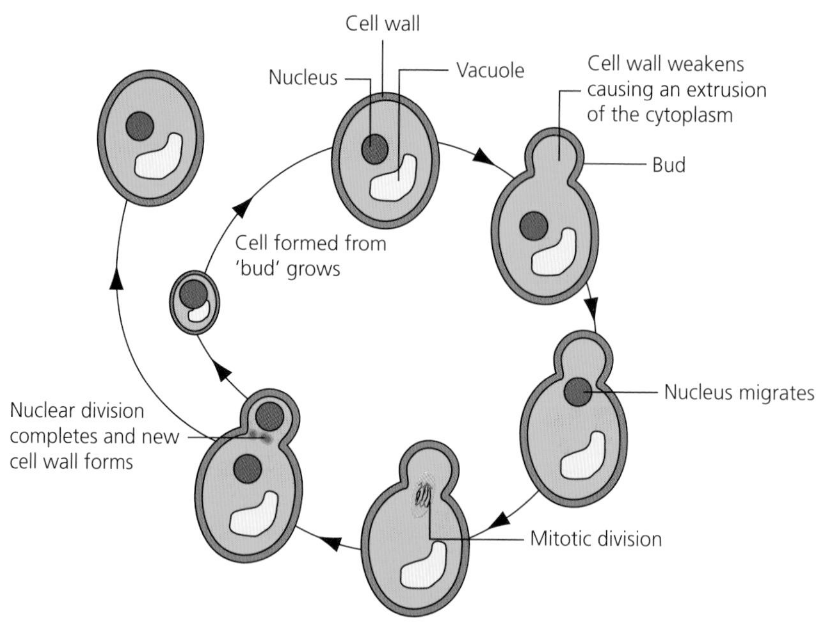

Figure 6.9 The budding process in yeast. (×4375)

Test yourself

1 Describe the events that occur in a cell during interphase.
2 Chromosomes become visible during prophase. Why are they not visible in interphase?
3 What type of structures form the spindle?

The budding process is shown in Figure 6.9. The stages are as follows:

1 DNA replicates inside the nucleus.
2 An area of the cell wall is weakened. Due to the turgor inside the cell, this causes the cytoplasm and cell wall to bulge out, forming a bud.
3 A ring of chitin forms at the junction of the cell and the bud. After the bud separates, this remains as a 'scar'.
4 The nucleus migrates to the region of the bud and undergoes a mitotic division, but without the nuclear membrane breaking up.

Test yourself

4 Identify, with reasons, the stage of mitosis shown in the photo.

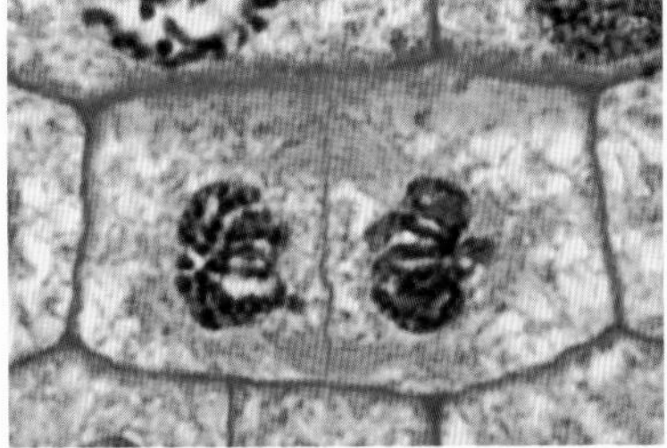

5 Define the term *cytokinesis*.
6 What is the difference between the process of budding in yeast and mitotic divisions in animal and plant cells?

5 One nucleus migrates into the bud along with organelles.
6 Cytokinesis is completed with the formation of a new cell wall between the mother and daughter cells.
7 The new daughter cell grows to full size before starting the budding cycle again.

As opposed to mitosis in animal and plant cells, in yeast the cytoplasm divides unequally, leaving one cell to grow while the other is already full size. In animal and plant cells, both new cells are equal in size and smaller than the cell that produced them, and so both grow to reach full size. The nuclear division in yeast is by mitosis. The process is similar to that described in animal and plant cells except that the nuclear membrane remains intact throughout.

Activity

Observing mitosis

Mitosis can be observed in root tips of garlic by squashing the tip, staining the chromosomes and observing them under the microscope. A measure of the extent of cell division in the tissue is the *mitotic index*.

Method

1 A garlic bulb was left overnight with its base just in contact with water (Figure 6.10), so that it would grow roots.
2 From the end of each root, 1–2 cm root tips were cut. These were placed in a small volume of ethanoic acid on a watch glass and left for 10 minutes. Ethanoic acid is a fixative that stops all living processes in the cells.
3 The tips were washed in cold water for four to five minutes then dried with filter paper.
4 The tips were then transferred into 1 M hydrochloric acid at 60 °C and left for five minutes; this breaks down the bonds between the cell walls in the root.
5 The tips were again washed in cold water for four to five minutes and dried on filter paper.
6 Two tips were transferred to a clean microscope slide.
7 The final 2 mm (approximately) was cut from each tip and the rest discarded.
8 A drop of ethano-orcein stain was added and left for two to three minutes. (This stains the chromosomes.)
9 The root tissue was broken up using a mounted needle.
10 A coverslip was placed over the root tips and they were squashed by dropping the blunt end of a pencil onto the middle of the coverslip about 20 times, from a height of around 5 cm.
11 The tips were then observed under the microscope. The number of cells observed at each stage of mitosis and at interphase was recorded in Table 6.2.

Figure 6.10 Growing garlic roots.

Table 6.2

Stage	Number of cells seen
Prophase	54
Metaphase	6
Anaphase	2
Telophase	4
Interphase	934

A further experiment was done to investigate the mitotic index of root tip cells at different times of the day. At each time, 1000 cells were counted. The results are shown in Table 6.3.

Questions

1 Suggest a reason why hydrochloric acid was used to break the connections between the cell walls in Step 4.
2 Calculate the mitotic index of the root tissue in the first experiment.
3 Calculate the percentages of the **dividing** cells that were at each stage of mitosis. (Remember that interphase is not a stage of mitosis.)
4 Suggest a reason why the percentage of dividing cells was different at different stages of mitosis.
5 In the second experiment, it was suggested that the time of day influenced the number of cells undergoing division. Assess the strength of evidence for this hypothesis, using the results given.

Table 6.3

Time	Mitotic index/%
12 midnight	5.1
4 am	5.4
8 am	6.1
12 noon	7.5
4 pm	6.6
8 pm	6.7

Homologous chromosomes and meiosis

Key terms

Homologous pair Two chromosomes with the same sequence of genes on them.
Haploid Containing one set of chromosomes.
Diploid Containing two sets of chromosomes.
Allele/Gene variant A type or version of a gene for a given characteristic.

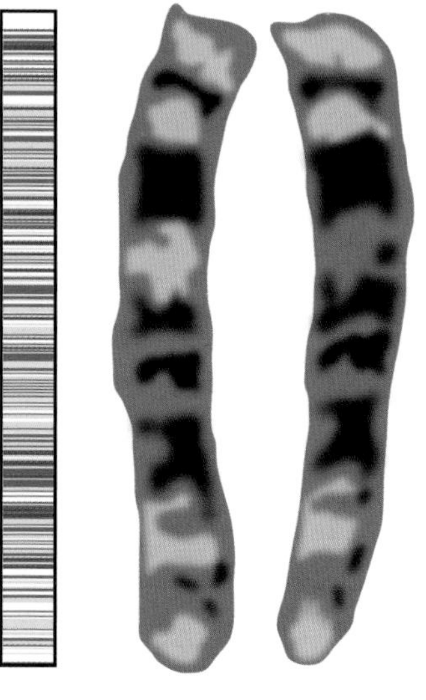

Figure 6.11 The gene sequence of homologous chromosomes (chromosome 10 of the malarial parasite, *Plasmodium falciparum*). The technique used analyses the gene sequence and then represents it as bands of colour. Note that the gene sequences of the two chromosomes are identical, but there may be differences in the alleles.

A full set of chromosomes is made up of a number of **homologous pairs**. In humans, for instance, there are 46 chromosomes and 23 homologous pairs. So, what is a 'homologous pair'?

When gametes fuse in sexual reproduction, their nuclei join together into one nucleus in the zygote. For this reason, each gamete must only have one set of chromosomes (the **haploid** number) so that the zygote nucleus then has the same number of chromosomes as the parents (the **diploid** number). However, the chromosomes passed into a gamete cannot be a random selection from the parents' set. If that was the case, then certain essential genes might not be passed on. In fact, since there are homologous paired chromosomes, each gene will be present on both chromosomes. These copies are not necessarily identical; for instance, a human has two copies of the eye colour gene, but one may be a 'blue' **allele** and the other a 'brown' allele. The two copies of a gene are found in the same position on two chromosomes, and all the other genes on those two chromosomes are also matching pairs (see Figure 6.11). The two chromosomes with the same sequence of genes on them are called a homologous pair. Within each pair, one has originated from the father's sperm (sometimes called the paternal chromosome) and the other (the maternal chromosome) from the mother.

Homologous chromosomes look the same and have the same staining properties because they have the same basic structure. Therefore, it is possible to pair them up from images taken through a microscope. This has been done for human chromosomes in Figure 6.12. Note that exceptions to this are the X and Y sex chromosomes in a male, which have different structures.

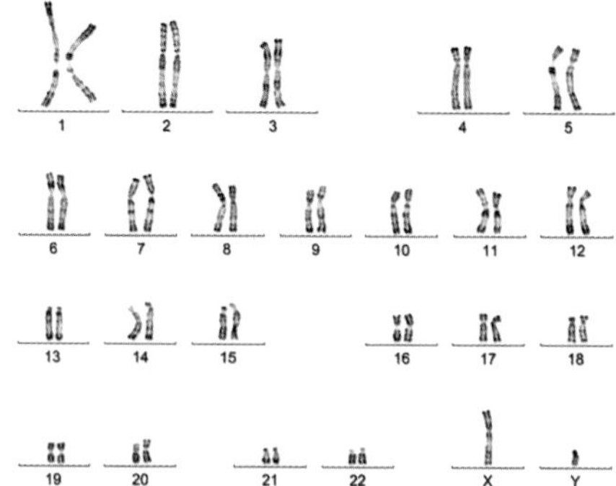

Figure 6.12 Human chromosomes of a male, arranged in homologous pairs. This is known as a karyogram.

Stages of meiosis

Meiosis consists of two successive divisions, the first of which halves the number of chromosomes in each cell. The two haploid cells then divide again so that a total of four cells are formed. Each of the divisions consists of stages that are comparable with those seen in mitosis and have the same names but with a number after them to indicate which division is referred to (for example prophase I, telophase II, etc.). The events of meiosis are shown in Figure 6.13 and summarised in Table 6.4 below.

Table 6.4 Events of meiosis.

Stage	Events
Prophase I	Chromosomes condense as sister chromatids (the DNA having replicated in interphase). Homologous chromosomes pair up as four chromatids (known as a **tetrad**). Crossing over may occur. Centrioles move to the poles of the cell.
Metaphase I	The tetrads align at the equator of the spindle.
Anaphase I	The chromosomes of the tetrad separate and move to the poles of the cell. The sister chromatids remain together.
Telophase I	The cell now has a haploid number of chromosomes at each pole. In some cells, the chromosomes may decondense, but more usually they remain condensed. Nuclear membranes may reform.
Cytokinesis I	The cytoplasm divides to form two haploid cells. No replication of DNA occurs.
Prophase II	The nuclear membrane breaks up again (if it was reformed).
Metaphase II	The chromosomes (consisting of two sister chromatids) align at the equator of the spindle of each cell.
Anaphase II	The sister chromatids separate and move to the poles of the cells.
Telophase II	The chromosomes decondense and nuclear membranes reform.
Cytokinesis II	The cytoplasm divides, creating four haploid cells.

Prophase I

- chromosomes condense so that they become shorter, thicker and visible in the light microscope
- homologous chromosomes pair to form bivalents (maternal chromosomes are blue, paternal are pink)
- chiasmata (singular: chiasma) form to hold chromosomes together; non-sister chromatids join, break and exchange parts in crossing over
- nuclear membrane breaks up into small sacs of membrane which become part of the endoplasmic reticulum; centrioles replicate and move to opposite poles and form spindle microtubules

Metaphase I

- bivalents move to the equatorial (or metaphase) plate across the centre of the cell
- the paternal and maternal chromosomes in each bivalent position themselves independently of the others
- microtubules attach to the centromere of each chromosome

Anaphase I

- chromosomes (each with two chromatids) are pulled by the shortening of microtubules towards the poles
- (chromatids would be pulled to opposite poles if this was anaphase of mitosis)

Telophase I

- chromosomes reach opposite poles
- nuclear membrane reforms to make two daughter nuclei that have half the number of chromosomes of the parent cell – these nuclei are haploid
- cytokinesis occurs – the cell surface membrane 'pinches in' leaving small cytoplasmic bridges between the cells

An interphase may occur between the divisions of meiosis I and II in which case the chromosome uncoil.

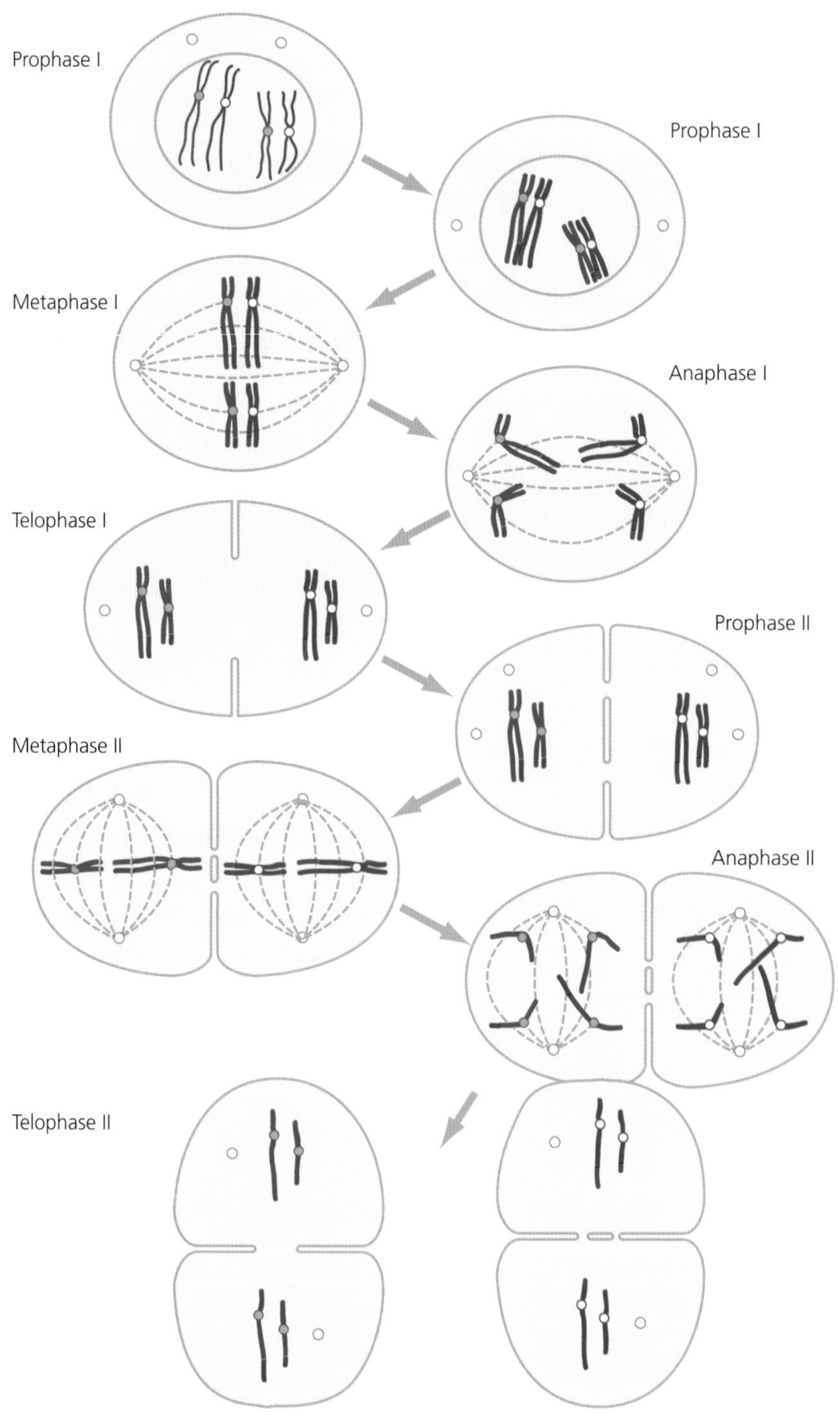

Figure 6.13 The process of meiosis.

Prophase II

- centrioles replicate and move to poles that are at right angles to those in meiosis I
- nuclear membrane breaks up

Metaphase II

- individual chromosomes align on the equator with their chromatids randomly arranged (important if crossing over has occurred in meiosis I)
- microtubules attach to the centromeres

Anaphase II

- sister chromatids break apart at the centromere and move to opposite poles

Telophase II

- nuclear membranes reform
- cells divide to give four haploid cells that are genetically different to one another and from the parent cell

The second division of meiosis is exactly the same as mitosis, but the first division shows key differences.

- In meiosis, the homologous chromosomes pair up and then separate into different cells. This does not occur during mitosis.
- The two chromatids formed from a single chromosome remain together during the first division of meiosis, but separate in the second division. The second division of meiosis is similar to mitosis in that chromatids separate so that the number of chromosomes remains the same.

Independent assortment, crossing over and genetic variation

One advantage of reproducing sexually using gametes produced by meiosis is that the gametes (and therefore the new individuals formed from them) show genetic variation. The significance of this will be dealt with later.

Homologous chromosomes make it possible for the cell to pass on a set of chromosomes to gametes during meiosis, while making sure that all the genes are present in the new individual. When meiosis occurs, one of each homologous pair of chromosomes is transferred to the new gamete. Which of any two homologous chromosomes goes into a particular gamete is completely random; it is said that there is **independent assortment** of the chromosomes (see Figure 6.14). When fertilisation occurs, not only is the diploid number restored, but the homologous pairs are also reinstated. In this process, either of the two chromosomes in a homologous pair may be passed on. Remember that the homologous pairs have the same genes but not necessarily the same alleles. Therefore, depending on which chromosome from each pair is passed on, each gamete has a different genetic make-up.

Independent assortment is illustrated in Figure 6.14 in a parent cell with two pairs of homologous chromosomes (four bivalents). The more bivalents there are, the more variation is possible. In humans, for example, there are 23 pairs of chromosomes giving over 8 million combinations.

Tip

The letters A/a and B/b represent the alleles (gene variants) of two different genes. These genes are on different chromosomes. This diagram shows how independent assortment leads to variation in the gametes produced by meiosis.

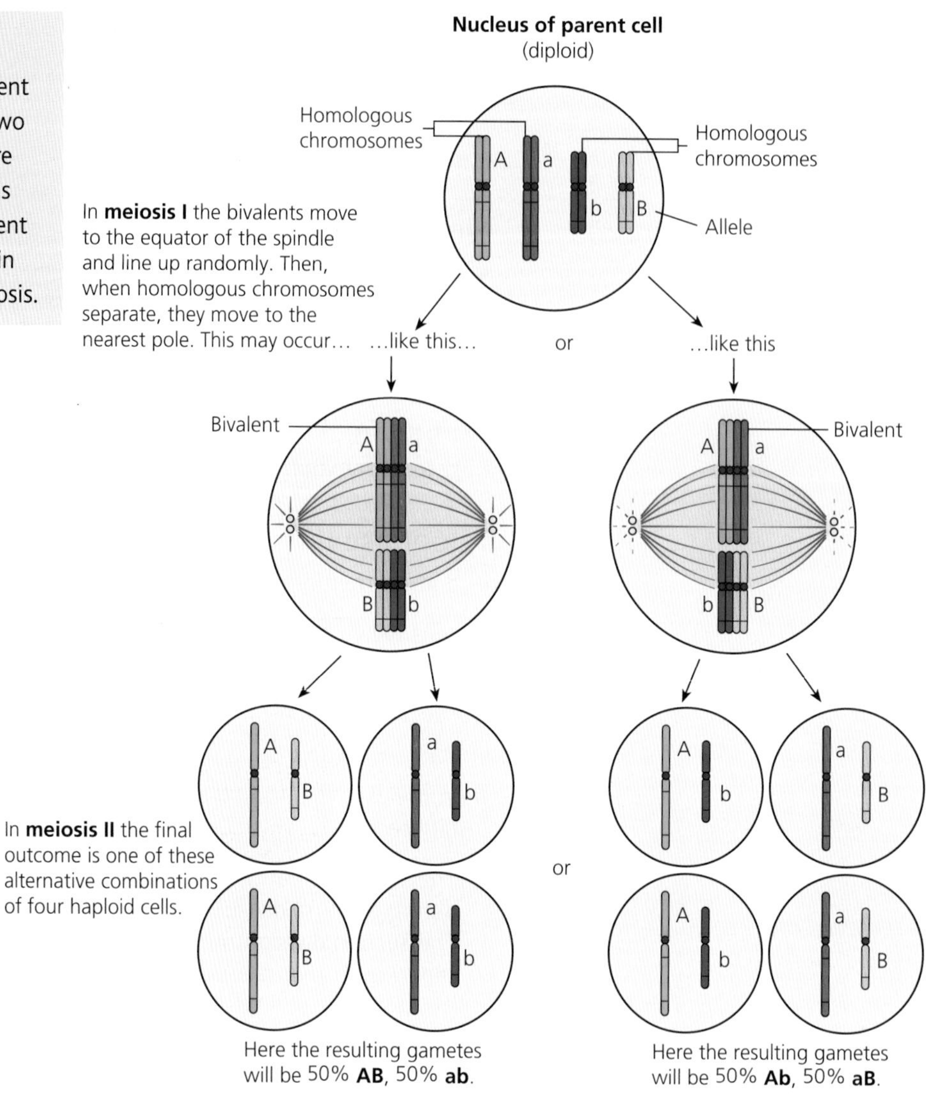

Nucleus of parent cell
(diploid)
Homologous chromosomes
Homologous chromosomes
A
a
b
B
Allele
In **meiosis I** the bivalents move to the equator of the spindle and line up randomly. Then, when homologous chromosomes separate, they move to the nearest pole. This may occur…
…like this…
or
…like this
Bivalent
Bivalent
In **meiosis II** the final outcome is one of these alternative combinations of four haploid cells.
or
Here the resulting gametes will be 50% **AB**, 50% **ab**.
Here the resulting gametes will be 50% **Ab**, 50% **aB**.

Figure 6.14 Genetic variation due to independent assortment.

Key terms

Synapsis When two homologous chromosomes come together.

Chiasmata The places where chromatids cross over.

The variation is increased by the process of crossing over, which occurs during prophase 1 of meiosis. Crossing over is a process by which the two chromosomes of a homologous pair exchange part of their genetic material. The two homologous chromosomes come together (a process called synapsis), and breaks appear at equivalent points on two of the non-sister chromatids at places where the chromatids cross over, called chiasmata (singular: chiasma). The broken sections are then exchanged. Because the homologous chromosomes have identical gene sequences, the exchange of broken sections does not disrupt the genetic information but does rearrange the alleles from the original paternal and maternal chromosomes. The second division of meiosis is similar to mitosis in that chromatids separate so that the number of chromosomes remains the same. The process is shown in Figure 6.15.

In humans, with 23 chromosomes, the chance that genetically identical gametes will occur is 1 in 20 million. Males produce over 400 billion sperm cells in a lifetime, so some of those are genetically identical. However, women release only about 450 eggs in total, so each egg cell is almost certain to be genetically unique. The chances of two fertilisations occurring with genetically identical sperm cells and genetically identical egg cells is one in 400 000 billion. This explains why all humans (apart from identical twins) have a unique genetic make-up.

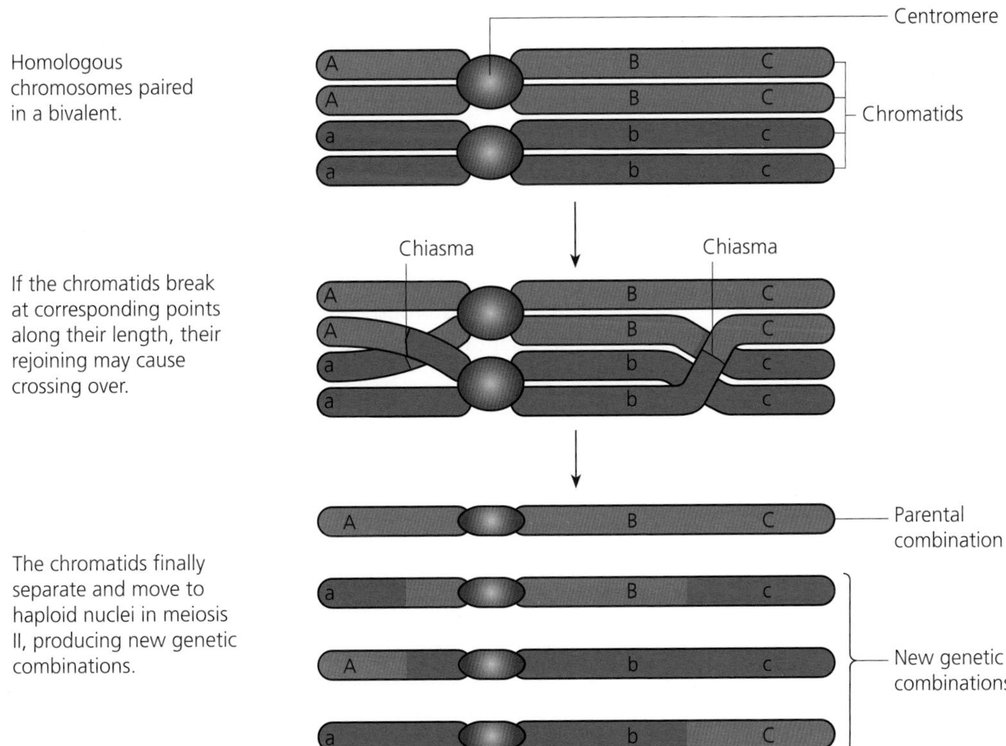

The effects of genetic variation are shown in one pair of homologous chromosomes.
Typically, two, three or more chiasmata form between the chromatids of each bivalent at prophase I.
Centromere
Homologous chromosomes paired in a bivalent.
A
A
a
a
B
B
b
b
C
C
c
c
Chromatids
Chiasma
Chiasma
If the chromatids break at corresponding points along their length, their rejoining may cause crossing over.
Parental combination
The chromatids finally separate and move to haploid nuclei in meiosis II, producing new genetic combinations.
New genetic combinations

Figure 6.15 Genetic variation due to crossing over between non-sister chromatids.

Differentiation

All multicellular organisms have stem cells, which are **undifferentiated** (i.e. unspecialised) cells. These divide by mitosis and undergo a process of differentiation during which they become more specialised for a particular function. Stem cells can also divide to produce more stem cells. Some examples of differentiation are given on the next few pages.

Erythrocytes

Erythrocytes are red blood cells. Their main function is to transport oxygen around the body, although they also play a role in the transport of carbon dioxide. Erythrocytes are formed from stem cells in the bone marrow. Because a mature red blood cell has no nucleus, it cannot divide, so new cells have to be constantly formed from stem cells in order to maintain the erythrocyte count in the blood. This process is called **erythropoiesis**. The stages of erythropoiesis are shown in Figure 6.16.

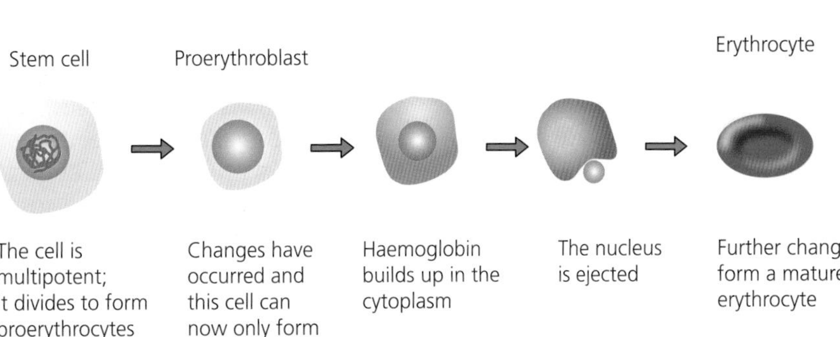

Stem cell
Proerythroblast
Erythrocyte
The cell is multipotent; it divides to form proerythrocytes
Changes have occurred and this cell can now only form an erythrocyte
Haemoglobin builds up in the cytoplasm
The nucleus is ejected
Further changes form a mature erythrocyte

Figure 6.16 Differentiation of red blood cells by erythropoiesis.

Key term

Multipotent Capable of forming a number of different types of body cell.

The stem cells of the bone marrow are multipotent – that is, they are capable of forming a number of different types of body cell. Fairly early on in the differentiation process, however, changes occur that mean that the cells formed by mitosis from the stem cell can only become erythrocytes. **Haemoglobin**, the oxygen-carrying pigment of red blood cells, then accumulates in the cytoplasm, and after a while the nucleus is ejected from the cell. Even at this stage, the cell is not a fully formed erythrocyte. It has to undergo further changes, which include a change in shape to give the characteristic biconcave disc shape of the mature erythrocyte.

Structure and function in erythrocytes

Differentiation involves changes that adapt the structure of the cell to its function. Red blood cells (Figure 6.17) transport oxygen around the body. The adaptations of red blood cells are as follows:

- The shape of the cell is a biconcave disc. This shape has a greater surface area, due to the double indentation. Because oxygen is absorbed through the surface, the greater area allows more gas to be taken in.
- The cytoplasm contains lots of haemoglobin (about 280 million molecules). This pigment binds with oxygen and only releases it when oxygen concentrations are low.
- The lack of a nucleus (and other organelles such as mitochondria, endoplasmic reticulum and Golgi apparatus) allows more haemoglobin to be packed into the cytoplasm, so increasing the oxygen-carrying capacity of the cell.
- An elastic membrane allows the red blood cell to change shape, which is important when it squeezes through narrow capillaries.

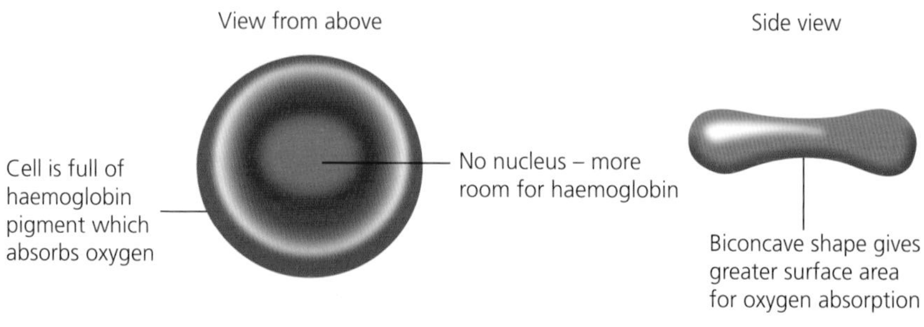
View from above
Side view
Cell is full of haemoglobin pigment which absorbs oxygen
No nucleus – more room for haemoglobin
Biconcave shape gives greater surface area for oxygen absorption

Figure 6.17 Adaptations of erythrocytes.

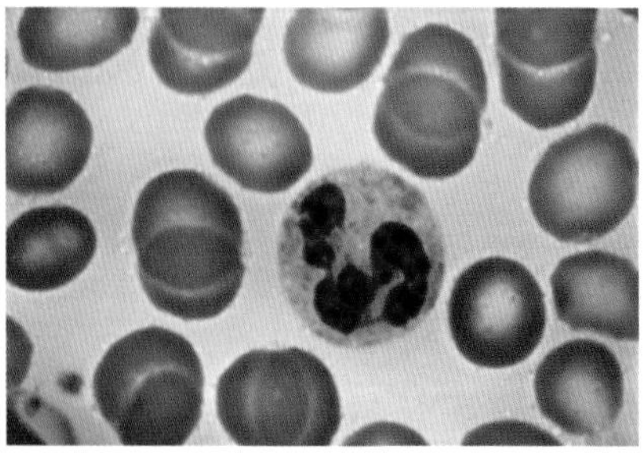

Figure 6.18 A neutrophil among red blood cells (×1100).

Neutrophils

The same stem cells in the bone marrow that produce erythrocytes also produce **neutrophils**, a type of white blood cell. A neutrophil is shown in Figure 6.18. The changes from the original stem cell are less dramatic than in the erythrocyte, but are significant nevertheless. The main changes are the indentations of the nucleus, giving it its characteristic lobed structure, and the accumulation of numerous granules. These granules do not stain darkly with acidic dyes or basic dyes and this is why they are called neutrophils. The granules are lysosomes, which contain hydrolytic enzymes.

Structure and function of neutrophils

The neutrophils are the first white blood cells to arrive at the site of an infected wound. They squeeze through gaps in the capillary walls, collect around any foreign bodies that have penetrated the tissues, and destroy them by engulfing them by phagocytosis and secreting enzymes. Their adaptations are as follows:

- Their shape is very flexible, which allows them to penetrate between the junctions of the cells of the capillary wall and to form pseudopodia (projections of the cytoplasm) to engulf microorganisms.
- They contain many lysosomes, which produce digestive enzymes to destroy invading cells.
- The nuclear membrane is more flexible than normal, which presumably helps the cell to squeeze through the tiny gaps between the capillary cells. The advantages of having the lobed nucleus are unclear, but it may be that the flexibility of the membrane results in the lobed shape.

Xylem and phloem

In plants, the **xylem vessels** and **phloem sieve tubes** that form the plant's transport systems are formed from stem cells in the tissue between them, which is called the **cambium** (see Figure 6.19). In both stems and roots, the cambium divides to form phloem cells on the outside and xylem cells on the inside. The cambium is an example of a **meristem** – undifferentiated plant tissue that can give rise to new cells. There are meristems at the shoot tips and root tips.

Tip

There is more about the structure of xylem and phloem in Chapter 9.

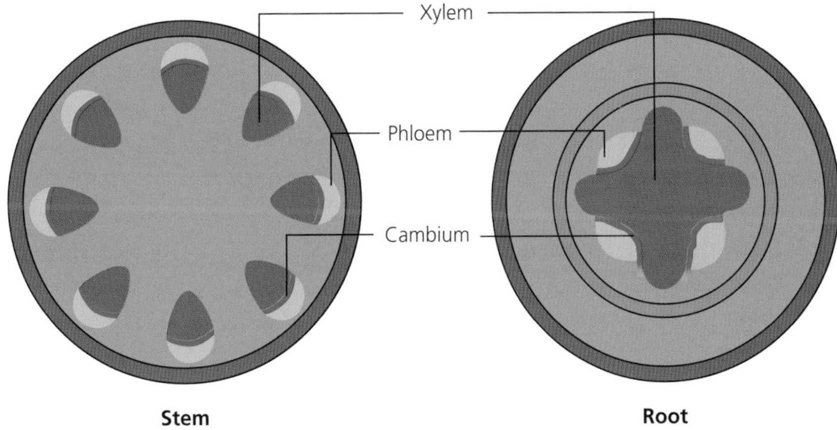

Figure 6.19 General distribution of the cambium, xylem and phloem in a young stem and root. (The precise arrangement varies from species to species.)

Key term

Hormone An organic substance produced in one part of an organism and transported to other parts, where it has specific effects.

The production of xylem and or phloem from the cambium cells is stimulated by hormones, and the balance of the different hormones can shift production between xylem and phloem. Although the cells in the cambium look the same, the ones towards the outside only produce phloem cells, and the ones on the inside only xylem. The cells that become xylem lose their cytoplasm and deposit lignin in their cell walls. The end cell walls may be lost. In the formation of phloem, there is some loss of cytoplasm and organelles, together with the development of sieve plates at the ends of the cells.

Specialised cells

> **Tip**
>
> See page 115 for more examples of specialised cells.

Many cells are specialised in ways that adapt them to their functions. For example, you have already seen how red blood cells differentiate to carry oxygen, and neutrophils differentiate to enable them to fight infection. Some other specialised cells in animals and plants are shown in Figure 6.20.

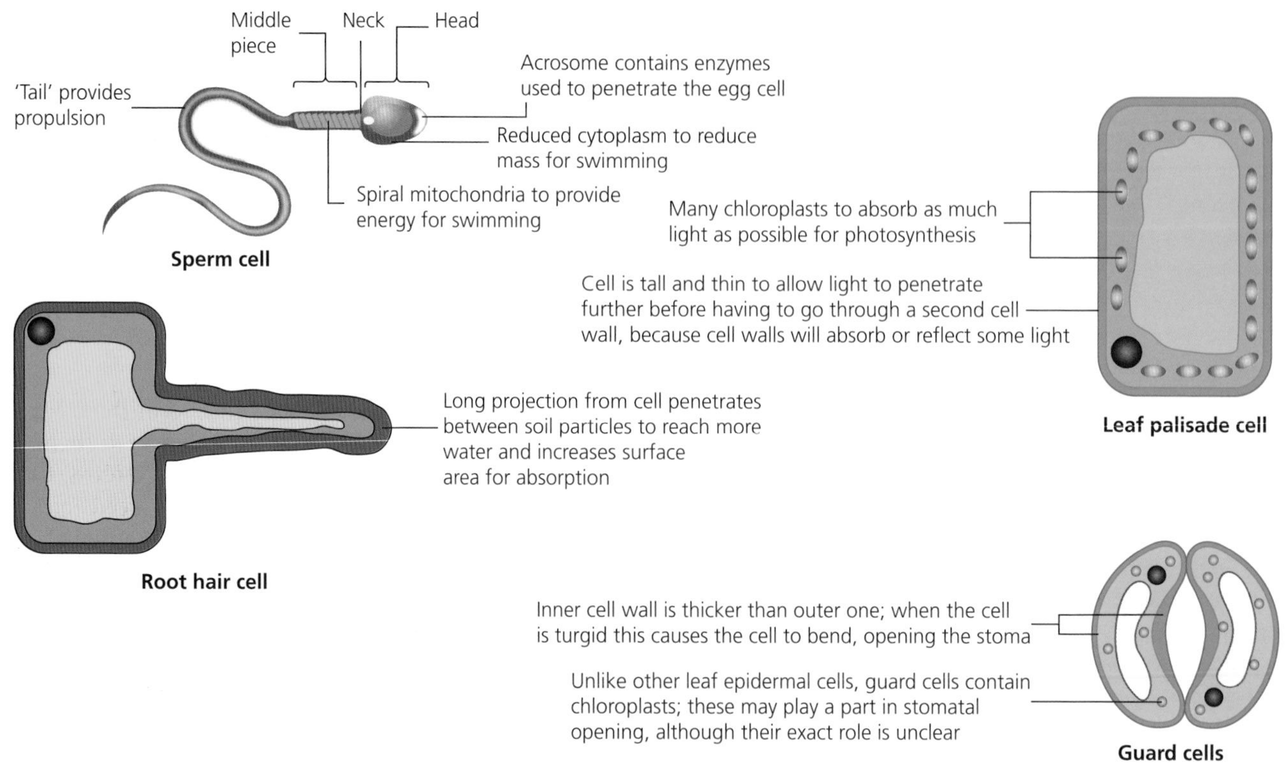

Figure 6.20 Examples of specialised cells in animals and plants (not to scale).

> **Test yourself**
>
> **7** Define the term *homologous pair* in relation to chromosomes.
>
> **8** What causes gametes to show genetic variation?
>
> **9** Where are erythrocytes formed in a mammal?
>
> **10** How is the structure of a neutrophil related to its function?
>
> **11** State the name of the tissue in plants that forms the xylem and phloem.

Stem cells and medicine

Stem cells are classified into three types, listed in Table 6.5.

Scientists have discovered that they can induce stem cells to form a specific type of cell according to the conditions in which the cell is cultured.

Table 6.5

Totipotent	These are stem cells that can differentiate into any type of body cell and can also form the extra-embryonic cells that make up the placenta and umbilical cord. The only cells that are totipotent are the zygote and those in the very early embryo.
Pluripotent	These cells can form any type of body cell but cannot form extra-embryonic cells. Embryonic stem cells (after the first couple of divisions of the zygote) are pluripotent.
Multipotent	These stem cells can form more than one cell type, but not all types. Adult stem cells such as those found in the bone marrow are multipotent.

Potentially, this allows a certain type of cell to be grown and used to repair damaged cells in the body. In addition, if an individual's own stem cells can be used to treat them, there is much less chance of rejection.

Tip

Stem cells and medicine is a fascinating area of research; find out more about stem cell treatment for the conditions listed here.

Research is underway into using stem cells to treat the following conditions:

- Alzheimer's disease: Growing stem cells into nerve cells with the potential to repair neurological conditions such as Alzheimer's disease. This is challenging, however, because the damage to the brain in dementia is widespread rather than restricted to a certain area.
- Parkinson's disease: Replacing the dopamine-producing brain cells that die in people with this neurological condition. Without the dopamine that these cells normally produce, people's movement is affected, causing shaking, stiffness and slowness of movement.
- Age-related macular degeneration: Growing retinal cells to treat this common condition that can cause loss of vision in the over-50s.
- Spinal injuries: Introducing stem cells into the site of spinal injuries to try to repair the damage.
- Blood diseases (e.g. sickle cell disease): Using bone marrow stem cells to treat a number of blood diseases.
- Type 1 diabetes: Protecting pancreatic beta cells (which form insulin) in newly diagnosed patients with type 1 diabetes.
- Heart attack: Using stem cells to repair damaged heart tissue following a heart attack.
- Chronic obstructive pulmonary disease: Repairing lung tissue.

Stem cells are also valuable for the study of developmental biology to discover how the body develops from a fertilised egg. Such investigations could provide useful information about the behaviour of cells and how problems may arise during development.

Embryonic stem cells are particularly useful in all these studies because of the wider range of cell types they can form, but the use of human embryos (from a surplus from *in-vitro* fertilisation clinics) is controversial. Although the embryos used are the waste embryos from *in-vitro* fertilisation treatment, they have the potential to develop into new individuals, and some people have ethical objections to their use.

Example

Parkinson's disease is a long-term degenerative disease of the central nervous system that impairs movement and speech due to the insufficient production of the cell-signalling compound dopamine. Human adult stem cells from the endometrium in the uterus were shown to differentiate *in vitro* into neurones (nerve cells) with the ability to make dopamine.

In an investigation, a neurotoxin that specifically kills dopamine-secreting cells was injected into a group of mice so that they showed symptoms of Parkinson's disease. The mice were divided into two groups, A and B.

Mice in group A were given a transplant of the stem cells; mice in group B were given no further treatment. A third group of mice that had received neither of these treatments was used as the control group, C.

After five weeks, the concentrations of dopamine and DOPAC, a breakdown product of dopamine, were determined in the brains of all the mice.

The results are shown in the charts.

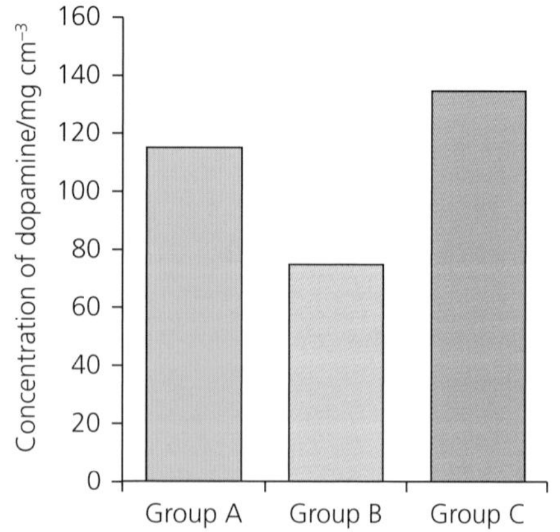

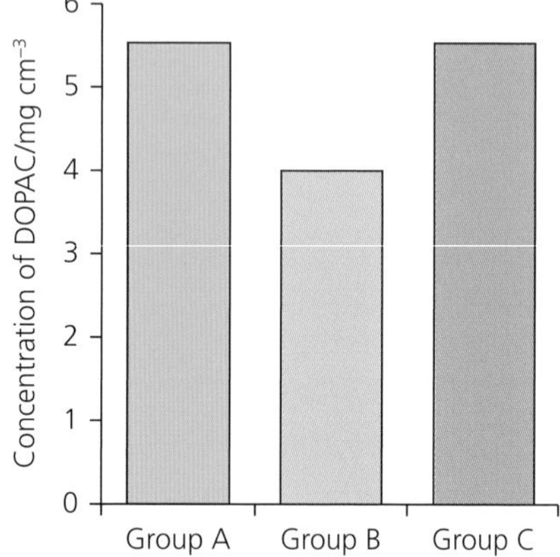

Figure 6.21

1 Explain the terms *cell-signalling compound* and *in vitro*.
2 Explain why the researchers used the mice in groups B and C in their investigation.
3 Suggest why the researchers measured the quantity of DOPAC as well as dopamine.
4 Use the information in Figure 6.21 to describe the effects of the neurotoxin and the transplanted stem cells.
5 Discuss the ethical issues involved in this study.
6 Suggest what further work should be done to assess the effectiveness of using stem cells in the treatment of Parkinson's disease.

Answers

1 A cell-signalling compound is a compound that is released by one cell and travels to another (target) cell. It binds to a receptor on the cell membrane or in the cytoplasm of the target cell and stimulates a change in the target cell. *In vitro* means taking place outside a living body, in an artificial environment.
2 This is to make sure that the results can be compared and to check their validity:
   - Group B shows whether stem cells produced any dopamine. If there were no dopamine-secreting cells in the mice then any dopamine produced would come from the stem cells.
   - Group C shows the normal concentrations of dopamine secreted in animals kept in identical conditions to groups A and B.
3 DOPAC can only be produced from dopamine. Dopamine produced by the differentiated stem cells may have been broken down after five weeks or by the end of the investigation. DOPAC could only be produced by the breakdown of dopamine from differentiated stem cells.
4 For both dopamine and DOPAC, the concentration is higher in group A than in group B (which has the lowest concentrations of both dopamine and DOPAC). The concentration of dopamine is highest in the control group – group C. The concentration of DOPAC is the same in groups A and C. (It is fine to compare using figures as long as you use the correct units. For example, the dopamine concentration in group A is 115 mg cm^{-3} and in cells treated with neurotoxin is 75 mg cm^{-3}.) The results for group B show that neurotoxin did not kill all the dopamine-secreting cells, but the dopamine concentration in mice treated with stem cells (group A) is higher.
5 Animals have been killed in the study and they also are likely to have experienced pain and stress. The investigation has been carried out in mice, so the results may not be relevant to humans. It is also not known whether there are likely to be any benefits to people with Parkinson's disease. However, scientists cannot use untried methods like this on humans. If the study offers a method of treatment to relieve human suffering then it may be justified.
6 It needs to be shown that these differentiated cells survive for longer than five weeks for the treatment to be of any use, and that the increase in dopamine has an effect on alleviating symptoms of Parkinson's disease in mice. It would also need to be shown that there are no side effects of the treatment and that stem cells do not give rise to tumours or cancer. Clinical trials would need to be carried out on human volunteers.

Tip

This question is challenging and is more like an A Level question than an AS Level question.

Tissues, organs and organ systems

In multicellular organisms, cells are organised into tissues. A tissue is a group of cells that work together to perform a particular function. It is important to understand that although the cells in a tissue are often described as 'similar', a tissue can be made up of more than one cell type. For example, blood is a tissue, yet there are clear differences between red cells and the various types of white cell. Some examples of tissues are discussed below.

Key terms

Epithelium A sheet of cells that forms an outer layer, i.e. a covering or lining.

Basement membrane A thin, delicate membrane of protein fibres and polysaccharides separating an epithelium from underlying tissue.

Squamous epithelium

Squamous **epithelium** is the name given to any type of epithelium that consists of a layer of flattened cells. Squamous epithelium is found in many places in the body of animals, and sits on a **basement membrane**. There are simple squamous epithelia that are made of a single layer. The cells are relatively unspecialised but they are flattened. Because squamous epithelial cells form a surface covering, this thin cross section helps by reducing the distance that substances have to travel to pass through it, shortening the diffusion pathway.

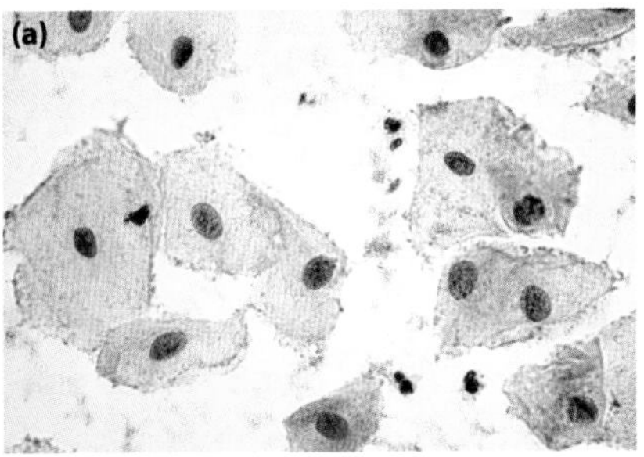

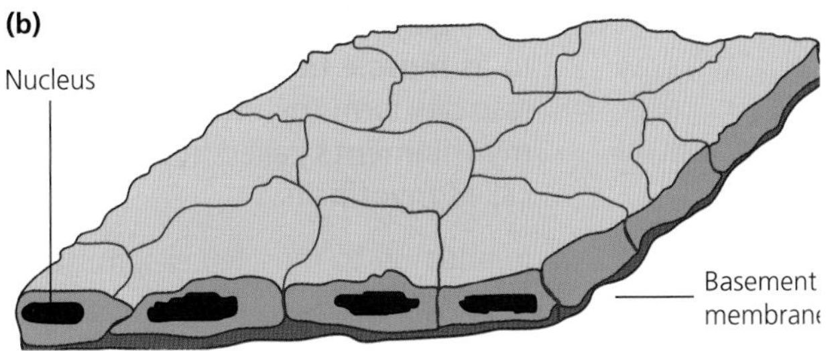

Figure 6.22 (a) A photomicrograph of squamous epithelium (×1500) and (b) a diagram of squamous epithelium.

Key term

Connective tissue A tissue that connects, supports, binds, or separates other tissues or organs.

Ciliated epithelium

As the name suggests, a ciliated epithelium is made up of cells with a lot of cilia on their surfaces. It is found in animals in places where something has to be moved across the surface (e.g. in the trachea, to move mucus, and in the oviduct to move the ovum). The movement of the cilia shifts material along the surface of the epithelium, and so these cells are also specialised for their function.

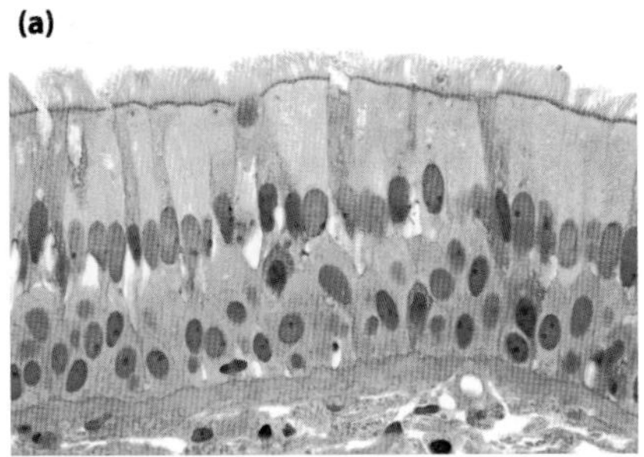

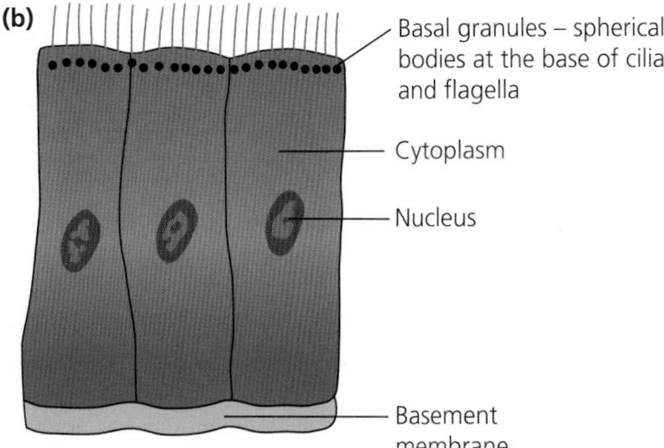

Figure 6.23 (a) A photomicrograph of ciliated epithelium (×160) and (b) a diagram of ciliated epithelium.

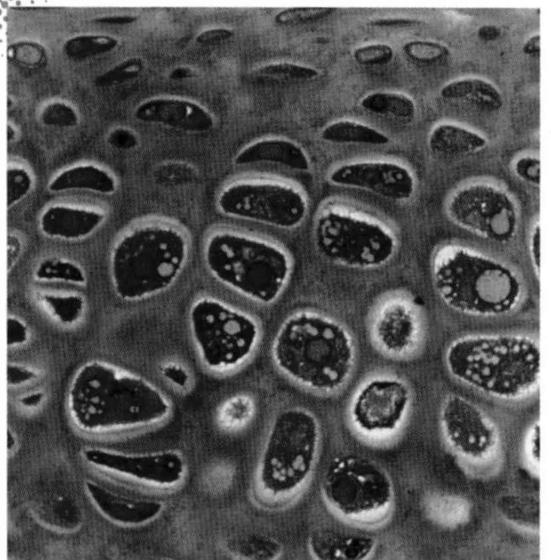

Figure 6.24 The structure of cartilage (×300).

Cartilage

Cartilage is an example of a connective tissue. Cartilage is composed of specialised cells called chondrocytes that produce an extracellular matrix consisting of collagen fibres (which stiffen and strengthen the tissue) and elastin fibres (which give it flexibility). The function of cartilage is to protect and strengthen. Cartilage is found in mammals in the nose, the ear, the rib cage, the trachea and bronchi, and in joints. The structure of cartilage is shown in Figure 6.24.

Key term

Connective tissue A tissue that connects, supports, binds, or separates other tissues or organs.

Muscle

Muscle is a highly unusual tissue in that its function is to move parts of the body. A muscle, such as the bicep muscle in the arm, is an organ made mostly of muscle tissue with nerve tissue, blood and connective tissue. The organisation of muscle tissue into muscles is complex and is shown in Figure 6.25.

Mammals have three types of muscle: skeletal (voluntary), smooth (involuntary) and cardiac (only found in the heart). Skeletal muscle cells are highly specialised and are multinucleate. They are commonly referred to as **muscle fibres**. Groups of fibres form a **fascicle**, and groups of fascicles form the muscle. The muscle fibres contain contractile protein **myofilaments** called **actin** and **myosin**, which are formed into structures called **myofibrils**.

In addition to muscle cells, muscles contain nerve fibres, blood vessels and connective tissue.

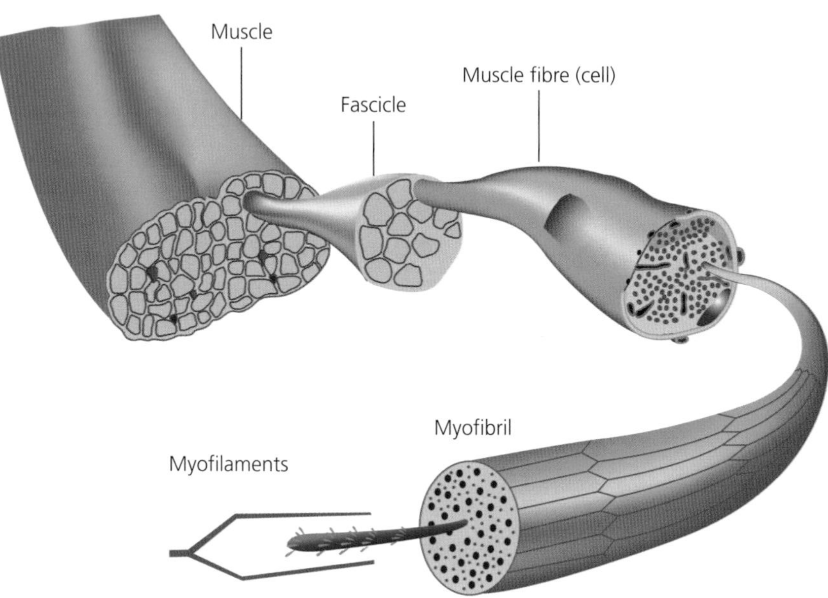

Figure 6.25 Structure of a skeletal muscle.

Xylem

Xylem is a plant tissue that transports water and minerals up the plant stem and also strengthens it. It consists of up to four different types of cell (the exact composition varies in different species): **vessel elements**, **tracheids**, **fibres** and **parenchyma**.

Parenchyma cells are not specialised and simply form a packing tissue between the other cells.

The xylem vessel elements are the main water-transporting cells, and as such have a wide lumen and their end walls are perforated or completely absent. The pits allow water to be moved laterally, which can be helpful for avoiding blockages. The vessels are aligned so they are continuous and stretch the length of the plant.

The tracheids also transport water, but in angiosperms they are also important as a strengthening tissue. They do not have as wide a lumen as vessels, but still have perforated end walls and pits where the cell wall is very thin and water can move easily between cells.

> **Key term**
>
> **Angiosperms** A large group of plants that have flowers and produce seeds.

Fibres do not transport water; their function is purely support. Fibres have hardly any lumen and are more or less strips of lignin.

The walls of vessel elements, tracheids and fibres are all thickened with **lignin**, which strengthens the walls and allows the cells to perform their support function. This lignin thickening can be present as rings, spirals or a more or less continuous sheet perforated by pores (see Figure 6.26).

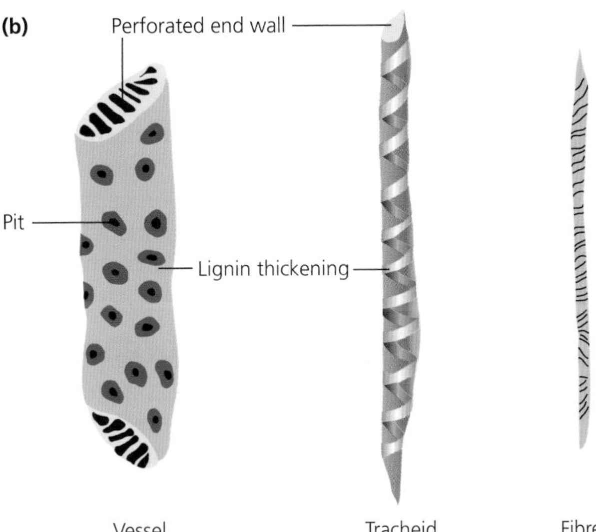

Figure 6.26 (a) Photo of longitudinal section of xylem tissue (×20); (b) diagram of xylem cell types.

Xylem cells are all dead and have no cytoplasm (which is unnecessary for strengthening, and would tend to obstruct water transport).

Phloem

Phloem is the other tissue involved in transport in plants, transporting organic nutrients up and down the plant. Like the xylem, it consists of more than one cell type: sieve tube elements and **companion cells**, with **phloem parenchyma** packed in between (see Figure 6.27).

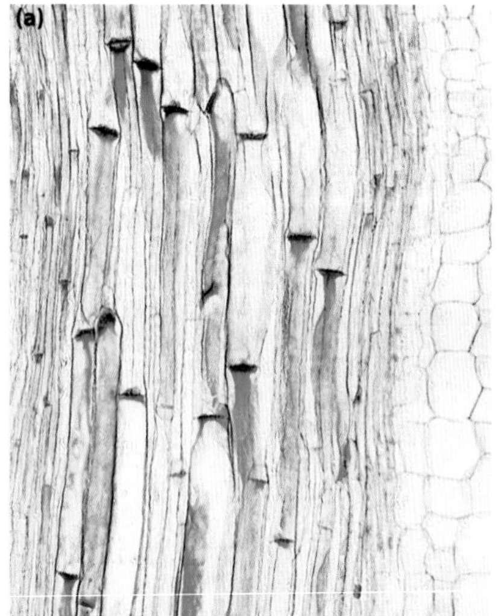

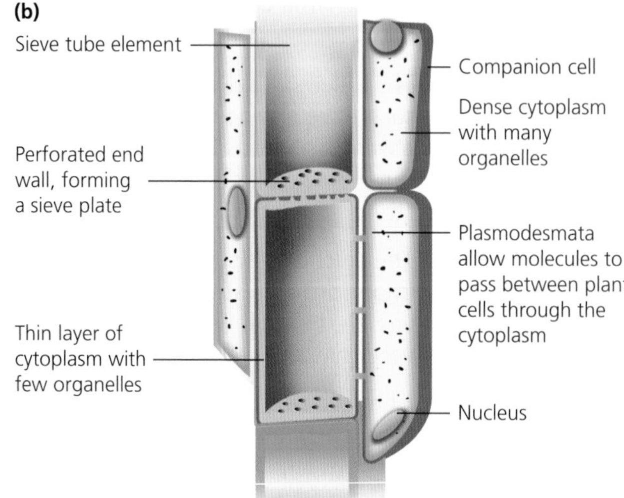

Figure 6.27 (a) Photo of a longitudinal section of phloem tissue (×65); (b) diagram of phloem cell types.

The sieve tube elements are the cells that transport nutrients. In order to do so, they have a reduced amount of cytoplasm, few organelles, and their end walls form sieve plates, allowing the cytoplasm of the cells to connect. The importance of these modifications is dealt with in Chapter 9.

The reduction in cytoplasm in the sieve elements means that the cells cannot support themselves without the help of a companion cell. The exact role of the companion cells is still the subject of research, but their large number of mitochondria and ribosomes suggests that they play an active part in transport in phloem.

Organs and organ systems

An organ is defined as a group of different tissues that work together. In animals and plants, specialised organs have a particular function or functions. Often, these functions have different aspects that require different tissues to carry them out. Examples of organs include the following.

- The **heart** pumps blood around the body. Cardiac muscle requires nutrients and oxygen for respiration, and these are supplied by the blood. Finally, the beating of the heart requires coordination, and this is done by nervous tissue.
- The **leaf** is a plant organ that carries out photosynthesis and gas exchange. It has within it a number of tissues: xylem, which brings water to the leaf; phloem, which transports organic nutrients away; palisade and spongy parenchyma, which carry out photosynthesis; epidermis, which waterproofs the leaf and also has stomata within it to allow for diffusion of gases into and out of the leaf.

In animals, organs often work together to form organ systems. An example is the **digestive system**, which involves many organs, including the oesophagus, stomach, small and large intestines, liver, pancreas and gall bladder.

Test yourself

12 What is a stem cell?
13 Give a reason why embryonic stem cells have a wider variety of uses than adult stem cells.
14 Give an example of how a ciliated epithelium is adapted to its function.
15 How does the absence of end cell walls aid the functioning of a xylem vessel?
16 Why is a lung defined as an organ?

Exam practice questions

1 What is the name of the stage in the mitotic cell cycle in which replication occurs?

A cytokinesis
B interphase
C prophase
D telophase *(1)*

2 Which describes a homologous pair of chromosomes?

A any two chromosomes from a diploid nucleus
B chromosomes in a mammalian sperm cell that have the same genes
C chromosomes that pair together during prophase of mitosis
D chromosomes that pair together in meiosis I. *(1)*

3 Which correctly describes the changes that occur during anaphase of the second division of meiosis (meiosis II) in an animal cell? *(1)*

	Distance between poles of the spindle	**Distance between centromeres and poles of the spindle**	**Distance between centromeres of sister chromatids**
A	increases	decreases	stays constant
B	decreases	stays constant	increases
C	stays constant	decreases	increases
D	increases	increases	stays constant

4 Plants and animals have cells specialised for specific functions.

a) Outline what happens to a stem cell when it differentiates into a specialised cell. *(4)*

b) Explain how the following cells are specialised to carry out the functions given:

i) squamous cell – exchange of substances *(2)*
ii) ciliated epithelial cell – movement of mucus *(3)*
iii) palisade mesophyll cell – photosynthesis *(3)*
iv) guard cell – diffusion of gases into and out of a leaf. *(3)*

5 The drawings show stages in a mitotic cell cycle.

A
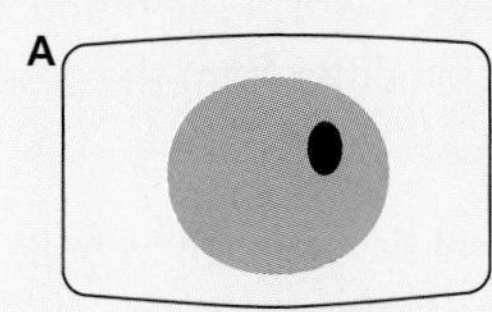
B
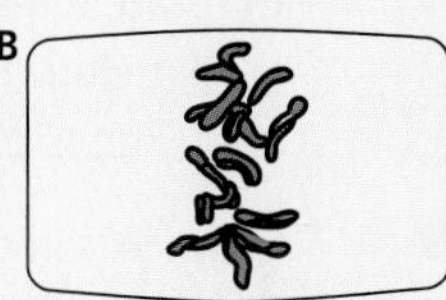
C
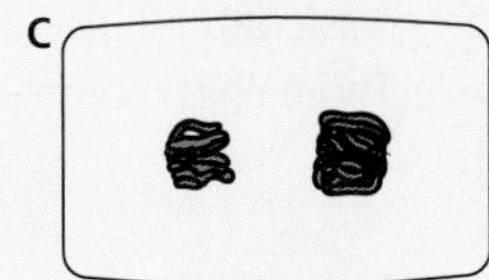
D
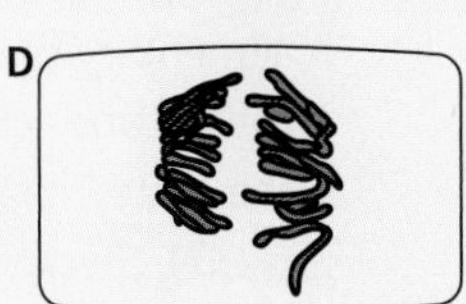
E
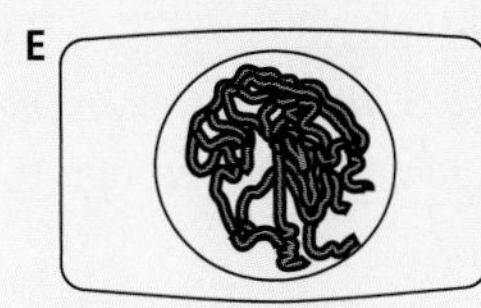

a) List the letters in the order in which the stages of mitosis occur, starting with stage A. *(1)*

b) Outline what happens to DNA in a cell during stage A. *(4)*

c) Explain the significance of mitosis in the life cycle of a plant. *(5)*

d) Describe how mitotic cell division in plant cells differs from this type of division in animal cells. *(4)*

6 A flowering plant has a diploid number of four. The diagram shows drawings of the chromosomes from this plant.

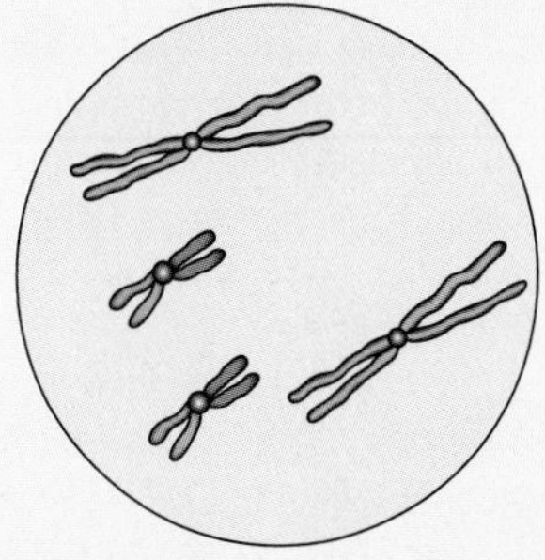

a) Make an annotated diagram to show the arrangement of these chromosomes in a cell during the following stages of mitosis

i) metaphase *(3)*
ii) anaphase. *(3)*

b) Make an annotated diagram to show how these chromosomes appear in anaphase I and anaphase II of meiosis. *(6)*

c) State three ways in which the products of a meiotic division differ from the products of mitosis. *(3)*

7 Stem cells in animal tissues such as bone marrow, the skin and the epithelium that lines the alimentary canal progress continuously through the mitotic cell cycle. Three checkpoints control the cycle. These are:

- G1 checkpoint
- G2 checkpoint
- metaphase checkpoint.

Each of these checkpoints determines whether the cell cycle should continue through to the next stage of the cycle.

a) G1 and G2 are two stages during the cell cycle.

  i) State when each of these stages occurs during the cycle. *(2)*

  ii) Outline what happens in each stage. *(4)*

b) Suggest what happens to cells that do not progress past the checkpoints at:

  i) G1 *(2)*

  ii) G2 *(2)*

  iii) metaphase. *(2)*

c) Stem cells in the bone marrow give rise to mature red blood cells that are specialised for the transport of oxygen. Outline the changes that occur in a stem cell as it differentiates into a red blood cell in bone marrow. *(6)*

d) Explain why stem cells are useful in research into developmental biology. *(4)*

Stretch and challenge

8 Explain how the behaviour of chromosomes during meiosis I contributes to genetic variation.

9 *Allium triquetrum* is a flowering plant in the same family as onion and garlic. A root tip squash preparation of this species showed that each cell contains 18 chromosomes.

a) Outline a procedure that you could use to confirm that cells of *A. triquetrum* have 18 chromosomes.

b) *A. triquetrum* produces flowers for sexual reproduction. Meiosis occurs in the anthers inside the flower. Cells known as pollen mother cells divide by meiosis to form pollen grains. Each pollen mother cell forms four pollen grains.

Soon after formation, some pollen grains from the anthers of *A. triquetrum* were sectioned, stained and photographed with a light microscope. A few of the photos showed chromosomes in metaphase of mitosis. The chromosome number of these cells was nine.

Explain fully why:

  i) there are only nine chromosomes in the pollen grain

  ii) only a few of the cells were in metaphase.

c) Discuss the advantages of studying nuclear division (mitosis and meiosis) with a light microscope rather than a transmission electron microscope.

d) What extra advantage is gained by using a laser scanning confocal microscope to study cell division?

Chapter 7

Exchange surfaces and ventilation

Prior knowledge

Before you start, make sure that you are confident in your knowledge and understanding of the following points.

- Most living things need oxygen in order to release energy from food in the process of respiration.
- Small organisms take in oxygen through their surfaces, which are moist and permeable.
- Larger, more complex organisms have special gas exchange organs such as lungs and gills.
- Increasing the surface area of a gas exchange surface increases the rate of gas exchange.
- The human gas exchange system includes the trachea, bronchi, bronchioles, lungs, ribs, pleural membranes, intercostal muscles and diaphragm.
- The alveoli form the gas exchange surface in human lungs.
- Breathing consists of the processes of inspiration and expiration.
- Breathing is not the same as respiration.
- In the lungs, oxygen moves into the blood and carbon dioxide leaves it. In the tissues, these gases travel in the reverse direction: oxygen moves out of the blood and carbon dioxide moves into the blood

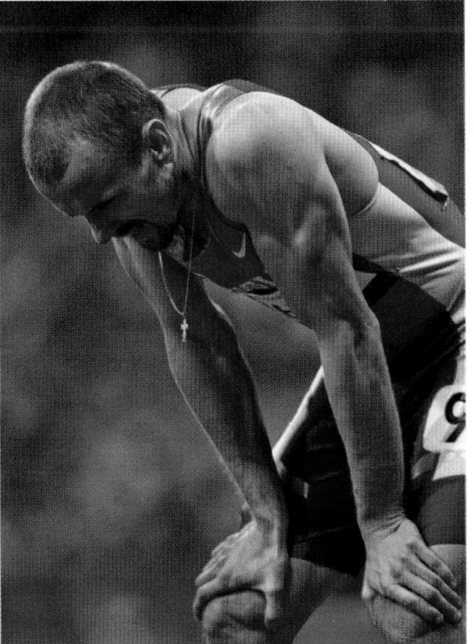

Figure 7.1 A sprinter gasping for breath at the end of a race.

Test yourself on prior knowledge

1. Breathing is not the same as respiration. Explain the difference.
2. Why are gas exchange surfaces often moist?
3. Explain why large animals need specialised gas exchange organs.
4. What process causes oxygen from the air to move into the blood, and causes carbon dioxide to move in the opposite direction?

Most living things ultimately get their energy from the Sun, but it is released by the process of aerobic respiration. This requires oxygen and releases carbon dioxide as a waste product. Organisms have to get sufficient oxygen into their cells for their needs, and the waste carbon dioxide must be released. You see this when you perform strenuous activity and you start breathing deeply as your body attempts to take in enough oxygen to keep your muscles working properly.

When sprinting, you cannot take in enough oxygen. Muscle cells can release energy for a short time without oxygen, but afterwards the oxygen deficit needs to be made up.

The rest of this chapter deals with gas exchange, but you must remember that other things are exchanged through surfaces, and many of the adaptations for gas exchange are also found at other exchange sites.

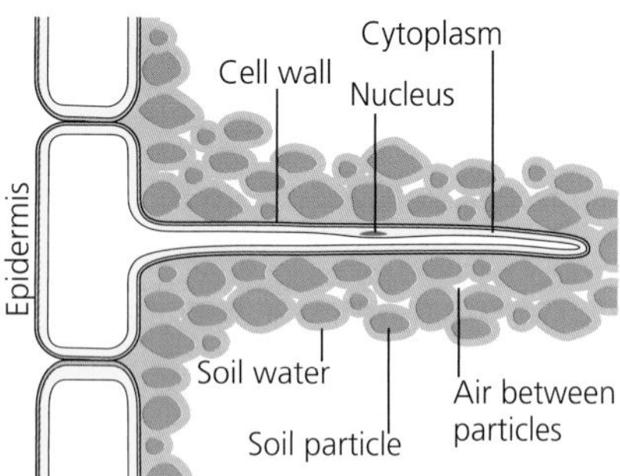

Figure 7.2 Root hair cells in the epidermis of plant roots increase the surface area for intake of water. (a) Photo of a root tip showing root hairs; (b) diagram showing the shape of a root hair cell.

For example: plant roots have root hair cells to increase the surface area for the absorption of water (see Figure 7.2), and cells in the endothelium layer that lines the gut have microvilli on the surface to increase the area for absorption of digestive food products.

Tip

Remember that 'breathing' is not gas exchange – it is a specific process involving taking air into the body, exchanging gases and then expelling the air again. It is incorrect to talk about plants, fish and many invertebrates 'breathing'. Ventilation is the correct term.

The need for specialised gas exchange surfaces

Over the history of evolution, organisms have become more complex and have developed specialised organs and organ systems. Organisms that respire aerobically need oxygen. As a result, there is also a need to get rid of the waste carbon dioxide produced, particularly in animals; plants, however, can make use of some carbon dioxide for photosynthesis.

In unicellular and very small organisms, the process is straightforward: because oxygen is used for respiration, the concentration of oxygen within the organism is lower than in the surroundings, and so oxygen diffuses in, down a concentration gradient, as long as it can penetrate the organism's surface. Carbon dioxide is produced in respiration, therefore its concentration gradient is favourable for movement out of the cell. In order for this to work, the organism's surface has to be **permeable** to the two gases, and it also has to be moist – oxygen and carbon dioxide can only get through membranes in solution, not in their gaseous form. The moist surface allows the gases to dissolve and then pass through. Simple organisms usually live in water, so this is not a problem. Diffusion is a slow process, but because the distance between the centre of a small organism and the outside is small, diffusion is adequate for the purpose.

As organisms become more complex and often larger, problems arise. These organisms tend to be more active and have an increased demand for oxygen to supply all their cells (i.e. they have increased metabolic activity). The distances that the gases have to travel become much larger, and oxygen would take so long to diffuse into the centre of the organism that the central cells would die before the oxygen reached them. Figure 7.3 gives an indication of this problem, although diffusion rates and sizes of the species do vary, so the times are approximate. A further problem is that organisms that evolved to colonise land needed to develop a waterproof surface to avoid excessive water loss, and such surfaces are impermeable to oxygen and carbon dioxide. Finally, there is the issue of surface area to volume ratio.

Gases diffuse through the surface of small animals, so the greater the surface area, the higher the quantity of gases that can diffuse through. The number of cells (and therefore the volume of the organism) is an indication of the demand for oxygen (and the production of carbon dioxide). As animals get larger, their surface area increases, but their volume increases

Single-celled organism (*Paramecium caudatum*)

Maximum distance for diffusion = 50 μm

Time taken = 8 seconds

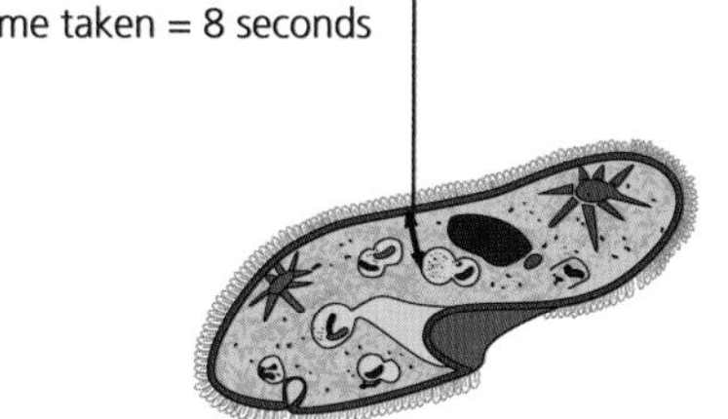

Maximum distance for diffusion = 15 cm

Time taken = 7 hours

Figure 7.3 The problem of increasing size related to rate of diffusion.

by a greater factor. In other words, the **surface area : volume ratio** gets smaller, and demand for oxygen outstrips supply. In addition, the volume of carbon dioxide produced exceeds the capacity of the surface to remove it. This principle can be shown using cubes of different size (see Figure 7.4). Animals have a more complex shape than a cube, but they have compact bodies and the use of a cube is valid as a model.

Length of side/cm	Surface area/cm^2	Volume/cm^3	SA : V ratio
0.5	1.5	0.125	12 : 1
1.0	6.0	1.0	6 : 1
2.0	24	8.0	3 : 1

Figure 7.4 Decrease of surface area : volume ratio in cubes of increasing size.

Gaseous exchange system in mammals

In mammals, the gas exchange surface is the **alveoli** (singular: **alveolus**) (air sacs) within the lungs. The human gas exchange system is shown in Figure 7.5.

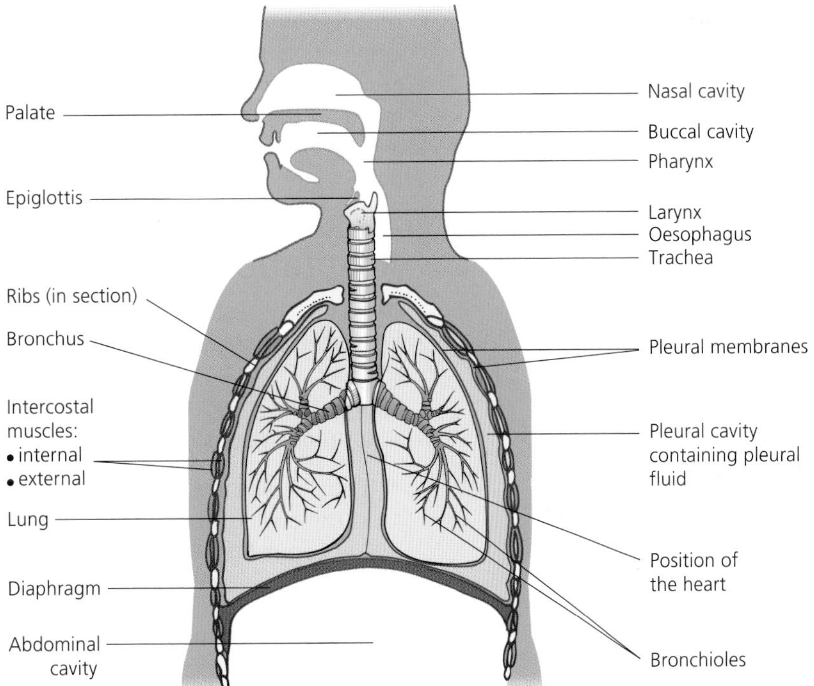

Figure 7.5 The human gas exchange system.

Air is taken in through the mouth and the nose and enters the **trachea**. The trachea branches into two **bronchi** (singular: **bronchus**), one going to each lung. Each bronchus then divides into smaller tubes called **bronchioles**, which eventually end in a group of air sacs or alveoli. This is the gas exchange surface, and is shown in Figure 7.6.

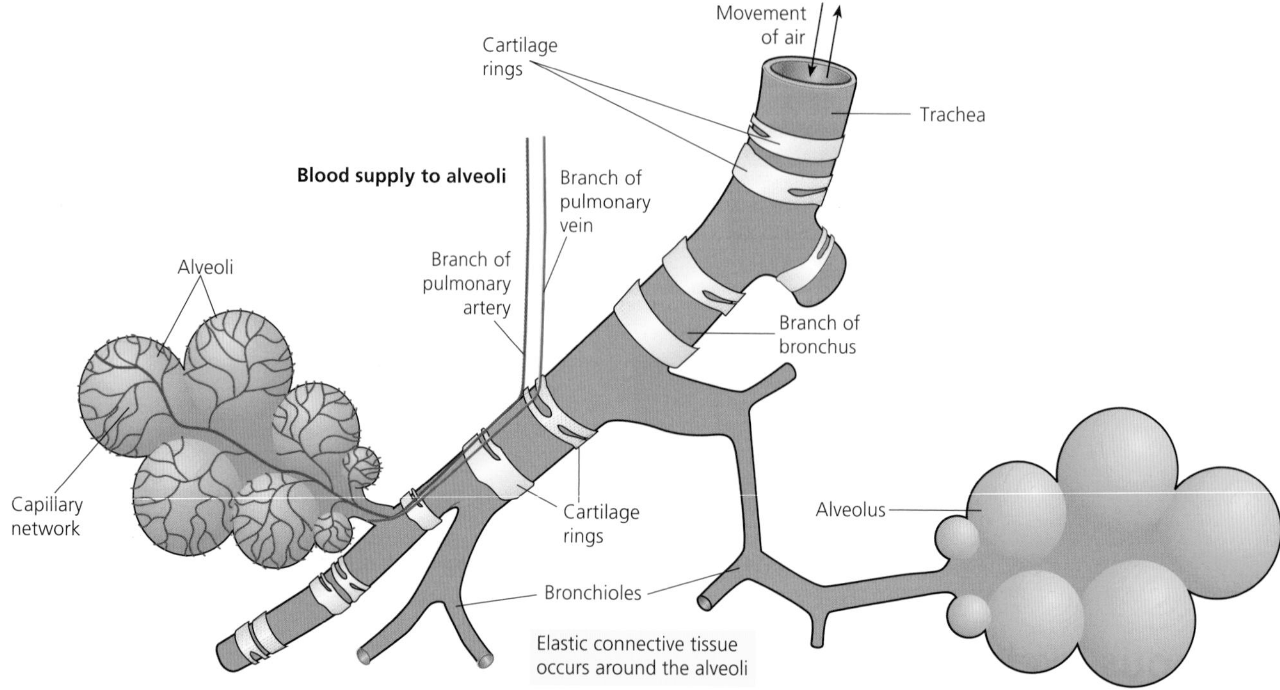

Figure 7.6 The airways of the mammalian breathing system with alveoli shown (not to scale).

The lungs are situated within the **rib cage**, which protects them, in the **thoracic cavity**. Between the ribs are the interior and exterior **intercostal muscles**, which move the rib cage in and out during deeper breathing. The lungs are surrounded by a double membrane, the **pleural membranes**, between which, in the **pleural cavity**, is a fluid. This fluid provides lubrication between the lungs and the rib cage. At the base of the thoracic cavity is a sheet of muscle, the **diaphragm**, which is also part of the ventilation mechanism.

Adaptations for gas exchange

Like all body systems, the mammalian gas exchange system has structural features that are suited to its function and make gas exchange more efficient. These adaptations are common to many animal groups, even though they may involve different structures (e.g. gills in fish). The adaptations are as follows:

- **The surface area for gas exchange is very large**. Gas exchange occurs in the alveoli, and it has been estimated that an adult human being has about 500–700 million alveoli in total. This gives a surface area of 70–100 m^2.
- **The gas exchange surface is very thin**. The alveolar walls are only one cell thick, composed of simple squamous epithelium, as are the capillary walls that the gases travel through to get into or out of the blood. This means that there is a short diffusion pathway. Diffusion is a slow process, so it is important that the gases do not have to diffuse very far.

- **There is a good blood supply**. The alveoli are well supplied with capillaries, which will absorb the oxygen and deliver the carbon dioxide. In addition, the circulation of the blood means that oxygen is taken away and fresh carbon dioxide is brought in, maintaining the **concentration gradients** of both gases.
- **There is a ventilation mechanism**. Inside the alveoli, the concentration gradient is also maintained by breathing, which is a ventilation mechanism. Breathing constantly removes the air containing the waste carbon dioxide and reduced levels of oxygen, and replaces it with fresh air containing higher levels of oxygen and lower levels of carbon dioxide.
- **The gas exchange surface is moist**. Oxygen molecules dissolve in the water and diffuse across the cells into the blood.

The gas exchange mechanism in an alveolus is shown in Figure 7.7.

Tip

When discussing the gas exchange surface, the surface area, not the surface area : volume ratio, is important. When discussing the whole animal, the surface area : volume ratio, not the surface area, determines the need for a gas exchange system.

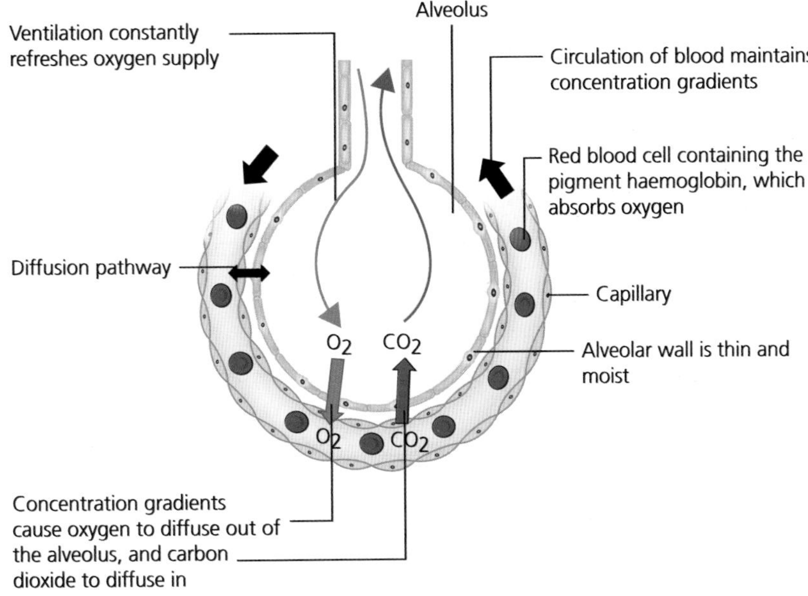

Figure 7.7 Gas exchange in an alveolus.

Tip

You will sometimes see the term *diffusion gradient* used instead of *concentration gradient*. The terms are interchangeable and you can use either term in examination answers.

Test yourself

1. How does living on land pose a problem for animals when absorbing gases through the body surface?
2. Large animals have a much greater surface area than small animals. Why are they less able to supply their need for oxygen, despite this?
3. Large animals have specialised gas exchange systems, but large plants do not have a circulatory system for gases and rely on direct diffusion from the air. Suggest a reason for this.
4. State how the alveolar epithelium is adapted for efficient gas exchange.
5. Explain how a ventilation mechanism makes gas exchange more efficient in mammals.

Tissues of the gas exchange system

Each of the different tissues making up the various organs of the gas exchange system performs a specific purpose. The structure and arrangement of the tissues ensures efficient working of the system as a whole. In order to understand this adaptation of structure to function, you first need to consider the individual tissues present.

Cartilage

Cartilage is a form of connective tissue that provides strengthening and support. It is composed of cells surrounded by material consisting of mucopolysaccharides, which are complex polysaccharides containing amino groups. Cartilage is resistant to both tension and compression; it is not as strong or rigid as bone (which can be formed from it in certain parts of the body), but is more flexible.

In the gas exchange system, the function of cartilage is to keep the larger tubes open (the trachea, the bronchi and the larger bronchioles). These tubes have a relatively large diameter and thin walls, and without the strengthening supplied by the cartilage they would collapse.

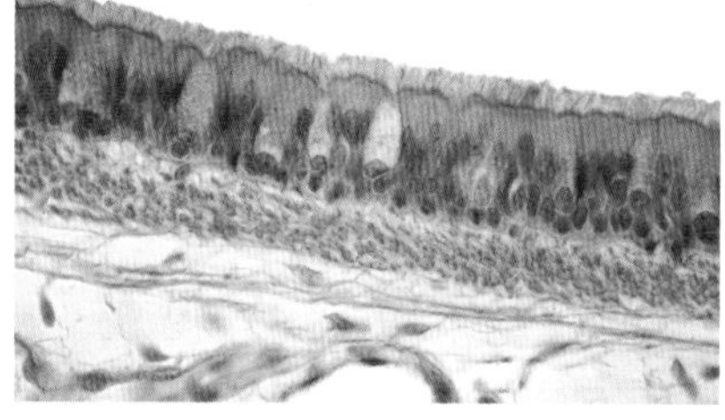

Figure 7.8 Ciliated epithelium (×90).

Ciliated epithelium

'Epithelium' is the name given to any layer of cells that forms a covering or lining. The distinctive feature of ciliated epithelium, as the name suggests, is the presence of cilia. The airways are coated with mucus, which traps dust and bacteria that enter with the air. It is important that this mucus is then disposed of, and this is done by moving it to the top of the trachea, which is close to the opening of the oesophagus. The mucus can then be swallowed and is eventually eliminated via the digestive tract. The cilia of the ciliated epithelium beat constantly, moving the mucus towards the top of the trachea (see Figure 7.8).

Goblet cells

Goblet cells produce the mucus that lines the trachea, bronchi and larger bronchioles. They get their name from their goblet shape, with the 'bowl' of the goblet containing the mucus, ready to be secreted. These cells are found between the ciliated cells in the epithelium (see Figure 7.9).

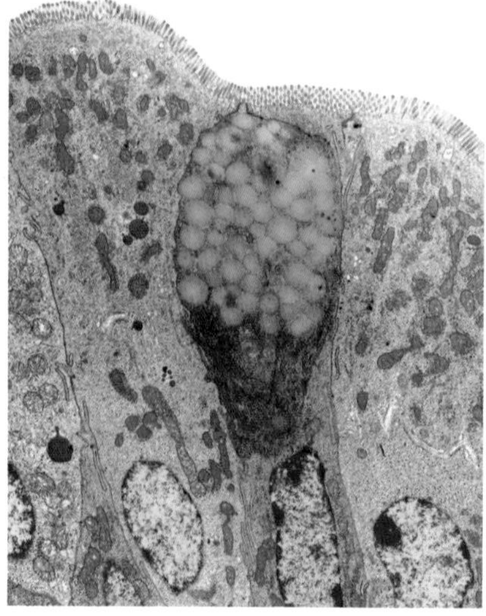

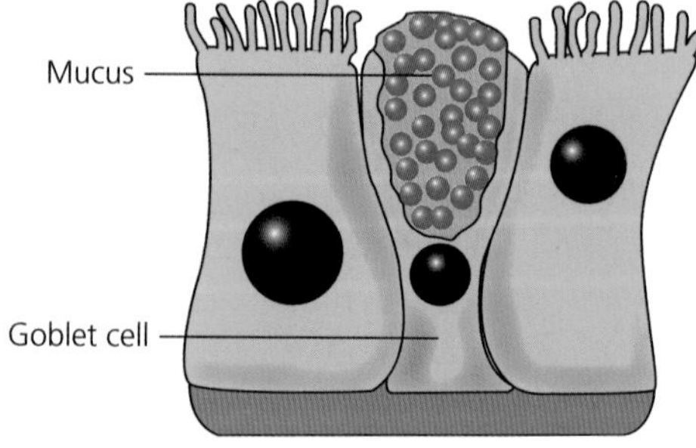

Figure 7.9 (a) A transmission electron micrograph of a goblet cell (×720); (b) a diagram of a goblet cell.

Smooth muscle and elastic fibres

Smooth muscle is found in the trachea, bronchi and the large bronchioles. It maintains the tone in the airways and also allows expansion in conditions where extra oxygen is needed (e.g. during exercise). **Elastic fibres** are found in all the lung tissues, including the alveoli. The elasticity of the lungs is vital, because **expiration** is mainly a passive process due to the recoil of the lungs. Elasticity provided by the fibres is necessary for this recoil.

Gas exchange system components

The trachea

The trachea is the widest tube in the gas exchange system. No gas exchange takes place in the trachea, the bronchi or the bronchioles, and the tracheal wall therefore does not need to be especially thin. Neither does the tracheal wall need to be particularly thick, because it has no active role in ventilation, simply being a channel that allows air to reach the lungs. The only really important thing for the trachea is to be able to keep open at all times to allow air to pass through. Thin-walled tubes can collapse quite easily, but rather than develop a uniformly thick wall (which would need a lot of cells and therefore a lot of resources), the trachea has evolved rings of cartilage that provide a form of 'scaffolding' to keep the tube open. These rings are distributed at intervals along the trachea, so the softer tissue in between retains flexibility. The oesophagus runs very close behind the trachea and expands when food passes down it. This could create friction if the oesophagus wall rubbed against the semi-rigid rings of cartilage, so the rings are C shaped, with a gap at the back so the oesophagus wall does not rub against them. The ends of the cartilaginous rings are joined by tissue containing smooth muscle and elastic fibres.

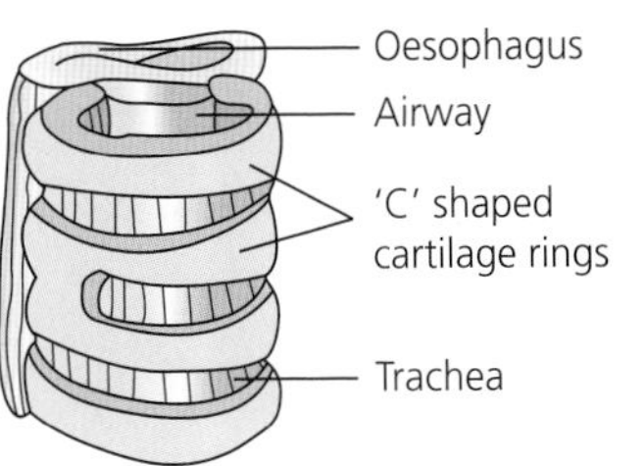

Figure 7.10 Structure of the mammalian trachea.

The trachea is lined by ciliated epithelium that contains goblet cells, which secrete mucus. In addition, below the epithelium there are **mucous glands**. A good covering of mucus is particularly important in the trachea, because it is the first part of the gas exchange system to encounter the air. It is best if most of the dust particles and bacteria are trapped by the mucus here, where they can be moved the fairly short distance up the trachea and then into the oesophagus. The structure of the trachea is shown in Figure 7.10.

Bronchi

The bronchi are very similar in structure to the trachea, but they have a smaller diameter and thinner walls, and there is one other important difference: the bronchi have complete rings of cartilage rather than the C-shaped ones in the trachea. There is no longer a need for an incomplete ring, because the bronchi do not lie against the oesophagus.

Bronchioles

There are many bronchioles, and they vary in size and structure as they get smaller towards the alveoli. Although the walls of the bronchioles are thin, narrow tubes tend to be more self-supporting; the bronchioles, which are around 0.3–5.0 mm in diameter, therefore no longer have rings of cartilage, because they can support themselves without it.

There are also no mucous glands, although there are goblet cells in the larger bronchioles. The smaller bronchioles have no goblet cells and also no cilia on their epithelium. The smallest bronchioles are deep in the lungs, and there has been plenty of opportunity to trap and remove any irritant particles.

The walls of the larger bronchioles contain elastic fibres and smooth muscle, and the muscles can adjust the diameter of the airways to increase or reduce air flow. The smallest bronchioles lack muscle fibres but retain elastic fibres.

Test yourself

6 Why is cartilage needed in the trachea and bronchi, but not in the bronchioles?
7 What is the function of the goblet cells found in the ciliated epithelium?
8 Smoking paralyses the cilia. Explain why this may lead to irritation of the lungs.
9 Explain the role of the elastic fibres in the alveoli.
10 State one difference in structure between the trachea and the bronchi.

Alveoli

The alveoli are arranged in groups at the end of the smallest bronchioles. The wall of an alveolus consists of a single layer of epithelium, but there is also an extracellular matrix that contains elastic fibres, which allow the alveoli to expand during inspiration and recoil during expiration.

Many capillaries surround the alveoli. The capillaries absorb the oxygen from the alveoli and deliver waste carbon dioxide for excretion by the lungs. The watery fluid lining the alveoli facilitates the diffusion of gases but also produces a surface tension.

Comparison of structures in the lungs

Table 7.1 compares the various airways in the lungs.

Table 7.1 Features of the various airways in the lungs.

Structure	Cartilage	Smooth muscle	Elastic fibres	Ciliated epithelium	Goblet cells	Mucous glands
Trachea	✓	✓	✓	✓	✓	✓
Bronchi	✓	✓	✓	✓	✓	✓
Larger bronchioles	✗	✓	✓	✓	✓	✗
Smaller bronchioles	✗	✓	✓	✗	✗	✗
Smallest bronchioles	✗	✗	✓	✗	✗	✗
Alveoli	✗	✗	✓	✗	✗	✗

The mechanism of ventilation

In order for gases to be exchanged in the alveoli, there must be concentration gradients of oxygen (the concentration of which must be higher in the alveoli than in the blood) and carbon dioxide (which must be lower in the alveoli than in the blood). These concentration gradients are maintained by constantly refreshing the supply of air in the alveoli – in other words, by breathing. Ventilation results from the action of the intercostal muscles between the ribs and the diaphragm, which is a sheet of muscle that seals off the bottom of the thoracic cavity. The elasticity of the lungs also plays a key role, although the lungs simply respond to the events around them and only play a passive part in the process of drawing air in (**inspiration**) and moving it out again (**expiration**).

The movements of the rib cage and diaphragm that result in ventilation are shown in Figure 7.11. In order to understand how these movements result in breathing, you need to know a little about how these cause air pressures in the thorax.

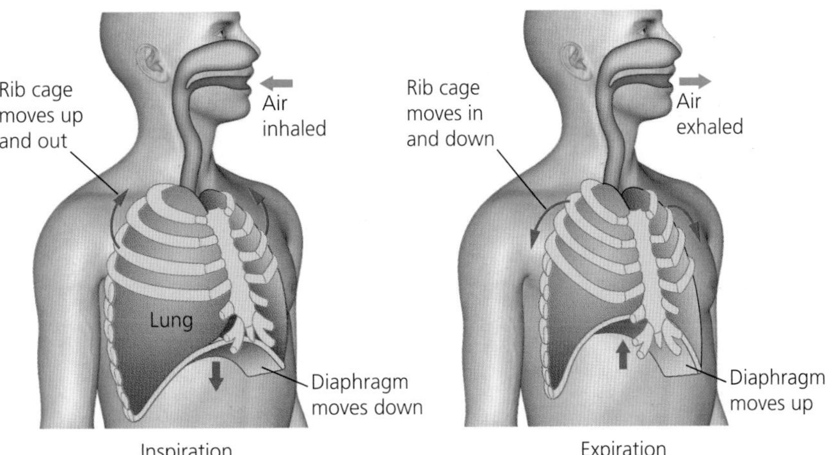

Figure 7.11 Ventilation movements.

The pressure in the thoracic cavity can be increased or decreased by altering the cavity's volume. Movements of the diaphragm and the rib cage alter the volume of the thoracic cavity. There is an inverse relationship between volume and pressure in any air-filled cavity: an increase in volume decreases the pressure, and a decrease in volume increases it.

When the volume of the thoracic cavity increases during inspiration (due to the rib cage moving up and out and the diaphragm moving down), the pressure inside the lungs reduces to below atmospheric pressure, causing air to move in from the outside. Air moving into the lungs causes them to expand. The process is reversed during expiration. The natural elasticity of the lungs, which causes them to recoil, helps expiration.

The sequence of events during one breath in and out is shown in Figure 7.12. Note that an expiratory centre in the brain controls another set of muscles, the internal intercostal muscles. When these muscles contract, they pull the rib cage down and in. This is not necessary in normal ventilation, when the elasticity of the lungs and the relaxation of the external intercostal muscles is enough to cause expiration. However, the internal intercostal muscles are used when there is forced expiration (e.g. blowing) and during strenuous exercise, when extra pressure on the lungs is needed.

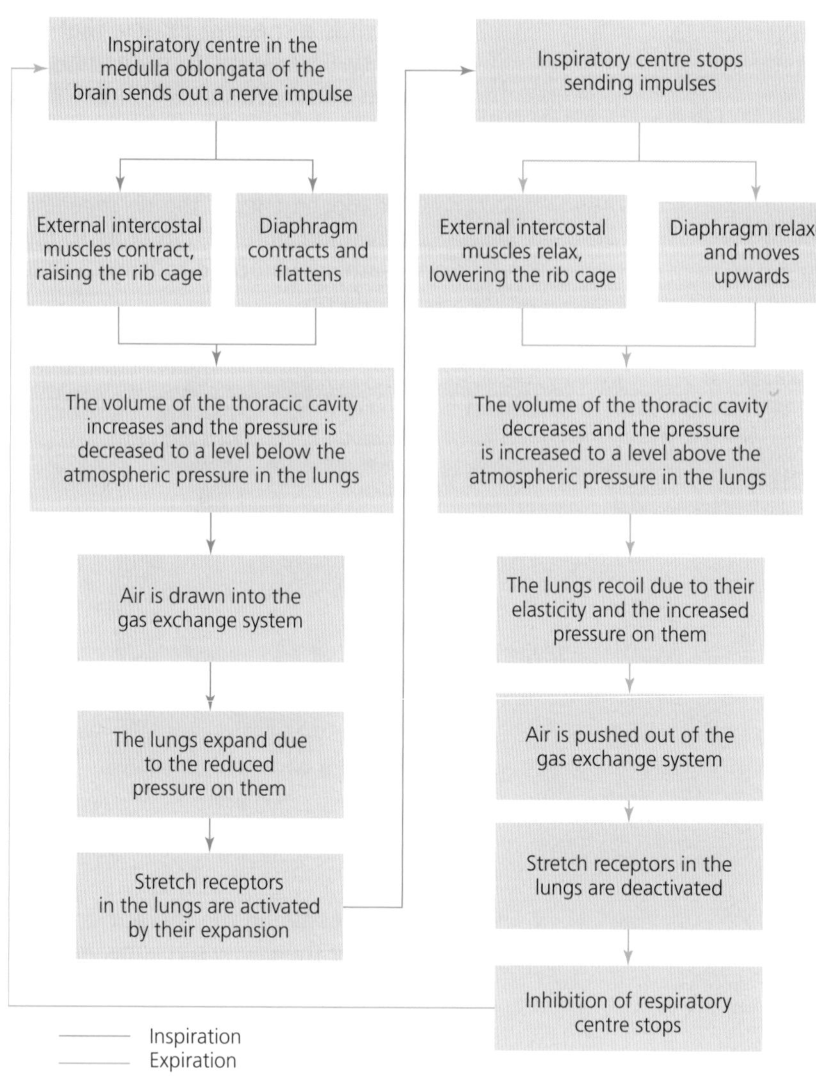

Figure 7.12 Flowchart showing the mechanism of ventilation in humans.

Breathing measurements

A variety of measurements related to breathing can be made using a piece of equipment known as a **spirometer**. A spirometer is shown in Figure 7.13. The spirometer produces a trace on a revolving drum (a kymograph) or on a screen.

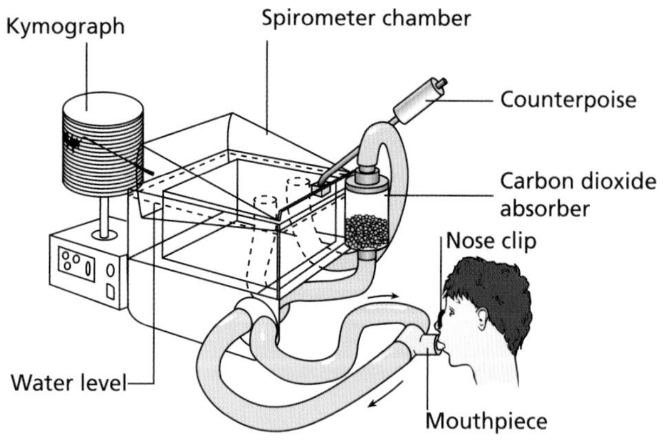

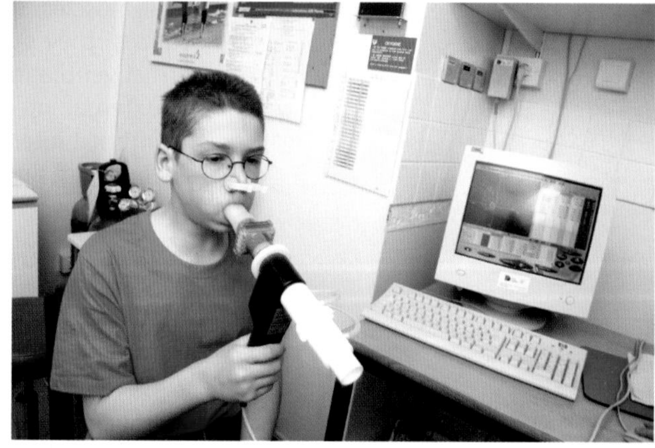

Figure 7.13 (a) A mechanical spirometer; (b) an electronic spirometer.

> **Key terms**
>
> **Vital capacity** The maximum volume of air that can be breathed in or out in a single breath.
>
> **Tidal volume** The volume of air that is normally breathed in or out at rest.
>
> **Breathing rate** The number of breaths per minute (a breath is taking air in and breathing it out).
>
> **Oxygen uptake** The amount of oxygen consumed by the subject. (The spirometer removes carbon dioxide from the re-breathed air, therefore the volume of air gradually decreases as oxygen is extracted from it. The change in volume is a measure of oxygen uptake.)

The subject breathes in and out through the spirometer. The carbon dioxide is absorbed from the re-breathed air by soda lime to prevent respiratory distress, which may be caused if air containing high levels of carbon dioxide is breathed in.

A spirometer can measure the **vital capacity**, the **tidal volume**, the **breathing rate** and the **oxygen uptake**.

Figure 7.14 shows a read-out from a spirometer, indicating the different quantities measured. A small volume of air is always retained in the lungs, and this residual volume cannot be measured using a spirometer.

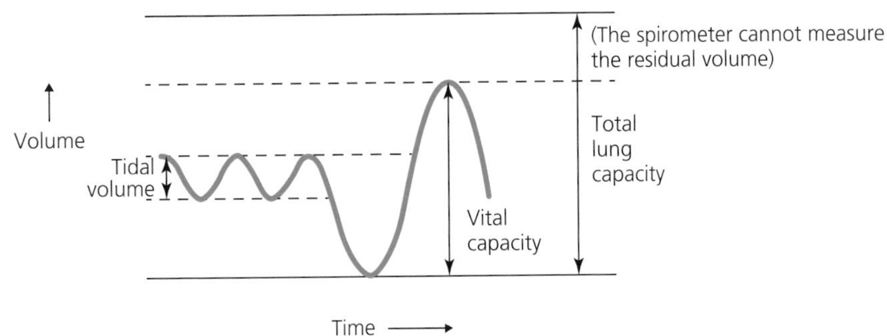

Figure 7.14 The features of a spirometer trace, showing various measurements of lung volume.

Analysing spirometer data

Look at the spirometer trace in Figure 7.15.

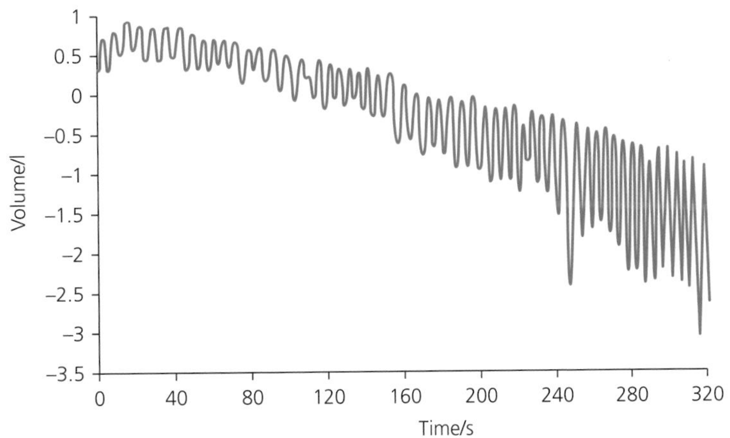

Figure 7.15 Spirometer trace for a subject at rest.

Calculating breathing rate

On the trace in Figure 7.15, the breathing rate changes with time. The calculation of breathing rate is straightforward if you use a 30-second or one-minute period, because the breathing rate will be expressed in breaths min^{-1}.

In the first two minutes, there are 18 breaths, giving a breathing rate of 9 breaths min^{-1} $\left(\frac{18}{2}\right)$.

Remember that in one breath the trace goes up **and** down. There is some confusion with a breath at around 100 seconds, where the trace seems to give an extra 'twitch', which may or may not be an extra breath.

The breathing rate speeds up and between 240 and 320 seconds (again a 'convenient' time span of two minutes) there are 14–15 breaths, meaning a breathing rate of 10.8–11.5 breaths min^{-1}.

Calculating tidal volume

Calculating the tidal volume in any time period is more difficult than calculating breathing rate, because tidal volume varies to some extent from breath to breath as well as over the period of the study. The mark scheme in an exam allows a greater degree of tolerance for tidal volume than breathing rate.

You can find the tidal volume by measuring from the top to the bottom of the trace. In the early period, tidal volume is about 0.4–0.5 litres. In the last minute of the study, it is roughly 1.5 litres but is sometimes as high as 2 litres.

Further analysis

When using a spirometer to give a trace like this, the air is re-breathed over and over again. Although the carbon dioxide is removed, a drop in the oxygen concentration occurs and, in response, the subject eventually starts taking deeper breaths and breathing faster in an attempt to get more oxygen. This explains the increase in breathing rate and tidal volume. For this reason, this experiment was curtailed after 320 seconds.

These traces always slope down from left to right. This is because, as the air is re-breathed, the carbon dioxide is absorbed by the spirometer. This causes a steady decrease in the total volume of the air, hence the slope of the trace.

Activity

A group of students used a spirometer similar to the one shown in Figure 7.13. They performed their experiment during their lesson after lunch.

The spirometer trace for one of the students while sitting down is shown in Figure 7.16.

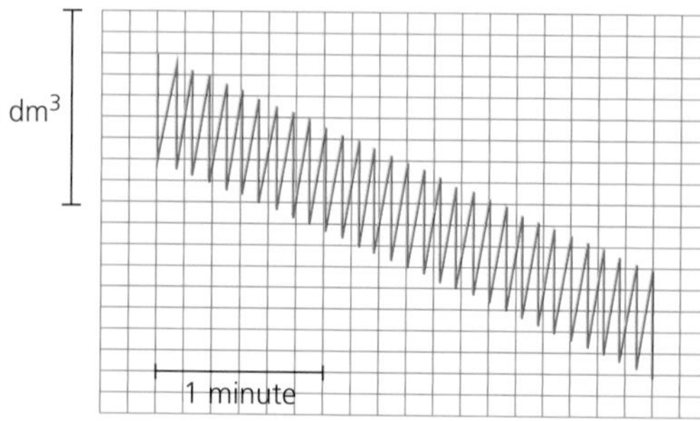

Figure 7.16 Student's spirometer trace.

1 Use the spirometer trace to calculate:
   a) mean tidal volume **b)** rate of breathing.
   In both cases show your workings.
2 Basal metabolic rate is the energy released in 24 hours when the body is completely at rest and no food has been eaten for at least 12 hours. Humans release 21.2 kJ of energy for each 1 dm^3 of oxygen consumed.
   a) Use this information and the spirometer trace to calculate the student's metabolic rate in kJ h^{-1}. Show your workings.
   b) Explain why this result is not a good estimate of the student's basal metabolic rate.

Gas exchange in other animals

Tip

As part of your course, you should have the opportunity to dissect a bony fish and/or an insect to examine their gas exchange system, and examine microscope slides of a range of gas exchange surfaces.

The adaptations of structure related to function that are seen in mammalian lungs are seen in the gas exchange surfaces of other animals as well. The gas exchange system always has a large surface area to exchange as much gas as possible. The surface is always moist (no special adaptation is needed in fish, of course, because they live in water). There is always a short diffusion pathway for the oxygen to enter, and there is always an ample blood supply, except in insects, which have a different type of gas exchange system that does not rely on blood to transport the oxygen around the body.

Gas exchange in bony fish

There are two groups of fish, distinguished by the composition of their skeleton. Bony fish, as the name suggests, have a skeleton made of bone. This group is known as the teleosts. The other group (the elasmobranchs) contains fish that have a skeleton made of cartilage.

The structure of a teleost's gas exchange system is shown in Figure 7.17. The gills are the gas exchange organs, and their surface area is increased by the division into many **gill filaments**. In addition, each gill filament has its surface area further increased by the presence of many **gill lamellae** (also called **gill plates**). The lamellae are well supplied with blood vessels and are very thin, making the diffusion pathway of oxygen from the water to the blood very short. Because fish live in water, there is no need for any special modification to keep the gills moist.

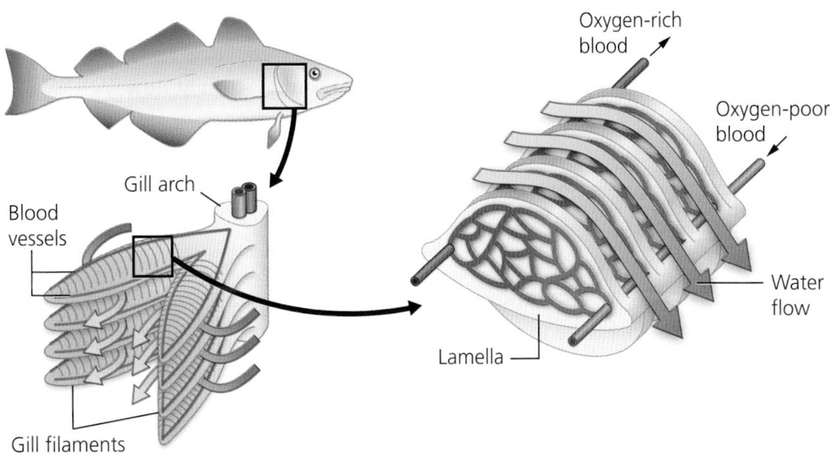

Figure 7.17 Structure of the gills in bony fish, showing adaptations for gas exchange.

Key term

Operculum A flap of tissue that covers the gills in bony fish.

Tip

The way a fish moves water across its gills for gas exchange is all about pressure differences. Remember, water always flows from an area of higher pressure to a lower one. If you are asked about the ventilation mechanism in fish, always explain it in terms of pressure differences.

The fish has a ventilation mechanism that constantly pushes water across the surface of the gills (Figure 7.18). With the mouth open, the fish lowers the floor of its **buccal cavity** and so increases the volume of the cavity, which in turn lowers pressure in the cavity. Because the pressure in the buccal cavity is now lower than the outside pressure, water flows in. The fish then closes its mouth and raises the floor of the buccal cavity. The increased pressure pushes water into the **gill cavity** (which has a lower pressure). Pressure now builds up in the gill cavity; this in turn forces open the operculum, and water is pushed out. The cycle is repeated, and when the buccal cavity floor is lowered again, this also has the effect of 'sucking' the operculum shut.

This ventilation mechanism ensures that the gills are constantly supplied with water containing oxygen, which helps to maintain the oxygen concentration gradient between the water and the blood.

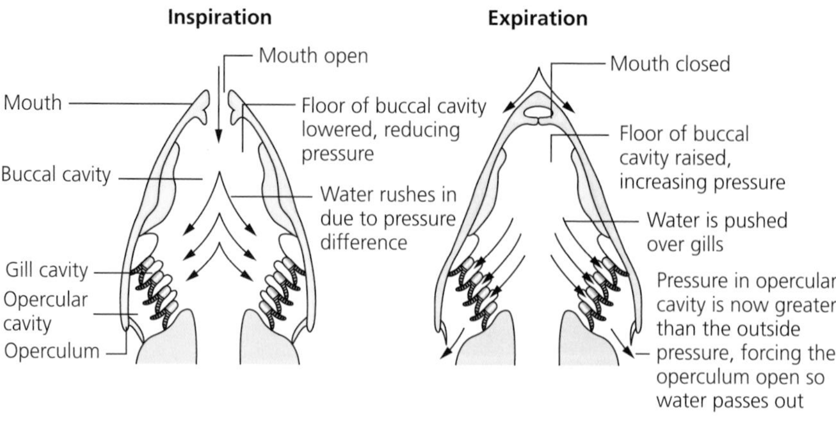

Figure 7.18 Structure of the gas exchange system in a bony fish, showing the mechanism that forces water over the gills (shown in red). The diagrams show the fish head in horizontal section.

The diffusion gradient between the water and the blood is also maintained because the flow of blood through the gill lamellae is in the opposite direction to the flow of water over the gills. This is called **counter-current flow**. The counter-current flow ensures that, all the way across the gill lamella, blood constantly encounters water with a higher oxygen concentration than it has, so maintaining the diffusion gradient at all times. This maximises the oxygen that can be extracted from the water.

In cartilaginous fish there is a concurrent flow, where the blood flows through the gills in the same direction as the water does. With concurrent flow, an equilibrium is reached part way across the gill; at this point, the amount of oxygen in the water and the blood is the same, so the blood no longer takes up oxygen. This is far less efficient than counter-current flow, as is shown in Figure 7.19.

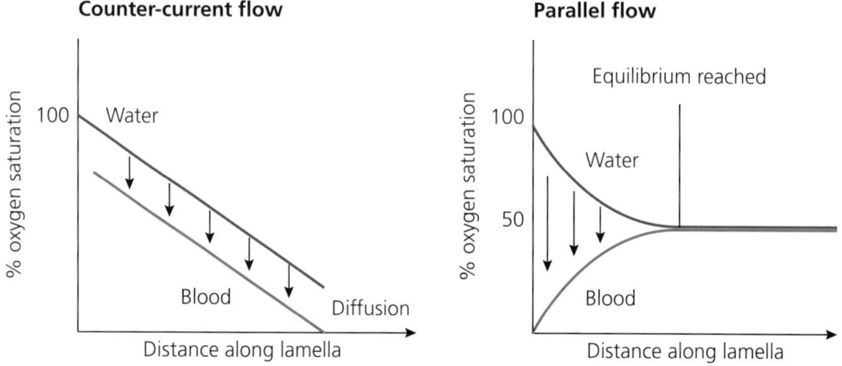

Figure 7.19 Comparison of counter-current and concurrent flow. Oxygen diffuses from the water into the blood as long as the concentration in the blood is lower than in the water. In counter-current flow, this is true across the whole gill lamella. In concurrent flow an equilibrium is reached part way across.

As always with respiratory systems, carbon dioxide travels in the opposite direction to the oxygen, leaving the tissues and being expelled from the body.

Test yourself

11 Describe the action of the diaphragm during inspiration.
12 An increased pressure in the thoracic cavity is one of two causes of expiration. What is the other?
13 Suggest why an injury that punctured the thoracic cavity would make it difficult for the lung on the side of the injury to function.
14 To calculate the volume of air breathed in and out per minute, which two quantities do you need to know?
15 List the features of the mammalian gas exchange system that make for efficient gas exchange.
16 Why is tracheal fluid necessary at the end of each tracheole?

Gas exchange in insects

Insects' gas exchange systems are very different from those in vertebrates. Insects do not rely on the blood system to transport oxygen. Instead, air containing oxygen is delivered directly to all the tissues by a system of tubes called **tracheae** (singular: trachea). The structure of the insect gas exchange system is shown in Figure 7.20. There are series of openings called **spiracles** along each side of the insect's abdomen. Each segment has a pair of spiracles. The insect makes pumping movements with its thorax and abdomen. When the body expands, air is sucked in through the spiracles and into the tracheae. The air then passes down a series of smaller and smaller tubes called **tracheoles** to reach the respiring tissues. The walls of the larger tracheae are strengthened by a series of rings of **chitin**, which keep the tracheae open rather like the C-shaped rings of cartilage which keep the mammalian trachea open.

At the end of the each tracheole there is a small amount of tracheal fluid. This fluid allows the oxygen from the air to dissolve and then diffuse into the tissues. Carbon dioxide diffuses from the tissues into this fluid and then passes into the air.

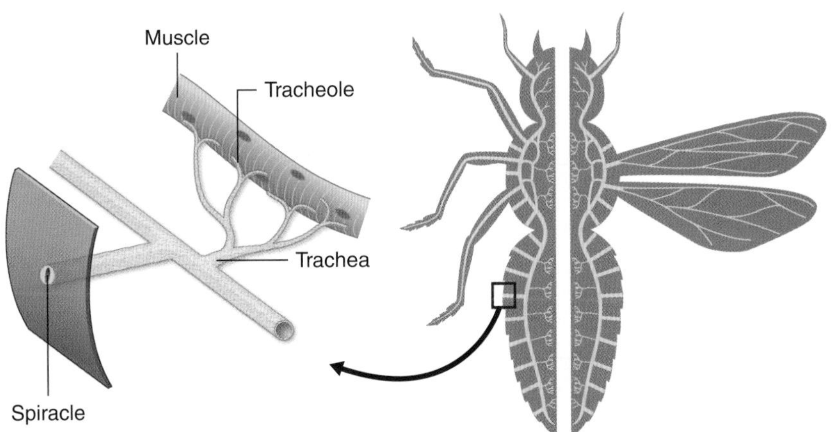

Figure 7.20 Insect gas exchange system. The tracheoles are the site of gas exchange.

Histology of gas exchange surfaces

Gas exchange surfaces in different organisms show similarities due to the similarities in their functions, but differ in detail. Three different exchange surfaces are shown in Figure 7.21.

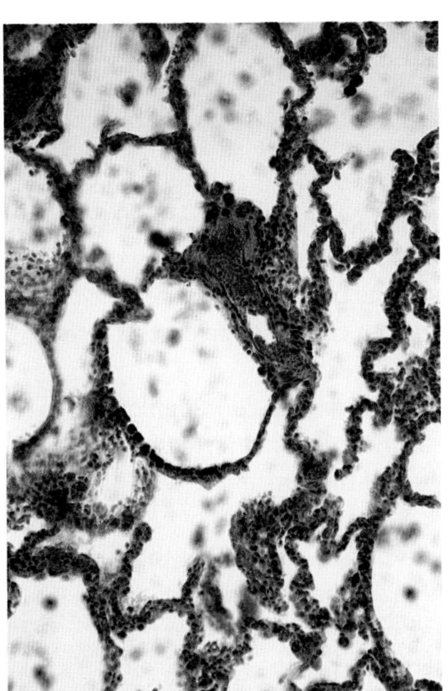

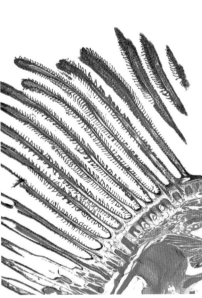

(a) Section of the lung showing alveoli (×75). The alveolar epithelium is very thin to provide a short diffusion pathway.

(b) Section of gill lamellae in a fish (×77). Once again, the exchange surface is very thin.

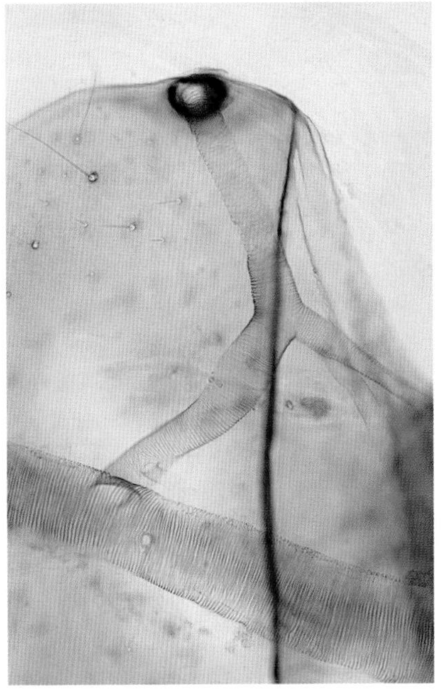

(c) Low-power image of the tracheae of an insect (×25). The tracheae are thin (as indicated by their transparency) and the rings of chitin that help to keep the larger tubes open are evident.

Figure 7.21 Gas exchange surfaces.

Exam practice questions

1 Which is a gas exchange surface?

A body surface of a single-celled organism
B bronchioles in mammals
C gill filaments in bony fish
D tracheae in insects *(1)*

2 The table shows the activity of the diaphragm, intercostal muscles and rib cage in a human during breathing.

	Diaphragm	Intercostal muscles		Rib cage
		External	Internal	
A	contracts	contracts	contracts	moves downwards
B	contracts	contracts	relaxes	moves upwards
C	relaxes	relaxes	contracts	moves downwards
D	relaxes	relaxes	relaxes	moves upwards

Which occurs during forced expiration during strenuous exercise? *(1)*

3 The diagram below was drawn from a microscope slide of human lung tissue.

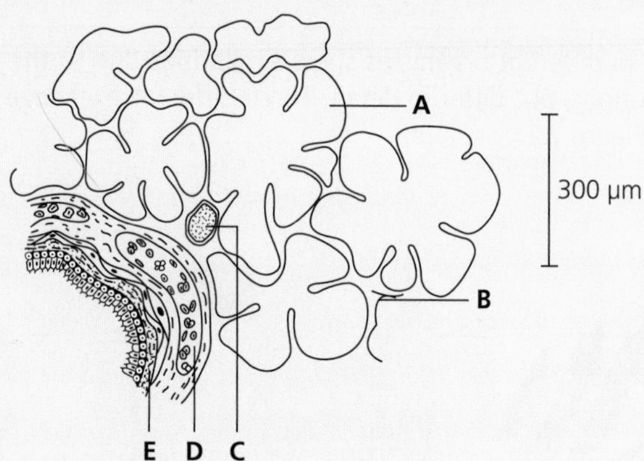

a) Calculate the magnification of the drawing. Show your working. *(2)*
b) Name structures A, C and D. *(3)*
c) i) Give the label of an area that contains elastic fibres. *(1)*
ii) Describe the function of elastic fibres in the lungs. *(3)*
d) Explain how the lungs are protected from damage and infection. *(4)*
e) The concentration of oxygen in the air-filled spaces in the lungs fluctuates between 13 and 14%. Explain why this is less than the oxygen concentration in the atmosphere. *(3)*

4 The rate of oxygen consumption was determined for seven animals.
The table shows the results.

Animal	Gaseous exchange surface	Activity	Oxygen consumption / $mm^3 g^{-1} h^{-1}$
Sea anemone	body surface	resting	13
Earthworm	body surface	resting	60
Octopus	gills	resting	90
Butterfly	tracheoles	resting	600
		flying	100 000
Frog	body surface and lungs	resting	150
Sparrow	air capillaries in lungs	resting	6 700
Mouse	alveoli in lungs	resting	2 500
		running	20 000

a) Explain why the results are all expressed as $mm^3 g^{-1} h^{-1}$. *(1)*
b) i) The percentage increase in oxygen consumption for the mouse when it became active was 700%.
Calculate the percentage increase for the butterfly when it started to fly.
Show your working. *(2)*
ii) Explain why the oxygen consumption increases when an animal runs or flies. *(3)*
iii) Explain how the butterfly achieves a high rate of oxygen consumption. *(3)*
c) Attempts to stimulate the sea anemone, earthworm and octopus to increase their rates of oxygen consumption above those in the table all failed.
Explain why the oxygen consumption of these animals does not increase to the levels of the bird and mammal. *(3)*
d) State and explain three features that make the gas exchange surface of the mouse much more efficient than that of the earthworm. *(3)*

5 The diagram shows part of the gill of a bony fish.

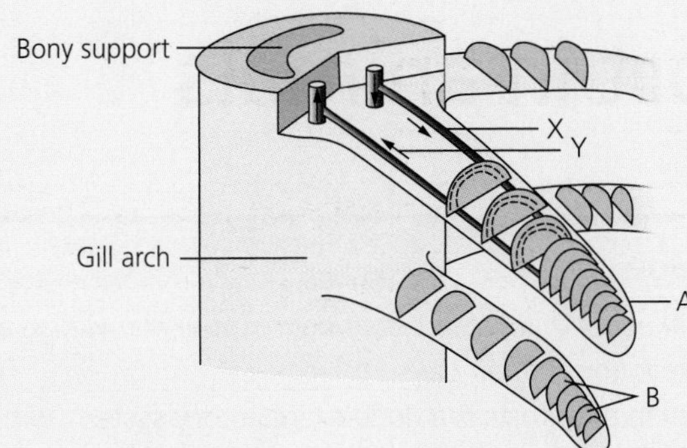

a) Name the structures labelled A and B. *(2)*

b) Blood in artery X has a concentration of oxygen of less than 5 cm^3 100 cm^{-3} of blood, whereas the concentration of oxygen in artery Y is 15 cm^3 100 cm^{-3} of blood. Explain how the structure of the gill shown in the diagram ensures an efficient exchange of gases. *(6)*

c) During ventilation of the gills, the pressure of water in the buccal and opercular cavities fluctuates. At its highest the pressure in the buccal cavity is about 0.098 kPa greater than the pressure in the opercular cavity. Explain how the gas exchange surface of the fish is ventilated. *(5)*

Stretch and challenge

6 The diagram shows apparatus that can be used to investigate the effect of different concentrations of gases on the breathing rate of a locust. Carbon dioxide, oxygen and nitrogen from gas cylinders can be mixed together in different combinations before passing through the apparatus shown below.

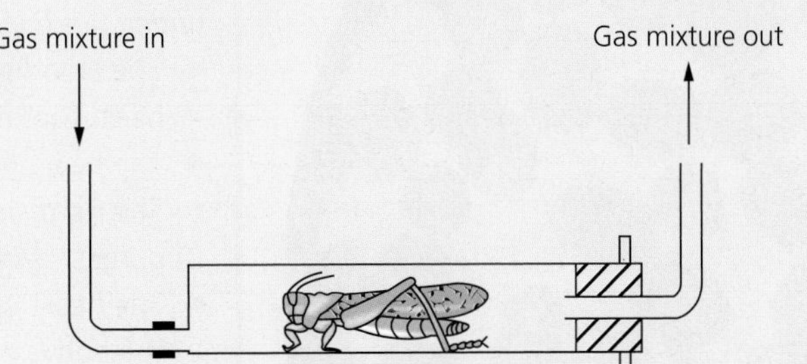

Is the breathing rate of locusts more sensitive to changes in oxygen or to changes in carbon dioxide?

Plan an investigation using the apparatus to investigate this question.

Chapter 8

Transport in animals

Prior knowledge

Before you start, make sure that you are confident in your knowledge and understanding of the following points.

- The blood system in mammals is a double circulatory system, with a heart that is divided vertically into a right and left side.
- The heart has four chambers: two atria and two ventricles.
- The right side of the heart pumps deoxygenated blood and the left side pumps oxygenated.
- The heart contracts to put pressure on the blood and force it around the body.
- Blood moves around the body in arteries, veins and capillaries, which have different structures.
- The blood in arteries has a higher pressure than the blood in veins.
- The blood transports oxygen around the body in the red blood cells, using a pigment called haemoglobin.
- Haemoglobin combines with oxygen in the lungs to form oxyhaemoglobin.
- Blood contains red and white blood cells and platelets in a liquid plasma.

Test yourself on prior knowledge

1 What are the main differences in structure between arteries and veins?
2 In what type of blood vessel does exchange of materials take place?
3 What happens to oxyhaemoglobin in tissues other than the lungs?
4 What are the functions of the red blood cells, white blood cells and platelets?

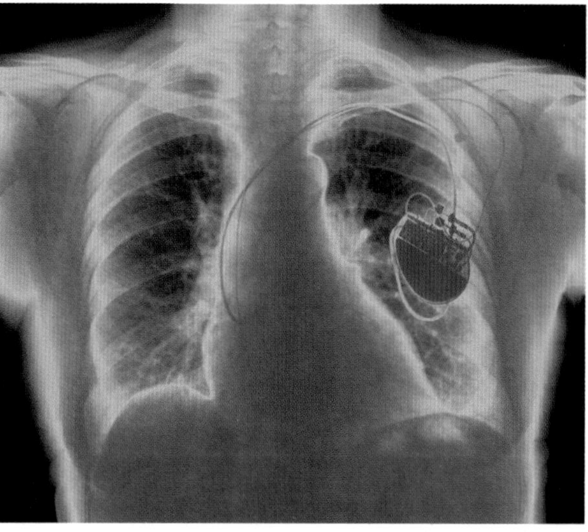

Figure 8.1 A coloured X-ray of a patient's chest, showing a fitted heart pacemaker. The pacemaker will detect any abnormalities in heart rate and restore normal heart function by direct electrical stimulation.

As animals have evolved to become larger, the increased distances between body parts and the development of specialised organs resulted in the need for transport systems for oxygen, nutrients and wastes. **Circulatory systems** have evolved in which blood is transported through blood vessels, propelled by contractions of a muscular heart. Within these systems, different types of blood vessel perform different functions, and red blood cells are adapted to transport oxygen effectively.

The heart is vital for the functioning of the circulatory system, and has features that ensure efficient one-way flow of blood through it and to and from the organs. In humans, a rough estimate is that the heart beats 2.4 billion times in an average lifetime. The importance of a healthy heart has led to the need for techniques to monitor its action and ensure that, where possible, faults are remedied.

Reasons for a transport system

Living cells need a supply of nutrients and oxygen from their environment, and they need to get rid of waste products.

Single-celled organisms and very small animals get their oxygen and nutrients by diffusion through their body surface. The distance the substances have to travel is small, so the slow speed of diffusion is not a problem. The **diffusion gradients** are favourable, and because these animals are small and have relatively low levels of activity, their demand for nutrients and oxygen is not great.

In contrast, multicellular animals have developed and increased in size, and this poses certain problems when obtaining essential materials and disposing of wastes.

Increasing transport distances

As an organism increases in size, the furthest distance that oxygen has to reach (if absorbed at the surface) gets larger. Diffusion is a slow process, and the time taken for oxygen to reach the innermost cells would be so long that the cells would die before they received it.

Tip

The change in surface area : volume ratio can be shown using cubes of different sizes (look back at Figure 7.4 on page 123). Animals have a more complex shape than a cube, but they have compact bodies and the use of a cube is valid as a model.

Surface area : volume ratio

Oxygen diffuses through the surface of small animals, so the surface area is a measure of the *supply* of oxygen. The number of cells (and therefore the volume of the organism) is an indication of the *demand* for oxygen. As discussed in Chapter 7, larger animals have a smaller surface area : volume ratio than small animals.

Increasing level of activity

Large-sized animals tend to be more active. Even fairly sluggish species have a higher level of activity than their smaller counterparts, because of the increased number of cells in their body. This means that the demand for oxygen and nutrients is increased, and the larger number of cells also produce more waste.

Larger animals face several problems. Diffusion through the body surface is not efficient enough to provide oxygen and nutrients and get rid of waste products, such as carbon dioxide. Diffusion is also too slow a process for the movement of substances between cells in the body and the surface. To solve these problems animals have mass flow transport systems that propel substances around the body. Each system is in one area of the body but gathers materials from, and delivers products to, the rest of the body via the circulatory system.

Key term

Mass flow The movement of materials. In biology, this means directed movement, with some sort of force causing it (e.g. the contraction of a heart).

Test yourself

1. Explain why a favourable diffusion gradient exists for oxygen in a small freshwater animal.
2. Why is diffusion alone an inadequate process for supplying the tissues of a large animal with nutrients?
3. Apart from considerations of distance and surface area : volume ratios, suggest another reason why terrestrial (land) animals would have difficulty using their body surface for gas exchange.

Single and double circulatory systems

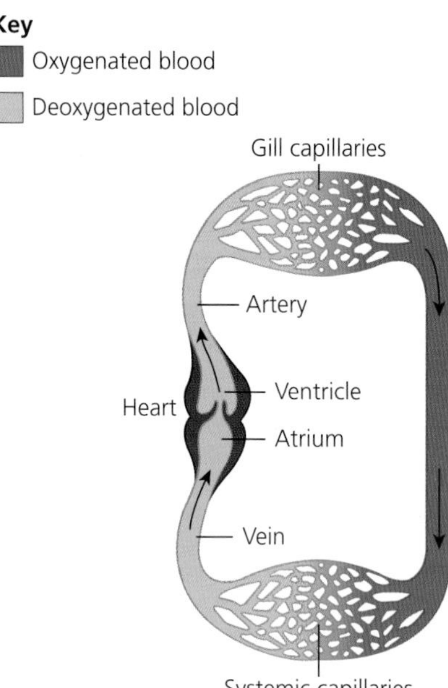

Figure 8.2 Single circulatory system in fish.

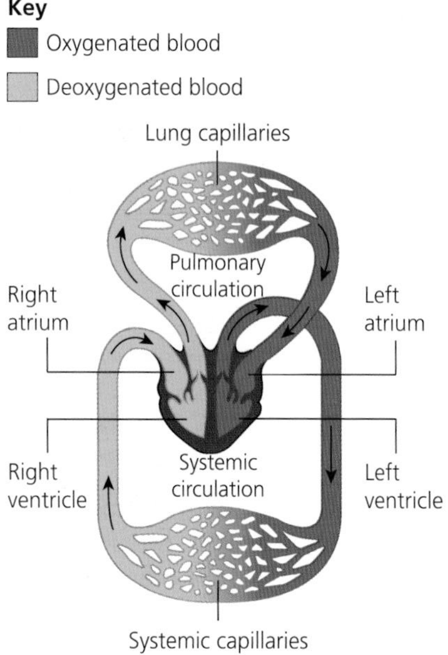

Figure 8.3 Double circulatory system in mammals.

Animals have circulatory systems to carry oxygen, nutrients and waste. In vertebrates, there are two main models of circulatory systems: **single circulatory systems** and **double circulatory systems**. The 'single' and 'double' in their name refers to the number of times the blood passes through the heart in one complete circuit of the body.

Single circulatory systems

In single circulatory systems, the blood passes through the heart once during each circuit of the body. Fish have single circulatory systems, and the structure of the system is shown in Figure 8.2.

Deoxygenated blood is pumped by the heart to the gills, where it absorbs oxygen and carbon dioxide is excreted. The **oxygenated blood** then travels to the other areas of the body, passing through the organs in capillaries, from which it delivers oxygen, before returning to the heart. Because there is only a single circulation, the heart has a single atrium and a single ventricle.

Double circulatory systems

Mammals have a double circulatory system, in which the blood passes through the heart twice during each circuit of the body. This is shown in Figure 8.3.

The heart is divided into a left and right half, the left side dealing with **oxygenated blood** and the right side dealing with **deoxygenated blood**. The blood leaving the right side of the heart travels to the lungs, where it is oxygenated. It then returns to the left side of the heart and is pumped to the other organs, before returning to the right side of the heart again.

It is important to understand that, in general, the oxygenated blood that travels through an organ goes directly back to the heart, not to another organ. The only exception is blood going to the gut, which then goes on to the liver via the hepatic portal vein before returning to the heart.

Advantages of a double circulation

In evolutionary terms, fish are older than mammals and it is reasonable to assume that the double circulatory system evolved from the single, and is therefore an improvement on it.

The advantages of a double circulation relate to what happens to the blood when it goes through a capillary network. As blood travels through capillaries, its pressure and speed both drop. In a double circulatory system, the blood only travels through one capillary network before returning to the heart, whereas in a single circulation it has to travel through two. The double circulatory system therefore maintains a higher blood pressure and a higher average speed of flow, which in turn helps to maintain steeper concentration gradients and makes exchange of materials more efficient.

Open and closed circulatory systems

Insects have very different circulatory systems from those of fish and mammals. Vertebrates have a closed circulatory system, in which the blood circulates entirely within blood vessels. Insects have an open circulatory system, in which the blood is not always confined within blood vessels, but also enters the body cavity (Figure 8.4).

Tip

The gas exchange system in insects is explained in Chapter 7.

Test yourself

4 What is the defining difference between a single circulatory system and a double circulatory system?
5 Why is a mammalian heart divided into right and left sides, when the heart of a fish is not?
6 Describe the advantages of a double circulatory system over a single circulatory system.
7 Suggest reasons why an open circulatory system would not be very efficient if it had to deliver oxygen to the tissues of insects.

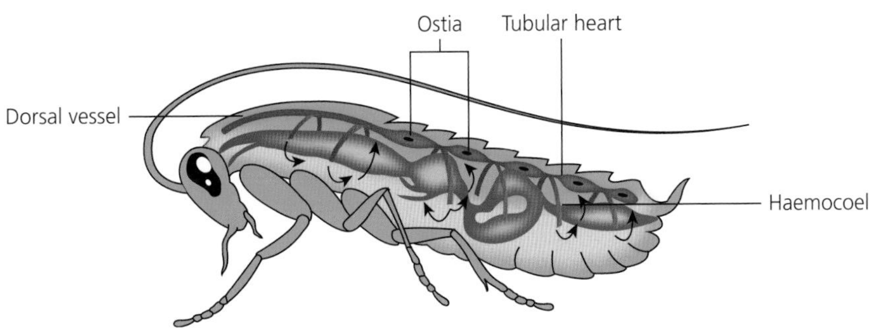

Figure 8.4 Circulatory system of an insect.

Insects (and other arthropods) have one main blood vessel, the **dorsal vessel**, which delivers blood (called **haemolymph** in insects) into the **haemocoel** (the body cavity) due to the pumping of a **tubular heart** in the abdomen. The haemolymph bathes the organs and re-enters the heart through openings with one-way valves called **ostia** when the heart relaxes.

Haemolymph is not specifically directed to any organ. Insects can function with a circulatory system that is so haphazard probably because it does not transport oxygen and carbon dioxide. In insects, oxygen is delivered directly to the tissues by a system of tubes (called tracheae) that connect to the outside. This reduces the importance of having a very efficient circulatory system.

The mammalian heart

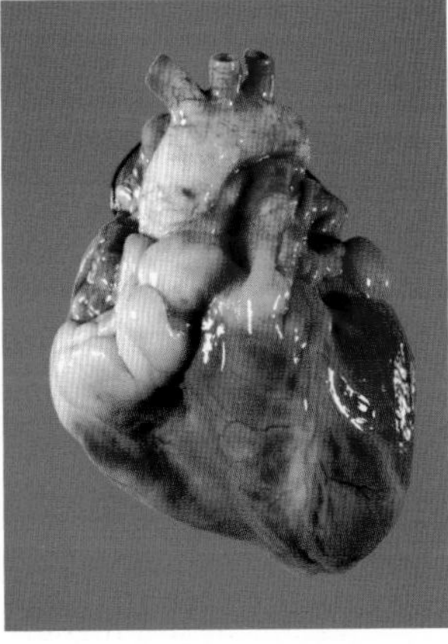

Figure 8.5 External structure of the mammalian heart.

The heart is a muscular organ that pumps blood around the body by means of regular contractions, the speed of which can be adjusted to suit the circumstances. This continual pumping requires a lot of energy and, although the heart is full of blood, the muscular walls are so thick that a separate blood supply to the outside of the heart is needed. The external structure of the heart is shown in Figure 8.5.

The blood vessel which divides into three above the heart is the **aorta** which carries oxygenated blood at high pressure to the body. The blood is pumped into the aorta by the left ventricle which is on the right hand side of the heart in this photo. Just below the aorta and to the right is the **pulmonary artery** which carries deoxygenated blood to the lungs. Blood is pumped into this artery by the right ventricle which is directly underneath it in this photograph. The outside of the heart is supplied with oxygenated blood by the **coronary artery**, which sends branches across the surface. This blood is returned in the **coronary vein**. Fatty tissue is often deposited on the heart, which can make the coronary vessels more difficult to see.

Internal structure of the heart

The internal structure of the mammalian heart is shown in Figure 8.6.

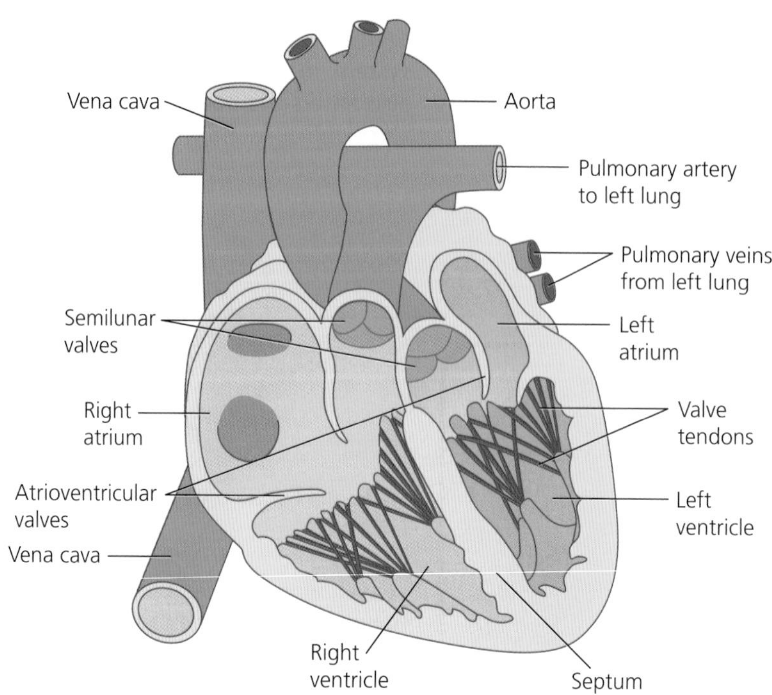

Figure 8.6 Internal structure of the mammalian heart.

The heart consists of four chambers, with two **atria** at the top and two **ventricles** at the bottom. The atria and ventricles are arranged more or less symmetrically, so that the heart is divided into right and left halves. Deoxygenated blood from the body is in the right half of the heart. This deoxygenated blood enters the atrium, passes through to the ventricle, and then leaves via the pulmonary artery to the lungs. The oxygenated blood that returns from the lungs enters the left atrium and then the left ventricle, before being pumped to all regions of the body (except the lungs).

The thickness of the muscular walls of the various chambers reflects their function. The thickest wall is in the left ventricle, which pumps the blood to everywhere in the body except the lungs. The right ventricle has a thinner wall because less muscle is required to pump blood through the pulmonary circuit. The pulmonary circuit has less resistance than the circuit from the left ventricle, which goes to the rest of the body. This is because the lung is a spongy organ, being full of air, and has fewer arterioles, which provide most of the resistance. The distance to the lungs is short compared with that to the rest of the body. The atria, which only pump blood to their adjacent ventricles, have the thinnest walls of all.

Valves are found between the atria and the ventricles (the **atrio-ventricular valves**) and at the base of the arteries leaving the heart. The latter are the **semi-lunar valves**, which are sometimes referred to as the pulmonary and aortic valves, to indicate which is which. The ventricles produce considerable pressure when they contract, so the flaps of the atrio-ventricular valves are attached to the ventricle walls by tendons that prevent the valves being blown inside out.

Activity

Heart dissection

Figure 8.7 shows photos taken during a heart dissection.

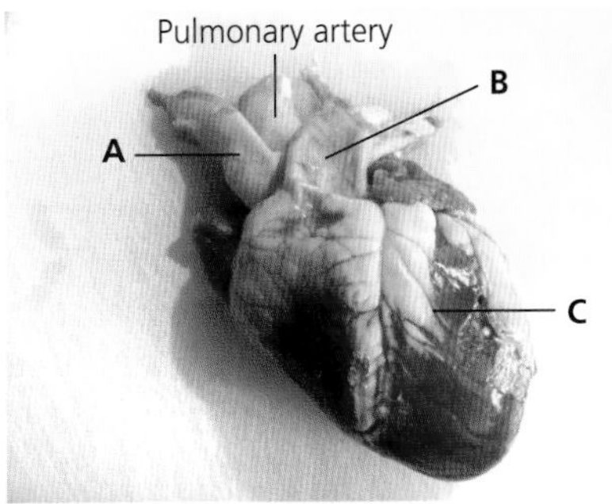

(a) External features

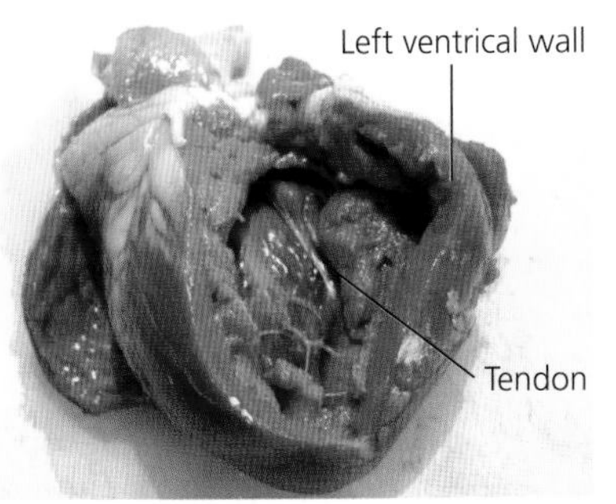

(b) Dissection of the left ventricle

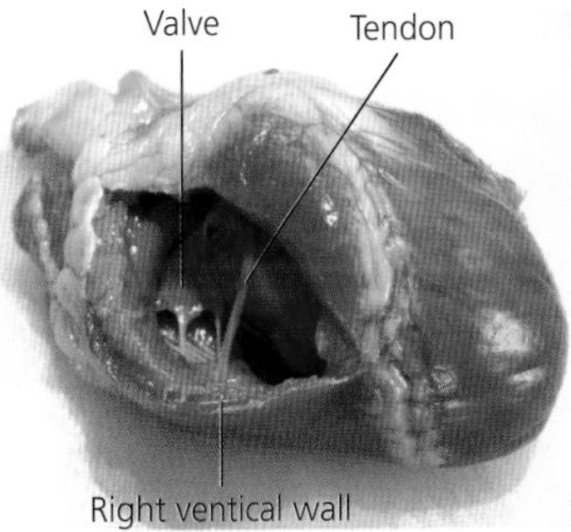

(c) Dissection of the right ventricle

Figure 8.7 Dissection of a mammalian heart.

1 On a copy of the photo of the external view, identify vein A and arteries B and C.
2 Artery C supplies the outside of the heart with oxygenated blood. Why is this necessary, when the heart chambers are filled with blood?
3 Copy and complete Table 8.1 below.

Table 8.1 Structure and function of parts of the transport system.

Structure	Function
	chamber that receives oxygenated blood from the lungs
aorta	
	pumps blood to the lungs
	brings blood back to the heart from the body
aortic semi-lunar valve	

4 Photos (b) and (c) show that the wall of the left ventricle is much thicker than that of the right ventricle. What is the reason for this?
5 Identify the valve shown in photo (c).
6 What is the purpose of the tendons shown in photos (b) and (c)?

The cardiac cycle

The heart beats and relaxes in a regular cycle. This is called the **cardiac cycle**. The atria contract, followed by the ventricles, and then there is a short period of relaxation before the cycle starts again. The valves in the heart maintain one-way flow in the necessary direction during the contractions of the heart chambers. The contraction phase is called **systole**, and the relaxation phase is called **diastole**.

When the heart muscles contract, they 'squeeze' the blood and so put pressure on it, raising the blood pressure. The cardiac cycle can be followed by monitoring the pressures in the different parts of the heart; this is shown in Figure 8.8 for one cycle. The data shown relates to the left side of the heart, but similar changes take place in the right side.

At the beginning of the cycle, both chambers are relaxed and the pressures are low. The changes in pressure then show that the following events are taking place.

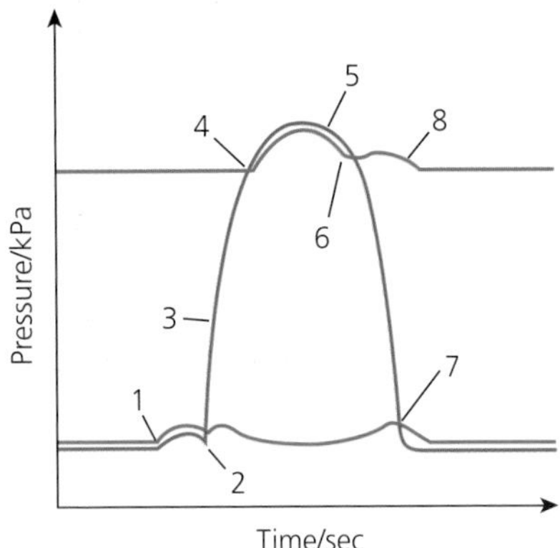

Figure 8.8 Changes of pressure in parts of the left side of the heart during the cardiac cycle.

Tip

In the cardiac cycle, remember that blood always tends to flow from an area of higher pressure to one of lower pressure. Whenever lines on the cardiac cycle graph cross, something happens to a valve. If blood *should* flow from the higher pressure area to the lower pressure area, the valve will open. If it should not, the valve will close.

1 The atrium starts to contract, and the pressure in it increases. This increase is relatively small, because the atrium has relatively thin muscular walls. There is also a small increase in the pressure of the ventricle, caused by the blood flowing into it.

2 The atrio-ventricular valve closes. The atrium relaxes and the ventricle starts to contract. When the pressure in the ventricle goes above that of the atrium, the atrio-ventricular valve is forced closed.

3 The ventricle contracts. Continued contraction of the ventricle, which has a thick muscular wall, increases the pressure in the ventricle considerably. The small rise in the atrial pressure is due to the closed atrio-ventricular valve bulging into the atrium and putting a little more pressure on the blood.

4 The semi-lunar valves open. When the pressure in the ventricle goes above that of the aorta, the semi-lunar valves open and blood flows into the aorta. This increases pressure in the aorta.

5 The ventricle starts to relax, and the pressure in it starts to decrease.

6 The semi-lunar valves close. When the pressure in the ventricle drops below that of the aorta, the semi-lunar valves close.

7 The atrio-ventricular valves open. When the pressure in the ventricle drops below that of the atrium, the atrio-ventricular valve opens. Blood has been flowing into the atrium from the body, causing a gradual rise in pressure that is released when the valve opens and blood can flow into the ventricle.

8 The aorta walls recoil. When blood flows into the aorta, its walls (which are highly elastic) stretch and expand. When the blood flow slows, the elastic tissue recoils and puts extra pressure on the blood. This gives the blood an extra 'push', which means that the blood still flows from the heart during the period when both the atrium and the ventricle are relaxed. Note that this is passive **elastic recoil**, and should not be referred to as 'contraction'. Because it is a passive process, it would also be incorrect to say that the aorta 'pumps' the blood.

Coordination of the heartbeat

Key terms

Autonomic nervous system The part of the nervous system that controls involuntary activity (e.g. heart rate, change in size of blood vessels, peristalsis). It consists of two parts: the parasympathetic and sympathetic nervous systems, which have opposite effects.

The heart is myogenic, meaning that the cardiac muscles can contract without stimulation from outside nerves. The stimulus for the heart muscles to contract is created entirely within the heart; the contraction needs no outside stimulus. Impulses from the parasympathetic and sympathetic nerves of the **autonomic nervous system** can adjust the heart rate to suit different circumstances. The autonomic nervous system is the part of the nervous system that controls involuntary activity (e.g. heart rate, change in size of blood vessels, digestion). The autonomic nervous system consists of two parts: the parasympathetic and sympathetic nervous systems, which have opposite effects. The parasympathetic nerves decrease the heart rate, and the sympathetic nerves increase it.

The heartbeat originates from a patch of specialised muscle tissue in the upper wall of the right atrium, known as the **sino-atrial node**. The sino-atrial node is often referred to as the heart's pacemaker. The nodal tissue can generate electrical impulses when it contracts. The nerve impulses set off a wave of muscular contraction across both atria. Non-conductive tissue between the atria and ventricles prevents the spread of the impulse

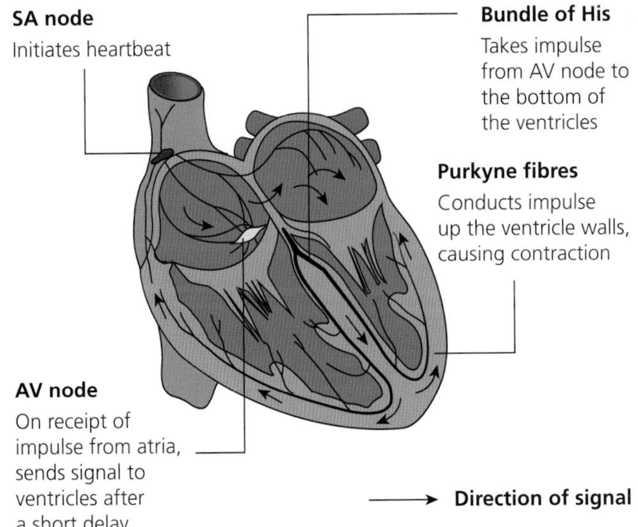
SA node
Initiates heartbeat
Bundle of His
Takes impulse from AV node to the bottom of the ventricles
Purkyne fibres
Conducts impulse up the ventricle walls, causing contraction
AV node
On receipt of impulse from atria, sends signal to ventricles after a short delay
Direction of signal

Figure 8.9 Coordination of the heartbeat.

to the ventricles, except at the **atrio-ventricular node**. This node delays the signal for a short time (which allows the atria to complete their contraction), and then sends an impulse into the ventricles. Because blood has to be forced out of vessels at the top of the heart, it is best if the contraction of the ventricles starts at the bottom and moves upwards. To achieve this, the impulse travels from the atrio-ventricular node down specialised muscle fibres known as the **bundle of His** to the bottom of the ventricles, and then back up via another set of muscle fibres, the **Purkyne fibres** (sometimes known as the Purkinje fibres). The impulse travelling along the Purkyne fibres causes contraction as it moves up the walls of the ventricles.

The coordination of the heartbeat is shown in Figure 8.9.

Detecting heart activity

Because electrical activity in the muscle cells causes the heartbeat, this electricity can be monitored to detect heart activity and identify certain problems. The process is called **electrocardiography** (usually abbreviated to ECG) and it involves placing electrodes on the skin that can detect electrical signals to produce an **electrocardiogram** (usually abbreviated to ECG) (see Figure 8.10). An electrocardiogram of a healthy heart is shown in Figure 8.11.

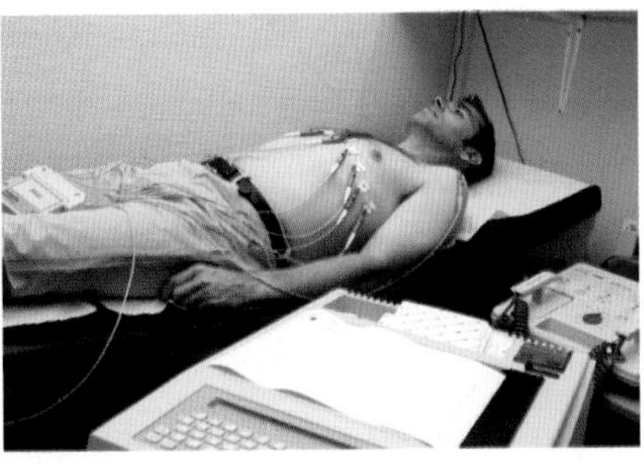

Figure 8.10 A patient having an electrocardiogram.

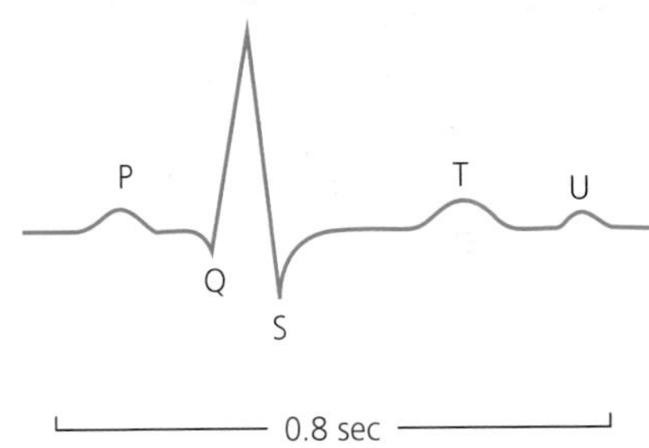
R
P
T
U
Q
S
0.8 sec

Figure 8.11 An electrocardiogram showing one beat of a healthy heart.

The electrocardiogram shows a number of distinctive waves caused by heart activity:

- The **P wave** is caused by **depolarisation** of the atria, resulting in their contraction.
- The **QRS complex** results from ventricular depolarisation/contraction.
- The **T wave** is caused by **repolarisation** of the ventricles (causing relaxation).
- The cause of the **U wave** is uncertain.

Tip

Polarisation is a state in which something has different properties in different directions. In the example of nerves and muscles, polarisation refers to different electrical charges on either side of membranes. **Depolarisation** is the balancing or reversal of these charges, and **repolarisation** is their restoration.

Heart problems and ECGs

If a doctor suspects that a person has a heart problem, the doctor will send the person to have an electrocardiogram (ECG). Some examples of ECGs are shown in Figure 8.12.

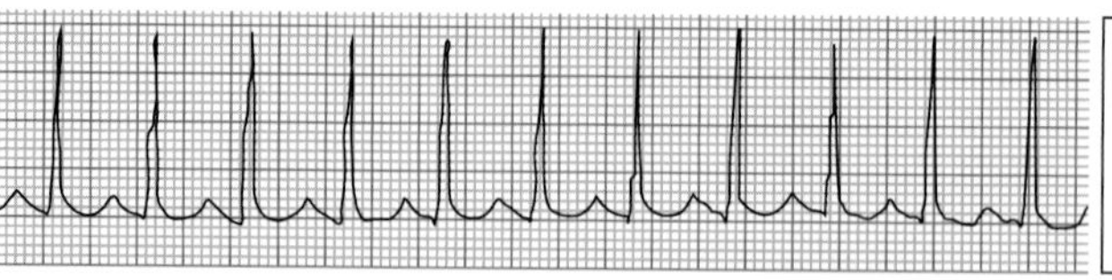

Tachycardia: The heart is beating too rapidly (resting heart rate > 100 bpm) – the peaks are too close together.

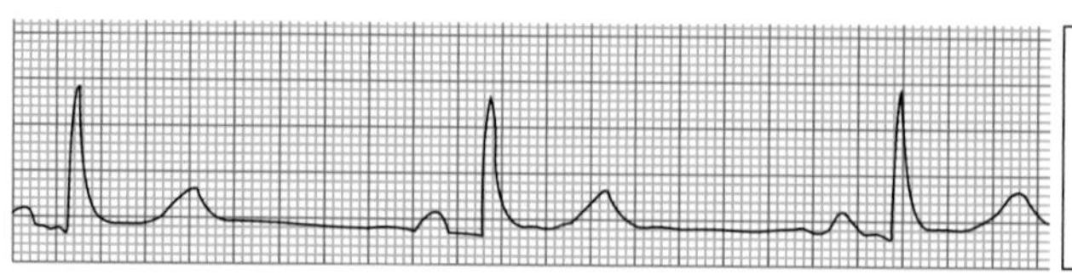

Bradycardia: The heart is beating too slowly (less than 60 bpm) – the peaks are too far apart.

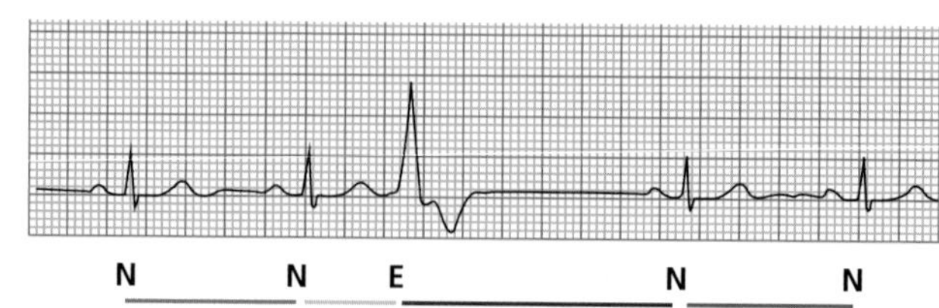

Ectopic heart beat: The heart beats too early (E), followed by a pause. This is common and usually requires no treatment. N = normal heart beat.

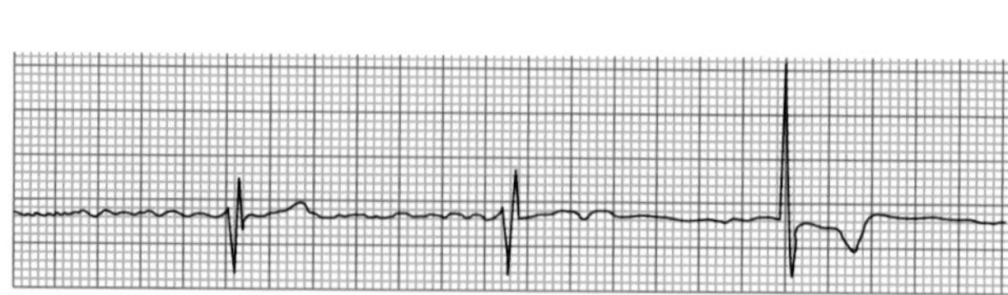

Fibrillation: The heartbeat is irregular and has lost its rhythm. Severe fibrillation is very dangerous and can result in death.

Figure 8.12 ECGs of faulty heartbeats.

Test yourself

8 Why is the wall of the left ventricle thicker than that of the right?

9 In patients who need heart valve replacements, it is much more common for the left atrio-ventricular valve to need replacing than the right. Suggest a reason for this.

10 In the atria, the wave of contraction spreads directly through the muscle, whereas in the ventricles it is channelled down the bundle of His and up the Purkyne fibres. Suggest a reason for this difference.

11 In an ECG, suggest why the peak during the QRS complex is higher than the P wave.

Blood vessels

As the blood travels around the body from the heart, through the other organs and back again, it passes through three different types of vessels: **arteries**, **veins** and **capillaries**. The smaller branches of the arteries and veins are given the names **arterioles** and **venules**. The differences in structure in the three types of blood vessel are related to their different functions.

The arteries carry blood away from the heart. This blood is under high pressure, which puts a strain on the blood vessel walls. As the blood moves through the arteries, the pressure drops; by the time it reaches the capillaries, it has decreased considerably. No exchange of materials takes place between the blood and the tissues as the blood travels along the arteries and arterioles, but in the extensive network of capillaries, which takes blood very close to the individual cells, a variety of materials are exchanged. Oxygen and nutrients leave the blood, and carbon dioxide and nitrogenous wastes enter.

When the blood then enters the veins to return to the heart, the exchange of gases, nutrients and waste stops. The movement from the narrow capillaries to the much wider veins speeds up the flow of blood, but the pressure remains low. Despite this low pressure, the veins must return the blood, often against the force of gravity, to the heart. Each of the blood vessels therefore has a specific task to carry out, with its own problems, and its structure is adapted to achieve that task and overcome those problems.

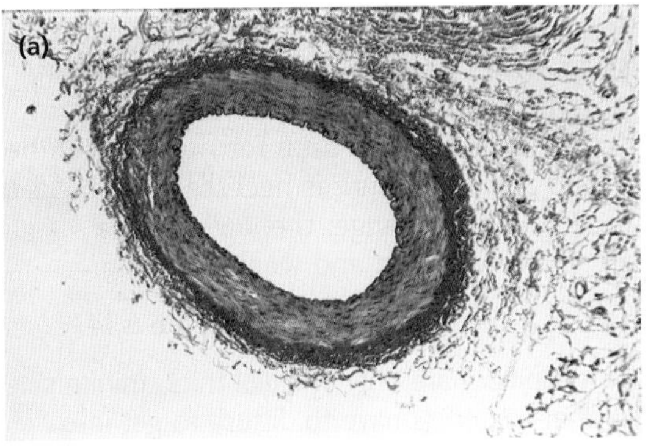

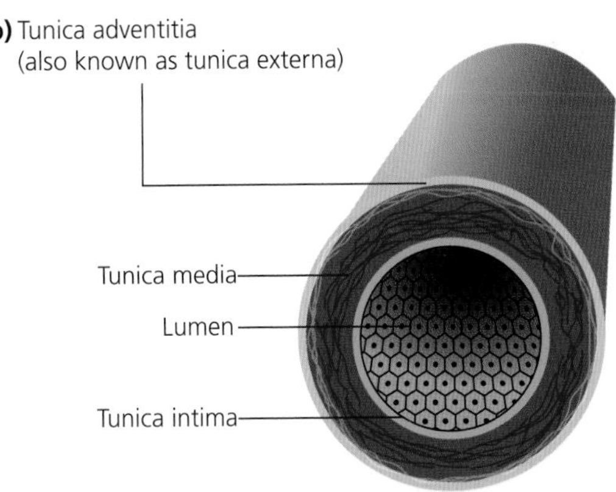

Figure 8.13 Structure of an artery, showing the three-layered wall. (a) A photomicrograph (×24) and (b) a diagram.

Arteries

The function of the arteries is to carry blood at high pressure to the tissues. The walls of the arteries have a three-layered structure, shown in Figure 8.13.

The **tunica intima** is the innermost layer, lining the whole of the blood system, and consists of an **endothelium**, which is one cell thick and lines the **lumen**, plus a network of connective tissue and a layer of elastic fibres. The endothelium has a very smooth surface, which reduces friction with the blood and so allows it to flow freely through the vessel.

The **tunica media** is a thick layer – much thicker in arteries than in veins. The tunica media consists of smooth muscle cells and a thick layer of **elastic tissue**. These two tissues perform different functions. The elastic layer allows the blood vessel to stretch, which helps it to maintain the pressure in the arteries. As the elastic tissue stretches, it lowers the pressure slightly, then as it recoils it increases the pressure again, so evening out fluctuations. The muscle tissue strengthens the artery and so resists the high blood pressure in the vessel, and can contract to reduce the flow of blood. The body constantly directs more blood to where it is most needed, while restricting flow to areas that are less important at that moment. The contraction and relaxation of the muscles in the arterioles brings about this control of flow.

The **tunica adventitia** covers the outside of the artery, and mainly consists of the protein **collagen**. This is a tough material that prevents the blood vessel from over-stretching, which could result in damage to the artery wall.

Arterioles

As the blood moves further from the heart, the arteries branch and get smaller. The smallest arteries, which lead into the capillaries, are called **arterioles**. The arterioles are very important in regulating blood flow. Their walls contain a high proportion of smooth muscle, and when this contracts it narrows the arteriole and restricts the flow of blood. The autonomic nervous system controls the contraction of these muscles, although they will also respond to local factors such as pH and levels of oxygen and carbon dioxide.

At the junction of the arteriole and the capillaries, there are rings of smooth muscle known as pre-capillary sphincters. Contraction of these sphincters will prevent blood flowing into the capillary bed, and so the blood can completely bypass a given capillary bed if necessary.

Capillaries

The role of capillaries is to exchange materials with the cells. It is important that the capillaries penetrate the tissues and reach areas close to all the cells; therefore, capillaries are numerous and have a small cross section (i.e. a small lumen). The small lumen causes friction with the blood, which slows it down and lowers its pressure. Both of these are important: the slow flow allows a longer time for materials to be exchanged, and a low pressure is necessary because, in order to carry out exchange, the walls of the capillary are extremely thin and weak.

The structure of a capillary is shown in Figure 8.14.

The walls of a capillary are only one cell thick, minimising the distance that materials have to diffuse to get into and out of the vessel. White blood cells are capable of squeezing through the intercellular junctions, which allows them to enter the tissues to combat infections.

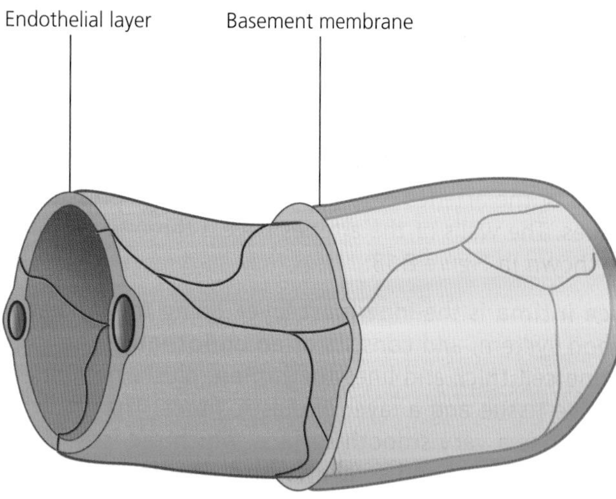

Figure 8.14 Structure of a capillary.

Venules

Blood flows out of the capillaries into the smallest veins, called **venules**. The smallest of these consist only of an endothelium, but the larger ones have a structure which is similar to veins (see Figure 8.15), but scaled down and with few or no elastic fibres.

Veins

The function of the veins is to return the blood to the heart. This blood is under low pressure, and the pulse that was present in the arteries has now disappeared. The veins have the same three layers that are present in arteries, but the tunica media (with its muscular and elastic tissue) is much thinner. The structure of a vein is shown in Figure 8.15.

The differences seen between the structure of a vein and that of an artery relate to the vein's function:

- The large lumen compensates for the absence of a pulse. It is important that the blood returning to the heart in the veins travels as fast as the blood leaving the heart in the arteries (otherwise the heart would 'run out' of blood). Although the rate of flow is slower in the veins than in the arteries, the large lumen means that the volume of blood delivered per unit of time is the same. The large lumen also means that much of the blood is not in contact with the walls, so reducing friction.
- The blood is at low pressure, and so there is no need for a thick muscular layer to resist a high pressure, and elastic tissue is reduced because the vessels are not subjected to much stretching force (because of the low pressure) and the flow of blood is even with no pulse.
- The tunica adventitia is *relatively* thicker in veins than in arteries, because of the thinner tunica media. In actual thickness, however, it is much the same in veins and arteries.
- Veins have **valves** along their length, which ensure one-way flow towards the heart. In arteries, valves are not necessary, because the pulse pushes the blood in the right direction.

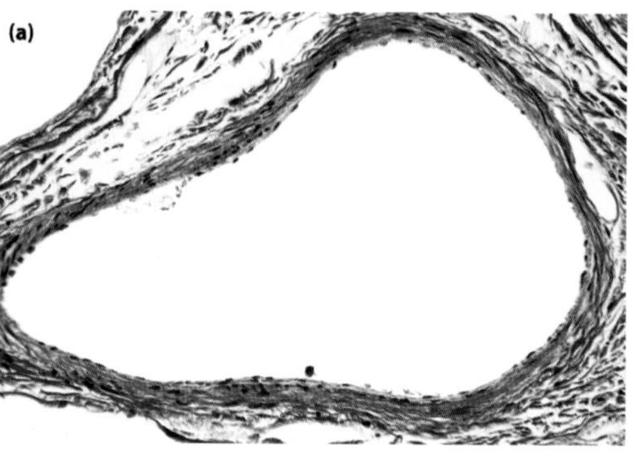

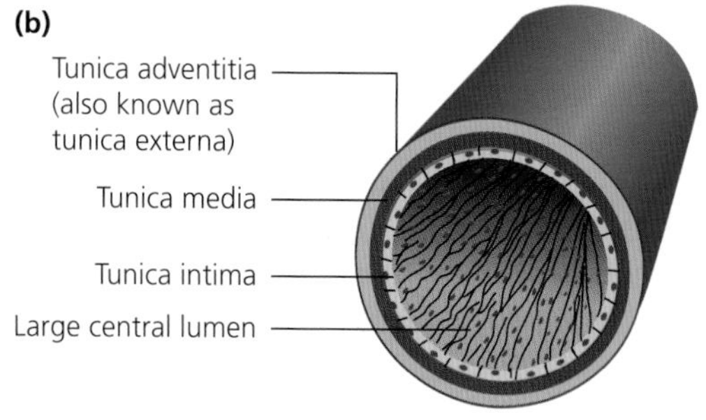

Figure 8.15 Structure of a vein. (a) A photomicrograph (×20) and (b) a diagram.

- The force needed to push the blood along the veins is mainly produced when the muscles around the vein contract during their normal activity. The thin vein wall means that this force is not resisted by the vessel.

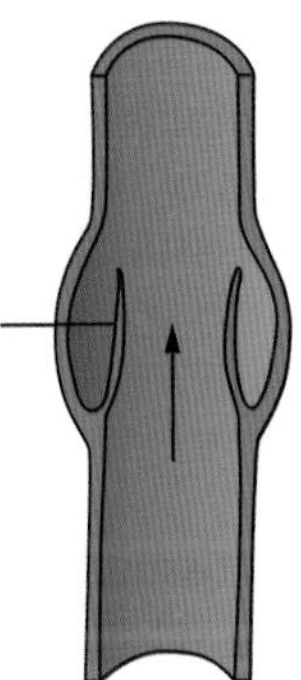

Blood flow in the correct direction pushes the pockets of the valves flat against the vein wall and blood flows through

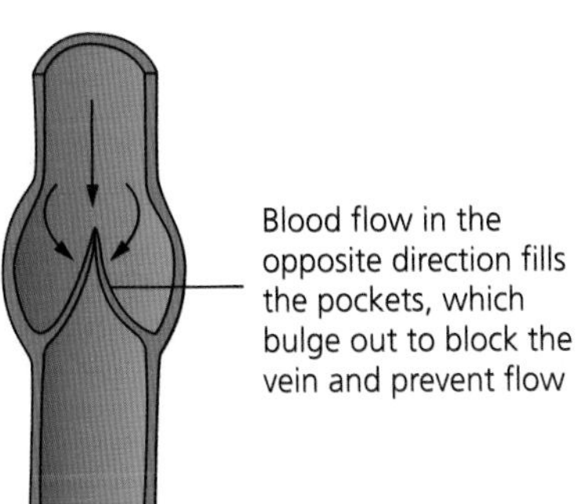

Blood flow in the opposite direction fills the pockets, which bulge out to block the vein and prevent flow

Figure 8.16

Summary of differences between arteries and veins

Table 8.2 shows some of the main differences between arteries and veins.

Table 8.2

Feature	Arteries	Veins
Direction of flow	away from heart	towards heart
Blood pressure	high	low
Pulse	present	absent
Thickness of tunica media	thick	thin
Size of lumen	narrow	wide
Valves	absent	present

Tip

Note that direction of flow, blood pressure and the presence of a pulse are not **structural** differences.

Blood, tissue fluid and lymph

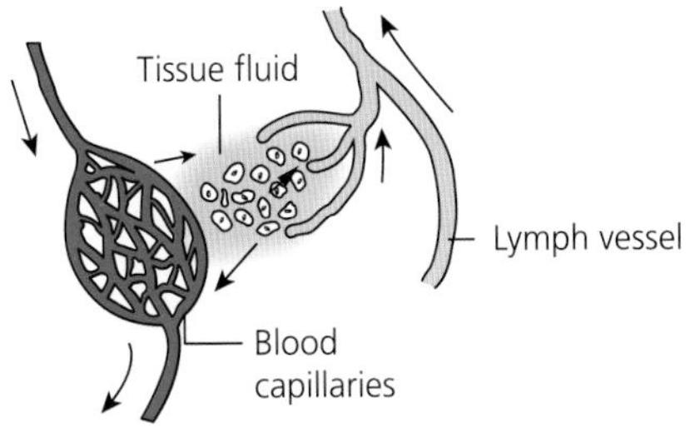

Figure 8.17 Relationship of blood, tissue fluid and lymph.

All the substances needed by the cells must be in solution so that they can be absorbed through membranes. Therefore, at all stages, a liquid medium is needed. The blood carries chemicals around the body, but between the blood vessels and the cells the transport is done by **tissue fluid**, which bathes and surrounds the cells. Excess tissue fluid is returned to the blood by a third fluid, **lymph**, which also has a role in immunity. The relationship between blood, tissue fluid and lymph is shown in Figure 8.17.

Blood

The components of blood are as follows:

- **Red blood cells (erythrocytes)**: These are responsible for the transport of oxygen and also play a role in the transport of carbon dioxide. These functions are carried out by the red pigment **haemoglobin** in the cells.
- **White blood cells (leukocytes)**: Different types of white blood cell play different roles in the immune system.
- **Platelets**: These are cell fragments that are important in the clotting process.
- **Plasma**: The liquid medium of the blood, which transports dissolved substances (including amino acids, sugars, fatty acids, hormones and other proteins, clotting factors and carbon dioxide).

The cellular components of blood are shown in Figure 8.18.

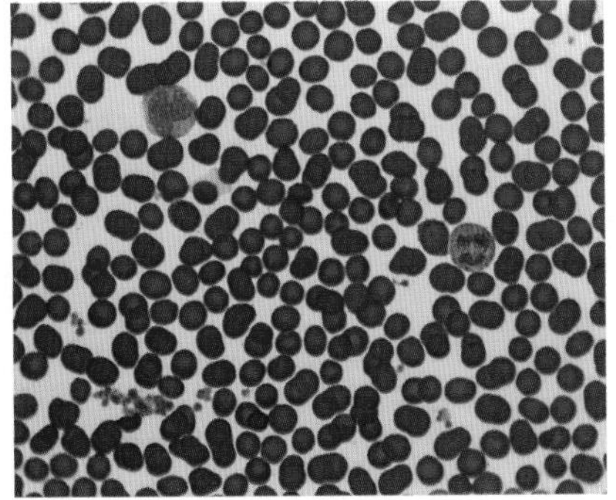

Figure 8.18 Components of blood seen in a blood smear (×300). You can see the numerous red blood cells, a few white blood cells with nuclei stained purple and some platelets.

Tissue fluid

Tissue fluid is formed from plasma and contains many of the solutes that are found in the plasma, with the exception of the larger protein molecules, which cannot escape from the capillaries. Tissue fluid contains some white blood cells, but no red cells or platelets.

Lymph

Of the fluid leaving the blood and forming tissue fluid, around 90% is returned to the capillaries. The remainder is returned to the bloodstream as lymph, via the **lymphatic system**. Lymph is therefore very similar to tissue fluid, but contains more white blood cells. These cells collect in large numbers in the lymph nodes to fight infections, and so tend to be more common in lymph as a whole compared with in tissue fluid.

Lymphatic system

The lymphatic system is a network of vessels and organs that connect to the circulatory system. Swellings called lymph nodes are found at intervals throughout the system. The tonsils, spleen and thymus gland are part of the lymphatic system.

A comparison of the contents of blood, tissue fluid and lymph is shown in Table 8.3.

Table 8.3

Component	Blood	Tissue fluid	Lymph
Red blood cells	yes	no	no
White blood cells	many	few	many
Proteins	many	few	more than in tissue fluid (antibodies added)
Dissolved solutes	yes	yes	yes

As the tissue fluid is formed from blood that has passed through gaps in the capillary walls, it has effectively been filtered, and so larger molecules and structures do not get through. The cells do not go through, except for some white blood cells which actively 'squeeze' themselves through (particularly when tissue is damaged and/or infected). Lymph is effectively tissue fluid on its way back to the blood system, but along the lymphatic system there are lymph nodes that contain lymphocytes (B and T cells) and plasma cells. As the lymph flows through the lymph nodes, some of these cells, together with antibodies produced by the plasma cells are added to the fluid, which explains the increased numbers of white cells and proteins found in the lymph.

Formation of tissue fluid

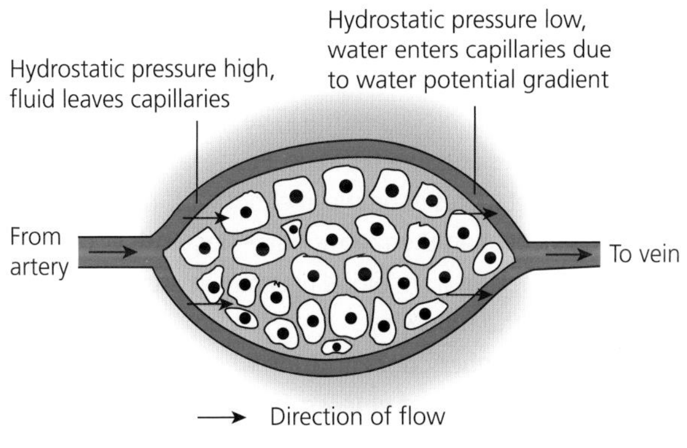
Hydrostatic pressure high, fluid leaves capillaries
Hydrostatic pressure low, water enters capillaries due to water potential gradient
From artery
To vein
Direction of flow

Figure 8.19 Formation and recovery of tissue fluid.

The capillaries are very leaky: about 20 litres of fluid leaves the blood and enters the tissues every day. This fluid must be returned or else the entire blood volume would be lost within an hour or two. The formation and recovery of tissue fluid is shown in Figure 8.19.

Two factors control the movement of fluid to and from the blood capillaries: the **hydrostatic pressure** of the blood (i.e. blood pressure), which tends to force fluid out, and an **oncotic pressure** gradient, which tends to draw water in by osmosis. The water potential of the blood is lower than that of the tissue fluid. This is mainly because of the presence of plasma proteins, most of which do not enter the tissue fluid.

Key terms

Hydrostatic pressure The pressure exerted by a fluid, for example the blood.

Oncotic pressure A form of osmotic pressure exerted by proteins, notably albumin, in a blood vessel's plasma, that usually tends to pull water into the circulatory system.

The hydrostatic pressure of the blood drops considerably as the blood passes through the capillaries. When blood enters the capillary network, its hydrostatic pressure is relatively high; this overcomes the water potential gradient to force fluid through the capillary walls. No blood cells or large protein molecules leave, because they are too large to get through the 'gaps' in the capillary walls. At the venous end of the capillary bed, the water potential gradient is largely unchanged, and the drop in hydrostatic pressure means that hydrostatic pressure is now less that the pressure due to the water potential gradient. As a result, water re-enters the blood. About 90% of the fluid lost at the arterial end of the capillary bed is recovered at the venous end; the remaining 10% is collected by lymph vessels, which eventually return it to the blood.

Example

1 Explain why multicellular animals such as mammals and fish need a transport system.
2 Explain why the walls of arteries are thicker than the walls of veins.
3 Describe the role of the pre-capillary sphincter muscle shown in Figure 8.20.
4 Describe how tissue fluid is formed as blood flows through capillaries.
5 Describe three ways in which the composition of plasma differs from the composition of tissue fluid.
6 Name the fluid in the vessel A in Figure 8.20.

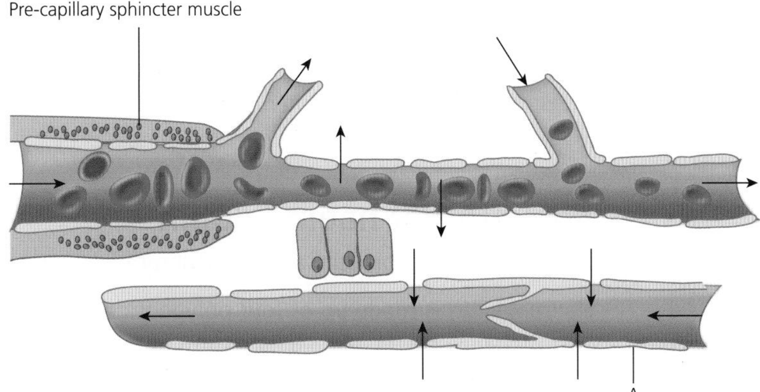
Pre-capillary sphincter muscle
A

Figure 8.20 The blood vessels supplying a tissue.

Answers

1 These are large animals with small surface area : volume ratios, so there are long distances between sites of exchange with the environment and cells, tissues or organs (in the body) for the transport of oxygen, carbon dioxide, water and nutrients. The distances involved are too large for diffusion to be effective. Animals are also active, so need a good supply of oxygen.

2 Arteries have more elastic fibres or collagen fibres, and more smooth muscle to withstand high(er) blood pressure. These recoil to propel blood along, so maintaining the blood pressure, and blood arrives at organs at only a slightly lower pressure than when it left the heart.

3 The pre-capillary sphincter muscle contracts to control blood flowing through capillaries.

4 The hydrostatic pressure of blood is greater than the hydrostatic pressure of tissue fluid. (You might also say that the hydrostatic pressure of blood is greater than the oncotic pressure of blood plasma.) The higher pressure in the blood causes pressure filtration. Fluid leaves the blood plasma through pores in between endothelial cells. Water, ions and glucose can then move out. Pinocytosis occurs across endothelium from blood plasma to tissue fluid.

5 Any three of the following.
Plasma has a: higher concentration of plasma proteins; higher concentration of glucose; higher concentration of fatty acids and glycerol; higher concentration of amino acids; lower water potential; lower carbon dioxide concentration; higher oxygen concentration.
Tissue fluid has a higher concentration of any substance secreted by the cells, for example hormones such as insulin.

6 Vessel A contains lymph.

Test yourself

12 Explain why arteries have a much thicker layer of muscle in their walls than veins have.

13 Explain why veins have a relatively larger lumen than arteries have.

14 State how lymph differs from tissue fluid.

15 Explain why fluid leaves the capillaries at the arterial end of the capillary bed, but return to the capillaries at the venous end.

Haemoglobin and the transport of oxygen and carbon dioxide

The solubility of oxygen in water is quite low, and dissolved oxygen in the plasma would never be enough to supply the body's needs. Red blood cells have a red pigment, haemoglobin, which readily absorbs oxygen and transports it around the body. Haemoglobin also has a role in the transport of carbon dioxide.

Structure of haemoglobin

Haemoglobin is a protein, and its structure is shown in Figure 8.21.

Haemoglobin is made up of four polypeptide chains: two α chains and two β chains. Each chain has an iron-containing **haem group** associated with it, and these haem groups allow the molecule to bind oxygen (one oxygen molecule per haem group, so a total of four in all). Haemoglobin exists in two forms: **oxyhaemoglobin**, when oxygen is bound to it, and **deoxyhaemoglobin**, when it is not.

Tip

Look back at Chapter 3 to remind yourself about the structure of haemoglobin and quaternary structure of proteins.

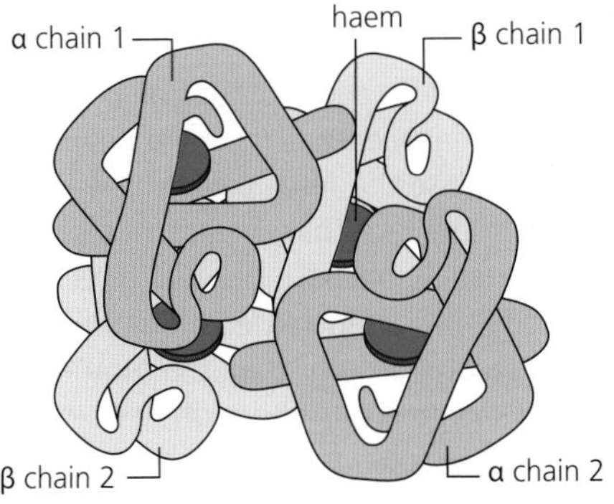

Figure 8.21 Structure of haemoglobin molecule.

Key terms

Affinity A natural attraction. In evolution, it can also mean a relationship or resemblance between species that suggests a common origin.

Partial pressure The pressure exerted by one component of a mixture of gases, usually expressed in mm Hg or kPa.

Transport of oxygen

The **affinity** of haemoglobin increases with increasing **partial pressure** of oxygen in the surroundings. In addition, when one haem group absorbs oxygen, it alters the structure of the haemoglobin molecule so that it becomes easier to absorb a second molecule. This increase in affinity continues when the second and third molecules of oxygen are absorbed.

The changing affinity for oxygen shown by haemoglobin at different partial pressures of oxygen is expressed by an oxygen dissociation curve. The dissociation curve for adult human haemoglobin is shown in Figure 8.22.

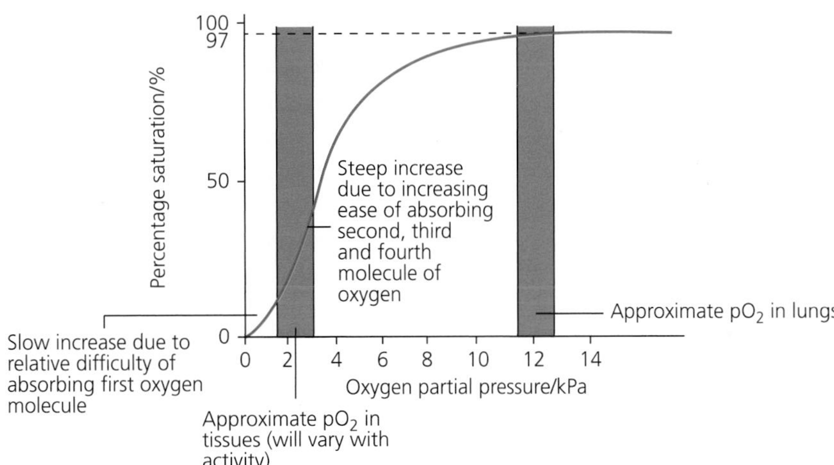

Figure 8.22 Oxygen dissociation curve of adult haemoglobin.

The properties of haemoglobin shown by the dissociation curve make it perfectly suited to the body's needs. The pigment needs a high affinity for oxygen in order to absorb as much as possible from the air, yet a high affinity also makes haemoglobin unlikely to give up that oxygen. The fact that the affinity varies with the partial pressure of oxygen gets around that problem. In the lungs, the affinity is high, so oxygen uptake is maximised. In the tissues, however, where the partial pressure of oxygen is low and there is a need for the gas, the affinity is low, causing the oxyhaemoglobin to disassociate.

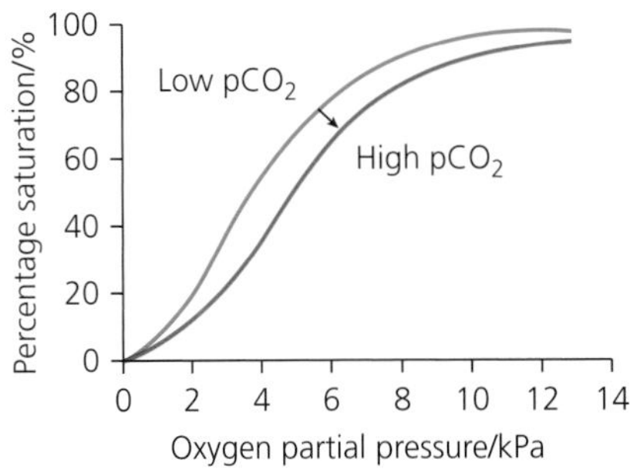

Figure 8.23 Effect of high carbon dioxide concentrations on the oxygen dissociation curve of haemoglobin – the Bohr effect.

Bohr effect

The functionality of haemoglobin is further enhanced by the fact that carbon dioxide concentrations modify the dissociation curve. In actively respiring tissues, which have the greatest demand for oxygen, carbon dioxide is produced. Increasing levels of carbon dioxide lower the affinity of haemoglobin for oxygen, so that even more oxygen is released. This is known as the **Bohr effect** (after its discoverer, Christian Bohr) and is shown in Figure 8.23.

The Bohr effect is caused by both the carbon dioxide and by the decrease in pH that it creates (see page 155).

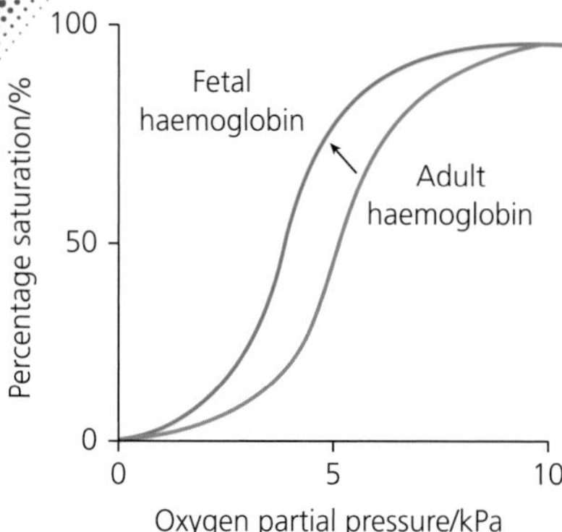

Figure 8.24 Oxygen dissociation curve of fetal and adult haemoglobin.

Fetal haemoglobin

In mammals, the developing fetus gets its oxygen from the mother's blood via the placenta. The partial pressure of oxygen in the placenta is nowhere near as high as in the lungs of an adult, which could cause the fetus problems with absorbing sufficient oxygen. This is overcome by the fact that the haemoglobin in the fetus is slightly different from that of the adult, because the two beta chains are replaced by two gamma chains, and it has an increased affinity for oxygen. Because the fetal haemoglobin has a greater affinity than the mother's haemoglobin, the fetal blood is able to extract oxygen from the mother's blood. In the first few months after birth, the fetal haemoglobin is replaced by the adult form. The dissociation curve for fetal haemoglobin is shown in Figure 8.24.

Activity

Investigating physiological changes at high altitude

A study was carried out on 50 soldiers who were stationed at high altitudes for two years. At high altitude, the partial pressure of oxygen is lower than at sea level. At regular intervals before, during and after the soldiers' stay, blood samples were taken. The red blood cells were counted and the concentration of haemoglobin in the blood was measured.

The mean results are shown in Figure 8.25. Note that the scale on the *x* axis is not linear, in order to present the data on the page.

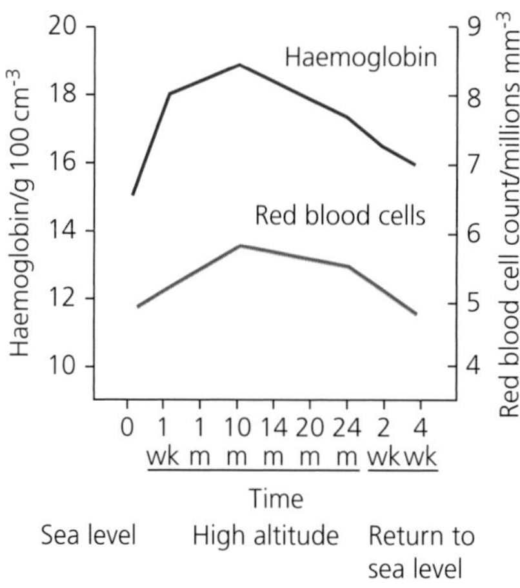

Figure 8.25

Additional measurements showed changes in pulse rate, breathing rate and blood pressure, all of which increased at high altitude.

1 Describe the effect of high altitude on haemoglobin concentration and red blood cell numbers, as shown in the graph.
2 The volunteers were all male and had a good level of fitness. Their height, mass and age could not be precisely controlled. Suggest why the inability to control these factors does not significantly affect confidence in the conclusions.
3 Suggest how the changes shown in the graph would benefit people living at high altitudes.
4 Suggest a reason for the trends seen in the graphs after 10 months at high altitude.

Tip

This activity is using this knowledge in a different context; you don't need to know about the effects of altitude.

Carbon dioxide transport

Carbon dioxide is transported in the blood in three forms:

- Dissolved carbon dioxide: Carbon dioxide is much more soluble in water than oxygen is, and about 5% of the total carbon dioxide transported is dissolved in the plasma.
- Hydrogen carbonate ions (HCO_3^-): This is the main form of transport (70–90%). Hydrogen carbonate ions are formed in the red blood cells but transported in the plasma. This process is shown in Figure 8.26.
- Bound to haemoglobin: Carbon dioxide is bound to haemoglobin in the form of carbaminohaemoglobin, in which carbon dioxide combines with the terminal amino group of the haemoglobin molecule.

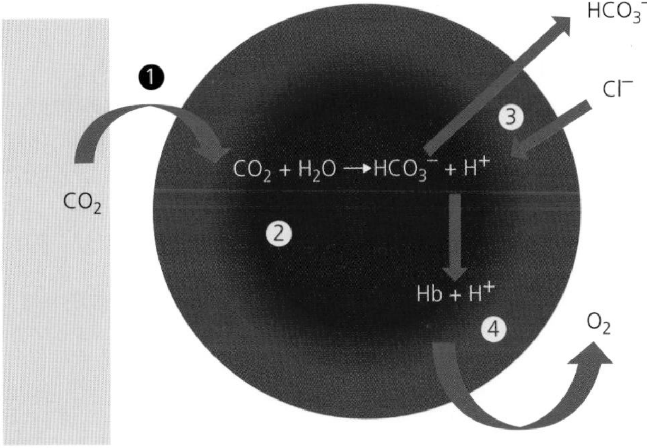

Figure 8.26 The transportation of carbon dioxide by red blood cells.

1 Carbon dioxide diffuses into the red blood cell.

2 The carbon dioxide is converted to carbonic acid, catalysed by the enzyme **carbonic anhydrase**. The carbonic acid dissociates to form hydrogen (H^+) and hydrogen carbonate (HCO_3^-) ions.

3 The HCO_3^- ions diffuse out of the red blood cell down a concentration gradient and combine with sodium ions in the plasma to form sodium hydrogen carbonate. In order to maintain electrochemical neutrality in the cell, the negative hydrogen carbonate ions are replaced with chloride ions (Cl^-), which diffuse in from the plasma. The exchange of hydrogen carbonate ions with chloride ions is referred to as the chloride shift.

4 The H^+ ions combine with haemoglobin (Hb) to form **haemoglobinic acid**. This enhances oxygen release (the Bohr effect, described above) and also becomes the substrate for carbamino formation, which binds carbon dioxide to haemoglobin. It also has a buffering effect, preventing the H^+ from causing a decrease in pH in the cell.

Test yourself

16 What level of protein structure is shown by haemoglobin? (You will need information from Chapter 3 to help you answer this.)

17 What part of the haemoglobin molecule combines with oxygen?

18 Describe how the dissociation of haemoglobin suits its function of oxygen transport.

19 What is the Bohr effect?

20 Why do chloride ions need to go into red blood cells when HCO_3^- ions leave?

Exam practice questions

1 Which describes the circulatory system of insects?

A closed
B double
C open
D single *(1)*

2 The drawing shows a cross section of an artery.

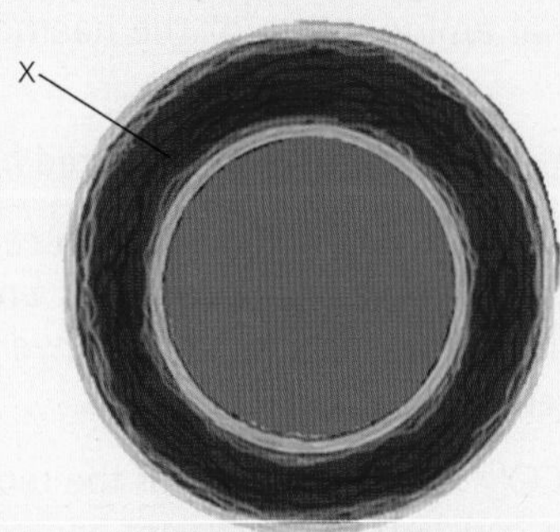

What type of fibres are at X?

A collagen
B elastic
C fibrin
D keratin *(1)*

3 **a)** Define the term *double circulatory system.* *(1)*

The cardiac output is the volume of blood pumped out by each ventricle in a minute. The diagram shows the events that occur in the cardiac cycle over a period of two seconds. During this time the volume of blood pumped out by each contraction of the left ventricle (stroke volume) was 70 cm^3.

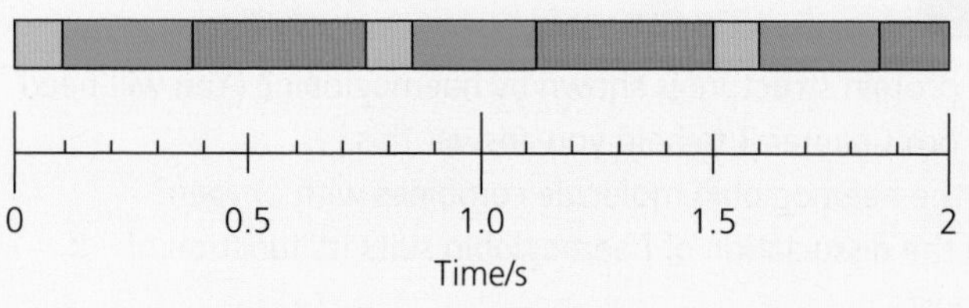

Key

Atrial systole | Ventricular systole | Diastole

b) Use the diagram to calculate
i) the heart rate *(1)*
ii) the cardiac output. *(1)*

c) Explain why the walls of the ventricles are not the same thickness. *(3)*

d) Describe the roles of the following in the action of the heart:
i) atrio-ventricular septum
ii) atrio-ventricular valves
iii) papillary muscle
iv) semi-lunar valve at the base of the aorta. *(6)*

4 **a)** Explain the roles of the sino-atrial node, atrio-ventricular node and Purkyne fibres in coordination of the heart. *(6)*

The diagram shows three ECGs. **ECG1** is a normal ECG.

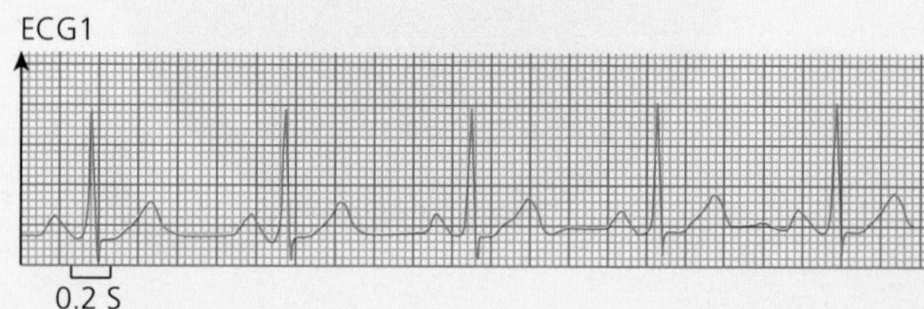

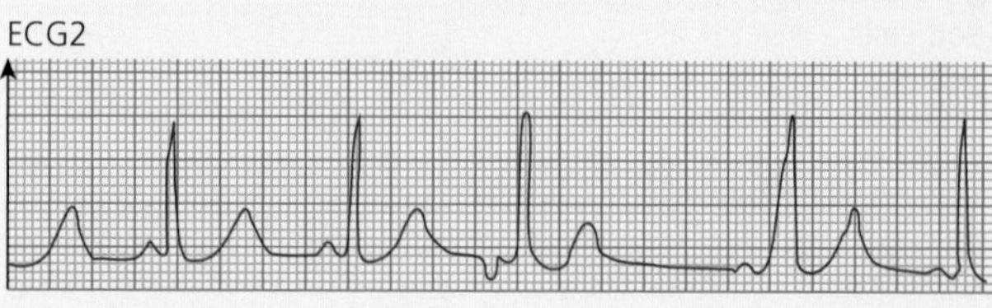

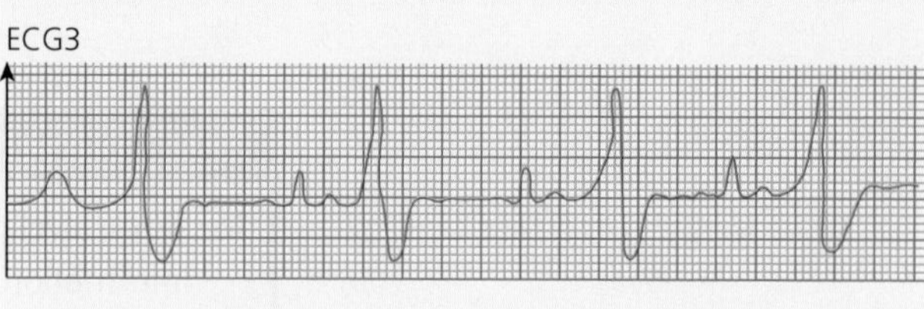

b) Calculate the heart rate from **ECG1**. Show your working. *(2)*

ECG2 and **ECG3** are from people with ectopic heartbeats.

c) Describe how **ECG2** differs from **ECG1**. *(3)*

d) Describe how **ECG3** differs from **ECG1**. *(3)*

e) Explain the causes of the ectopic heartbeats shown in **ECG2** and in **ECG3**. *(4)*

5 Oxygen is transported in the blood and delivered to respiring cells. Carbon dioxide is transported away from respiring cells.

a) Describe how oxygen molecules travel from red blood cells to respiring cells. *(3)*

b) An enzyme in red blood cells catalyses the reaction between carbon dioxide and water as blood flows through respiring tissues.

$$CO_2 + H_2O \xrightarrow{\text{enzyme}} H_2CO_3 \longrightarrow H^+ + HCO_3^-$$

i) Name the enzyme that catalyses this reaction. *(1)*

ii) Explain why the reaction shown above is important in the transport of carbon dioxide in the blood. *(3)*

The graph shows the effect of increasing the carbon dioxide concentration on the oxygen haemoglobin dissociation curve.

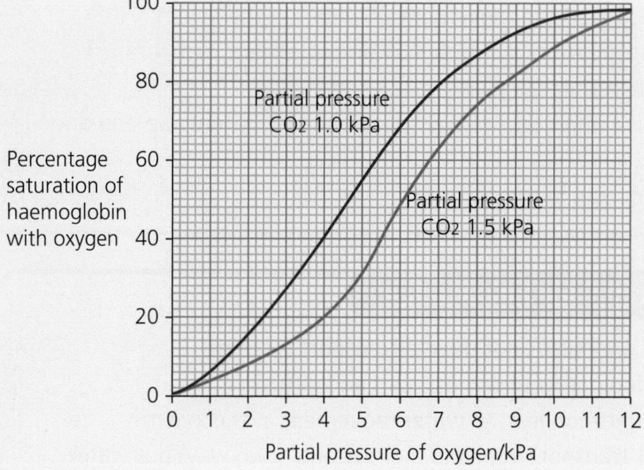

c) State the percentage saturation of haemoglobin with oxygen at a partial pressure of 5 kPa of oxygen when the partial pressure of carbon dioxide is 1.0 kPa and when it is 1.5 kPa. *(2)*

d) The graph shows that the percentage saturation of haemoglobin with oxygen decreases as the partial pressure of carbon dioxide increases.

i) Explain how this happens. *(3)*

ii) Name the effect shown by the graph. *(1)*

iii) Explain the importance of this effect of carbon dioxide on oxygen transport as shown in the graph. *(4)*

Stretch and challenge

6 A student dissected a sheep's heart and took the following measurements.

Structure	Width of wall/mm
left ventricle	20.0
left atrium	5.0
right ventricle	8.5
right atrium	4.5
aorta	3.5
superior vena cava	1.5
pulmonary artery	3.0
pulmonary vein	4.5

Comment on the figures given in the table.

Chapter 9

Transport in plants

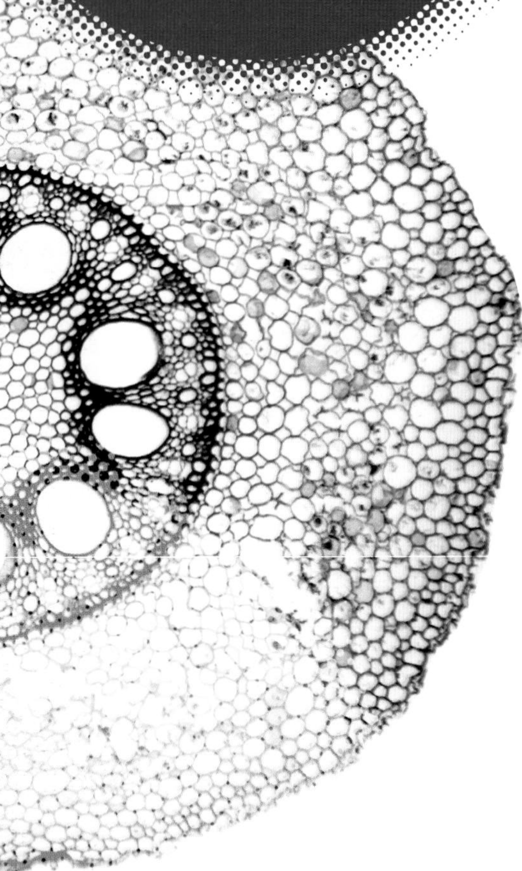

Prior knowledge

Before you start, make sure that you are confident in your knowledge and understanding of the following points.

- As multicellular organisms became more complex, they need a transport system.
- Diffusion and osmosis play an important part in the transport of water in plants.
- Active transport plays an important part in the transport of mineral ions in plants.
- Water and mineral ions are transported in the xylem and sucrose is transported in the phloem.
- Plants may have root hair cells to increase the surface area of the roots for the absorption of water and mineral ions.
- Transpiration is the process of evaporation of water from leaves.
- Transpiration leads to the movement of water up the xylem.
- The rate of transpiration is affected by air movement, temperature, humidity and light intensity.
- Translocation is the process of transportation of nutrients up and down a plant.
- Translocation occurs in the phloem.

Test yourself on prior knowledge

1 Why do simple organisms not need a transport system?
2 What is the difference between osmosis and diffusion?
3 How does transpiration lead to water movement in the xylem?
4 Why is nutrient transport in plants bidirectional (two way) but water transport is unidirectional (one way)?
5 How is water lost through the leaves?
6 Why does air movement affect transpiration?

Figure 9.1 Giant sequoias, *Sequoiadendron giganteum,* are the largest trees and living things as well as some of the oldest living organisms on Earth.

The oldest known sequoia is thought to have been 3500 years old. Giant sequoias are found on the western side of the Sierra Nevada, California. They reach, on average, a height of 85 m and a diameter of 8 m. Because of the species' extreme size, scientists have studied its water uptake. They found that sequoias have aerial roots that take in water from the air (fog) to supplement water from the soil.

Just as in animals, large and complex plants need transport systems. Flowering plants have specialised regions for absorbing water and ions from the soil (the roots) and for collecting light for photosynthesis (the leaves, held in place by the stem). Because light comes from above and water from below, the plant develops to become tall and thin. Transport systems become necessary to transport water and ions to the leaves and nutrients to the roots, which may be a considerable distance.

The need for a transport system

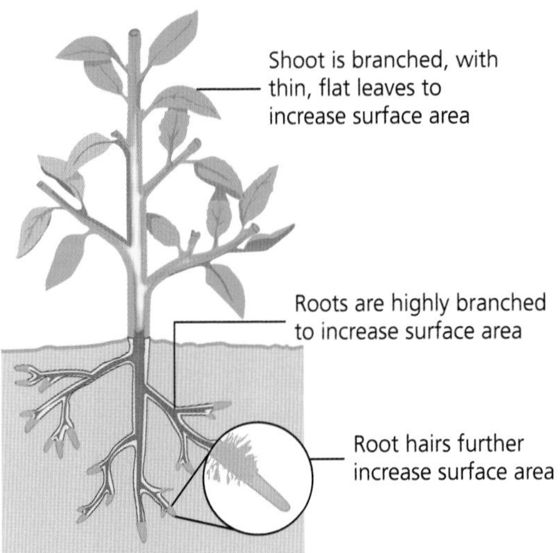

Figure 9.2 Plant structure – adaptations to increase surface area : volume ratio.

Flowering plants take in water and mineral ions through their roots and make glucose in their leaves by photosynthesis. Water, mineral ions and glucose are needed by all the cells in the plant and so need to be transported from one area to another. Glucose is transported as sucrose. Flowering plants have two separate transport systems: the xylem for transporting water and mineral ions, and the phloem for transporting nutrients, such as photosynthates and amino acids.

Unlike most animals, plants have a branching body shape, which helps to give a very high surface area to volume ratio, even in large plants. The leaves, which absorb light, are thin and flat and so maximise their surface area : volume ratio, and the roots have root hairs to greatly increase their surface area (see Figure 9.2).

Unlike larger animals, plants do not have any system for transporting oxygen and carbon dioxide. This is because of their very large surface area : volume ratio, the fact that the leaves and stems have chloroplasts that can generate their own oxygen and use up carbon dioxide, and the low metabolic rates of plant tissues and hence a low demand for oxygen for aerobic respiration.

Transport system structure

Key term

Dicotyledonous plant A flowering plant with two embryonic seed leaves (cotyledons), which usually emerge from the seed at germination.

Plants have a **vascular system**, but it is not actually a single system but two. One system transports water and mineral ions up the plant from the roots through a system of tubes called **xylem vessels**. The other transports assimilates and minerals up and down the plant and is made up from another set of tubes called the **phloem sieve tubes**. The location of the xylem and phloem tissue in roots, stems and leaves is shown in Figures 9.3 and 9.4.

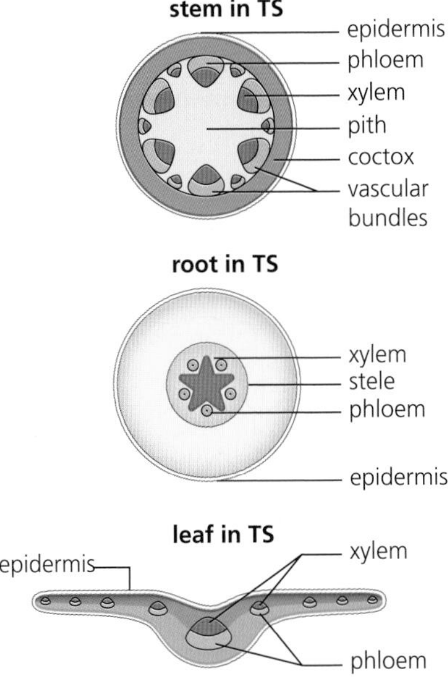

Figure 9.3 Location of xylem and phloem tissue in the stem, root and leaf of a dicotyledonous plant. Note that the shape of the xylem in the root varies a little between species.

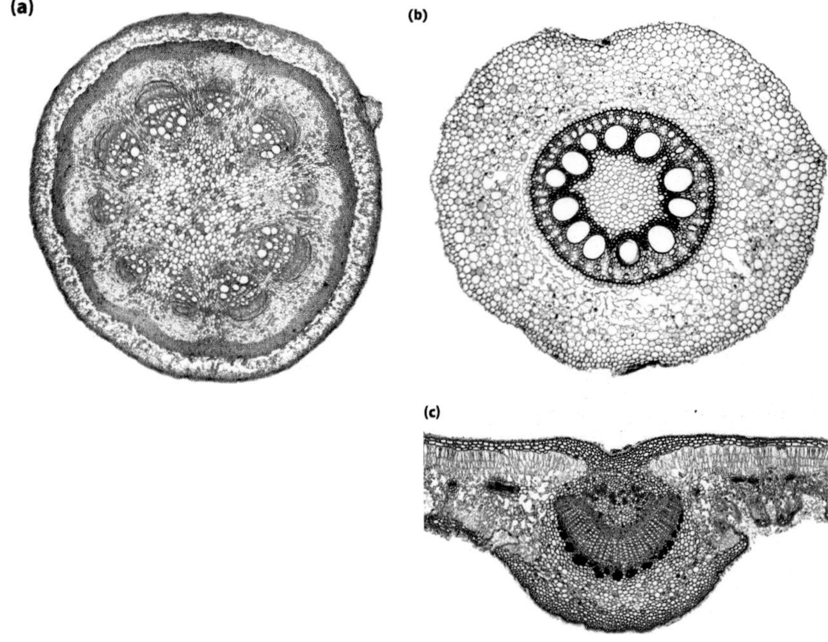

Figure 9.4 Images of sections of the (a) stem (×50), (b) root (×24) and (c) leaf (×55) of a dicotyledonous plant, showing the distribution of xylem and phloem.

Xylem

The xylem consists of several types of cell – vessels, tracheids, fibres and parenchyma – but the cells that are primarily responsible for transport are the **xylem vessel elements**. These are aligned end to end to form continuous xylem vessels. The structure of xylem vessel elements adapts them well for transporting water and dissolved ions, as well as for their secondary function of supporting and strengthening the plant.

The structure of a xylem vessel is shown in Figure 9.5.

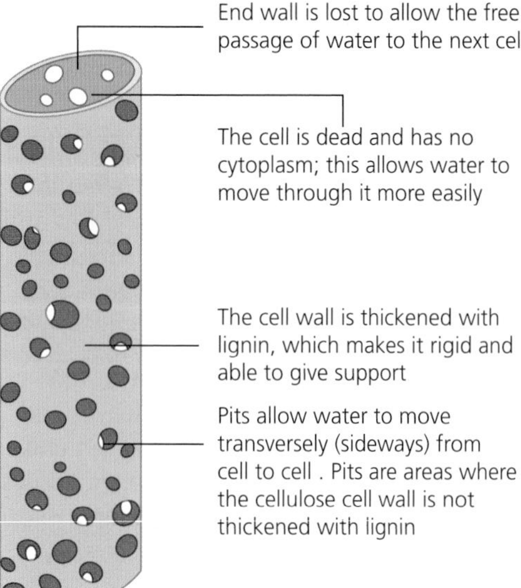
End wall is lost to allow the free passage of water to the next cell
The cell is dead and has no cytoplasm; this allows water to move through it more easily
The cell wall is thickened with lignin, which makes it rigid and able to give support
Pits allow water to move transversely (sideways) from cell to cell . Pits are areas where the cellulose cell wall is not thickened with lignin

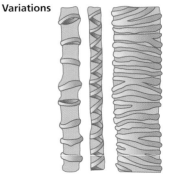

Variations

The end wall may be perforated rather than completely missing

The pattern of lignin thickening may be in rings, coils or strips

Figure 9.5 Structure of xylem vessels.

Activity

Celery

The celery stalk that you eat is actually a petiole, because it supports the leaf.

Method

1 Some students investigated xylem tissue using celery. They cut the ends off two celery petioles with a sharp razorblade and placed them in a beaker of water with some drops of red food dye, and left them for 24 hours.
2 The students cut transverse sections 1–2 mm thick, placed them on a slide and examined the cut surface for the presence of dye.
3 The students then cut sections of the stem at the base of the petiole, the middle of the petiole, and just below the leaf. The students used the technique shown in Figure 9.6 to cut several very thin cross sections.
4 The students took the thinnest section, mounted it on a slide and stained it with iodine. They observed it under a microscope and looked for some thick-walled cells in about the same position as they found the dye previously.

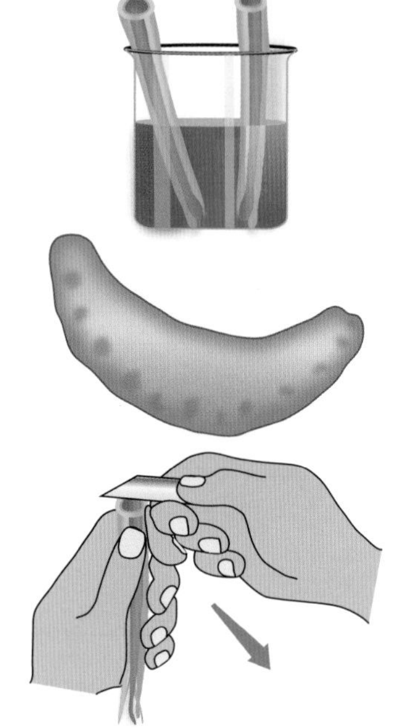

Rest blade on top edge of celery, then pull blade towards you with a single slice in direction of arrow. Repeat rapidly several times.

Figure 9.6

Questions

1 What sort of cells do you think these were and why?

They then cut a 1 cm thick slice from the remaining undyed celery, cut out a small piece of the vascular bundle (see Figure 9.3) and put this on a microscope slide with a drop of iodine. They teased this piece of tissue apart, added a coverslip, and gently squashed the bundle.

2 What cell types are shown in Figure 9.7?

3 Xylem vessel cell walls are thickened by lignin. The circles and spirals in Figure 9.7 are also made of lignin. Suggest the advantage of both these features for a plant.

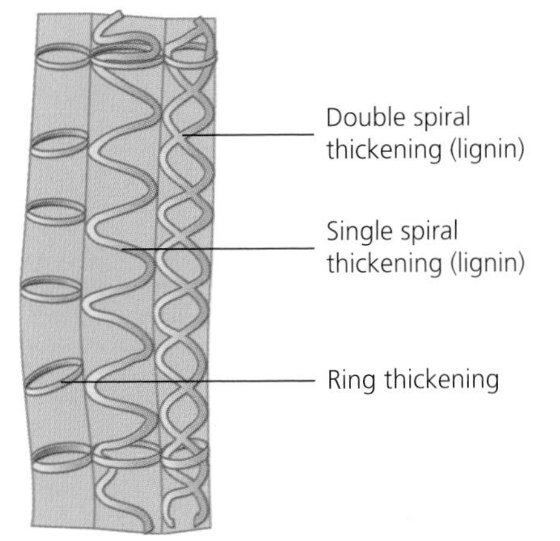

Figure 9.7

Phloem

Unlike the xylem, the phloem consists of living cells. Phloem cells are of several types, but the two that are concerned with transport are the **sieve tube elements** and the **companion cells**.

> **Key term**
>
> **Plasmodesma (plural: plasmodesmata)** A microscopic channel through plant cells walls, connecting the cytoplasm of two cells. These are lined with plasma membrane.

As the name implies, the sieve tube elements form into continuous tubes, and each of them has an associated companion cell. The sieve tube elements, although alive, have no nucleus, few organelles and little cytoplasm; this may aid their ability to transport assimilates, but it means that they are probably not self-sufficient. The relationship between the sieve tube element and the companion cell is complex, but a main role of the companion cell seems to be as a sort of 'life support system' providing materials to keep the sieve tube element alive. The sieve cell and the companion cell are connected by plasmodesmata, which allow the transport of molecules.

> **Tip**
>
> Revise what you learned about plasmodesmata in Chapter 1. Look back at Chapter 6 to remind yourself about the structure of the phloem.

The structure of a sieve tube element and its associated companion cell is shown in Figure 6.28b.

> **Test yourself**
>
> 1 What are the two functions of the xylem?
> 2 State three differences between a xylem vessel and a sieve tube.
> 3 Why is it useful that some areas of the xylem vessel wall are *not* thickened with lignin?
> 4 What is the function of the companion cells in the phloem?

Absorbing water

Water potential and water movement

You saw earlier (Chapter 2) that the movement of water between plant cells depends upon the **water potential** of the two cells, and the same applies to movement between cells and the atmosphere. Water always moves from an area of higher water potential to one of lower water potential. In the case of transpiration, water usually moves out of the leaf because the water potential inside the leaf is greater than that outside the leaf. Water potential gradients are also involved in the absorption of water in the roots. As water enters the roots and leaves via the leaves, there is a **water potential gradient** from the roots (which have the highest water potential) up the stem to the leaves (which have the lowest water potential).

Water enters the roots from the soil down a water potential gradient, assisted by the increased surface area provided by the root hair cells (see above). It then needs to move across the root cortex and through the endodermis to reach the xylem, in which it can be transported up the stem. This movement is down a water potential gradient and can occur via several pathways (see Figure 9.8).

Key terms

Apoplast The 'non-living' extracellular space that surrounds the symplast. It consists of cell walls and spaces between cells.

Symplast A continuous network of interconnected plant cell protoplasts (the cytoplasm of the cells, linked by plasmodesmata).

The apoplastic pathway goes through the cell walls. The cell walls are readily permeable, and so this pathway offers little resistance. Most of the water travels along this path.

The symplastic pathway goes through the cytoplasm, and from cell to cell via plasmodesmata.

The problem with the apoplastic pathway is that there is no opportunity to 'select' what comes into the plant and what is kept out – everything dissolved in the water travels with it. This problem is solved by the presence of the **Casparian strip** in the walls of the **endodermal cells** (see Figure 9.8).

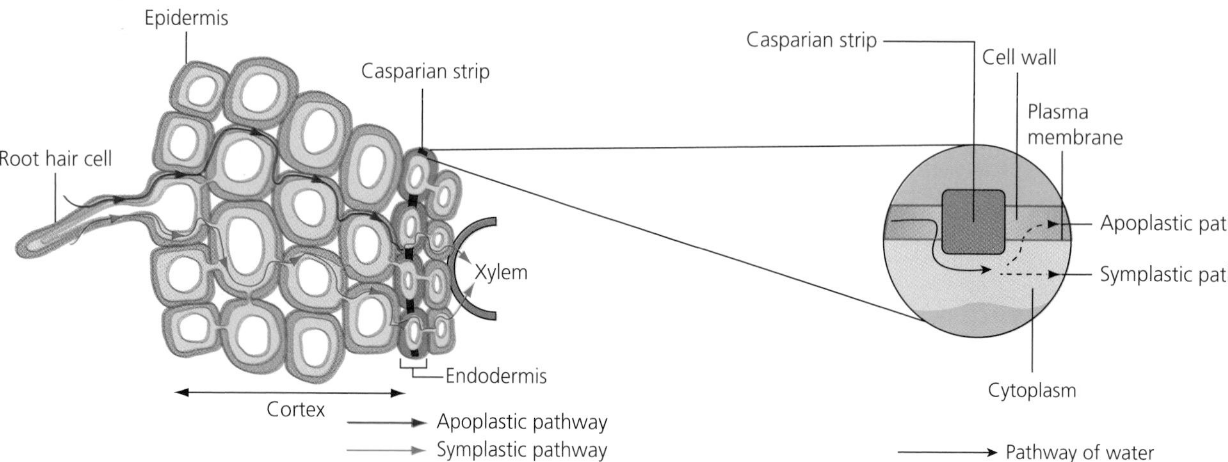

Figure 9.8 (a) Pathways taken by water across the root; (b) effect of the Casparian strip on the passage of water.

The Casparian strip is a continuous band that goes around the wall of the endodermal cells. It is made of a waterproof, waxy substance called **suberin**, and so it blocks the apoplastic pathway because water cannot get past it. The water and dissolved substances such as ions must then enter the cytoplasm and, in doing so, have to go through the plasma membrane, which is selectively permeable and so controls what enters. The membrane of the endodermal cells also contains a number of protein carriers that can regulate active transport. Once in the cytoplasm, the water can continue through the symplastic pathway or can return to the apoplastic pathway. It then continues into the xylem.

Mechanism of water transport

Once it has entered the xylem, water with its dissolved mineral ions travels up the plant. This can be quite a challenge. The tallest trees in the world are over 100 m high. Scientists have calculated that no tree can grow to more than around 130 m, possibly because there is a limit to how high the water column can be pulled up the xylem. Raising water any distance up the plant requires that gravity is overcome. Pressures of up to −600 kPa have been measured in the xylem. How are such suction pressures created?

Tip

See Chapter 2 to remind yourself about cohesion between water molecules.

Suction pressures are all down to transpiration. Evaporation of water from the leaves is responsible for pulling water up the stem. The most well-known theory that explains how the **transpiration stream** works is known as the **cohesion–tension theory**. This theory is based on the principle that water molecules are **cohesive** – that is, they stick together. This results from the fact that water is a polar molecule. The more negatively charged oxygen of one water molecule forms a hydrogen bond with a more positively charged hydrogen atom in the second water molecule (see Figure 2.2 on page 19).

Key term

Transpiration stream The uninterrupted stream of water and solutes that is taken up by the roots and transported via the xylem vessels to the leaves, where it evaporates into the sub-stomatal cavity.

If a water molecule moves from one place to another, it pulls other water molecules with it because of these cohesive forces. So, when water molecules evaporate from the cells inside the leaf, they pull other water molecules behind them to the interface between water and air in the cell wall, from which they then evaporate. These water molecules pull others from the xylem in the leaf, and so on. As a result, water is pulled all the way up from the xylem in the root. It is a similar principle to drinking through a straw, when sucking the drink into your mouth causes more liquid to be pulled up the straw. The 'tension' part of the theory's name comes from the fact that this cohesion of the water molecules creates a tension (i.e. a suction) at the top of the xylem column.

This water transport system only works if there is a continuous column of water with no breaks, because the molecules must be very close to each other to cohere. If air gets into the xylem, it causes an airlock – which stops water movement. This is not usually a problem, however, because water can move laterally from one xylem vessel to another (through the pits in the vessel walls) and so the air bubble can be bypassed. The blocking effect of air bubbles explains why it is so important to recut a stem underwater above the air bubble when setting up a potometer or receiving a bunch of flowers. Air will draw up into the xylem when a stem is cut because the tension of the water column is suddenly broken.

The cohesion–tension theory is summarised in Figure 9.9.

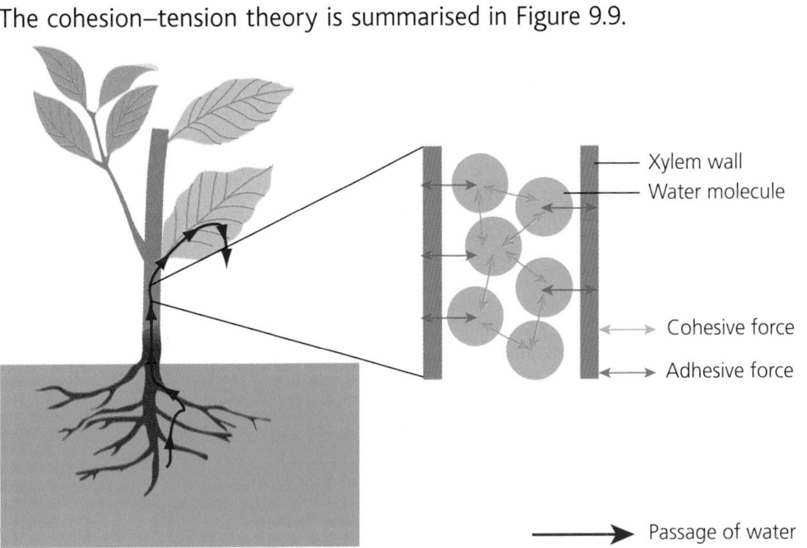

Figure 9.9 The transpiration stream and cohesion–tension theory.

The cohesion forces are not the only ones involved in water transport, however. Water molecules also **adhere** to the walls of the xylem. This helps to maintain the water column in position even when transpiration is not taking place (e.g. at night). Without this adhesion, the force of gravity could cause the water column to drop.

Transpiration

Key term

Stoma (plural: stomata) A pore in the epidermis of the leaf, surrounded by a pair of guard cells.

Tip

Note that transpiration is the main cause of the movement of water up the stem, but the term is restricted to the process of evaporation of water, not water movement.

In plants, the necessity to exchange gases to provide for photosynthesis has led to the evolution of stomata, so that carbon dioxide and oxygen can move in and out. When the stomata are open, however, water loss is inevitable from the moist surfaces of the cells inside the leaf. There is relatively little water in the air around the leaf so water vapour diffuses from the air spaces in the mesophyll, where the water potential is high, through the stomata to the lower water potential outside. This is called **transpiration** and is a consequence of gas exchange for photosynthesis. The internal surface area of the leaf is considerable, so large amounts of water can be lost. Most of the water vapour evaporates when the stomata are open. Water loss is therefore an inevitable side effect of opening the stomata to allow for gaseous exchange. During darkness, when carbon dioxide is not needed for photosynthesis, the stomata close to reduce water loss.

Factors affecting transpiration

The rate of transpiration depends upon the extent of the diffusion gradient of water vapour between the inside of the leaf and the surrounding air. The larger the gradient, the faster the rate of diffusion (see Figure 9.10).

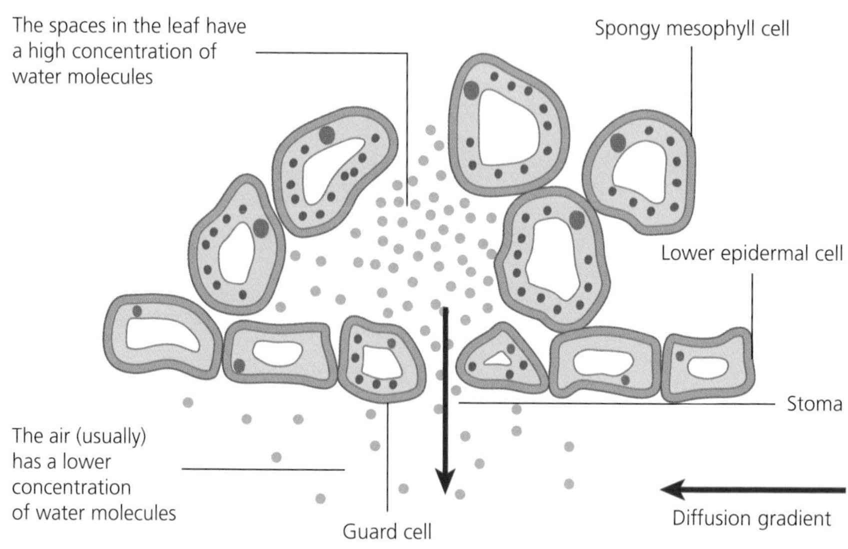

The spaces in the leaf have a high concentration of water molecules
Spongy mesophyll cell
Lower epidermal cell
Stoma
The air (usually) has a lower concentration of water molecules
Guard cell
Diffusion gradient

Figure 9.10 Diffusion of water vapour through a stoma of a leaf (the whole leaf structure is not shown).

A number of factors affect the diffusion of water vapour out of the leaf and, therefore, the transpiration rate (see Figure 9.11).

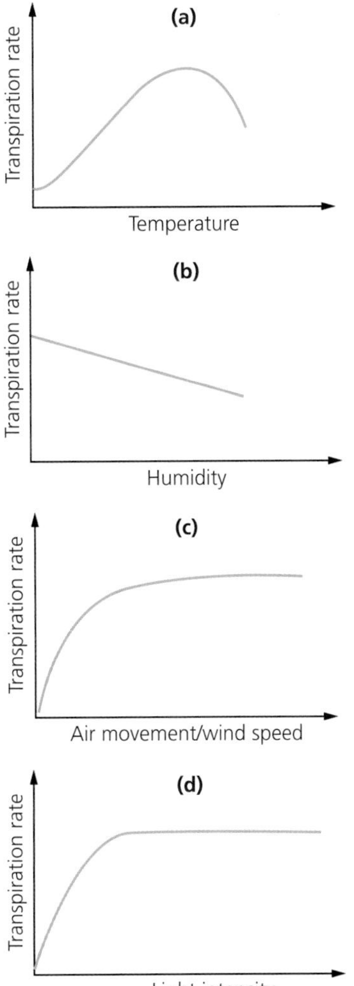

Figure 9.11 Effect of (a) temperature, (b) humidity, (c) wind speed and (d) light intensity on the rate of transpiration.

- **Temperature** (see Figure 9.11a): Increasing temperature increases the kinetic energy of all molecules, so an increase in temperature will increase the transpiration rate (assuming the concentration of water molecules in the air is lower than in the leaf). Very high temperatures cause the stomata to close, and transpiration then decreases.
- **Humidity** (see Figure 9.11b): As the humidity of the air surrounding the plant increases, the concentration of water molecules in the air rises, and so the diffusion gradient between the inside of the leaf and the surrounding air decreases and the transpiration rate decreases; eventually an equilibrium is reached and there is no net water vapour loss from the leaf.
- **Air movement** (see Figure 9.11c): In still air, the water molecules are not taken away in the wind so accumulate close to the leaf surface, giving an area of high humidity. The concentration gradient is low or non-existent. Air currents move the water molecules away from the surface of the leaf and so maintain the concentration gradient. Air movement therefore increases the transpiration rate, up to a maximum level.
- **Light** (see Figure 9.11d): In the dark, the stomata close and the transpiration rate is drastically decreased. Once there is enough light to cause the stomata to open, increasing the light intensity any further has no effect. The intensity of light that is needed for the stomata to open is quite low, and the stomata remain open even on the cloudiest days.

Measuring transpiration

Transpiration is difficult to measure directly, but it is possible to measure water uptake using a **potometer**. It is assumed that water uptake is directly proportional to water vapour loss by transpiration.

One design of potometer is shown in Figure 9.12. Designs vary, but there is always a calibrated tube connected to a plant in such a way that there is a continuous column of water.

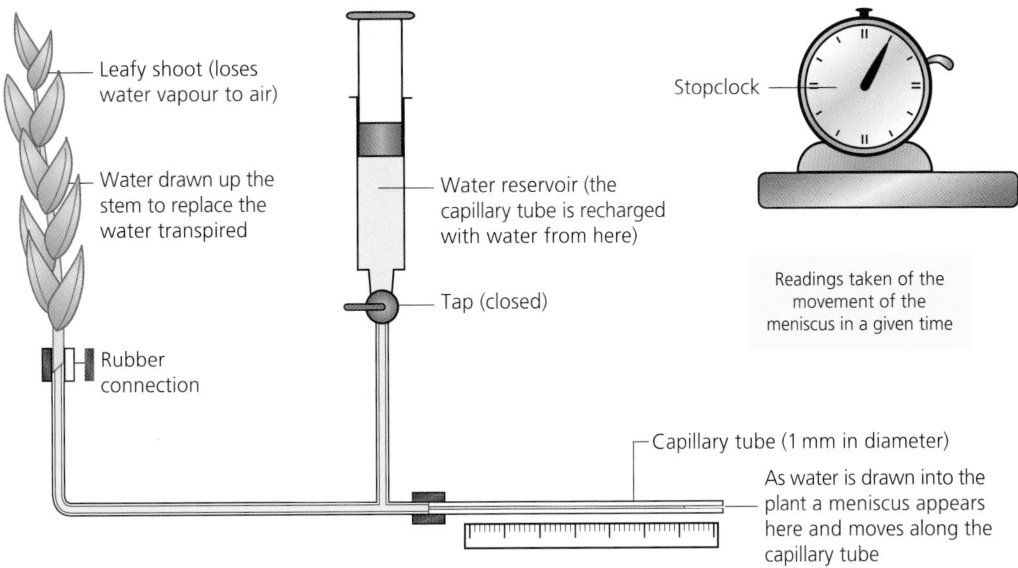

Figure 9.12 A potometer.

When setting up a potometer, it is important to ensure that no air gets into the apparatus (apart from the bubble in the capillary tubing) or else an airlock will form, preventing water movement. The plant stem must be cut underwater to prevent air entering the xylem, because the transpiration stream pulls water upwards and sucks air into the cut xylem if the stem is not in water. The apparatus must then be assembled underwater, ensuring all the joints are airtight, because bubbles in the apparatus also stop water movement. If the leaves of the plant get wet during set-up, they should be dried before taking readings, otherwise the water will create a humid atmosphere around the leaf, which may reduce transpiration. The potometer should be left until the rate of movement of the bubble is more or less constant before starting any experiment.

Tip

You should know how to set up a potometer, why it is necessary to do so underwater, and why the apparatus must be airtight.

Activity

The effect of air movement on transpiration rate

An experiment was carried out as follows:

1 A potometer was set up as shown in Figure 9.12.
2 Readings of the movement of the bubble were taken every minute for five minutes, and the mean distance travelled per minute was calculated. This was recorded as 'no air movement'.
3 The readings were repeated with a fan set at different distances from the plant. Air movement was recorded as $\frac{1}{\text{distance}}$ of the fan.

All trials were carried out at room temperature. The results are shown in Table 9.1 and Figure 9.13.

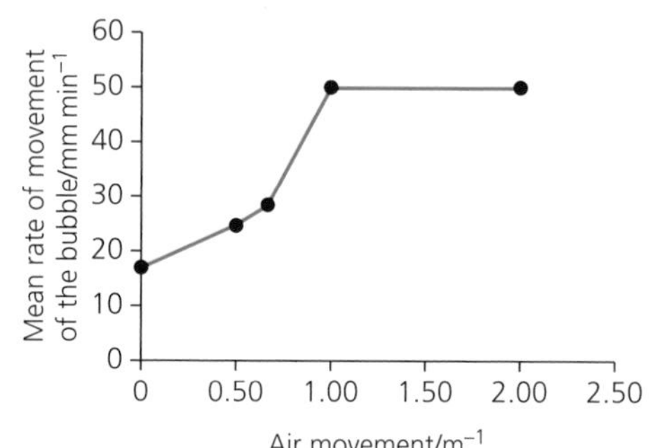

Figure 9.13

Table 9.1 Effect of air movement on the movement of the bubble in a potometer.

Air movement/m^{-1}	Distance bubble moves in one minute/mm					Mean rate of movement of the bubble/mm min^{-1}
	1 min	2 min	3 min	4 min	5 min	
0.00	21	16	19	15	14	17.0
0.50	31	24	22	19	26	24.4
0.67	29	29	27	30	26	28.2
1.00	47	51	48	54	52	50.4
2.00	53	50	49	51	48	50.2

Questions

1 Describe the effect of air movement on the mean rate of movement of the bubble.
2 Suggest an explanation for the shape of the graph
 a) between 0 and 1.00 m^{-1}
 b) between 1.00 and 2.00 m^{-1}.
3 It is assumed that water uptake can be used to estimate the transpiration rate. Suggest reasons why that assumption may not be completely valid.
4 Temperature was controlled in this experiment. List three other variables that should be controlled.

Calculating water uptake rates

In order to calculate the water uptake rate of a plant using a potometer, it is necessary to have the glass tubing through which the bubble moves calibrated in standard units (usually millimetres) and to know the cross-sectional area of the bore of the tubing. This allows the conversion of the distance moved in a given time into a rate of uptake, using the following formula:

$$\text{volume of water absorbed (mm}^3) = \pi r^2 \times \text{distance moved (in mm)}$$

$$r = \text{radius of the bore of the tubing} \left(\text{i.e. } \frac{\text{bore diameter}}{2}\right)$$

$$\text{rate of water uptake (mm}^3\ \text{sec}^{-1}) = \frac{\text{volume of water absorbed (mm}^3)}{\text{time (s)}}$$

Tip

See Chapter 18 for more information about calculating rates.

Example

A student used the potometer shown below to measure the rate of water uptake of a cut shoot of the sycamore tree, *Acer pseudoplatanus*. The student placed the potometer on a balance to measure the rate of water vapour loss.

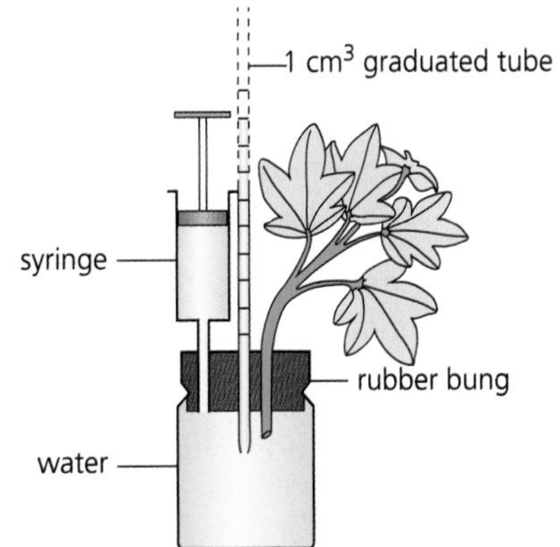

Figure 9.14 The student's potometer.

The table shows the student's results.

Table 9.2

Time/min	Volume of water in the graduated tube/cm³	Mass of potometer and shoot/g
0	10.00	376.8
4	9.90	376.6
8	9.75	376.4
12	9.65	376.2
16	9.50	376.0
20	9.40	375.8

a) i) State what precautions should be taken to ensure that there are no air locks in the cut end of the sycamore stem.

ii) Suggest the advantages of using potometers that measure both mass loss and water uptake.

b) The student saw that the rates of water vapour loss and water uptake were both constant over the 20 minute period. Calculate the rate of water vapour loss and the rate of water uptake per hour. Show your working.

c) State the assumption that is made in using mass loss as a measurement of water vapour loss.

Tip

If the rates were not linear then the student would draw a graph and use the graph to calculate the rates.

Answers

a) i) Put the stem underwater and cut it preferably with an oblique cut to give a larger surface for absorption of water than a straight cut. Keep the cut end underwater and insert it into the potometer underwater.

ii) Not all the water absorbed by the plant is lost in transpiration, so using the two methods allows measurement of water uptake and water vapour loss (as mass loss). It also allows measurement of the volume of water retained by the plant (difference between uptake and loss). You can also use this to check for leaks in the system.

b) decrease in mass = rate of water vapour loss = 376.8 – 375.8 = 1.0 g

water uptake = 10.00 – 9.40
= 0.60 cm³ / g (1 cm³ water = 1 g)

rate of water vapour loss = 1.0 / 20 × 60 = 3.0 g h^{-1}

rate of water uptake = 0.60 / 20 × 60 = 1.8 g h^{-1}

c) The assumption is that the mass lost or gained for other reasons (e.g. respiration, photosynthesis or maintaining turgidity of cells) is too small to make much difference.

Test yourself

5 What part of the leaf structure controls the rate of transpiration?
6 Plants wilt due to excessive water loss. Explain why this is more likely in warm weather.
7 Why do potometers need to be set up underwater?
8 Why is the transpiration rate lower during the night?
9 Explain why the transpiration rate is lower in wet weather.
10 Water loss is disadvantageous to the plant. Explain why plants carry out transpiration.

Reducing water loss

In any organism, losing a lot of water is a bad thing. Water is a substance that is vital to life, and it is best if it is not lost. Leaves have evolved to reduce this water loss wherever possible, and especially so in plants that live in dry conditions where water lost may not easily be replaced.

Most land plants have some means of restricting water vapour loss from their leaves. These include:

- The leaves have a waxy waterproof cuticle, so that water vapour loss is restricted over the whole surface of the leaf, apart from the stomata.
- Stomata are mostly or only on the lower surface of the leaf, which is usually cooler because it faces away from the Sun, so that diffusion is less rapid. (In water-living plants with floating leaves, the stomata are on the upper surface of the leaf, because water vapour loss is not a problem and the richest supply of gases is in the air rather than the water.)
- The stomata close in the dark, when it is not necessary to take up carbon dioxide, because there is no light for photosynthesis.

Plants that live in dry conditions have extra adaptations to reduce water vapour loss. Plants with such adaptations are called **xerophytes**. Common xerophytic adaptations are:

- The number of stomata may be reduced.
- The stomata may be sunken in pits or grooves. Water vapour leaving the stoma is therefore sheltered from air movement, so that high humidity builds up outside the stoma and reduces the diffusion gradient.
- The stomata may be surrounded by hairs, which trap the water vapour leaving the stoma and, again, lower the diffusion gradient between the inside and outside of the leaf.
- The waxy cuticle is often much thicker than in non-xerophytic plants, to further reduce water vapour loss through the leaf surface.
- The leaf may be rolled, with the lower surface inwards. This again allows high humidity to build up inside the 'coil', reducing the rate of diffusion from the stomata.
- The whole leaf may be reduced in size (e.g. the spines of a cactus) or the plant may lose some or all of its leaves, so that the surface area over which water vapour may be lost is drastically reduced.

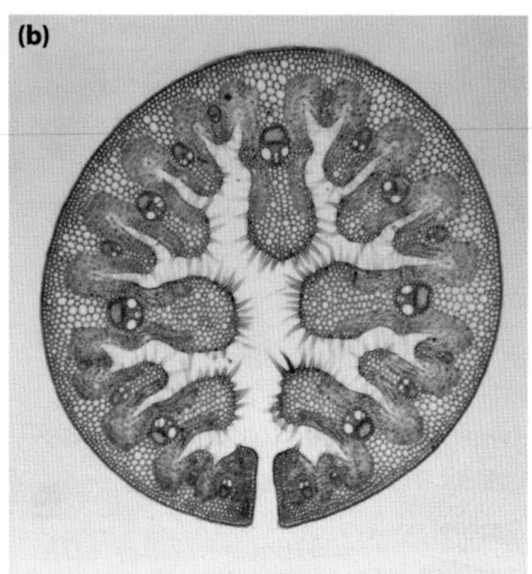

Figure 9.15 (a) Marram grass, *Ammophila arenaria,* growing in sand dunes; (b) transverse section of the leaf of marram grass (×25).

Figure 9.16 The leaves on this cactus have been reduced to spines, which greatly decreases the surface area over which water can be lost.

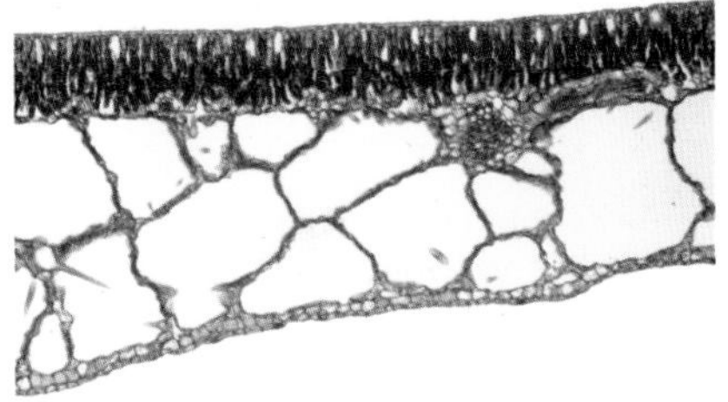

Figure 9.17 The structure of a water lily leaf (×25), showing hydrophytic adaptations. The stomata in the upper epidermis are too small to be seen.

Marram grass (*Ammophila arenaria*) is an example of a xerophytic plant. It lives on sand dunes, which do not retain water – so the plant's access to water is restricted. Marram grass has many of the modifications listed above, and is shown in Figure 9.15. Figure 9.16 shows a cactus – another xerophytic plant.

Plants that live in fresh water (with leaves beneath the surface or floating on the surface) have a different set of adaptations. Such plants are called **hydrophytes**. A typical hydrophyte is the water lily, and the structure of its leaves is shown in Figure 9.17. For animals that live in fresh water, excess water uptake can be a problem. In plants, however, the presence of cell walls prevents their cells absorbing too much water, but some issues remain. There is obviously less need for mechanisms to transport water from the roots or to prevent water loss. However, water contains less oxygen and carbon dioxide than air does, so obtaining sufficient carbon dioxide during the day and oxygen at night can be a problem. The structure of hydrophytes, and particularly their leaves, shows features to address these issues. Common adaptations are:

- In floating leaves, stomata tend to be mostly or entirely on the upper surface rather than the lower. This means that gas exchange can occur with the air rather than with the water.
- The leaves often have very large air spaces, which gives them buoyancy.
- Floating leaves are particularly thin and flat, which also gives them greater buoyancy.
- The waterproof waxy cuticle of the leaf is thin.
- The veins in the leaf are much reduced (especially the xylem). There is much less need for transporting water and for support, so the xylem is much less important.
- Hydrophytes have a greatly reduced root system.

Translocation

Translocation is the transport of dissolved photosynthetic assimilates in a plant. These assimilates are nutrients made by photosynthesis in the leaves, the main one being **sucrose**, although amino acids and a number of other substances are also transported. The transport requires energy, and so needs living cells, unlike the transport of water by the xylem.

In translocation, movement is from areas called **sources** to others called **sinks**. The sources of sucrose are the leaves, because they produce sucrose as a result of photosynthesis. A sink is any area that requires sugar (or other nutrients). The main sinks are the growing points (meristems) in the roots, stems and leaves, and the roots in general, particularly when they are the storage organs, for example, potato tubers. This is true when stores are being created, but these organs act as sinks when they are being used, for example, when the potato tuber starts growing into a new plant. The sources have a high concentration of sucrose (because it is produced there) and the sinks have a low concentration, because they use it up or, as in the case of storage organs, convert it to starch. Assimilates move down the concentration gradient from sources to sinks.

Test yourself

11 Describe the apoplastic and symplastic pathways across a root.
12 Explain the importance of the Casparian strip in the endodermis.
13 What causes water molecules to cohere?
14 Explain how hairs surrounding stomata can assist with water conservation.

Activity

Investigating phloem transport using radioactive tracers – assessing the evidence

One leaf of a plant is injected with sucrose solution that contains a radioactive isotope of carbon, ^{14}C. The radioactive sucrose is then transported around the plant. The position of the tracer can be established by putting the plant in contact with photographic film, which is 'fogged' by radioactivity. When the film is developed, it gives an image of the radioactivity. Some of the plants used in the experiment were ringed – that is, the phloem was removed. The method and results are shown in Figure 9.18.

1 Explain the distribution of radioactivity in the un-ringed plant.
2 Suggest a reason why this experiment does not need a large number of repeats.
3 Give your opinion of the strength of evidence from this experiment for the hypothesis that 'sucrose is transported both up and down the plant, by the phloem'.

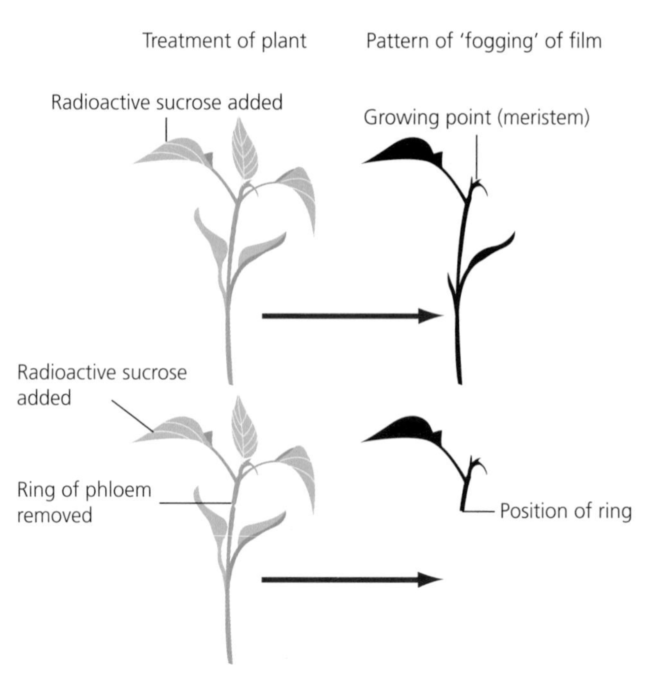

Figure 9.18 Using a radioactive tracer to investigate sugar transport.

Mechanism of translocation

The mechanism of transport in the phloem is not yet fully understood. The best supported theory is the **mass flow hypothesis**. The proposed mechanism involves **loading** of sucrose into the phloem in the leaves and then **unloading** it again in the sinks.

Loading of sucrose into the phloem in the leaves involves active transport, although it seems that some sucrose may be transported passively through the cytoplasm and plasmodesmata. The ATP needed for active transport into the phloem is supplied by the companion cells. As sucrose concentrations build up in the sieve tube elements, the reduced water potential draws in water from the neighbouring xylem. This creates a hydrostatic pressure in the phloem, which forces the sucrose away from the source towards the sink areas.

In the sinks, sucrose is moved out of the phloem and into the cells that need it. The exact mechanism is uncertain, but is thought to involve active transport. As a result, the water potential in the phloem is increased and water leaves the phloem by osmosis. At least some of this water re-enters the xylem. The pressure in the phloem is reduced, so creating a pressure gradient from the sources (where it is high) to the sinks (where it is lower). The mass flow hypothesis is summarised in Figure 9.19.

Test yourself

15 Define *translocation*.
16 Explain how different sucrose concentrations at the source and the sink lead to a pressure gradient.
17 Why is active transport needed to load the phloem at the source?

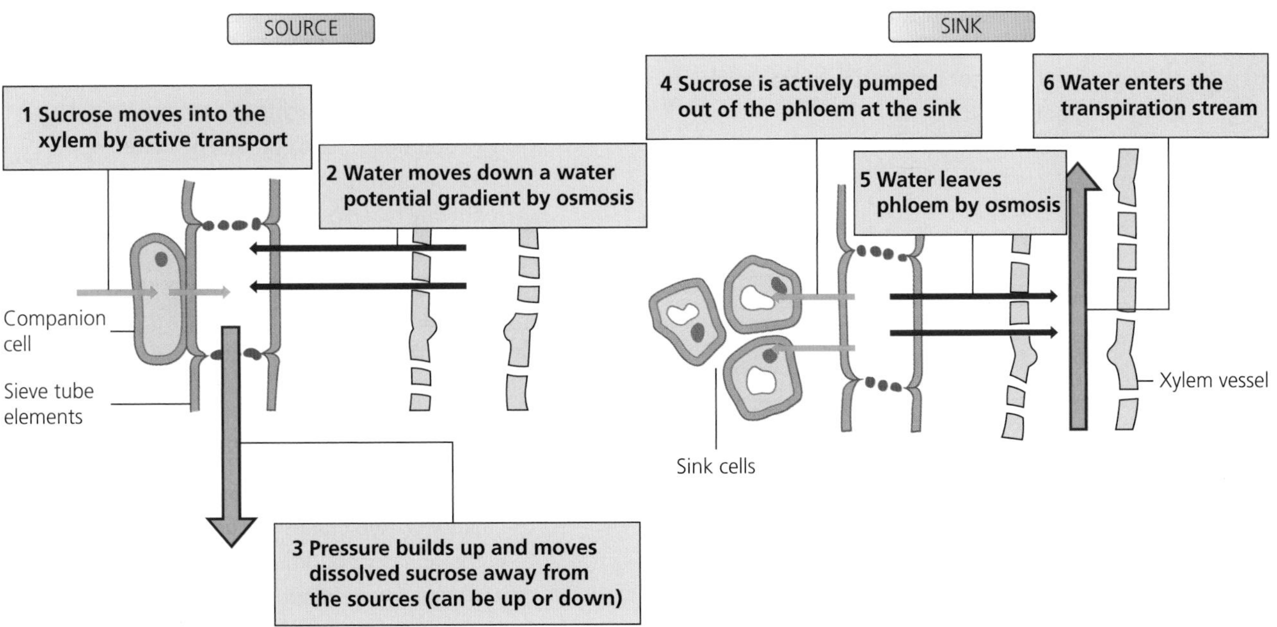

Figure 9.19 Summary of the mass flow hypothesis.

Exam practice questions

1 The photos show two multicellular organisms.

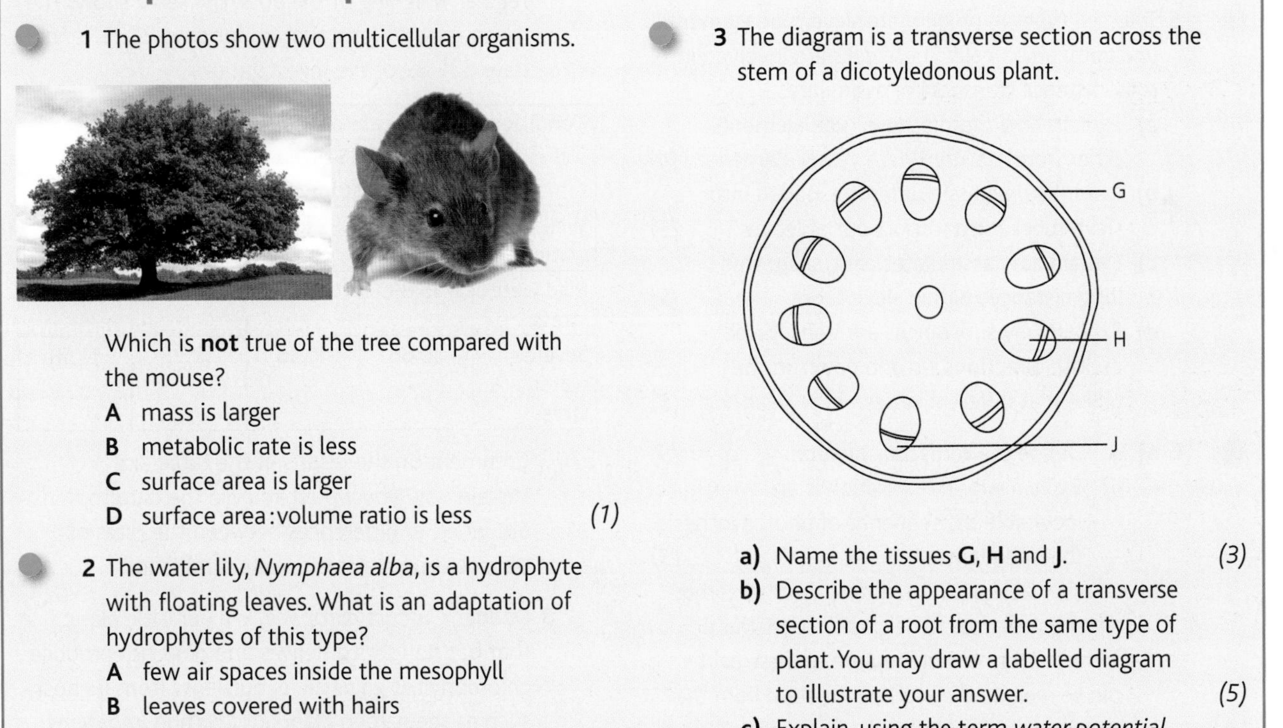

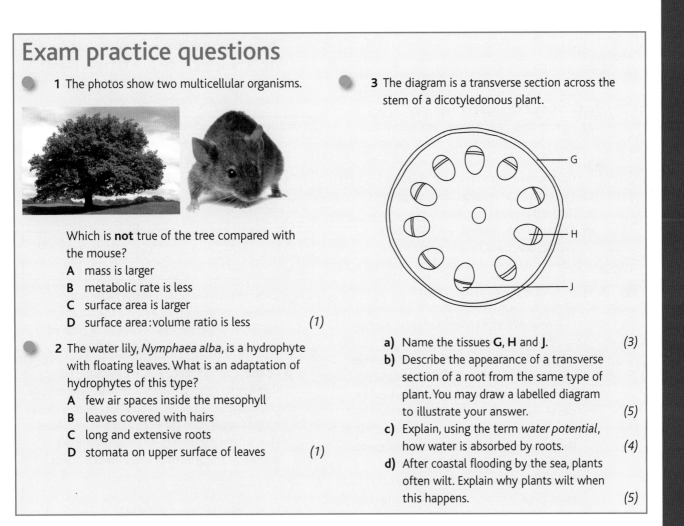

Which is **not** true of the tree compared with the mouse?

A mass is larger
B metabolic rate is less
C surface area is larger
D surface area : volume ratio is less *(1)*

2 The water lily, *Nymphaea alba*, is a hydrophyte with floating leaves. What is an adaptation of hydrophytes of this type?

A few air spaces inside the mesophyll
B leaves covered with hairs
C long and extensive roots
D stomata on upper surface of leaves *(1)*

3 The diagram is a transverse section across the stem of a dicotyledonous plant.

a) Name the tissues **G**, **H** and **J**. *(3)*

b) Describe the appearance of a transverse section of a root from the same type of plant. You may draw a labelled diagram to illustrate your answer. *(5)*

c) Explain, using the term *water potential*, how water is absorbed by roots. *(4)*

d) After coastal flooding by the sea, plants often wilt. Explain why plants wilt when this happens. *(5)*

4 The graphs show the effect of increasing an environmental factor on the rate of transpiration of crop plants.

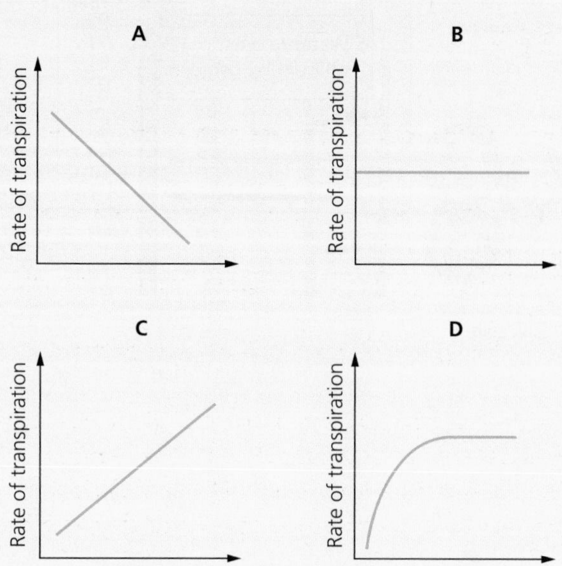

Which graph shows the effect of increasing humidity? *(1)*

5 Two cell types in phloem are sieve tube elements and companion cells. These cells are involved in long-distance transport of assimilates.

a) Explain how mature sieve tube elements differ from mature xylem vessel elements. *(4)*

b) Describe how assimilates are loaded into sieve tubes for transport out of leaves. *(5)*

c) Explain how assimilates are transported long distances within sieve tubes. *(5)*

d) Explain why substances are transported in both directions, up and down, in the phloem at different times of the year. *(5)*

6 **a) i)** Define the term *transpiration*. *(3)*

ii) Explain why transpiration is an inevitable consequence of photosynthesis for land plants. *(5)*

b) *Calamus laevigatus* is a rattan palm that climbs on rainforest trees to heights of 30 metres or more. The rate of water uptake of these palms can be measured by cutting the stem from the roots and placing it into a bucket of water. Palms like this with stems 25 mm in diameter absorb water at rates of up to 12 cm^3 per minute. Explain how plants such as *C. laevigatus* move large volumes of water over distances of 30 metres or more. *(6)*

c) Researchers investigated the structure of xylem vessels in *C. laevigatus* by cutting stems underwater and placing them in a solution of latex paint. They later took sections of the stem to follow the pathway taken by water.
Suggest how the researchers used this technique to locate the pathway taken by water through the stems of *C. laevigatus*. *(4)*

Stretch and challenge

7 Eucalyptus trees are grown as a crop for the timber trade. Researchers in Spain investigated the ability of two clones of *Eucalyptus globulus* to withstand dry conditions. Ninety seedlings of each clone were planted in waterproof pots and the soil covered by inert beads to form an impermeable covering. The seedlings of each clone were divided into three groups and exposed to one of three different watering conditions.

The transpiration rates were determined by regular weighing of the pots. The table shows the mean rates of transpiration for days 25–57 and days 62– 89 of the investigation.

Condition	Mean rate of transpiration/g plant^{-1} day^{-1}			
	Days 25–57		Days 62–89	
	Clone 1	Clone 2	Clone 1	Clone 2
Well watered	33.74	21.41	57.21	37.38
Mild water stress	15.89	16.46	14.37	7.88
Severe water stress	10.66	11.58	12.98	11.58

Comment on the results in the table and explain how you would analyse the data to find out whether differences between the rates of transpiration are statistically significant.

8 Dodder, *Cuscuta europaea*, is a parasitic plant that has no roots or leaves and does not produce chlorophyll. It gains all its nutrients from its host, such as the garden plants *Coleus* and *Impatiens*. Plan an investigation to find out the impact of dodder on the growth of these garden plants.

Chapter 10

Diseases of animals and plants

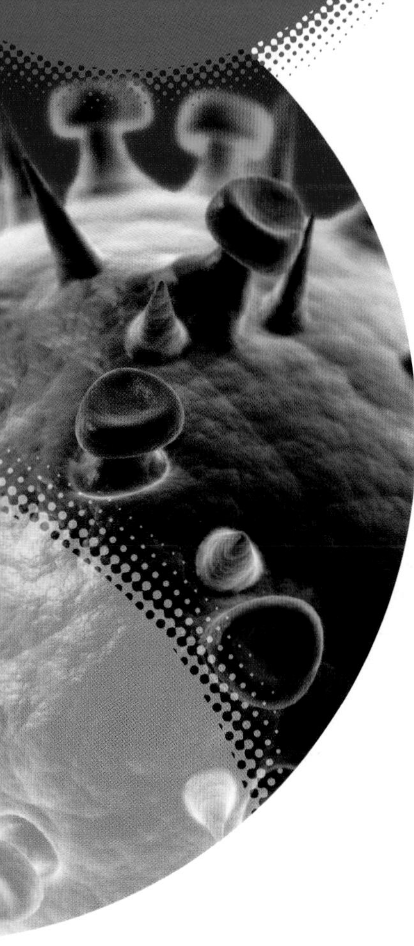

Prior knowledge

Before you start, make sure that you are confident in your knowledge and understanding of the following points.

- The different types of disease are: communicable (infectious) and non-communicable diseases.
- The causative agents of infectious diseases in humans are bacteria, viruses and fungi.
- Bacterial cells have a simple structure with a cell wall, cell surface membrane, a loop of DNA, small (70S) ribosomes and cytoplasm.
- Vascular tissue in plants is composed of xylem and phloem. Xylem transports water and ions, phloem transports assimilates, such as sucrose and amino acids.
- Stems, roots and leaves are plant organs. Tissues, such as epidermis, parenchyma, xylem and phloem, are distributed in different ways in each of these organs.

Test yourself on prior knowledge

1 Some bacterial parasites infect human white blood cells. State three ways in which the structure of a bacterial parasite differs from the structure of its host cell.
2 Why are all viruses parasitic?
3 State the difference between communicable (infectious) and non-communicable diseases.
4 Name four non-communicable diseases.
5 Name four communicable diseases that are common throughout the UK.

Life on Man

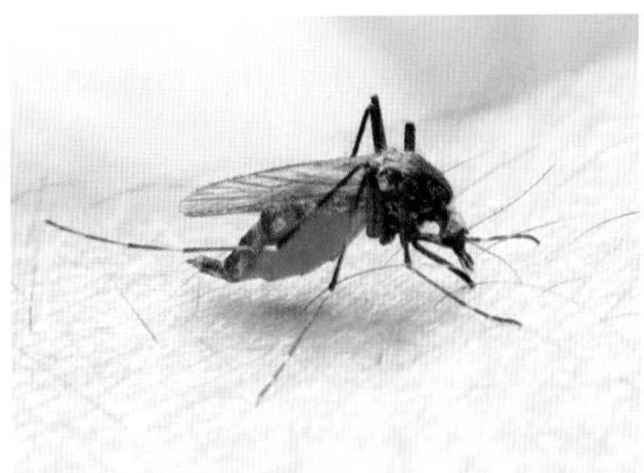

Figure 10.1 A female *Anopheles* mosquito feeding on human blood and transmitting the parasite that causes malaria.

Our bodies are a rich source of food for a variety of organisms. Some are temporary visitors, coming to feed on blood from time to time. Mosquitoes, midges and bed bugs are like this. People in South America have reported being fed on by vampire bats. Other external parasites, such as ticks, lice, fleas and mites, attach themselves almost permanently to the outside of the body.

We are not only a source of food, but a habitat that provides conditions such as warmth, moisture and a suitable pH. Our skin is covered in many bacteria and yeasts that survive on dead skin and various secretions. Others live inside our natural openings, particularly inside the mouth and nasal cavities. The lower part of the gastro-intestinal tract has many bacteria and protoctists, which form our gut flora. These organisms living on and in us depend on us for food, but none are harmful; in fact some are beneficial to us because their presence makes it more difficult for harmful organisms to grow.

Some parasites breach our defences and enter the bloodstream and then spread throughout the body. Some of these are extracellular parasites living on the surfaces of cells or in the spaces between cells. Many, including all the viruses that infect us, are intracellular parasites, since they can enter cells, survive and reproduce. Some of these organisms are obligate parasites, because they cannot exist as free-living organisms.

Pathogens and communicable diseases

Key terms

Parasite An organism that lives on or inside another organism known as the host. Parasites gain energy from their hosts and depend on this energy for their reproduction.

Pathogen A disease-causing organism; many are microorganisms such as bacteria, viruses, fungi and protoctists.

Communicable diseases Diseases caused by a pathogen that is transmitted from one host organism to another.

Non-communicable diseases Diseases not caused by a pathogen. (These diseases have numerous other causes.) Examples in this category include inherited (genetic), degenerative and deficiency diseases.

Disease transmission The transfer of a pathogen from an infected host to an uninfected host.

Parasites live inside or on the surface of another organism known as the host. They obtain their energy and nutrients from the host, as well as protection. Some of our parasites do not cause many problems; for example, several species of *Entamoeba* live in the human gut, and none of them are associated with any diseases. Microorganisms that cause disease are called **pathogens**.

All diseases that are caused by pathogens are **communicable diseases**, also commonly known as infectious diseases. All other diseases that are not caused by pathogens are **non-communicable diseases**; examples include genetic diseases, and deficiency diseases caused by poor diet.

The transfer of pathogens from an infected host to an uninfected host is **disease transmission**. If it is known how communicable diseases are transmitted then it is possible to devise suitable control methods to stop them spreading. Often this involves public expenditure to improve the infrastructure of a country, as with diseases spread through water contaminated by human gut pathogens, such as those that cause cholera and typhoid. In other cases, people have to take precautions such as sleeping under bed nets to avoid malaria (see Figure 10.2) and adopt good hygiene to avoid the spread of airborne pathogens and those passed by direct skin contact.

You will have seen fruit such as apples and oranges covered in mould fungi. These fungi are plant pathogens. Flowering plants are hosts to many different types of parasites, just as we are. Animals and plants are assaulted by pathogens of different types. This chapter will consider pathogens from the following four groups of microorganisms:

- bacteria
- viruses
- fungi
- protoctists.

Figure 10.2 A young girl helps a health worker demonstrate how to use insecticide-treated bed nets in Ambowuha in the Amhara region of Ethiopia.

Bacteria

Bacteria are prokaryotic organisms (see Chapter 1) with a huge diversity of different types, although the ones people tend to know most about are those that cause human diseases such as **tuberculosis** (TB) and the bacterial form of meningitis. Most cases of human TB are caused by *Mycobacterium tuberculosis* (see Figure 10.3), although the species that causes TB in cattle, *M. bovis*, can also infect humans. Table 10.1 on page 177 lists the different types of bacteria that cause **meningitis**; one of them, *Neisseria meningitidis*, is shown in Figure 10.4. Both *M. tuberculosis* and *N. meningitidis* infect human cells – they are intracellular parasites. Not all pathogenic bacteria infect cells; some remain in body spaces.

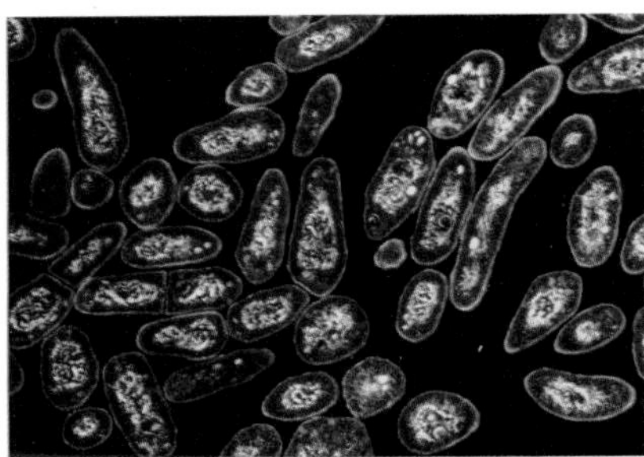

Figure 10.3 Colour-enhanced transmission electron micrograph of cells of *Mycobacterium tuberculosis* (x3000).

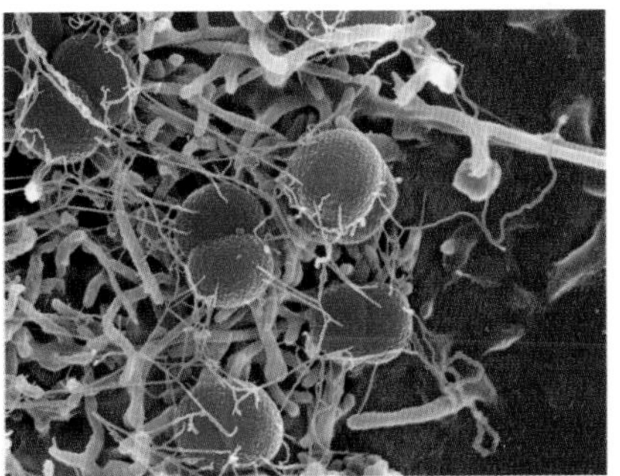

Figure 10.4 A scanning electron micrograph of bacteria that cause meningitis (*Neisseria meningitidis*) settled on the epithelium of human cells (x7000).

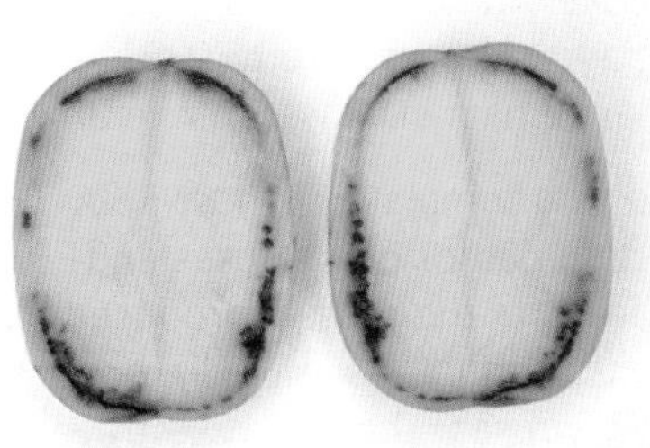

Figure 10.5 A potato tuber cut in half to reveal ring rot bacteria in the vascular tissue. Later in the infection, bacteria ooze out of the black area and then the whole potato goes 'cheesy'.

Figure 10.6 Model of the tobacco mosaic virus (×1000 000). The protein capsid is shown in blue. The capsid surrounds the genetic material of the virus. In TMV, the genetic material is RNA, which is shown in red.

The meninges are the tissues that surround the brain and spinal cord. They prevent the entry of most bacteria, but *N. meningitidis* is one of the few that can pass through this barrier to cause the disease known as bacterial meningitis. Some viruses also cross the meninges to cause a viral form of the same disease.

Bacterial pathogens of plants tend to kill their hosts and then feed on the dead and decaying tissues. There are fewer bacterial pathogens of plants than there are fungal and viral pathogens, but bacterial diseases cause huge losses of crops before and after harvest.

Many bacterial diseases of plants are called rots. If you have ever peeled potatoes and cut them up, you may have noticed that the outside and inside can go black. This is one of the rots of potato.

The bacteria that cause ring rot in potatoes infect vascular tissue (xylem and phloem) and block it so that less water reaches the leaves and they wilt. The bacteria spread through the vascular tissue in the stem and into the developing tubers. Vascular tissue in tubers is arranged as a ring, and that is why the disease first appears as a ring (see Figure 10.5).

Viruses

Tobacco mosaic virus (TMV) was the first virus to be discovered. It is a pathogen of many species, not just tobacco plants. Many other viruses cause mosaic diseases, which have the symptom of a yellowing of the leaves to give a mosaic pattern (see Figure 10.6, and also Figure 10.14 on page 181). TMV is one of the simpler types of viruses – it has a rod of protein surrounding a coil of single-stranded RNA. The RNA is the genetic material of the virus and codes for four proteins.

Viruses are the ultimate in parasitism: to be successful they have to infect cells, where they hijack the host cell's metabolism to make more of themselves. Virus particles like the one in Figure 10.6 have no cellular structure, so they cannot respire, make ATP or transcribe and translate the genes coded by the genetic material. Once a virus infects a cell, it hijacks the cell's machinery to make more copies of its genetic material and the proteins needed to make the coat, or capsid, that you can see in Figure 10.6.

Three types of virus cause **influenza**: influenza A, B and C. The most common form of influenza is caused by virus type A, which was responsible for the worldwide epidemics of 1918, 1957, 1968 and 2009 (see Chapter 11). Influenza A has a capsid that surrounds eight single-stranded molecules of RNA that between them code for 11 genes. This virus infects the cells lining the airways of the gas exchange system.

As virus particles leave their host cells they are enveloped in a phospholipid bilayer derived from the cell surface membrane. This membrane contains two types of glycoprotein, which you can see in Figure 10.7 – haemagglutinin and neuraminidase – which are involved in infecting new host cells and are coded by the genes in the viral RNA.

The **human immunodeficiency virus** (HIV) infects certain cell types, including brain cells and some in the immune system. Like influenza, it is an enveloped virus with RNA as its genetic material.

The protein capsid is surrounded by a matrix of viral protein and then by a phospholipid bilayer that is formed from the cell surface membrane of the host cell from which it emerged (see Figure 10.8). Inserted into this membrane are molecules of glycoprotein that fit into molecules of a cell surface protein of the cells that the virus can infect.

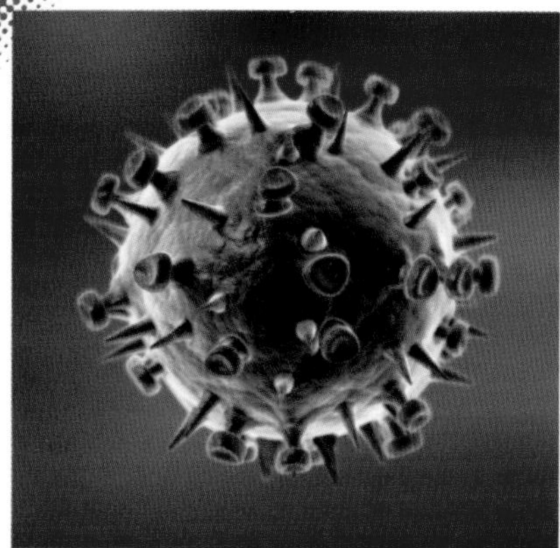

Figure 10.7 An artist's impression of the influenza A virus (×300 000).

Key term

Retrovirus A virus, such as HIV, with single-stranded RNA as its genetic material, not DNA. When a retrovirus invades a host cell, it uses its enzyme reverse transcriptase to make DNA from an RNA template. Other RNA viruses, such as influenza viruses, do not use their RNA as a template for making DNA.

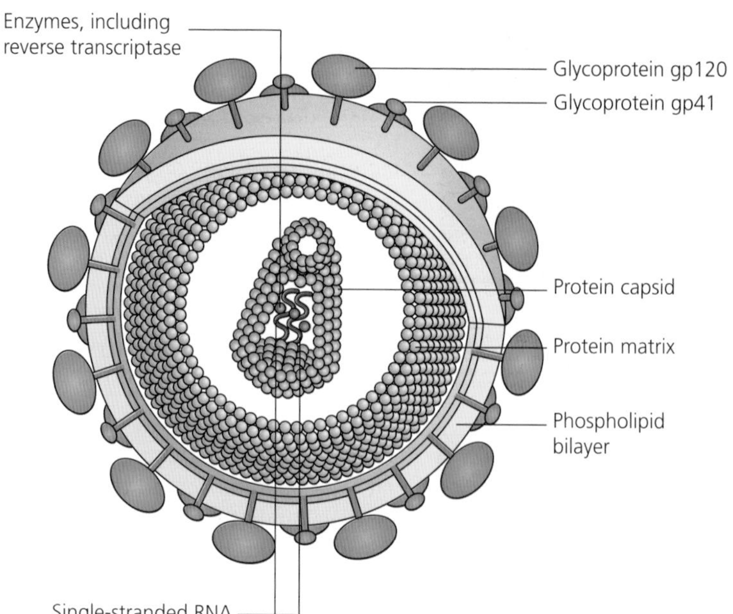

Figure 10.8 HIV is an enveloped virus. A protein capsid surrounds RNA and three enzymes. As the virus leaves a host cell, it is surrounded by host cell surface membrane that includes viral glycoproteins.

HIV is a **retrovirus**. Its RNA is used as a template to make DNA, which is the reverse of what normally happens in cells.

The enzyme reverse transcriptase uses the viral RNA as a template to make single-stranded DNA. This single-stranded DNA is replicated by DNA polymerase so that the DNA becomes double stranded. The DNA enters the nucleus, where the viral enzyme integrase attaches it to host DNA. This incorporated viral DNA is a provirus and may remain inactive for several years.

When activated, the DNA provirus is used as a template for host RNA polymerase to make RNA as the genetic material for new viruses and mRNA to make viral proteins. Viral protease cuts the protein produced on the host cell's ribosomes into short sections that are assembled around RNA to make the capsid and matrix of new viruses. These viruses travel to the cell surface membrane and leave surrounded by host cell membrane with HIV glycoproteins incorporated (see Figure 10.9).

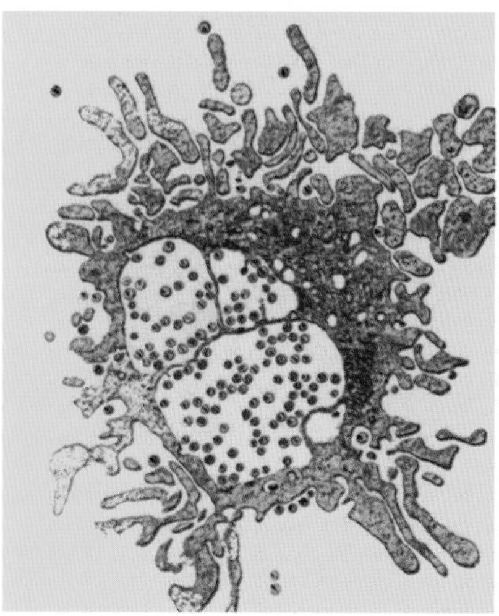

Figure 10.9 A transmission electron micrograph of a lymphocyte infected with HIV particles which will leave this cell to infect others (×25 000).

Protoctista

The parasite that causes malaria is classified with other unicellular eukaryotes in the kingdom Protoctista. There is more about the features of Protoctista in Chapter 14. Many mammals and birds are infected by these parasites. Human malaria is caused by several species of the genus *Plasmodium*, the most severe form of the disease by *P. falciparum*. The parasite goes through some of the stages in its life cycle in humans and other stages within the bodies of female *Anopheles* mosquitoes.

Fungi

Fungi are eukaryotic organisms that have a structure not unlike that of plants, with cell walls and large central vacuoles. Instead of being made of separate cells, many fungi are composed of filaments known as hyphae that form an extensive network throughout the soil or, in the case of

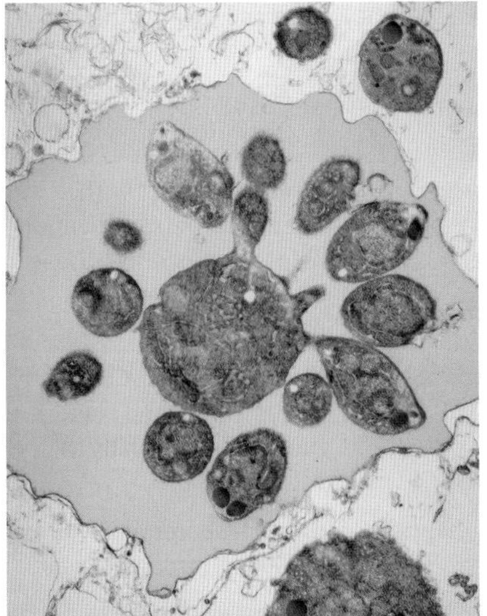

Figure 10.10 *Plasmodium falciparum* reproducing inside a red blood cell. The host cell has broken apart, or lysed, as new parasites (yellow) are released to infect new red blood cells. The green body in the centre is the remains of the original cell of *Plasmodium* (×2800).

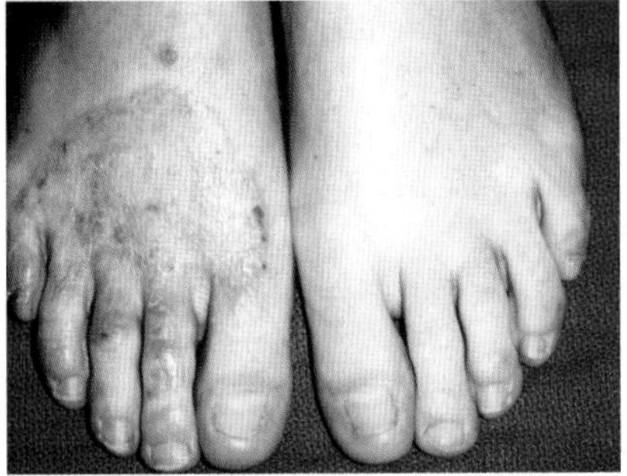

Figure 10.11 Athlete's foot – a fungal disease of the skin.

Figure 10.12 A banana leaf showing symptoms of black sigatoka. The fungus gets this name because the streaks on the leaves look similar to those of yellow sigatoka, which was discovered in the Sigatoka Valley in Fiji.

parasitic fungi, over the surface or within the body of their hosts. There is more about fungi in Chapter 14. Athlete's foot and ringworm of cattle are two diseases caused by fungi that grow over the surface of the skin.

There are many more fungal diseases of plants than there are of animals. Some of the most serious threats to our staple crops such as wheat, rice and maize are fungal diseases. Black sigatoka is caused by a fungus that causes black streaks in banana leaves. The fungus spreads through the leaf tissue, reducing the plant's ability to photosynthesise. As the disease spreads, the whole leaf dies – reducing the production of carbohydrate that is destined to be transported to the fruits.

Perhaps the most famous, or infamous, plant disease of all time is late blight of potatoes – more often just called potato blight. This disease was the cause of the potato famines in the nineteenth century, including the Irish potato famine of the 1840s. The pathogen that causes late blight of potatoes and tomatoes is *Phytophthora infestans*. *P. infestans* is classified as a protoctist although it shares some features with fungi. *P. infestans* has hyphae, like the mould fungi, but it also has features that are not associated with fungi: the walls of the hyphae are made of cellulose and not chitin, the main storage substance is starch not glycogen, and it has motile spores that have flagella to swim through water in the soil and on the surface of plants.

Question 5 on page 86 includes a diagram showing how late blight grows through a plant host and then spreads to other plants nearby.

Tables 10.1 and 10.2 are checklists of the animal and plant diseases described in this chapter.

Table 10.1 Checklist of animal pathogens, the diseases they cause and their methods of transmission.

Type of pathogen	Disease	Names of pathogens	Method of transmission
bacterium	tuberculosis (TB)	*Mycobacterium tuberculosis* *M. bovis*	**direct** – in droplets through the air
	bacterial meningitis	*Neisseria meningitidis*, *Haemophilus influenzae* (most often by type b, Hib), group B *Streptococcus pneumoniae*, and *Listeria monocytogenes*	**direct** – in droplets in the air and through exchange of fluids, e.g. during kissing
virus	HIV/AIDS	human immunodeficiency virus (HIV)	**direct** – contact between body fluids such as blood, semen and vaginal fluids; transmission from mother to child in breast milk
	influenza	three types of virus, e.g. influenza A, B and C	**direct** – in droplets through the air
protoctist	malaria	*Plasmodium falciparum*, *P. ovale*, *P. malariae*, *P. vivax*	**indirect** – by female *Anopheles* mosquito
fungus	cattle ringworm	*Trichophyton verrucosum*	**direct** – contact with infected cattle
	athlete's foot	*Epidermophyton floccosum*, *Trichophyton rubrum*, *T. mentagrophytes*	**direct** – e.g. contact with towels used by infected people

Figure 10.13 Late blight growing over tomato fruits and producing spores.

Tip

You should remember the four different types of pathogen (bacteria, viruses, fungi and protoctists) and be able to match the diseases of plants and animals given in Tables 10.1 and 10.2 with the types of pathogen that cause them. There is no need to remember the species names.

Table 10.2 Checklist of plant pathogens, the diseases they cause and their methods of transmission.

Type of pathogen	Disease	Host	Names of pathogens	Method of transmission
bacterium	ring rot	potato, tomato	*Clavibacter michiganensis*	**direct** – contact with infected tubers; cultivation helps to spread the disease as bacteria remain on machinery
virus	mosaic	tobacco	tobacco mosaic virus (TMV)	**direct** – contact with leaves of infected plants **indirect** - via aphids as vectors
fungus	black sigatoka	bananas	*Mycosphaerella fijiensis*	**direct** – spores are dispersed through the air
protoctist	late blight	potato, tomato	*Phytophthora infestans*	**direct** – swimming zoospores and aerial spores

Test yourself

1 What is the difference between an intracellular parasite and an extracellular parasite?
2 Define the term *pathogen*.
3 State three ways in which the structure of the influenza A virus differs from the structure of TMV.
4 State two bacterial diseases of humans.
5 Make a table to show the differences in structure between bacteria, viruses and the malarial parasite *Plasmodium*, which is a protoctist.
6 State two fungal diseases of mammals and two fungal diseases of crop plants.
7 Explain why *Phytophthora infestans* is described as fungus-like and is classified in the kingdom Protoctista and not in the kingdom Fungi? (You might like to look at information about fungi in Chapter 14.)

Transmission of pathogens

All communicable diseases are caused by pathogens. To be successful, any pathogen must transfer from one host to another. If it fails, it will become extinct with the death of its host.

The transfer of pathogens from an infected host to an uninfected host is **disease transmission** and involves a variety of methods. There are two main principles here. First, transmission occurs when a pathogen is transferred from an infected individual to an uninfected individual. Second, transmission is a risky business, and pathogens have mechanisms to produce large numbers of individuals to increase the chances of some of them finding a new host. The infective stages that are transmitted tend to be very small. There is no point in using much energy to produce a small number of large individuals unless they have a very good chance of finding a host.

Transmission occurs either **directly** from one host to another or **indirectly** via a second organism that is usually unaffected by the pathogen.

Direct transmission

Direct transmission may involve contact between two individuals; for example, in a field of tobacco plants, leaves infected with TMV may touch the leaves of an uninfected plant, so the virus particles are transmitted. When the spores of black sigatoka and late blight settle on a suitable host plant, they grow small tubes that penetrate their hosts, either through the cuticle or through stomata. TB bacteria, influenza viruses and the different species of bacteria that cause meningitis are breathed out in tiny droplets of water that may be breathed in by uninfected people.

HIV is not a very robust parasite and is only transmitted by direct contact involving some body fluids. Several ways in which this can happen are, if:

- a person who is HIV positive has unprotected vaginal or anal sex with a person who is uninfected
- blood in a needle or syringe that was used on an HIV-positive person is then transferred to someone who is not HIV positive; this can happen when a needle or syringe is not sterilised by a health worker following use or when a needle is shared between intravenous drug users
- blood from an HIV-positive person is used in transfusion or is a contaminant in blood products
- at birth, blood of an HIV-positive mother mixes with the blood of her baby
- a baby drinks the breast milk of an HIV-positive mother.

Tip

People who have tested positive for HIV are described as HIV positive (HIV+). Those who have tested negative are described as HIV negative (HIV–).
The condition that is ultimately caused by HIV is now known as HIV/AIDS, as if it is one disease. Remember that HIV is the pathogen that leads to a weakened immune system, which in turn allows many diseases to infect people who are HIV positive.

After transmission of HIV, there is a short incubation period of several weeks. Then there are mild flu-like symptoms that are often misdiagnosed. The infection is then symptomless for a fairly long time, until a variety of **opportunistic diseases** appear, including thrush, tuberculosis, a rare form of pneumonia and Kaposi's sarcoma (a rare form of cancer). These diseases develop because the number of lymphocytes has decreased, because they have been destroyed by HIV infection. The collection of opportunistic diseases associated with HIV infection is known as acquired immunodeficiency syndrome, or AIDS. This is not a single disease with a consistent set of symptoms. The symptoms shown by people who have AIDS vary according to the opportunistic diseases that they have.

Key terms

Spore A small reproductive structure. Spores are released into the environment, dispersed in wind or water, and start to grow when they reach a suitable food source. Some spores are produced following mitosis and others after meiosis. Spores can be haploid or diploid.

Vector An organism that transfers a pathogen from an infected host to an uninfected host. The vector is not usually harmed in any way by the pathogen.

A more complex method of direct transmission is the formation of spores. These are produced specifically for transmission via some medium, be it air or water. For example, *P. infestans* produces hyphae that grow out through stomata. These hyphae swell to produce sporangia, which are pear-shaped structures about 30 μm long. Sporangia may be blown by the wind to land on uninfected leaves; there they produce specialised hyphae that enter the plant to begin a new infection.

Indirect transmission

Malaria is spread by the female *Anopheles* mosquito. Mosquitoes of this type normally feed on plant sap, but when the female is ready to make eggs she needs a richer source of protein and gets this by taking blood meals. If a female mosquito takes a blood meal from someone infected with *Plasmodium*, she will also take in many reproductive forms of the parasite. These reproduce inside her gut and move to her salivary glands, ready to infect another human when she takes another blood meal. Transmission is complete when she takes a blood meal from an uninfected person, injecting her saliva to stop the blood clotting. The mosquito is the vector of the disease.

Tip

Remember that malaria is caused by a **protoctist**, which is **eukaryotic**, and is transmitted by an **insect vector**.

Very similar to transport by mosquitoes is the transfer of plant viruses by aphids. The peach potato aphid, *Myzus persicae*, is an important vector of plant viruses. Like other aphids, *M. persicae* inserts its stylets into phloem sieve tubes (see page 118 in Chapter 6). While feeding, viruses such as TMV attach to the stylets and, when the aphid flies to an uninfected plant, they will be transmitted. The potato leaf roll virus passes through the lining of the intestine of its vector and then via the blood to the salivary glands. This increases the chances that the virus will be transmitted when the aphids feed again. An individual aphid can carry potato leaf roll virus for a long time.

Anopheles and *M. persicae* are both insects. Many disease vectors are insects. The advantages of this for pathogens are that in suitable conditions, insects reproduce in very large numbers, which increases the chances of transmission. Aphids reproduce when plants are available, which is perfect timing for the transmission of pathogens.

Factors that influence transmission of communicable diseases

The factors that affect transmission depend on the type of transmission. The factors affecting the transmission of plant pathogens and animal pathogens are broadly similar. Some of the diseases from Tables 10.1 and 10.2 will be used to illustrate these factors. Some factors are more specific to the transmission of human pathogens, and some of these will be dealt with separately.

The first factor that affects the transmission of all diseases, however they are transmitted, is the presence of infected individuals. If the pathogen is not present in a population then no potential hosts can be infected. Diseases that are always present in a population are known as **endemic** diseases. Malaria is endemic in many tropical countries but not elsewhere; TB is endemic across the whole world. Typhus fever was introduced into the Americas by the Spanish explorers, where it caused many deaths among the native populations, and now persists in endemic form in Mexico, Central, and South America. The pathogen is always there ready to infect people during periods of social upheaval, when they are cold, dirty and hungry.

Tip

Do not confuse endemic with epidemic. An epidemic of a disease occurs when there is an increase in number of cases in a population. Endemic refers to a disease that is always present in a population even though numbers of cases may be very low at any one time.

Many pathogens have several different stages in their life cycles. Only some of the stages are infective. Of the diseases listed at the beginning of the chapter, only the malarial parasite, *Plasmodium*, has a complex life cycle. The only infective stage is the one found in the salivary glands of *Anopheles* mosquitoes (see Figure 10.1). If there are no infective stages then there can be no transmission. Disease outbreaks occur when new strains of diseases originate. There is more about this in Chapter 11.

Animals and plants are resistant to some diseases. This means that they have inherited genes that code for some mechanism that prevents infection by specific pathogens or prevents their spread within the body. People who are heterozygous for the sickle cell allele are resistant to malaria. This means that when such people are exposed to the disease for the first time, the parasite does not develop inside their bodies sufficiently to cause symptoms. Most humans are resistant to many diseases, including leprosy and the prion disease known as Creutzfeldt–Jakob disease. Some are even resistant to HIV infection.

Resistance is different to immunity. As you will see in Chapter 11, animals develop immunity to pathogens. This means that they develop symptoms the first time they are infected by a specific pathogen, but are very unlikely to develop symptoms when infected by this pathogen on any subsequent occasion.

The proportion of resistant and/or immune individuals in a population is also a factor that influences transmission. The higher the proportion, the lower the chances of transmission occurring.

Factors affecting direct transmission

Diseases that are spread by droplet infection are dependent on potential hosts being in close proximity. Influenza spreads to infect more people if people live in high densities. For example, there are often higher infection rates in schools than in the general population, with infection rates higher in boarding schools than in day schools. People are more likely to be infected with TB bacteria if they live in close proximity to infected individuals who breathe out, cough or sneeze droplets of water containing TB bacteria. TB transmission is more likely to happen among people who sleep in close proximity, for example in poor housing where the whole family may sleep in the same room, or in shelters provided for the homeless.

Figure 10.14 Tobacco plants showing symptoms of tobacco mosaic virus, a disease that is easily spread when infected leaves of one plant touch the leaves of an uninfected plant.

Crop plants are usually grown in monocultures, at high densities, to ensure good use of the available land and for maximum absorption of light energy. Under these circumstances, leaves infected with TMV can easily touch uninfected leaves. Very soon the virus will spread throughout a crop, causing the characteristic yellow blotches on the leaves (see Figure 10.14).

Factors affecting indirect transmission

The transmission by vectors such as mosquitoes and aphids is influenced heavily by factors that affect their biology. The population of these insects is influenced by climate and by weather.

Anopheles mosquitoes need small areas of water in which to breed. They therefore tend to breed more frequently in wet conditions than in dry. The transmission of malaria is much higher during wet, rainy seasons than during dry seasons. The development of the malarial parasite within mosquitoes is temperature-dependent. If the temperature decreases below 20 °C, the parasites cannot complete their stages within mosquitoes. Aphids have a telescoped life cycle in which females give birth to live females (not eggs) every few days. The ideal conditions for aphids to reproduce in are those with plenty of food and with warm weather (temperatures just over 20 °C).

Factors affecting transmission of human diseases

The effect that migration has had on transmission of human diseases has already been mentioned. Many of our diseases are diseases of poverty. For example, water-borne diseases such as cholera, typhoid and polio are spread when human faecal wastes contaminate drinking water. This can happen in areas of poor housing where there is little or no sanitation, sewage treatment or provision of piped water that has been chlorinated to kill bacteria. Poor housing may also be unhygienic, so the transmission of other diseases is possible, especially if people are not able to cook food thoroughly.

Diseases are transmitted rapidly when they are brought into new populations with little (or no) natural resistance and no immunity. The Spanish conquistadores took diseases from Europe, such as smallpox, to the Americas, where people had no immunity and death rates were very high. In return, the first sailors to the Americas brought back syphilis, which caused an epidemic throughout Europe in the sixteenth century.

Increased movement of people around the globe has increased the chances of disease transmission. The first epidemic of the H1N1 strain of influenza A spread around the globe in 1918 in about a year. The 2009 pandemic took about three months to spread from Mexico to Ghana. Similarly, the viral disease known as severe acute respiratory syndrome (SARS) spread across three continents within weeks in 2003. Toronto in Canada became a focus for SARS, because a person travelled there from Hong Kong, where she had caught the virus. Someone who is infected in one country can travel, without showing any symptoms, to a major centre like London or New York by the following day, with the risk of infecting thousands of people.

People's behaviour can put them at risk of communicable diseases. For example, the risk of being infected with sexually transmitted diseases such as HIV increases with the number of sexual partners.

Activity

Identifying the pathogen that causes a disease

Figure 10.15 shows an apple with what looks like a fungal disease.

How do we know that a specific pathogen causes a disease? Some students followed the procedure used by scientists to confirm the identity of pathogens.

Method

1 The students described the symptoms of diseased apples like the one on the left of Figure 10.15. The most obvious symptoms are the brown colour and the blue mould on the surface.
2 The students then isolated some of the diseased tissue with a sterile scalpel and placed it on to malt agar in sterile Petri dishes. They incubated the dishes at 25 °C for seven days. When they checked the agar plates after seven days, only one type of fungus had grown.
3 Then the students took nine undamaged apples with no signs of disease and divided them into three equal groups and followed these instructions:
   - Group A – make a small incision through the skin with a sterile scalpel and use sterile forceps to transfer some of the fungus from the agar into the apple tissue.
   - Group B – treat as A but use a sterile pipette to insert sterile distilled water instead of the fungus.
   - Group C – leave untreated.
4 The apples were left at 25 °C for a further seven days and then examined. Fungus was growing on all the apples in group A, but not on the other apples. The symptoms were the same as in the diseased apple in Figure 10.15.

Figure 10.15 The apple on the left looks as if it is infected by a fungus.

5 The students removed some of the fungus from the apples in group A and put it on sterile malt agar. The students next used a microscope to compare the structure of the fungi that they grew on agar with that of the fungi removed from the apples in group A. The two samples were identical.

Questions

1 Explain why the students used sterile apparatus to remove the fungus from the apple.
2 Explain the reasons for including groups B and C in the investigation.
3 Suggest a way, other than by visual appearance, to make sure that the fungi isolated from the apples and grown on the agar belong to the same species.
4 Suggest how this disease of apples is transmitted.
5 Explain how the procedure described shows that the fungus is the cause of this disease of apples.

Plant defences against pathogens

Although plants seem very prone to infection by pathogens, and the damage caused by pathogens that cause ring rot, late blight and black sigatoka is devastating, plants do have many mechanisms to protect themselves against infection. These mechanisms are divided into passive and active.

Passive defence mechanisms are those that are present all the time. Some of these are physical barriers; others are chemicals that deter the growth of pathogens. Examples of passive defences are:

- physical barriers
  - waxy cuticle over leaf epidermis
  - bark
  - cellulose cell walls
  - Casparian strip in the endodermis of the root (see page 162 in Chapter 9)
  - stomata that close to stop pathogens from entering
- chemical defences
  - secreting compounds to support growth of microorganisms that compete with pathogens
  - secretion of compounds toxic to pathogens
  - secretion of inhibitors of enzymes, e.g. cellulose used by pathogens to break down cell walls to gain entry to cells
  - receptor molecules on cell surface membranes that detect pathogens and activate plant defences
  - sticky resins in bark that prevent the spread of pathogens.

Physical barriers such as waxy cuticles reduce the chances of plant pathogens gaining entry or spreading very far inside a plant. Viruses and bacteria, for example, cannot enter through the cuticles unless there is a wound on the surface of a leaf or stem. Wounds are made by herbivores such as caterpillars. Many fungi grow into roots and penetrate as far as the endodermis, but cannot grow any further because of the impenetrable Casparian strip.

Plants produce chemicals to prevent the growth of pathogens. By secreting substances on to the surface of leaves and roots they can make the environment too acidic for the growth of the pathogen. Leaves also secrete nutrients that support a community of microorganisms such as yeasts that are harmless to the plant but compete successfully with any pathogens that try to establish themselves on the leaf surface. Humans have a similar defence mechanism – see page 189. Plants also make substances that are directly toxic to pathogens, such as catechol. They also secrete enzyme inhibitors such as tannins.

Unlike animals, plants cannot produce cells that roam all over the body searching out pathogens and destroying them. Plant cells have cell walls, and those cell walls are fixed in one place. Plants have **active defence mechanisms** that are activated when pathogens invade. The most drastic of these is hypersensitivity – the almost immediate death of tissues surrounding the site of infection by a pathogen. This seems an extreme

Key term

Callose A polysaccharide that has 1,3 glycosidic bonds and some 1,6 glycosidic bonds. It is made by plant cells in defence against attack by pests and pathogens to block sieve pores in phloem sieve tubes, reduce the width of plasmodesmata and thicken cell walls.

measure but is highly effective, because many pathogens require living host tissues to survive and spread. If the cells die then there are no nutrients and there is no energy for continued growth. Plants also respond by making physical barriers. When bacteria and fungi attempt to penetrate cell walls, this often stimulates the cell to produce compounds such as callose and lignin to thicken and reinforce cell walls to make them harder to penetrate.

Callose is deposited between the cell surface membrane and the cell wall. Callose is a polysaccharide that forms a matrix in which antimicrobial compounds can be deposited, such as hydrogen peroxide and phenols, that kill any pathogens attempting to enter the cell. Callose also reduces the diameter of plasmodesmata, the membrane-lined channels that run through cell walls linking the cytoplasm of neighbouring plant cells. By narrowing these channels, callose reduces the spread of viruses from cell to cell. Some pathogens move through plants inside the vascular tissue. Tyloses are ingrowths into xylem vessels to prevent the movement of pathogens within the vessels. The cytoplasm of adjoining living cells grows into the xylem vessels and forms a wall often made of callose. Callose is also deposited in sieve pores to block the transport of phloem sap and impede the movement of pathogens.

Cell signalling plays a large part in plant responses to disease. Bacterial and fungal pathogens secrete cellulases to digest a pathway into cells. The breakdown products of cellulose hydrolysis act as signals that are detected by receptors on the surfaces of cells. Once detected the breakdown products stimulate the production of **phytoalexins**, which are defence chemicals that act in a variety of different ways and:

- disrupt the cell surface membranes of bacteria
- stimulate the secretion of chitinases that break down the cell walls of the hyphae of fungal pathogens
- disrupt metabolism in the pathogen
- delay reproduction of the pathogen.

Signalling molecules such as salicylic acid travel through plants to activate defence mechanisms in uninfected areas, giving them protection against many pathogens for some time after the original infection. This long-term response to infection is called **systemic acquired resistance**. Ethylene is another signalling compound. When secreted by plants under attack from pathogens, ethylene vapourises to stimulate other leaves of the same plant and also other plants in the immediate surroundings.

Example

1 Describe the transmission of
   a) tobacco mosaic virus (TMV)
   b) the pathogen that causes tuberculosis in humans.
2 Describe how plants are protected against infection by pathogens.
3 Following infection of a single leaf by a fungal pathogen, other leaves on the plant have a greater level of resistance to infection to all other pathogens. Explain how this happens.

Test yourself

8 Malaria literally means 'bad air'. In the late 1890s, Ronald Ross discovered that the malarial parasite is transmitted by mosquitoes, not directly through the air or through water. What were the consequences of Ross's discovery for control methods for this disease?

9 List the factors that influence the spread of communicable diseases that are transmitted directly between hosts.

10 List the factors that influence the spread of communicable diseases that are transmitted indirectly between hosts.

11 Explain why people living in affluent areas in the tropics are unlikely to be infected by water-borne pathogens but are at risk of vector-borne diseases.

Answers

1 a) TMV is transmitted directly from one plant to another by leaf contact. Viruses enter leaves through wounds or breaks in the leaf surface.

b) *Mycobacterium tuberculosis* is also transmitted directly from an infected host to an uninfected host. The bacteria are present in water droplets that are breathed out, coughed or sneezed by the infected person. The uninfected person breathes the droplets in and the bacteria settle inside the lungs.

2 Plants have physical barriers so that pathogens cannot enter. Leaves are surrounded by epidermis that is covered in a waxy cuticle. This waxy cuticle is impossible for viruses and bacteria to penetrate, although some fungi can penetrate it. Bark is much thicker than cuticle and even more difficult to penetrate. Stomata close in response to invasion by fungal pathogens. The endodermis in the root is protected by the Casparian strip, which is made of suberin, which pathogens cannot penetrate. This prevents pathogens from entering the xylem and being carried throughout the plant.

3 Infected tissues release cell-signalling compounds that travel in the vascular tissue. They also release ethylene, which is a volatile substance that travels through the air. These signalling compounds prompt cells in other leaves to produce phytoalexins to protect against the spread of the pathogen. Phytoalexins can be directly toxic to the pathogen or inhibit the enzymes that it needs to digest its way into the cell to gain nutrients.

Exam practice questions

1 The pathogen that causes malaria is a

A bacterium

B mosquito

C protoctist

D virus *(1)*

2 An organism that benefits from living on or inside another is known as a

A bacteruim

B microorganism

C parasite

D virus *(1)*

3 Tuberculosis (TB) is a communicable disease that is transmitted through the air in droplets of water.

a) State the type of pathogen that causes TB. *(1)*

Records have been kept of the numbers of cases of TB notified to medical authorities in England and Wales since 1913. The graph shows the number of these cases of TB between 1913 and 2012.

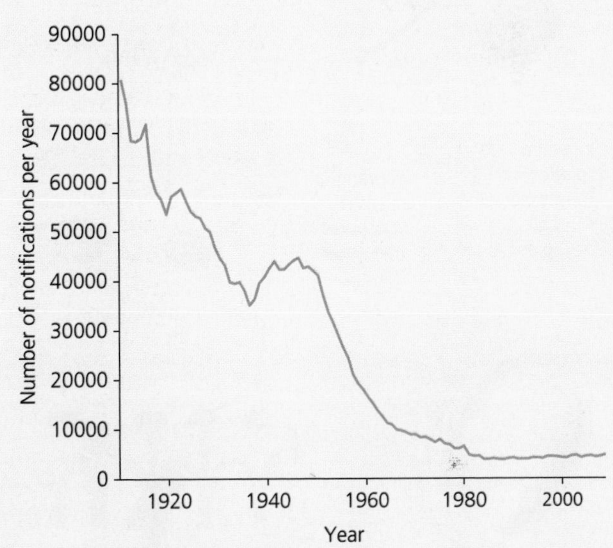

The number of cases of TB in 1913 was 80 788 and in 2012 it was 4563.

b) Calculate the percentage decrease in numbers of notified cases of TB between 1913 and 2012. Show your working. *(2)*

c) Describe the change in numbers of notified cases of TB between 1913 and 2012 as shown in the graph. *(5)*

d) The BCG vaccine for tuberculosis was introduced into England and Wales in 1953. It has been suggested that this had little effect in reducing the number of cases of TB in England and Wales.

i) State the evidence from the graph that supports the suggestion that the BCG vaccine had little effect in reducing the number of cases of TB. *(3)*

ii) Suggest and explain the social factors that influence the number of cases of TB. *(4)*

4 A fungal pathogen begins to grow through its plant host. Which defence is activated when the pathogen invades?

A callose deposition in cell walls

B formation of the Casparian strip in the endodermis

C production of a waxy cuticle over the epidermis

D secretion of resins *(1)*

5 The diagram shows part of the life cycle of *Phytophthora infestans*.

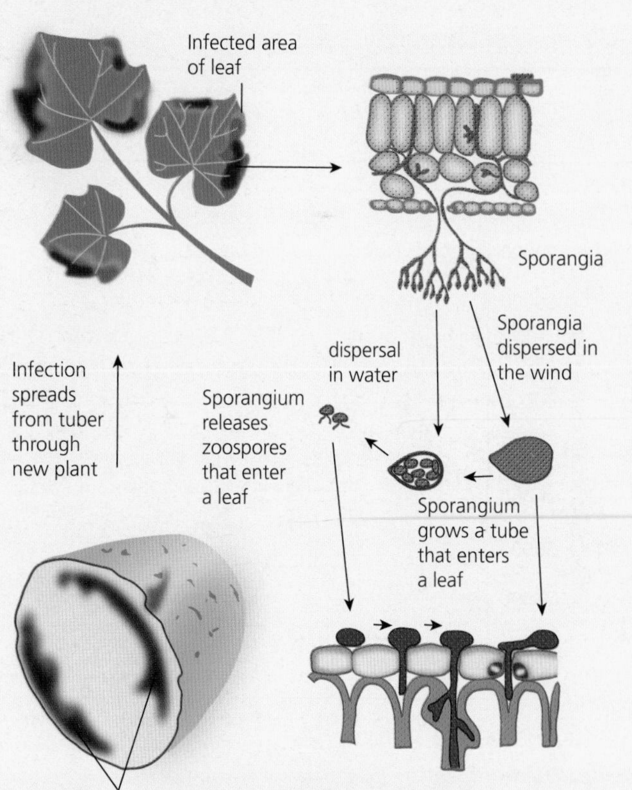
Infected area of leaf
Sporangia
Infection spreads from tuber through new plant
dispersal in water
Sporangia dispersed in the wind
Sporangium releases zoospores that enter a leaf
Sporangium grows a tube that enters a leaf
Infection in potato tuber left in the ground over winter
Infection spreads through previously uninfected plant

a) *P. infestans* shares some features with fungi, but it is classified in the kingdom Protoctista. State two features, visible in the diagram above, that *P. infestans* shares with fungi. *(2)*

b) Use the diagram to explain how zoospores infect a new host. *(3)*

c) *P. infestans* spreads very quickly at certain times of the year. Suggest the weather conditions that are most likely to favour the spread of late blight. *(2)*

d) Describe ways in which plants defend themselves following infection by pathogens such as *P. infestans*. *(5)*

6 The drawing shows an aphid that feeds on phloem and is a vector of the different viruses that cause barley yellow dwarf disease (BYD).

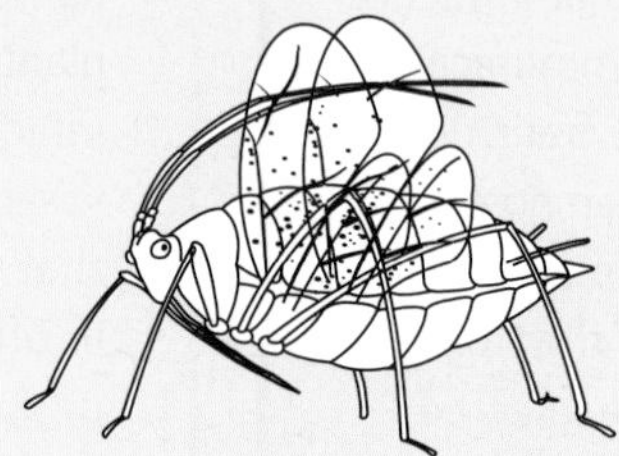

a) Aphids feed on phloem sap without using energy to make the sap flow into their mouthparts. Explain how aphids are able to feed in this way. *(2)*

b) Explain why aphids are effective vectors of plant diseases such as BYD. *(3)*

c) Some types of virus travel through the gut lining to reach the aphid's salivary glands.

i) Some of these viruses are absorbed selectively through the gut lining and by the cells of the salivary glands. Suggest how this is possible. *(3)*

ii) Describe the pathway taken by viral particles once they have been absorbed by the gut so that they reach the salivary glands. *(2)*

d) i) The RNA within each plant virus particle acts as mRNA in infected cells. Outline what will happen once RNA is released inside infected cells. *(4)*

ii) BYD viruses accumulate only in phloem cells and lead to the death of the tissue. Explain the effects this has on infected plants. *(3)*

Stretch and challenge

7 Irrigation schemes in Africa often involve constructing small dams to hold water for local agriculture. It is thought that these dams may influence the number of cases of malaria. A study was carried out in Tigray, northern Ethiopia, into the incidence of malaria in children in villages within 3 km of these dams and in villages 8–10 km from the dams.

The researchers used their results to calculate a risk ratio to compare the risk of catching malaria in the two communities. The results of this study are shown in Table 1.

Table 1

Months		Jan–Feb	Apr–May	Jul–Aug	Oct–Nov
Villages within 3 km of dams	Number of cases of malaria in the villages	7	19	33	43
	Number of cases of malaria per 1000 children	17.9	42.3	71.7	111.9
Villages 8 to 10 km from dams	Number of cases of malaria in the villages	2	2	1	3
	Number of cases of malaria per 1000 children	7.2	6.1	3.4	12.0
Risk ratio		2.5		21.1	

Table 2 shows the number of cases of malaria and the number of deaths resulting from malaria among people of all ages, for two African countries, in the years 2000 and 2012.

Table 2

Country	Botswana		Malawi	
Year	2002	2012	2002	2012
Total population	1 809 000	2 004 000	11 927 000	15 906 000
Total number of reported cases of malaria	1588	193	2 784 001	1 564 984
Total number of reported deaths from malaria	23	3	5775	5516

a) Calculate the missing risk ratios in Table 1.

b) Comment on the data in both tables.

c) Comment on the validity of the data in Tables 1 and 2.

Chapter 11

Disease prevention

Prior knowledge

Before you start, make sure that you are confident in your knowledge and understanding of the following points.

- Definitions of the terms *pathogen* and *disease transmission.*
- The different types of pathogens that cause disease in humans.
- The roles of the skin, the ciliated epithelium of the airways, and the blood in defence against invasion of the body by pathogens.
- The structure of cell surface membranes, including the structure and position of glycoproteins in membranes.
- The roles of organelles in protein synthesis.
- The four levels of protein structure.
- The principles of cell signalling, including the roles of signalling molecules and cell surface receptors.

Test yourself on prior knowledge

1 State four types of organism that are human pathogens.
2 State the roles of phagocytes in defence against pathogens.
3 Explain how the ciliated epithelium lining the airways protects the lungs against infection.
4 Explain the term *intracellular parasite*. Give two examples.
5 Why are viruses such as HIV described as *obligate parasites*?
6 Define the term *vector of disease*. Give an example.

Every town and village in Britain has a memorial to members of the armed forces who died in the First World War. Immediately after the First World War ended there was arguably an even greater catastrophe: the pandemic that came to be known as Spanish flu. This infection of influenza A was carried around the world by soldiers returning home after the war ended in 1918 (see Figure 11.1). One-third of the world's population were infected, and it is thought that as many as 50 million people died.

Researchers think that the world is overdue for another pandemic of influenza. There have been several since 1918, but nothing on the same scale.

There are few drugs for treating influenza: antibiotics are useless against viruses, although they can be used to treat secondary bacterial infections such as pneumonia. The alternative is prevention. One way to protect people against the disease is to use vaccination. At present, vaccines for flu are changed every year as the virus itself changes.

Years after the Spanish flu, survivors of the disease were found to produce antibodies against the strain of influenza A: a 2008 study, published in the scientific journal *Nature*, found that 90 years after the 1918 pandemic, survivors had antibody-producing lymphocytes that produced antibodies with the ability to block infection (in laboratory animals) by the 1918 strain of the flu. So developing a vaccine that is effective for longer is a possibility.

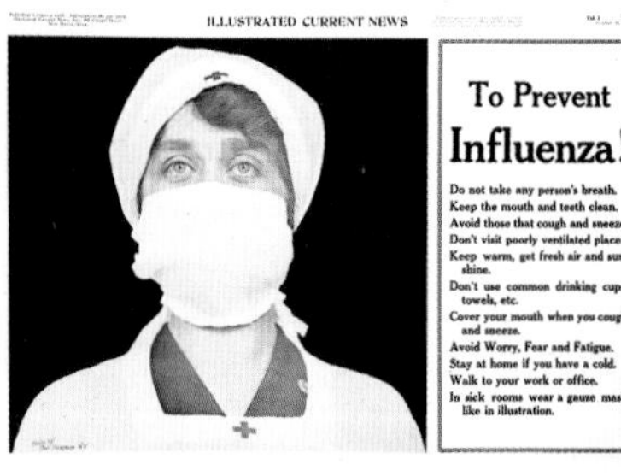

Figure 11.1 A poster printed in the USA during the influenza pandemic of 1918–1920, otherwise known as the Spanish flu.

Defences against infection in animals

Vertebrate animals have a sophisticated set of defences against invasion by pathogens. There are four main ways in which animals defend themselves against infectious diseases:

- Physical: Tissues provide barriers that pathogens cannot pass through unaided, for example skin and the mucous membranes that line the alimentary canal, the gas exchange system and the urinogenital tract.
- Cellular: Cells alert the body to the presence of pathogens, produce substances that provide protection and ingest and digest pathogens.
- Chemical: Substances secreted by the body provide inhospitable environments for pathogens, trap them, cause them to burst, stop them reproducing and/or growing, and stop them entering cells.
- Harmless bacteria and fungi: such harmless organisms (known as commensal organisms) that live on us and inside the alimentary canal and urinogenital system compete with pathogenic organisms to prevent infections.

Our defence system consists of three lines of defence, with each one consisting of a mixture of cellular and chemical defences.

Key terms

First line of defence Physical and chemical defences that prevent the entry of pathogens: the skin, mucous membranes, and chemicals such as lysozyme, sebum and hydrochloric acid.

Expulsive reflex Coughing and sneezing, which expel irritants that may include pathogens from the upper part of the gas exchange system, including the throat and nasal passage.

Tip

See pages 115 and 126 for further information about the structure and function of ciliated epithelial cells and goblet cells in the gas exchange system.

First line of defence

The skin and mucous membranes form a very effective **first line of defence**. The outer layer of skin consists of dead cells filled with keratin, which is a tough fibrous protein. This surface layer of dead, hardened cells is relatively dry, and skin secretions make the surface somewhat acidic. When sweat evaporates, salt is left behind on the skin. These conditions of low moisture, low pH and high salinity prevent most microorganisms from growing and multiplying on the skin.

The gut, airways and reproductive system are lined by mucous membranes that consist of epithelial cells interspersed with mucus-secreting cells, for example goblet cells. Mucus is a sticky, slimy substance full of glycoprotein molecules, each of which has long carbohydrate chains to make it sticky. In the trachea, bronchi and bronchioles, small particles in the air, such as bacteria, viruses, dust and pollen, are trapped and then moved upwards by cilia to the back of the throat (see Figure 11.2).

Figure 11.2 A scanning electron micrograph of the surface of a bronchus (×1000) showing cilia in between mucus-secreting goblet cells.

Sometimes pathogens irritate the lining of the airways, prompting a sneeze or a cough. These sudden expulsions of air carry secretions from the respiratory tract with many foreign particles that have entered. They are **expulsive reflexes** that eject pathogens from the upper parts of the airways, possibly increasing the chances that the pathogens will be transmitted.

The enzyme lysozyme is secreted into many body fluids as an antibacterial agent. It catalyses the breakdown of bacterial cell walls and is in blood, sweat, tears and milk, among many other fluids. Fatty acids in sebum secreted onto the surface of the skin also have antimicrobial properties.

Hydrochloric acid secreted by cells in the lining of the stomach kills bacteria taken in with food.

Cells along the length of the gut secrete mucus to protect the lining against attack by acid, enzymes and pathogens.

These defences prevent the entry of pathogens into tissues and into the blood. They also prevent pathogens growing inside the lungs, gut and reproductive system. These chemical defences are not always completely effective, however. For example, some pathogens, such as hepatitis A virus and cholera bacteria, can pass through the 'acid bath' treatment in the stomach.

We play host to many microorganisms, including the yeast *Candida albicans* and bacteria such as *Escherichia coli* that live in and on the body. These microorganisms grow on the skin and in the mouth, intestines and other areas of the body, but do not cause disease, because their growth is kept under control by our defence mechanisms. *E. coli* is one species of bacterium that lives in the large intestine, forming part of our normal intestinal flora. These microorganisms benefit us by competing successfully with disease-causing organisms, preventing the latter from invading host tissues. When the growth of this normal flora is suppressed, as happens during antibiotic treatment, other opportunistic organisms that do not normally grow in or on the body may be able to infect us and cause disease.

Key terms

Second line of defence Phagocytic cells and antimicrobial proteins that act against any type of invading microorganism that has breached the first line of defence.

Non-specific defences Present from birth, these defences do not distinguish between different pathogens, and they give the same response each time a particular pathogen enters the body. They usually act very quickly but are not always highly effective.

Second line of defence

If pathogens penetrate the first line of defence then there is a **second line of defence** that responds quickly to any invasion. Tissues respond to the presence of foreign cells or other foreign objects such as splinters by dealing with the problem locally and recruiting assistance from elsewhere in the body. The components of this second line are **non-specific defences**, because they do not distinguish between the different types of pathogen, for example between different species of bacteria. The responses are the same whatever type of pathogen invades the body.

Blood clotting

Blood clotting is a fairly rapid response to wounding. If there is a break in the skin or mucous membranes, platelets and cells lining the blood vessels release compounds that start a complex chain reaction, or cascade, of chemical reactions that involve a large number of plasma proteins. Clotting stops the loss of blood, prevents the entry of pathogens and provides a protective surface for wound healing to occur underneath. The blood clotting cascade is outlined in Figure 11.4.

There are many steps in the cascade, so a small stimulus is amplified to produce a large quantity of fibrin very quickly to seal the wound (see Figure 11.3).

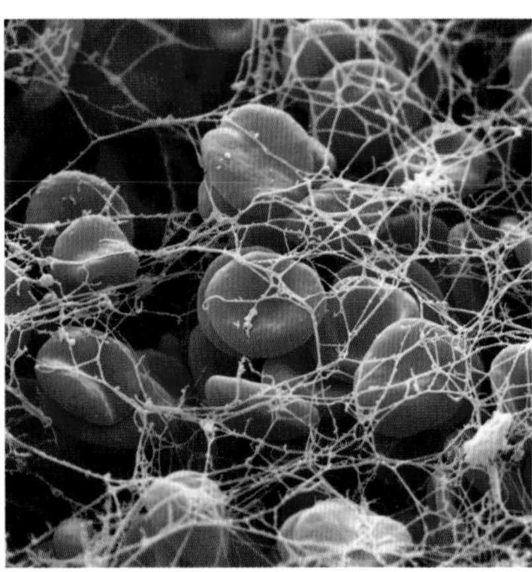

Figure 11.3 Blood clotting. Red blood cells trapped in a mesh of fibrin at the site of a wound (×2000).

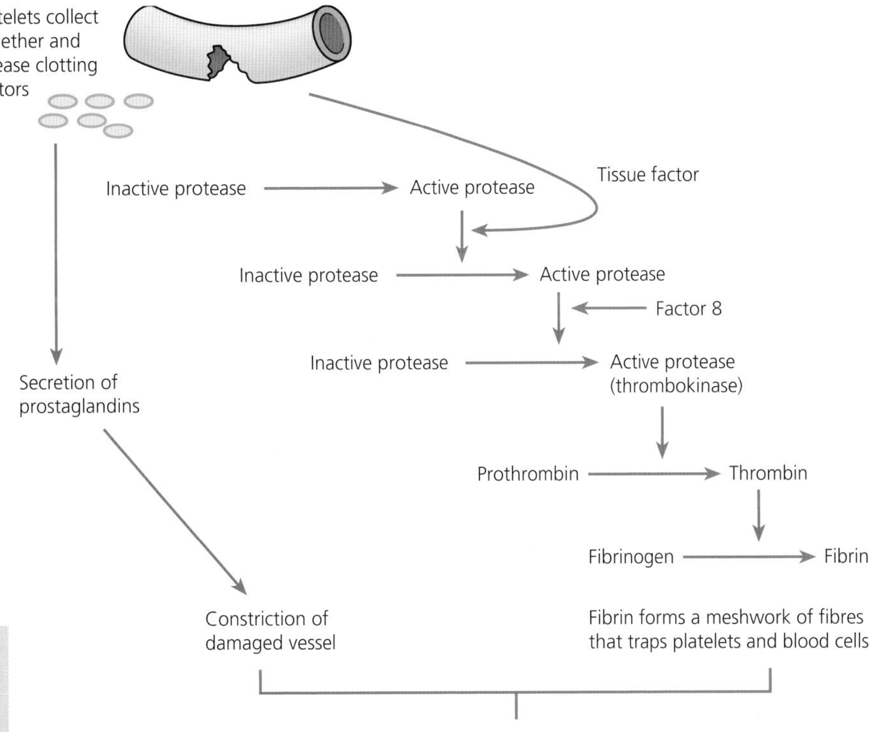

Figure 11.4 The blood clotting cascade.

Key terms

Inflammation A local response to tissue damage and infection, involving the release of chemical signalling molecules and resulting in increased blood flow and movement of phagocytes into the infected tissues.

Cytokines Small protein molecules that act as cell-signalling compounds. Many of them are involved in stimulating events that occur in response to infection, such as inflammation and the immune response. Interleukins are included in the group of compounds known as cytokines.

Test yourself

1 Explain how pathogens can get through the first line of defence.
2 What are the roles of blood clotting?
3 Outline what happens when a tissue becomes inflamed.

Inflammation

The area around a wound becomes red, swollen, hot and painful. These are the features of **inflammation** that we are all familiar with. What we cannot see at the site of a wound is the array of responses to tissue damage and infection as a result of the activity of cellular and chemical defences recruited from the blood.

Throughout the body, mast cells respond to tissue damage by secreting histamine. This cell-signalling compound stimulates a range of responses:

- Blood flow through capillaries increases as a result of vasodilation.
- Capillaries become increasingly 'leaky', so that fluid enters the tissues, leading to swelling.
- Phagocytes leave the blood, enter the tissues and engulf any foreign material.
- Some plasma proteins leave the blood.

Histamine stimulates cells to secrete **cytokines** to stimulate defences in the area of infection, and also some that have an effect on the whole body. Interleukin 1 (IL-1) and interleukin 6 (IL-6) are cytokines that promote inflammation. In addition, cytokines stimulate the liver to release acute phase proteins, some of which bind to the surfaces of bacteria and damaged host cells to promote phagocytosis by macrophages. The concentration of one C-reaction protein increases by over a thousand-fold at the beginning of an infection. Cytokines also have effects on other parts of the body: IL-1 stimulates the brain, causing fever and sleepiness.

Wound repair

Stem cells under a scab begin to divide to repair the wound by dividing by mitosis. Platelets secrete growth factors that attract cells to divide and grow. Wound healing involves the following (overlapping) stages:

1 The formation of new blood vessels.
2 The production of collagen.
3 The formation of granulation tissue that fills the wound, allowing further changes to occur underneath.
4 The formation of new epithelial cells by division of stem cells that migrate over the new tissue.
5 Wound contraction by contractile cells.
6 Death of unwanted cells.

Key terms

Neutrophil A short-lived phagocytic cell, produced in the bone marrow, that circulates in the blood. Of all white blood cells, 60–70% are neutrophils. They have lobed nuclei and granular cytoplasm.

Monocyte A larger cell than a neutrophil that circulates in the blood and leaves to remain as a long-lived macrophage in tissues such as the lungs.

Macrophage A large, long-lived phagocytic cell that remains in tissues. Macrophages process pathogens and present antigens to T lymphocytes.

Dendritic cell A large phagocytic cell with lengthy extensions that give a large surface area to interact with pathogens and with lymphocytes.

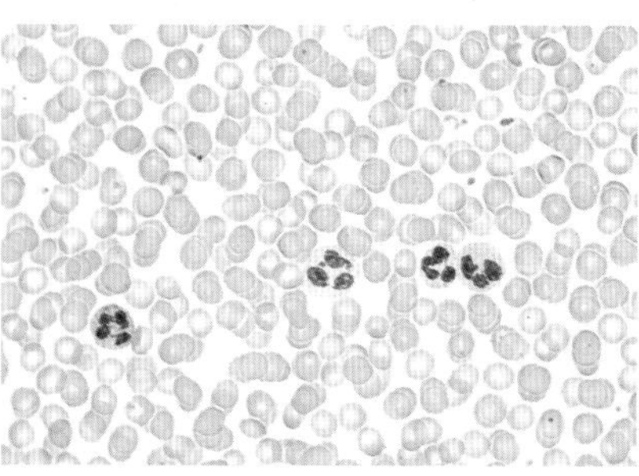

Figure 11.5 A blood smear with four neutrophils with their lobed nuclei stained blue (×300). Neutrophils form about 70% of all white blood cells.

Phagocytes

Phagocytes are cells that carry out phagocytosis to engulf pathogens and other foreign material (see Figure 11.6 and Chapter 5). There are three main types of phagocytes: neutrophils, monocytes/macrophages and dendritic cells.

Neutrophils circulate in the blood and spread into tissues during an infection. They are the 'rapid reaction force' of the immune system, responding quickly by rushing to an infected area and attempting to destroy any pathogens in the tissues. They do not last long. After engulfing bacteria and destroying them, neutrophils die and sometimes accumulate to form pus.

Monocytes pass out of the bloodstream and enter tissues where they form **macrophages** (literally: 'big eaters'). They are long-lived cells that have special roles to play in the specific defence system by processing and 'presenting' antigens to lymphocytes. Some migrate, others remain stationary in certain tissues, for example kidneys, lungs, brain, liver and especially the spleen and lymph nodes, where much antigen presentation occurs.

Dendritic cells have long processes that give them a large surface area with which to interact with pathogens and lymphocytes. Dendritic cells are found throughout the body, but once they have ingested some foreign material they migrate and take the material to the lymph nodes.

The number of neutrophils in the blood increases rapidly during an infection. The neutrophils pour out of the bone marrow, circulate in the blood and then leave through capillary walls into damaged tissues. Neutrophils do not live long after engulfing and digesting bacteria. More neutrophils are produced and released from bone marrow to replace those that die. During a lung infection, neutrophils leave the alveolar capillaries and digest their way through the lining of alveoli to reach bacteria.

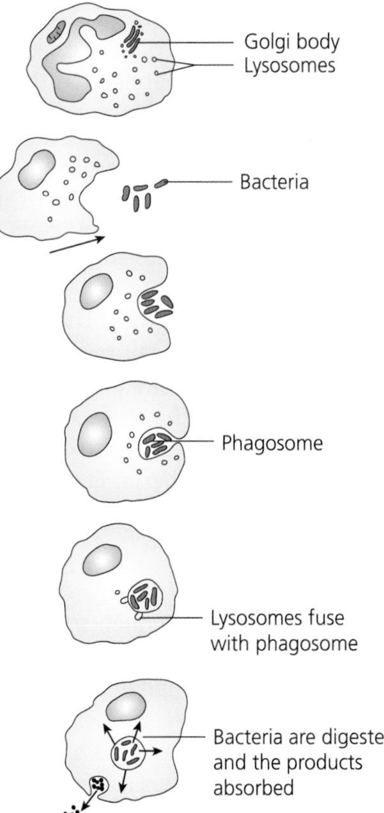
Golgi body
Lysosomes
Bacteria
Phagosome
Lysosomes fuse with phagosome
Bacteria are digested and the products absorbed

Figure 11.6 The stages in phagocytosis.

Key terms

Third line of defence A response to invading pathogens in which lymphocytes and antibodies act against specific pathogens.

Immune response The sequence of events that occurs as the specific immune system responds to a foreign antigen; this may involve activation of the cloning of B cells that have receptors specific to the antigen, or specific T cells as well.

B lymphocyte Type of white blood cell that matures in bone marrow; when B cells respond they become plasma cells.

T lymphocyte Type of white blood cell that matures in the thymus gland. When T cells respond during an immune response, they do not make and release antibodies.

Third line of defence

Phagocytes defend humans by destroying invading organisms, but they are not very effective on their own. We need a defence system that works specifically against the type of invading organism.

Lymphocytes and antibodies provide this and they are part of the third line of defence, often known as the specific defence system. This system recognises the different pathogens that enter the body. The responses are not the same each time pathogens breach the first line of defence: they differ according to the type of pathogen. Antibodies (see page 198) are produced and they are highly specific to molecules that stimulate their production; these are **antigens**.

Specific defences are highly effective but much slower than the non-specific defences. They are not present from birth. We are born, however, with the potential to produce specialised lymphocytes and antibodies to every strain of every type of pathogen that exists or is ever likely to exist. The lymphocytes involved need to be selected, activated and stimulated to produce larger numbers of themselves so that they can deliver an effective response. As humans, we do not have the space to store all the lymphocytes specific to all the various pathogens that we are likely to experience, so instead we have a small number of each. This is why specific defences are slower than non-specific defences.

The sequence of changes that occurs to select and increase the specialised lymphocytes is known as an immune response. It is an adaptive response, because it responds to a change in the environment (entry of a pathogen) and, if successful, helps us to survive in our environment if that pathogen invades again.

Once pathogens enter tissues, blood or lymph, they either remain in spaces between cells or enter cells. For example, parasitic worms can live in the gut, reducing the quantity of digested food we absorb, or in the blood absorbing nutrients from the plasma. Most of the pathogens that enter cells enter specific cells. Some, such as *Mycobacterium tuberculosis* and HIV, enter cells of the defence system itself. Removing pathogens from cells is difficult and often involves killing infected host cells.

Lymphocytes

Lymphocytes are small white blood cells. 'Blood cell' is not quite an accurate description of a lymphocyte since, like neutrophils and macrophages, they are active in tissues rather than in the blood. After all, the term 'lymphocyte' means lymph cell, not blood cell.

Lymphocytes mostly congregate in lymph nodes throughout the body, especially in areas most likely to be sites of infection, such as the throat, along the gut, in the lungs, and near the reproductive system. Lymphocytes simply use the blood to get from one place to another.

There are two groups of lymphocytes:

- B lymphocytes (B cells for short)
- T lymphocytes (T cells).

B cells originate from stem cells in the bone marrow, where they differentiate into mature cells that spread out through the body's lymphatic system.

T cells also originate in bone marrow, but migrate to the thymus gland in the chest where they mature. Once mature, T cells of different types also spread out and populate the lymphatic system.

Tip

Do not confuse antigen and antibody. Any substance that stimulates an immune response is an antigen. Antibodies are proteirs made as part of an immune response by plasma cells that differentiate from B lymphocytes.

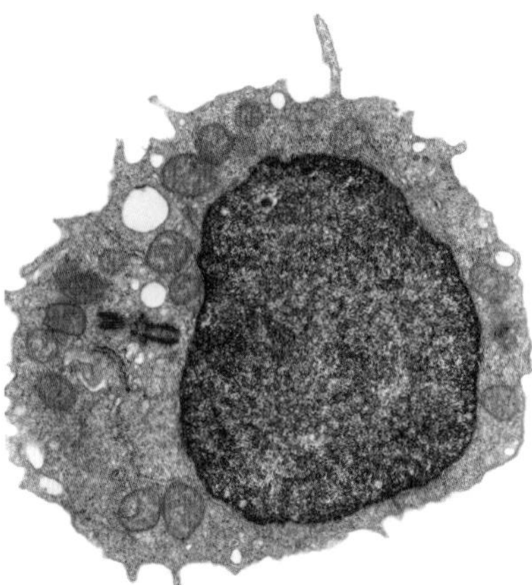

Figure 11.7 Transmission electron micrograph of a lymphocyte. The cytoplasm has many mitochondria but little rough endoplasmic reticulum (×25 000).

Key term

Clone A group of genetically identical cells. In the immune system, there are clones of B cells and clones of T cells. Each clone has its own unique cell surface receptor complementary to a specific antigen.

Both types of lymphocyte remain inactive until they come into contact with 'their' antigen. This leads to their activation and division by mitosis to form larger clones – groups of cells with the same specificity. When activated, B cells make and release antibodies. Active T cells do **not** make antibodies; instead they work in a variety of other ways, depending on the type of cell surface proteins that they have.

During the maturation process, both B cells and T cells gain cell surface receptors. These receptors are transmembrane glycoproteins that stretch across the membrane and have a specifically shaped extracellular region. Mature B cells have receptors that have exactly the same specificity as the antibody molecules that they are 'programmed' to secrete. Each type of B cell has a differently shaped B cell receptor (BCR) that identifies its specificity. Similarly, each type of T cell has T cell receptors (TCRs) that identify its specificity. This huge variation in BCRs and TCRs allows lymphocytes to recognise the molecules on the surface of invading microorganisms, and their cell products, such as the toxins released by bacteria that cause tetanus, cholera and diphtheria. Each lymphocyte has about 100 000 receptors on its cell surface – all with the same specificity.

It is estimated that there are as many as 10^{15} different types of antibody receptor that can be made by rearrangement of the coding sequences of genes for these receptors. It is thought that these cells have enough variation in their TCRs and BCRs to recognise *any* type of antigen that could ever exist. This gives lymphocytes the ability to distinguish between antigens that are only slightly different. Although there are many B and T cells in the body, there are only a small number of each type. Each group of identical B cells and each group of identical T cells form small clones. Among all these clones are some B cells and some T cells that have receptors specific to the same antigen, so that both groups of cells can respond together to the same threat. As you will see, B cells and T cells work together to defeat pathogens.

Lymphocytes circulate between blood, lymph, lymph nodes, the spleen and liver. In so doing, they come into contact with any pathogens, toxins, other foreign material and macrophages and dendritic cells that may be presenting antigens.

Immune response

The immune response is the series of events that happens once B cells and T cells detect the presence of 'their' antigen in the body. T cells can only detect antigens when the antigen is presented within a specific cell surface protein. B cells can interact with antigens without the need for other cells, but they do require the involvement of T cells to strengthen their response. Diagrams that show the immune response, such as those on the next few pages, always show the BCRs and TCRs as large structures on the cell surface. This is a highly diagrammatic way of showing what happens; just remember how many of these receptor molecules are on a cell surface!

Antigen-presenting cells

When neutrophils engulf anything foreign they destroy it completely. This does not happen in macrophages and dendritic cells. Instead they 'cut up' the pathogen and incorporate parts of the surface of bacteria and viruses within proteins that are then incorporated into the cell surface membrane. These proteins act to 'display' antigens, and in so doing they present the antigen to any lymphocytes that happen to be nearby.

Antigen presentation is the first stage in an immune response. However, antigen presentation is not unique to macrophages and dendritic cells; many cell types are able to do this, although they have a different type of cell surface display protein.

B cell activation

Key term

Plasma cell An activated B cell that makes and releases antibodies during an immune response.

B cells respond directly to large molecules with a repeated structure. Antigens like these are the large polysaccharide molecules on the surface of bacteria. These antigens interact with many BCRs on the surface of B cells. This multiple interaction is sufficient to activate the B cell to divide and differentiate into **plasma cells** (see Figure 11.8a).

B cells also respond to soluble antigens such as protein molecules that each interact with single BCRs. The antigens within the BCRs are taken into the cell by endocytosis. They are processed and attached to 'display' proteins and presented on the cell surface. T helper cells with TCRs complementary to the antigen are activated and secrete interleukin 2 (IL-2), which stimulates the B cell to divide and differentiate into plasma cells (see Figure 11.8b). This is a fail-safe mechanism to ensure that antibodies are not released unnecessarily.

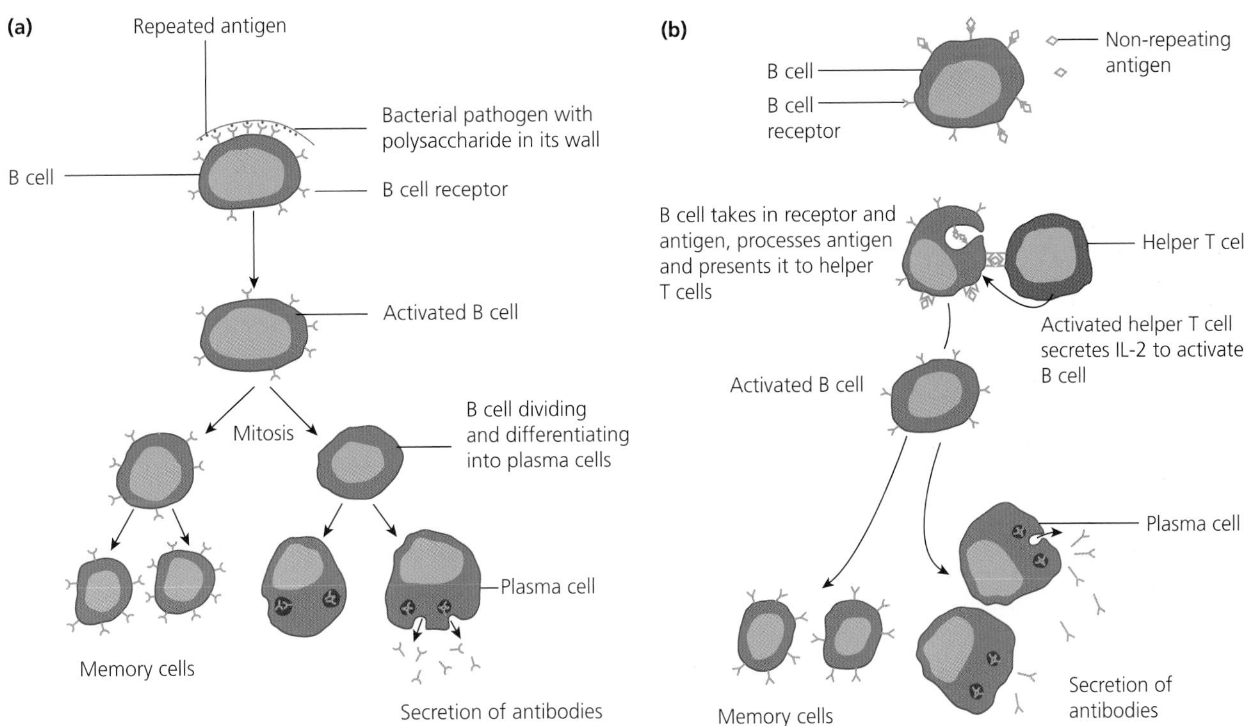

Figure 11.8 The selection and activation of B cells. An antigen binds to the BCR molecules on the cell surface. Only B cells with the complementary BCR to the antigen respond. The antigen is responsible for selecting the appropriate clone of B cells; (a) a B cell is activated by an antigen; (b) a B cell is activated by a T helper cell.

Some of the activated B cells form plasma cells that make many ribosomes and the membrane necessary for rough endoplasmic reticulum. Antibody genes are transcribed, the mRNA translated, and the polypeptides assembled to make antibody molecules. The antibody molecules are processed in the Golgi apparatus, packaged into vesicles and exported by exocytosis. Other cells in each clone do not become active and do not differentiate into plasma cells. Instead they remain in the body circulating between blood and the lymphatic system as long-lived memory cells. Their role will be discussed after the roles of T cells have been examined.

Key terms

T helper cells Type of T cell that coordinates an immune response by stimulating the responses of B cells and T killer cells by producing chemical signals such as interleukins.

Clonal selection The process during immune responses when specific clones of B cells and/or T cells interact with antigens displayed by APCs.

Clonal expansion The division by mitosis of small groups of identical B cells and T cells during an immune response. This gives rise to effector cells (e.g. plasma cells) and to memory cells.

T killer cell Type of T cell that kills infected host cells.

T regulator cell Type of T cell that stops immune responses and prevents T cells attacking the body's own cells and tissues.

Tip

The antibodies produced by an expanded clone of plasma cells will all have antigen binding sites that are identical. All of them are complementary in shape to the antigen.

T cell activation

Central to the immune response are **T helper cells** that respond to antigens presented by macrophages; this stimulates both B and T cells. The stages in an immune response involving helper T cells are as follows.

1 Antigen presentation – APCs (antigen presenting cells) (macrophages and dendritic cells) in lymph nodes engulf pathogens by endocytosis and then 'cut them up'. These macrophages process antigens from the surface of the pathogen and put them into special proteins within their own cell surface membranes.

2 **Clonal selection** – T cells with receptors complementary in shape to the antigens bind to the macrophage. These small groups of specific T cells are the clones that are selected by the macrophage. The APC secretes molecules of interleukin 1 (IL-1) to stimulate the selected T helper cells to become active.

3 **Clonal expansion** – Because very few T cells have the capability of destroying the invading pathogen, the cells in the clones divide by mitosis to form much larger clones. T helper cells release interleukin 2 (IL-2) to stimulate B cells to divide and differentiate into plasma cells. This coordinates the activity of the different lymphocytes as well as macrophages and other components of the immune system. IL-2 also has the action of self-stimulating T helper cells to become even more active as cytokine producers and produce another cytokine, interferon (IFN), which stimulates macrophage activity.

4 Antibody production – The stimulated B cells form plasma cells, which secrete the appropriate antibody.

Most of the antibodies produced in response to the first presentation of an antigen are large molecules each with ten identical antigen-binding sites. These super-large antibodies are good at sticking bacteria together.

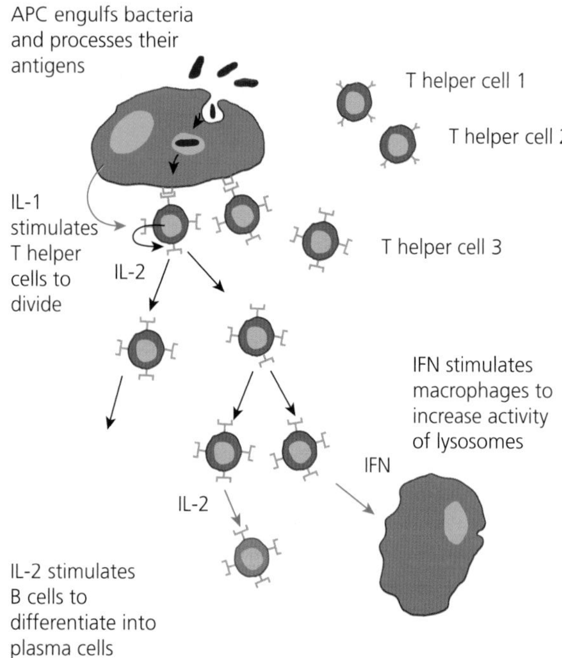
APC engulfs bacteria and processes their antigens
T helper cell 1
T helper cell 2
T helper cell 3
IL-1 stimulates T helper cells to divide
IL-2
IFN stimulates macrophages to increase activity of lysosomes
IFN
IL-2
IL-2 stimulates B cells to differentiate into plasma cells

Figure 11.9 The activation and role of T helper cells in an immune response.

Cells constantly cut up proteins and present them as short peptides within cell surface proteins, as if to show to patrolling **T killer cells** what they have inside them – which, on occasion, may be pathogens. When a T killer cell comes across an infected cell expressing a foreign antigen specific to its TCR, it becomes active. This only happens if the TCR is complementary to the antigen. Once activated by binding, T killer cells fix to the surface of the infected cells and secrete perforins, which are proteins inserted into the cell surface membranes of infected cells. Perforins form channels through membranes so that toxins such as hydrogen peroxide and nitric oxide can enter the infected host cells. The host cells die. This seems a drastic measure, but it is the only way to remove intracellular pathogens such as *M. tuberculosis* and viruses such as measles and influenza. Specific T helper cells are activated as well and promote the expansion of the appropriate clone of T killer cells by secreting IL-2.

The immune response cannot be allowed to continue unchecked. Another group of T cells, the **T regulator cells**, down-regulate the immune response. The T regulator cells shut down immune responses once the pathogens have been removed from the body. They also prevent the T cells from attacking uninfected host cells.

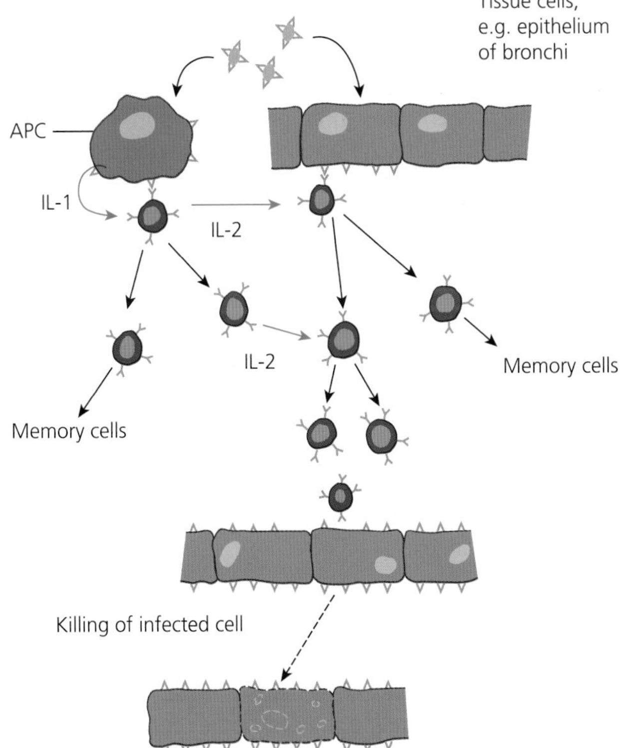

Figure 11.10 T killer cells are activated by an infected host cell.

Memory cells in long-term immunity

It takes about 10–17 days after an antigen enters the body for the first time for antibodies to appear in the blood. This is the **primary immune response**. This is why we are ill when we catch an infectious disease such as chicken pox. But, after a while, antibodies and activated T killer cells are produced that help to remove the infectious agent, and we recover.

Plasma cells do not live for long, and soon the antibody molecules that they make are broken down. However, when an antigen enters the body for a second time, the response is much faster. This is because during clonal expansion, B and T cells form memory cells. These memory cells remain circulating in the blood and lymph patrolling the body 'on the lookout' for the return of the same antigen. When the antigen returns, the memory cells respond more quickly because there are more of them to be selected than there were of the original clone before the first infection. The **secondary immune response** to an antigen occurs much faster than the primary immune response, and there are rarely any symptoms of the infection. In a secondary immune response it may take just a few days for the antibody concentration to increase as a result of the activation of memory cells. It is a greater response because more antibody molecules are produced and the response lasts for longer.

'Memory' is not a very good name for these cells because they have not 'learned' anything. They are just the representatives of what is now a much larger clone of cells.

Tip

Note that T cells differentiate into different classes, for example T helper cells, T killer cells and T regulator cells, when they mature in the thymus gland *before* taking part in an immune response; they do not differentiate into these different classes during or after an immune response.

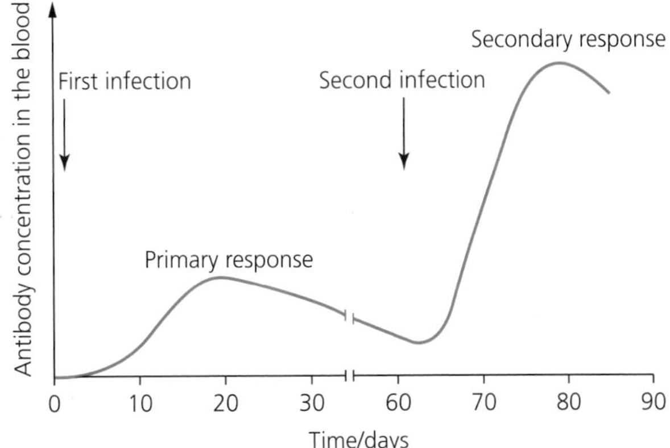

Figure 11.11 The change in the concentration of antibodies during the primary and secondary immune responses.

Tip

Notice in Figure 11.11 how the secondary response differs from the primary response: it is faster and reaches a higher antibody concentration in the blood.

Test yourself

4 Explain why people who have suffered from severe burns are at an increased risk of infectious disease.
5 Explain how the body's natural openings are protected against invasion by pathogens.
6 Describe the roles of phagosomes and lysosomes in defence.
7 Make a labelled drawing of the lymphocyte in Figure 11.7. Indicate its actual diameter.
8 Name two cell types that act as APCs.
9 Distinguish between T helper cells and T killer cells.

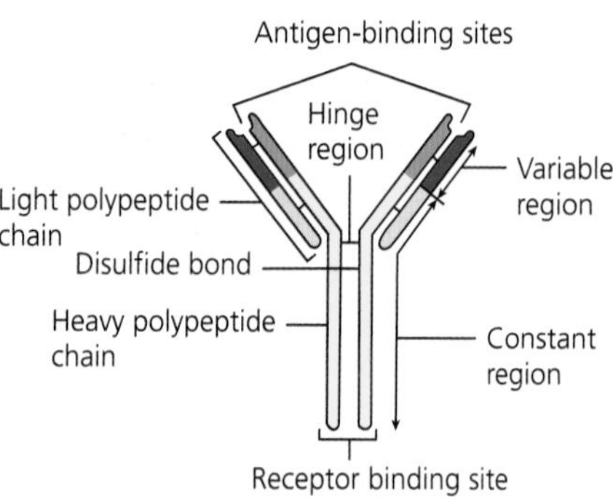

Figure 11.12

Tip

This is a good opportunity to revise protein structure. To explain how the structure of antibodies is related to their function, you need to use knowledge of the four levels of protein structure (see pages 42–7 in Chapter 3).

Antibodies: very special proteins

Antibodies are plasma proteins known as **immunoglobulins** (Ig for short). It helps here to recall your knowledge of protein structure. Proteins are formed from polypeptides, which each have three levels of organisation: primary, secondary and tertiary structure. All antibodies have quaternary structure because they are formed from four or more polypeptides. The simplest form of antibody molecule (Ig class G, or IgG) is composed of four polypeptides. Other classes of antibody are composed of more than four polypeptides. The largest class has 20 polypeptides.

Each IgG molecule is composed of two identical long polypeptides (or chains) and two identical short polypeptides. One region of the antibody molecule is the region that binds to antigens – the antigen-binding site. If you imagine an IgG molecule as Y-shaped, these binding sites are at the two ends at the top of the Y. The polypeptides are held together by disulfide bonds, with the hinge region giving the molecule some flexibility so that the two binding sites can make contact with antigens that may be separated by slightly different distances.

The constant region is the same for all antibodies of the same class. The constant regions of IgG molecules are all identical, whatever the specificities of the variable regions. These constant regions bind to receptors on the surfaces of phagocytes. This helps phagocytes to detect pathogens that have been 'labelled' by antibodies for destruction by phagocytosis.

In order to bind to its specific antigen, each type of antibody molecule has a different **antigen-binding site**. This is possible because amino acids can be arranged in many different sequences to give different three-dimensional shapes. Because these binding sites vary, these regions are also called **variable regions**. These regions are similar to the active sites of enzymes and the binding sites of receptors for cell-signalling molecules. All have a specific shape complementary to a binding agent.

The antigen-binding site of an antibody molecule is complementary in shape to the antigen with which it binds. We need many antibodies with different variable regions to 'fit around' the different antigens that may enter our bodies. The better the 'fit' between antigen and antibody, the more efficient the immune response at identifying and destroying pathogens.

Within each class of antibody, the antibodies function in different ways in defence; some of these are as **antitoxins**, **agglutinins** and **opsonins**:

- **Antitoxins** combine with toxins secreted by bacteria (e.g. tetanus and diphtheria toxins) to render them harmless.
- **Agglutinins** bind to two identical antigens on two or more pathogens, causing them to clump together. This makes it more difficult for them to spread and easier for phagocytes to engulf them. The largest class of antibody, with ten antigen-binding sites, is more effective at this than the antibody class with just two antigen-binding sites.
- **Opsonins** 'coat' pathogens to facilitate phagocytosis. Phagocytes have receptors for the constant regions of antibodies. By attaching to bacteria, these antibodies 'mark' them for destruction by phagocytes. Opsonisation is often compared to 'sugaring' pills to make them easier to swallow.

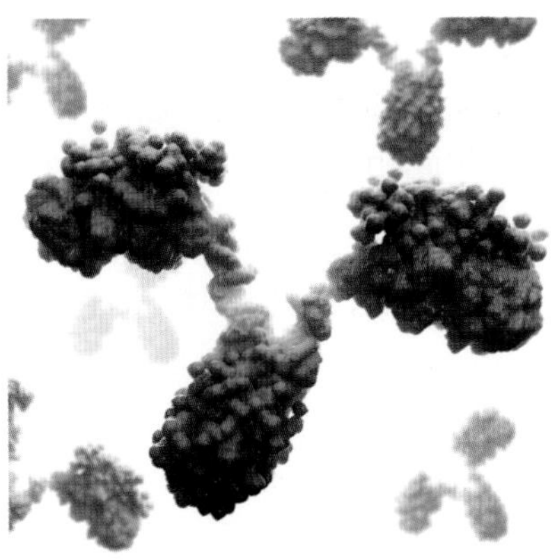

Figure 11.13 This computer model shows an antibody molecule (purple) attached to two antigen molecules (green), forming an antigen–antibody complex (×3000000).

Tip

Before answering a question like this, make a plan. You can write ideas at the top of the answer lines or in the margin. Here are some ideas for part 2(b):

- Antibodies are proteins ... made of polypeptides ...
- Polypeptides are made of amino acids ...
- Amino acids can be arranged in different sequences (primary structure).
- Polypeptides can fold into many different shapes.
- These shapes are specific ...

Tip

Always include a diagram if it helps you with your answer. It's a good idea to label and annotate any diagrams that you draw.

Example

Antibodies

The immune system produces antibodies in response to infection.

1 Name the type of cell that produces antibodies.

2 a) Explain why antibodies are said to have a quaternary structure.

b) Describe how the structure of an antibody molecule is related to its functions.

Answers

1 plasma cells

2 a) If a protein has quaternary structure, it has two or more polypeptides. Each antibody molecule is made from more than two polypeptides. (IgG antibodies have four.)

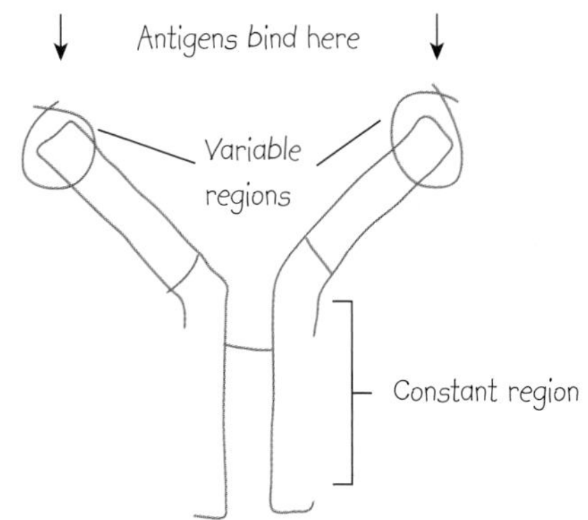

b) An antibody is a protein molecule that is composed of four polypeptides: two long (or heavy) polypeptides and two short (or light) polypeptides. The polypeptides are held together by disulfide bonds. The molecule has a Y shape. The two tips of the Y are the antigen-binding sites, which are also known as the variable regions.
Each type of antibody is folded into a specific shape for its antigen-binding site, so that it has a complementary shape to an antigen.
The antigen-binding sites are also known as the variable regions. The amino acid sequences of these regions vary to give structures that are complementary in shape to the antigens to which they bind. The projecting R groups give the shapes that are specific to the antigens. The hinge region between the R groups gives the antibody molecule some flexibility when it combines with antigens.
The other end of the molecule is the constant region that always has the same primary structure that is complementary to receptors of phagocytes. This helps phagocytes recognise pathogens and ingest them in phagocytosis.
Antibodies have two antigen-binding sites so they can attach to two antigens at the same time. If the antigens are on different cells or viruses then they can clump them together. This is called agglutination. By combining with viruses, antibodies prevent the viruses binding to cells – so the viruses do not infect the cells. Opsonins are antibodies that bind to pathogens, identifying them to phagocytes, which have receptors for the constant region.

Different types of immunity

So far, this chapter has considered what happens when an antigen enters the body. This is **active immunity**. It may happen naturally when you are infected, or it may happen artificially when you are given a preparation, or **vaccine**, containing an antigen. Active immunity always involves an immune response and protection is long term, often lasting a lifetime.

- **Natural active immunity** occurs when you are infected.
- **Artificial active immunity** is the result of being given a vaccine that contains one or more antigens.

It is also possible to become immune by receiving antibodies from another person. This is **passive immunity**. Here the body gains an antibody from another source. There is no contact with the antigen and no immune response occurs. Passive immunity may be gained naturally or artificially.

- **Natural passive immunity** occurs when antibodies cross the placenta during pregnancy; it also occurs when a child is breastfed by its mother. Breast milk is rich in IgA antibodies, but only against the diseases that the mother has had.
- **Artificial passive immunity** occurs when antibodies are injected into a person to give them instant immunity. People who are likely to have tetanus, rabies or diphtheria are often given antitoxin by injection as a precaution. In each case, the antitoxin neutralises the toxins released by each of the pathogens and prevents the damage that the toxins can cause.

Figure 11.14 Natural active immunity: the chickenpox virus has successfully invaded this child, but a primary immune response is busy making antibodies and memory cells.

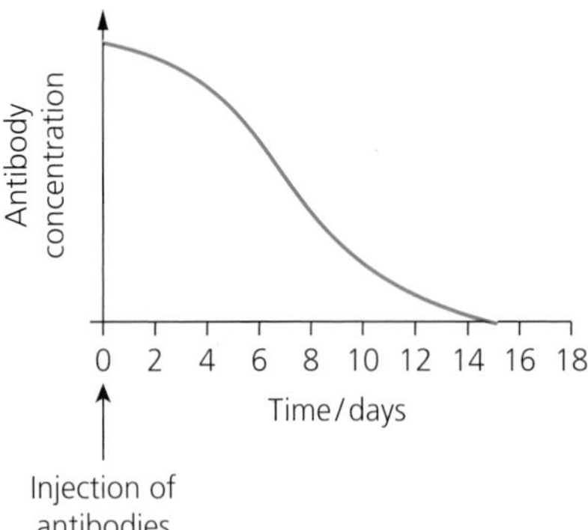

Figure 11.15 The change in antibody concentration in passive immunity. A person has been injected with antibodies. The concentration of antibodies decreases gradually because the antibodies are foreign to the body and are gradually removed. There are no activated lymphocytes (plasma cells) to secrete the antibody.

Test yourself

10 Make a table to compare the structure and mode of action of phagocytes and lymphocytes.

11 Distinguish between active immunity and passive immunity.

12 Make a table to compare the four different types of immunity: natural active, artificial active, natural passive and artificial passive.

Autoimmune diseases

The body sometimes attacks itself using antibodies and T cells that attack its cells. In the UK, about 5% of people have an autoimmune disease.

Autoimmune diseases occur because the immune system attacks one or more self antigens, usually proteins. Autoreactive T helper cells, T cytotoxic cells and B cells are involved as well. In some cases the attack is directed against one organ; in others it is directed against the whole body, as is the case with systemic lupus erythematosus (SLE), which is often known simply as lupus (see Figure 11.16).

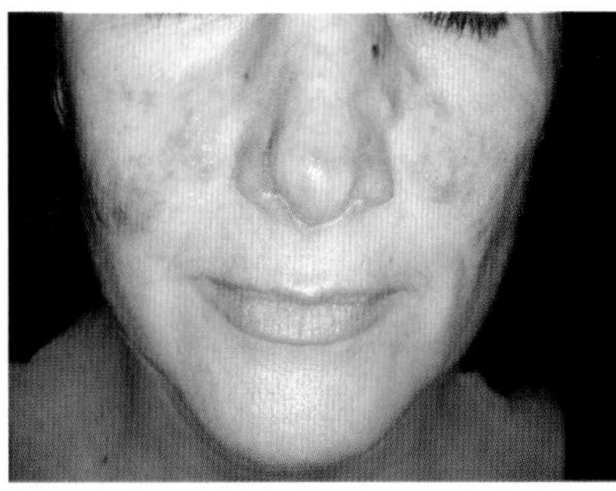

Figure 11.16 Systemic lupus erythematosus (SLE or lupus). The immune system attacks connective tissue, including the skin, joints and kidneys.

One of the common symptoms of lupus is the butterfly rash across the face. The disease affects many organs, including the heart, lungs and kidney, and causes pain in the joints. The symptoms of lupus are never quite the same when it flares up, which makes it a difficult disease to diagnose. Women are far more susceptible to lupus than men.

Rheumatoid arthritis is another long-term destructive process, this time occurring in the joints. It starts with the finger and hand joints and then spreads to the shoulders and other joints. There is constant pain and muscle spasm. Tendons become inflamed, and people with the disease are often lethargic. (Note that osteoarthritis is not an autoimmune disease.)

The causes of autoimmune diseases are not well known and are the subject of much research. Genetic factors are involved, because it has been shown that susceptibility is inherited. However, environmental factors are also important, as the increase in prevalence of these diseases in the developed world over the last 50 years suggests. Also, people who have moved from an area where these diseases are rare, such as Japan, to places where they are more common, such as the USA, have an increased chance of developing one of these diseases.

Control of infectious diseases

Tip

Immunisation refers to both artificial active immunity by vaccination and to artificial passive immunity. Some people think it means exactly the same as vaccination; it doesn't.

Vaccination controls the spread of infectious diseases

Vaccination programmes are an important part of the health protection offered by governments to their citizens. Infants and children are vaccinated against diseases that used to be common in populations and were responsible for much ill health and many deaths.

Table 11.1 Vaccination schedule for the UK.

Vaccine	Diseases protected against	Age/months				Age/years			
		2	3	4	12–13	2–3	just over 3	12–13	13–18
5 in 1 DTaP/ IPV/Hib	diphtheria, tetanus, pertussis (whooping cough), polio, *Haemophilus influenzae* type b (known as Hib)	✓	✓	✓			✓ (not Hib)		✓ (not Hib, not pertussis)
	pneumococcal (PCV) vaccine	✓		✓	✓				
	rotavirus vaccine	✓	✓						
	meningitis C		✓		✓				✓
MMR	measles mumps rubella				✓		✓		
	influenza					✓			
HPV	cervical cancer							✓ (girls only)	

Figure 11.17 A medical worker places drops of polio vaccine into a boy's mouth during National Immunisation Day in Bangladesh.

Many of these diseases are very rare in the UK. For example, the last case of polio that was naturally acquired in the UK was in 1984. But the disease still exists in the world and could be introduced into the UK by travellers from areas where it is still endemic. In 2013, there were 93 cases of polio in Pakistan, and as of 2014 the disease is endemic there and in Afghanistan and Nigeria, with some cases in seven other countries in Africa and the Middle East.

It is hoped that the World Health Organization (WHO) will soon announce the eradication of this disease. You can follow the progress of the campaign by searching online for 'polioeradication'.

During eradication programmes, vaccination is used in two ways: to achieve herd immunity or ring immunity.

Mass vaccination schemes give rise to **herd immunity**: as many people as possible are vaccinated; a pathogen cannot then easily be transmitted from an infected person to an uninfected person, because everyone, or nearly

Tip

Vaccination is the active way to immunise people. Injecting antibodies is the other way in which people can be immunised. Each has its advantages and disadvantages.

Tip

Look at the section on viral pathogens in Chapter 10 to remind yourself why viruses are obligate parasites.

Tip

Your knowledge of the immune response should tell you why a specific vaccine is required rather than any vaccine against influenza.

Test yourself

13 Sketch a graph to show what happens to the concentration of antibodies in a child who has a vaccination for tetanus at two, three and four months after birth.
14 Sketch a graph to show what happens to the concentration of antibodies if a girl has a vaccination for cervical cancer and then a vaccination for meningitis three weeks later.
15 Under what circumstances do people in the UK require vaccinations?
16 Who are those most at risk of influenza and who should have priority in receiving the vaccine during an epidemic?
17 List three infectious diseases for which no vaccine exists.
18 Girls are especially at risk of rubella. Why is it important to vaccinate boys against this disease?
19 Tamiflu and Relenza are enzyme inhibitors. Suggest how they may inhibit the enzyme that is used in the release of viruses from virally infected cells.

everyone, is immune. Vaccination programmes attempt to achieve nearly 100% coverage to achieve good herd immunity.

Ring immunity is used to vaccinate people living or working near someone who is infected (or their contacts) to prevent them catching the disease and then spreading it.

People who are vaccinated cannot harbour the pathogen and cannot pass it on to others. People who do not respond to vaccines are protected: the chances of them coming into contact with the disease are small, because most people around them have immunity and will not transmit the disease.

Problems with vaccines

It is difficult to develop vaccines against diseases that are caused by eukaryotic organisms, such as the malarial parasite *Plasmodium*, because they have many genes that code for cell surface antigens. Different strains of *Plasmodium* have different antigens. As *Plasmodium* passes through its different stages in liver and blood cells (see Chapter 10), it expresses different antigens. It also remains inside liver and red blood cells, where antibodies have no effect.

In 2008, the UK government announced that the greatest risk facing the population was an influenza pandemic. Cells infected with influenza virus become 'factories' for producing more viral particles, which are released to infect more cells and to be transmitted when infected people cough and sneeze. During the twentieth century, there were three major pandemics of influenza. The worst was the Spanish flu that followed the First World War (see page 188). The last pandemic was the 'swine flu' pandemic of 2009.

Viruses can change their surface antigens in one of three ways, by:

- **antigenic drift**, where there are small changes in the structure and shape of the antigens within the same strain of virus
- **antigenic shift**, where there are major changes in the antigens within the same strain
- different strains of viruses invading the same cell, with the formation of new viruses with antigens from different strains; this is known as 'cross-breeding'.

The virus that causes human influenza may cross-breed with viruses that cause similar diseases in animals, or a strain that is pathogenic in animals may cross the species barrier and infect people.

The WHO and national governments maintain a watch for new strains of the virus to which people have not been exposed. These strains may cause a pandemic because no-one has any immunity. Each year, the WHO issues guidance about the strains of influenza that are likely to spread. Vaccines against these strains are prepared and distributed. People who are at most risk of catching influenza are offered the vaccine in the UK. Herd immunity would be needed to prevent a pandemic on the scale of that in 1918, but would stocks of the appropriate vaccine be ready in time to vaccinate most of the population if a new strain emerged? Health authorities stockpile drugs for preventing and treating influenza. Two of these are oseltamivir (better known by its trade name of Tamiflu) and zanamivir (Relenza), which inhibit virally infected cells from releasing more viruses. It is unlikely that governments could stock enough of these drugs for those who fall ill during a pandemic. Who has priority if the vaccine and the drug are rationed?

The WHO estimates that 30 new diseases have emerged in the last 50 years. These include the viral diseases HIV/AIDS, SARS, ebola, West Nile disease and Lyme disease. Few drugs are effective against viruses.

Table 11.2

Bacterial process that is inhibited	Examples of antibiotics
Cell wall synthesis	penicillin, cephalosporin, vancomycin, cycloserine
Transcription	rifampicin
Translation	chloramphenicol, erythromycin, tetracycline, streptomycin
DNA synthesis	quinolones
Cell surface membrane function	polymixin
Synthesis of folic acid	sulfonamides trimethoprin

Antibiotics

An antibiotic is usually defined as 'any substance produced by a microorganism which, in dilute solution, harms or kills another microorganism'.

Antibiotics are compounds derived from some fungi, such as penicillin from *Penicillium*, but mainly from the actinobacteria, particularly *Streptomyces*. The actinobacteria are the source of most antibiotics – for example, streptomycin, erythromycin and tetracycline, which are used to treat bacterial infections. Many antibiotics and other drugs are semi-synthetic, in that they are modified chemically after production by microorganisms in fermenters, or they are produced entirely by chemical synthesis although they may originally have been discovered in organisms. The penicillins, for example, are extracted from cultures of *Penicillium chrysogenum* and then altered chemically.

Penicillin was one of the first antibiotics to be discovered and to be mass produced. Penicillin works by inhibiting the bacterial enzyme transpeptidase, which catalyses the synthesis of murein, from which bacterial cell walls are made (see Table 1.3 on page 15). Without their protective cell wall, the bacteria absorb water, burst and die. Penicillin is not effective against all bacteria. Antibiotics function only if they can enter the bacterial cells and if there is a target to inhibit.

Broad-spectrum antibiotics act on a wide range of bacteria. **Narrow-spectrum antibiotics** only work on a few.

Antibiotic resistance

Soon after antibiotics became available for widespread use, doctors found that some infections that had previously been cured with a course of antibiotics were no longer susceptible. The bacteria that caused these diseases, such as TB, had developed resistance to the specific antibiotic.

When someone takes an antibiotic for a bacterial infection, bacteria that are susceptible will die. In most cases if the dose is followed correctly, all the bacteria will die. However, if the dose is not followed, perhaps because people stop taking the antibiotic when they feel better as the symptoms disappear, then some susceptible bacteria may survive, and if any mutations occur these may give resistance. The next time there is an infection of this strain of bacteria, the antibiotic may not work, because there are some resistant bacteria among those that have infected the body.

Bacteria have one copy of each gene, because they only have a single loop of double-stranded DNA. This means that a mutant gene has an immediate effect on any bacterium possessing it. Bacterial cells with a mutant gene have a selective advantage: bacteria without this mutant gene will be killed, while those bacteria resistant to penicillin will survive and reproduce. This happens very rapidly: if there was initially only one resistant bacterium, it might produce 10 000 million descendants within 24 hours.

Bacteria can develop resistance by acquiring genes that protect them from the effects of the antibiotic. The genes are often found on plasmids, which spread easily from one bacterium to another – even from one species of bacterium to another. An example is a gene that codes for the enzyme penicillinase. This breaks down the antibiotic penicillin, making it harmless. Other examples are genes coding for:

- a modified enzyme for DNA synthesis, making quinolones useless
- a protein 'pump' in the bacterial cell membrane, through which antibiotics like tetracyclines are removed from the cell

Tip

Plasmids are small circular molecules of double-stranded DNA that are in prokaryotes and some eukaryotes. They replicate independently of the chromosomal DNA and carry genes that are not essential for normal metabolism, but can give organisms advantages such as antibiotic resistance.

- a modified ribosome structure to which antibiotics like erythromycin are unable to bind
- a modified cell wall structure, so that antibiotics like cycloserine are ineffective
- membrane components that reduce the inflow of antibiotics into the bacterial cell.

An alarming number of human pathogens have acquired genes to combat all the presently used antibiotics including vancomycin, which is used as an antibiotic of 'last resort' to treat infections that cannot be cured by other antibiotics. Vancomycin resistance took 30 years to develop from vancomycin's introduction. These multi-drug-resistant strains are particularly common in hospitals, where antibiotic use is heavy and the patients often have weakened immune systems. One bacterium that is causing particular concern is methicillin-resistant *Staphylococcus aureus*, or MRSA. This has caused deaths in hospital patients with suppressed immune systems, for example people taking immunosuppressive drugs following organ transplants.

S. aureus commonly colonises human skin and mucous membranes in the airways without causing any problems. It can cause disease, however, particularly if there is an opportunity for the bacteria to enter the body, for example through broken skin or a medical procedure. If the bacteria enter the body, illnesses that range from mild to life-threatening may then develop. These include skin and wound infections, infected eczema, abscesses or joint infections, infections of the heart valves, pneumonia and bacteraemia (bloodstream infection).

Most strains of *S. aureus* are sensitive to the more commonly used antibiotics, and infections can be treated effectively. Some *S. aureus* bacteria are more resistant. Those resistant to the antibiotic methicillin are termed methicillin-resistant *Staphylococcus aureus* (MRSA) and often require different types of antibiotic to treat them. Those that are sensitive to methicillin are termed methicillin-susceptible *Staphylococcus aureus* (MSSA). MRSA and MSSA only differ in their degree of antibiotic resistance; other than that, there is no real difference between them.

Another bacterium is *Clostridium difficile,* also widely known as *C. diff.* This is a gut bacterium that is usually kept under control by the presence of other bacteria. If many of these bacteria are killed when someone is taking a treatment of antibiotics, then *C. difficile* may increase in numbers and release toxins that interrupt the normal functioning of the epithelium of the intestine, resulting in diarrhoea and other symptoms such as fever. Symptoms can be particularly severe in the elderly.

Cases of MRSA and *C. difficile* increased significantly in the late 1990s and 2000s. In the USA, cases of infection by MRSA in hospitals halved between 2005 and 2011. In England over the same period the number of cases decreased by 80%. There have been similar decreases in the number of cases of *C. difficile*. The lower rates of infection are due to improvements in basic hygiene in hospitals, such as people washing their hands more frequently and cleaning surfaces thoroughly with bleach.

Sources of new medicines

No new class of antibiotic has been discovered since 1987. Scientists are searching for new antibiotics, but there are fears that old diseases that we thought were curable may return. Finding a way to cope with the development of antibiotic-resistant bacteria is an important challenge facing the scientists of the twenty-first century.

Tip

In 2013, Professor Sally Davies, the government's chief medical officer, warned that antibiotic resistance is a 'ticking time bomb' and should be ranked with terrorism on a list of threats to the UK. You should understand the dangers of antibiotic resistance to the health of the nation and steps that can be taken to reduce it. Early in 2015, researchers announced the discovery of teixobactin – a new type of antibiotic (see the Tip on page 292).

New drugs are discovered and developed in several ways, by:

- identifying likely compounds produced by organisms such as fungi, actinobacteria, plants and animals
- performing genetic analysis of organisms to search for likely genes that may code for potential drugs
- finding molecules that fit into drug targets, such as receptor molecules for hormones and neurotransmitters at synapses
- modifying existing drugs using computer modelling of the molecular structure of the drug and its target molecule.

Some examples of potential sources of new medicines are:

- Marine actinobacteria have been discovered to be a source of rifamycins – a group of antibiotics effective against bacteria because they inhibit transcription. Some actinobacteria live on or inside marine animals.
- *Calophyllum lanigerum*, a rare tree from the rainforest in Malaysia, is the source of calanolide A – a drug that stops HIV entering the nuclei of healthy T helper cells. This prevents the T helper cells producing new viruses and therefore decreases the spread of HIV throughout the body.

It takes a long time and a great deal of money to develop a drug successfully. Drugs go through a lengthy period of trialling and then approval by national regulatory authorities.

Plants used in traditional medicines are likely to make good potential medicines; many drugs in use today are derived from plants. Table 11.3 shows four such medicines.

Table 11.3 Four medicines derived from plants.

Medicine	Plant source	Use	Biological action
Artemisinin	*Artemisia annua*, sweet wormwood	treatment for malaria	kills the pathogen while it is inside red blood cells
Galanthamine	Amaryllis family: lilies, snowdrops, daffodils	treats Alzheimer's disease	competitive inhibitor of enzyme at synapses that breaks down the chemical transmitter acetylcholine
Teniposide	*Podophyllum peltatum* (mayapple)	treats childhood leukaemia	slows growth of cancer cells by inhibiting an enzyme involved in DNA replication
Quinidine	*Cinchona ledgeriana* (quinine tree)	treats fast heart rate – fibrillation	blocks channel proteins in cardiac muscle cells, so reducing conduction of impulses throughout the heart

Figure 11.18 Plants like sweet wormwood are one of the starting points in drug development, because they produce such a wide variety of compounds, many of which are difficult to synthesise artificially.

Cinchona ledgeriana is also the source of quinine, which was the first drug to be used against malaria and is still used to treat severe cases of *Plasmodium falciparum*.

Many of the drugs that we use originate from the study of organisms. Antibiotics are isolated from fungi and bacteria; anti-cancer drugs have been isolated from plants such as vinblastine from the Madagascan periwinkle, *Catharantus roseus*, and taxol from the Pacific yew tree, *Taxus brevifolia*. There is currently much interest in cataloguing plants used in Chinese medicines to see whether they can provide drugs such as artemisinin (see Table 11.3 above) that can be mass produced.

Of course it is important that all species at risk of extinction are conserved for many reasons other than their potential to provide drugs. But this utilitarian reason is a very powerful argument for those who cannot understand why we should be worried about the loss of biodiversity (see Chapter 13).

Activity

Drug-resistant tuberculosis

Strains of drug-resistant *Mycobacterium tuberculosis* were identified when treatment with antibiotics began in the 1950s.

Antibiotics act as selective agents, killing drug-sensitive strains of bacteria and leaving resistant ones behind. Bacteria can become drug-resistant as a result of mutation in their DNA. Mutation is a random event and occurs with a frequency of about one in every thousand bacteria. If three drugs are used in treatment then the chance of resistance arising by mutation to all three is reduced to one in a thousand million. If four drugs are used, the chance is reduced to one in a billion. This is the reasoning behind using combination therapy for treating tuberculosis (TB).

The four 'front-line' drugs for treating TB are isoniazid, pyrazinamide, rifampicin and ethambutol used in combination. This drug therapy cures 95% of all patients, is twice as effective as other strategies, and is helping to reduce the spread of multi-drug-resistant (MDR) strains.

If TB is not treated or the person stops the treatment before the bacteria are completely eliminated, the bacteria spread throughout the body; this increases the likelihood that mutations will arise, because the bacteria survive for a long time and multiply. Stopping treatment early can mean that *M. tuberculosis* develops resistance to all the drugs being used. People who do not complete a course of treatment are highly likely to infect others with drug-resistant forms of TB. It is estimated that one person can easily infect 10–15 others, especially if people live in overcrowded conditions.

The WHO promotes a scheme to ensure that patients complete their course of drugs, known as Direct Observation Treatment, Short Course (DOTS). This scheme involves health workers or responsible family members making sure that patients take their medicine regularly for six to eight months (see Figure 11.19).

Drug resistance is a problem in treating TB, especially the forms of *M. tuberculosis* that are resistant to two or more of the front-line drugs.

Figure 11.19 DOTS in action: TB patients in a community health centre in Chumpou Voan village, Kampot province, Cambodia, take medicines under the supervision of health workers.

There are two forms of drug-resistant TB:

- Multiple-drug-resistant strains of TB (MDR-TB) are resistant to at least the two main drugs used to treat TB – isoniazid and rifampicin.
- Extensively (or extreme) drug-resistant TB (XDR-TB) has also emerged as a serious threat to health, especially of those people who are HIV positive. These strains are resistant to first-line drugs and the drugs used to treat MDR-TB.

Resistant strains of TB do not respond to the standard six-month treatment with first-line anti-TB drugs and can take two years or more to treat with drugs that are less potent and much more expensive.

In 2012, an estimated 450 000 people developed MDR-TB in the world. It is estimated that about 9.6% of these cases were XDR-TB.

1 Describe how TB is transmitted.
2 Outline ways in which bacteria can become resistant to antibiotics.
3 Distinguish between MDR-TB and XDR-TB.
4 Explain why TB and many other infectious diseases are treated with combination therapy, not a single antibiotic.
5 Outline the consequences of drug resistance in combatting TB.
6 Describe the main features of the DOTS scheme.

Test yourself

20 Define the term *antibiotic*.

21 Why is there concern about antibiotic resistance?

22 List some likely sources of new antibiotics.

23 Suggest ways in which the spread of antibiotic resistance can be prevented.

Personalised medicines

Doctors have known for a long time that some medicines are effective on some of their patients, but not all. For example, it has been discovered that people treated with isoniazid for TB can be divided into those who metabolise the drug slowly, so find it is effective at treating *M. tuberculosis*, and those who metabolise it much faster and gain little benefit.

The reasons for certain medicines being more effective on some people than others should now be obvious to you from the information about genes, proteins and enzymes in Chapters 3 and 4. Proteins have complex structures, and a particular drug target may not be the same in everyone. By testing people's genomes it is possible to identify whether drugs will function on some people and not others. So instead of prescribing the same drug for everyone, doctors will be able to prescribe drugs that they know will be effective according to a person's own genome. People can be treated with **personalised medicines** that have been chosen as being suitable for them, rather than taking a 'one type fits all' drug as has been the case in the past. Now that detailed information about people's genomes is becoming available, it is possible for treatment to be far more precise and to know that a drug will not have adverse effects and will be much more effective.

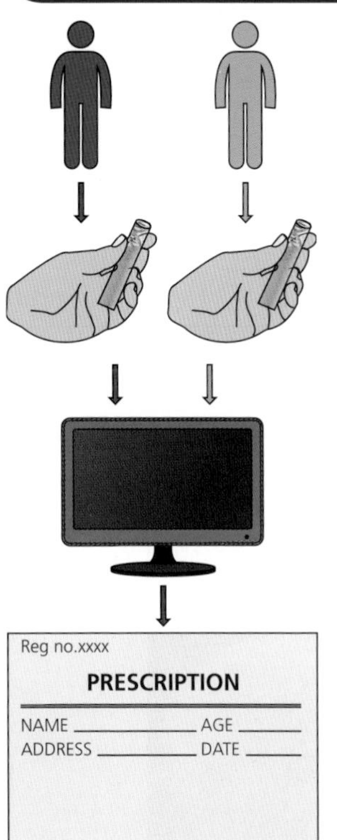

Figure 11.20 Personalised medicines. The population can be divided into different subgroups according to how they will respond to different drugs. Knowledge gained from genetic testing can make sure people receive a drug that will be most effective and have few if any side effects.

Most progress in personalised medicines has been made with drugs for treating cancers. For example, trastuzumab (Herceptin) is an antibody drug that interferes with a specific cell surface receptor and is used mainly to treat certain breast cancers. This drug is only used if a patient's cancer is tested and found to have large numbers of the receptor.

There is more about sequencing DNA and 'reading genomes' in Chapter 11 in OCR A Level Biology 2 Student's Book.

Synthetic biology

In May 2010, a group of scientists announced that they had assembled a complete genome of millions of base pairs, inserted it into an enucleated bacterial cell from which all the DNA had been removed, and caused that cell to start replicating DNA and dividing. These cells were called the first synthetic cells and were created at a cost of over US$40 million. This is more than just genetically engineering an organism to produce one or two products that it would not normally make; instead this is assembling a genome to operate a cell in ways that have never happened before. The genome can be assembled from existing DNA sequences and also new ones that have been written to produce specific proteins or to control existing genes in specific ways.

Synthetic biology, as it is known, holds the potential for developing new drugs that require complex methods of synthesis that are difficult to achieve by purely chemical methods that may be very expensive. One of the best-known applications of synthetic biology is genetically modifying *E. coli* and yeasts for the commercial production of a precursor of the antimalarial drug artemisinin.

Figure 11.21 A father gives his child an antimalarial drug based on artemisinin, which has been the first choice for treating malaria in South-East Asia for many years.

Recently, healthcare workers in Cambodia and Myanmar (Burma) have seen signs that *Plasmodium* is becoming resistant to artemisinin. If so, where will the next antimalarial drug come from?

Test yourself

24 Personalised medicine has been described as not being about creating medicines unique to each patient. If that is the case, what are personalised medicines?
25 Suggest the likely advantages of synthetic biology in the development of new medical drugs.
26 Make a table listing the named chemicals involved in defence against disease that have been included in this chapter. For each chemical, state its site(s) of synthesis and its role.

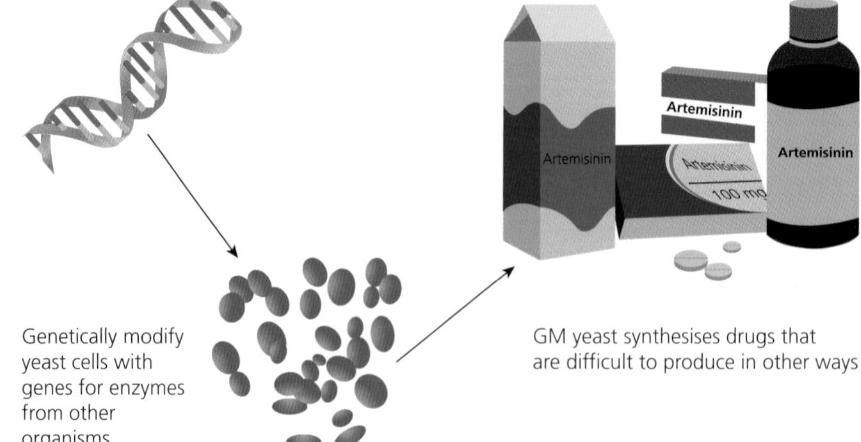
Genetically modify yeast cells with genes for enzymes from other organisms
GM yeast synthesises drugs that are difficult to produce in other ways
Artemisinin
Artemisinin
Artemisinin
100 mg

Figure 11.22 Synthetic biology involves reprogramming an organism to produce a particular product such as the antimalarial drug artemisinin.

Exam practice questions

1 Which results in long-term immunity?
- **A** antibodies passing across the placenta from mother to fetus
- **B** breastfeeding an infant
- **C** injecting serum into someone bitten by a dog with rabies
- **D** vaccinating a child with the rubella vaccine *(1)*

2 Which describes the difference between the responses of B cells and T killer cells to the invasion of a host cell by a bacterial pathogen? *(1)*

	B cells	T killer cells
A	divide to form antibody-secreting cells	make contact with host cells and kill them
B	make contact with host cells and kill them	kill host cells infected with viruses
C	provide active immunity	provide passive immunity
D	respond only during the primary immune response	respond only during the secondary response

3 The following are components of the body's defence system
- **1** ciliated epithelial cells
- **2** lysozymes
- **3** phagocytes

Which are in the first line of defence against infections?
- **A** 1, 2 and 3
- **B** Only 1 and 2
- **C** Only 2 and 3
- **D** Only 1 *(1)*

4 The primary defences of the body are non-specific.
a) Describe the roles of the bronchial epithelium as a primary defence against infection. *(4)*
The photo shows a neutrophil ingesting a cell of the yeast *Candida albicans*.

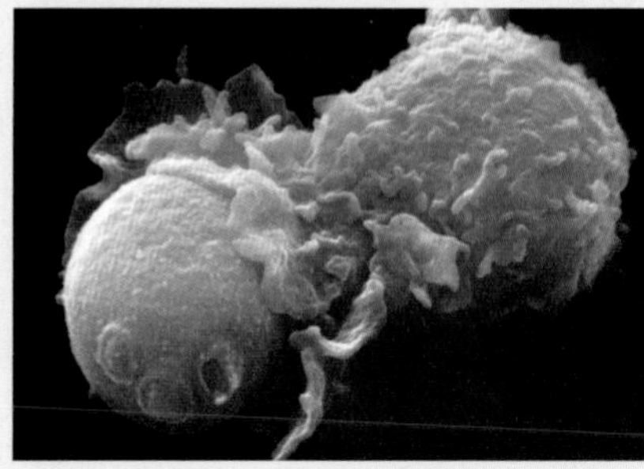

b) Explain how the neutrophil in the electron micrograph ingests cells such as those of *Candida albicans*. *(4)*
c) Describe what happens once the cell is inside the neutrophil. *(5)*
d) Outline how specific defences differ from non-specific defences. *(3)*

5 a) Explain the difference between antigens and antibodies. *(3)*
b) Measles is a viral disease. A child catches measles and is very ill. However, after a few weeks the child recovers.
Outline the events that happen in the specific immune system following infection by measles virus that help the child to recover. *(7)*

c) Explain why the child does not get measles when next in contact with someone with this disease. *(4)*

6 Bacteria that are resistant to the antibiotic penicillin may have the enzyme penicillinase. The figure shows the structure of the enzyme penicillinase. The arrow indicates the active site of the enzyme.

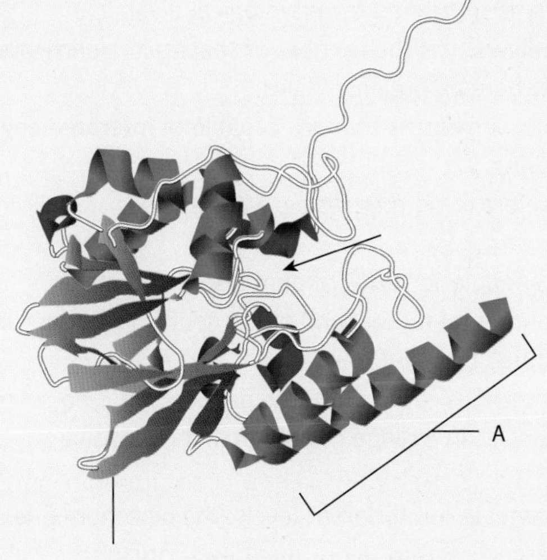

a) i) Identify the parts of the molecule labelled A and B. *(2)*

ii) Penicillin is an enzyme inhibitor. Explain why it does not inhibit this enzyme. *(2)*

b) Suggest how bacteria previously susceptible to penicillin could gain the ability to make penicillinase. *(3)*

c) The number of hospital-acquired infections caused by MRSA and *Clostridium difficile* (*C. diff.*) has increased in the UK. This is a great cause for concern, and as a result medical authorities have taken steps to reduce the number of cases.

i) State what MRSA stands for. *(2)*

ii) Explain the reasons for the concern over MRSA and *C. diff.* and outline the steps that are being taken to reduce the number of infections. *(5)*

Stretch and challenge

7 The causative organism of African sleeping sickness is *Trypanosoma brucei*, which is a protoctist that lives in the bloodstream. Tsetse flies are the vector of this disease.

Each cell is covered by a dense 'mat' of glycoproteins, which protects the cell surface membrane. *T. brucei* has over a thousand genes that code for these surface glycoproteins. The type of glycoprotein forming this mat can change with each generation of parasites.

The graph shows the changes in the numbers of parasites with four different glycoproteins that are identified as 1 to 4 over several weeks of an infection. The graph also shows the relative concentration of antibodies against each glycoprotein over the same period.

Use the information provided to comment on the likely progress of an infection by *T. brucei* beyond the period shown in the graph, assuming that no treatment is available.

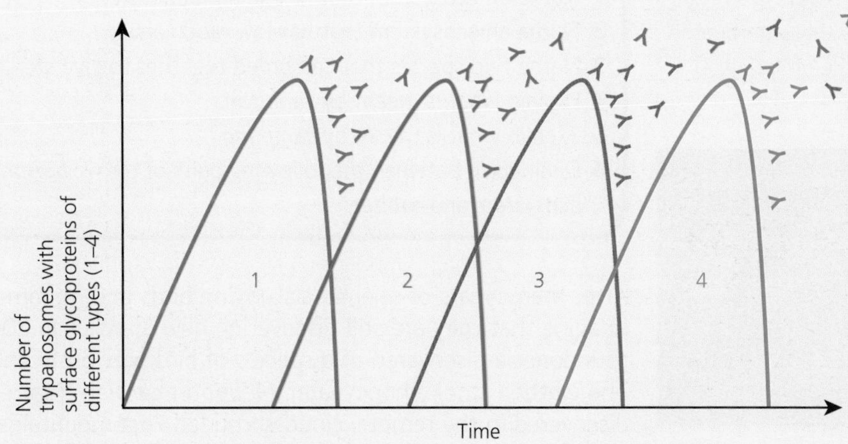

Chapter 12

Biodiversity

Prior knowledge

Before you start, make sure that you are confident in your knowledge and understanding of the following points.

- The term *biodiversity* is shortened from biological diversity.
- One measure of biodiversity is the number of species in a particular place.
- A habitat is a place where an organism lives.
- A species is a group of organisms that are capable of interbreeding to produce fertile offspring.
- The term *variation* refers to differences between species and differences within species.
- The distribution and population sizes of animals and plants are influenced by environmental factors and by competition for resources.
- Species can be identified by using keys.
- There are different methods to measure the distribution, frequency and abundance of species in a range of habitats. They involve taking samples.
- Biodiversity can be assessed at different levels: the biosphere (life on planet Earth), biomes, ecosystems, communities of organisms, habitats and populations of species.
- A gene is a length of DNA that codes for a polypeptide; alleles are alternative versions of genes that code for the same feature.
- The term *genotype* refers to the alleles of one or more genes of an individual; in diploid organisms, genotypes are *homozygous* if the two alleles of a gene are the same (AA/aa), and *heterozygous* if they are different (e.g. Aa).

Test yourself on prior knowledge

1 Name an ecosystem that has high biodiversity.
2 Name an ecosystem that has low biodiversity.
3 Name two species that can breed together to give sterile offspring.
4 Explain what is meant by *variation*.
5 Explain what is meant by *sampling*.
6 Distinguish between the following pairs of terms: *biosphere* and *biome*; *ecosystem* and *habitat*.

Key term

Species A group of organisms of common ancestry that interbreed to give rise to fertile offspring.

Figure 12.1 The wattled smoky honeyeater, *Melipotes carolae*.

After many years of scientists studying birds and mammals, it is hard to imagine that they are still discovering new **species**. In 2005, an expedition to Indonesia discovered of a species of bird previously unknown to science. The wattled smoky honeyeater, *Melipotes carolae* (Figure 12.1), was discovered in the remote, cloud-shrouded Foja mountains in the west of New Guinea. Co-leader of the expedition, Bruce Beehler, described the area as 'as close to the Garden of Eden as you're going to find on Earth'.

What is biodiversity?

Key term

Biodiversity The number of different ecosystems and habitats in an area, the number of species within those ecosystems, and the genetic variation within each species.

Rainforests, such as those of Papua New Guinea, are one of the most biodiverse ecosystems on the planet. At its simplest, **biodiversity** is a catalogue of all the living things in an area, a region, a country, an ecosystem or even the whole world.

Biodiversity has many different definitions, but the definition that this book will use is that given in the Rio Convention on Biological Diversity, which was presented at the Earth Summit of 1992:

'Biological diversity' means the variability among living organisms from all sources including (amongst others) terrestrial, marine and other aquatic ecosystems and the ecological complexes of which they are part; this includes diversity within species, between species and of ecosystems.

According to the Rio Convention definition, biodiversity is considered on three levels:

- the different ecosystems and habitats present in an area
- the different species present in the same area
- the genetic variation within each species.

The term biodiversity is used very loosely to apply to species diversity within the biosphere and more correctly to a very detailed analysis of a region with its various ecosystems and all the species present in those ecosystems. Various organisations publish information about biodiversity, for example County Wildlife Trusts and organisations such as Natural England and the National Parks. Most studies of biodiversity record details of ecosystems, habitats and species. They rarely include details on genetic diversity, except to highlight species that have very small populations and thus little genetic diversity. Biodiversity to most people means the variation in species, so that is where this chapter will start.

Tip

Conservation groups assess the genetic biodiversity of some groups, such as butterflies by looking for different phenotypic forms. You will see examples in Figure 12.3.

Species diversity

There are at least seven ways to define the term **species**.

One definition of a species is often given as:

A group of organisms that interbreed to give rise to fertile offspring and are reproductively isolated from other species. If the organisms do breed with other species, any offspring produced are usually sterile.

This definition describes a *biospecies*. However, it is often very difficult to apply this definition to many species. Scientists who discover and name new species rarely see evidence that individuals reproduce together, let alone check the fertility of the offspring by making sure they are able to interbreed. Many species have been described from preserved specimens collected during expeditions. Many species do not reproduce sexually; they only reproduce asexually. Extinct organisms are also classified into species. Obviously the biospecies concept cannot apply to them!

Most species are described using physical features such as morphology (outward appearance) and anatomy. So a different definition is used – a *morphological species*:

A group of organisms that share many physical features that distinguish them from other species.

Tip

Take care over the terms interbreeding and inbreeding. Interbreeding refers to organisms breeding together, whereas inbreeding refers specifically to breeding between individuals that are closely related. The opposite of inbreeding is *outbreeding*, where individuals mate with others of their species that are genetically different from themselves.

Increasingly, different methods of genetic analysis are used to distinguish between species.

Species diversity refers to the variety of species within an area. Such diversity can be measured in many ways, and scientists have not settled on a single best method. The number of species in an area is one measurement, but another measurement known as **taxonomic diversity** considers the relationship of species to each other. For example, an island with 20 species of bird and five species of lizard has a greater taxonomic diversity than an island with 20 species of bird but no lizards.

Genetic diversity

To see genetic diversity, look around at the human species. Many of the differences you can see between different people are due to the different gene variants or alleles in the human gene pool.

Humans make use of genetic diversity in selective breeding of plants and animals. You can see the results in the different breeds of domesticated animals such as cattle, sheep, dogs and cats, and the plants we grow for food and for their attractive appearance.

Figure 12.2 Fruit colour in these peppers of the species *Capsicum annuum* is an example of polymorphism (the existence of many forms within a species).

Figure 12.3 Genetic diversity in these *Heliconius* butterflies is expressed in the different colours and patterns of their wings.

Genetic diversity refers to the genetic variation within a species. This covers the genetic variation within distinct populations of the same species and within a single population. Genetic variation is high among the thousands of traditional rice varieties that are grown in India, whereas it is low among the populations of cheetahs, because they are all descended from a small number that survived near-extinction about 10 000 years ago.

Although different species may have many genes in common because they have similar cell structures and similar enzymes, for example those for respiration, each species has a unique combination of genes. All individuals of the same species have the same genes. The position that a gene occupies on a chromosome is called its **locus**, which is Latin for place and gives us the word location. In the explanation that follows, we will use the term gene locus or gene loci (plural) in describing the genetic diversity within a species.

Alleles (also known as gene variants) are the different forms of a gene. The **gene pool** consists of all the alleles of all the genes within a species. Some of the genes determine features that we can see, such as the colours and patterns of the wings of the *Heliconius* butterflies in Figure 12.3. But much genetic diversity does not have an effect on the appearance (morphology) of a species and has to be studied in other ways.

For example, genes that code for enzymes have alleles. Some of these alleles code for forms of the enzyme that function in slightly different ways.

These different forms of an enzyme have different primary sequences and are known as alloenzymes, or allozymes for short. Sometimes the different allozymes have an effect on the phenotype of individuals, and we can see it. The different allozymes present in individuals can be investigated by using the technique of electrophoresis (similar to chromatography) to separate them.

The amino acid sequences of the allozymes are different because the alleles have different sequences of nucleotides. The nucleotide sequences can be analysed using modern techniques of DNA sequencing. These techniques mean that now the genetic diversity within a species can be assessed by studying the different nucleotide sequences in a selected gene loci and even in all the genes in a species.

Genetic diversity is limited in populations that are very small. Inbreeding leads to many individuals being homozygous (e.g. AA or aa) for many of their genes, which reduces genetic diversity. Less genetic diversity can mean that genetic diseases that are caused by recessive alleles become more common – as is the case with genetic dwarfism in Californian condors and hip dysplasia in boxer dogs.

Genetic diversity is much greater in species that show outbreeding. Here the individuals that mate are more likely to be unrelated to each other and so give rise to offspring that show much variation.

Assessing genetic diversity

There are several ways to assess the genetic diversity in a population. These include assessing:

- the proportion of gene loci that have two or more alleles (gene variants) – that is, the proportion that show genetic polymorphism
- the proportion of the population that are heterozygous (e.g. Aa) for any particular gene locus
- the number of different alleles for certain genes – this is known as allele richness.

In each case, the measure of genetic diversity involves finding out whether there are different alleles of each gene locus. In some cases, all the alleles of a gene locus are expressed in the phenotypes of different individuals; for example, alleles that determine the fruit colour of peppers (Figure 12.2). Other genes are not expressed in a way that can be seen. In this case, it is necessary to examine the protein products of the different alleles or to sequence the DNA. However, in these cases it is not always known whether the genetic variation discovered is of any importance.

Calculating the proportion of polymorphic gene loci

Genetic polymorphism is the existence of different alleles of a gene, where the rarest allele has a frequency that is either more than 1% or more than 5% (depending on the judgement of the scientist carrying out the research). There is nothing biologically significant about 1% or 5%; these percentages have just been chosen arbitrarily. Scientists sample a population and determine the genotypes of the individuals. To calculate the percentage of alleles in the population, it is assumed that each individual contributes two alleles. Remember that when you write down a genotype for a diploid organism, you always write two letters, for example AA, Aa or aa.

Gene loci that do not have alleles are described as **monomorphic**. (In tables of data, the number of alleles may be recorded as 1, but strictly speaking there are no alleles because there are no alternative forms of the gene.) A **polymorphic** locus is one where the most common allele has a frequency that is less than 95% or less than 99% (depending on the level determined by the researcher). Scientists then find out how many genes are polymorphic and calculate the proportion of polymorphic gene loci (*P*) as:

$$P = \frac{\text{number of gene loci that are polymorphic}}{\text{total number of loci investigated}}$$

If, for example, investigators find that 50 out of a 100 gene loci are polymorphic, then the value of *P* is 0.5 (50%).

Within the species *Canis familiaris* (domestic dog) there are many pedigree breeds such as boxers, Chihuahuas and Pekinese. A study of the genetic diversity in pedigree dogs in the USA found that values of *P* were much lower in long-established breeds compared with those breeds that were established and registered with the American Kennel Club more recently. Researchers examined 100 gene loci in breeds of different types and determined whether the number of alleles per breed differed relative to the totality of alleles observed in all breeds. The total number of alleles observed for all breeds and loci was 1780. Within each breed, a range of 399–805 alleles per breed was found, with an overall average of 605 alleles. As a function of population size, the breeds with smaller populations had about 6% fewer alleles than the breeds with larger populations. The number of alleles observed per breed was lower for the older, longer established breeds by about 7%. This shows the effect that a certain degree of inbreeding can have.

However, studies have also shown that genetic diversity in *C. familiaris* as a whole is large – a reflection of the fact that many humans domesticated the grey wolf, *Canis lupus*, on more than one occasion, so that genetic diversity from the founding populations of wolves has persisted to the present day.

In another study of 18 blood proteins in dogs, all were found to be polymorphic (*P* = 1.0). The number of alleles per gene locus varied from 2 to 11, with a mean allele richness of 3.9 alleles per locus. This reveals the limitation of calculating *P* as a method of assessing genetic diversity.

Determining the amino acid sequence of the proteins reveals more genetic variation. This measure of genetic diversity is particularly suitable when investigating allozymes. Imagine a gene locus for an enzyme that is composed of one polypeptide. Proteins can be separated by using electrophoresis, a technique similar to chromatography. If the gene has three alleles that each give rise to change in one amino acid, then samples can be taken from many individuals to find out their genotypes. Table 12.1 shows the results for five variable gene loci of one such investigation of allozymes in a marine worm.

Tip

In Table 12.1, each gene locus is polymorphic at the 0.95 level if the most common allele has a frequency less than 0.95. To be polymorphic at the 0.99 level then the most common allele must have a frequency less than 0.99.

Table 12.1 Part of an analysis of variation in gene loci of 39 loci from the marine horseshoe worm, *Phoronopsis viridis*.

Gene locus	Proportions of alleles (most common to least common)						Polymorphic	
	1 allele	2 alleles	3 alleles	4 alleles	5 alleles	6 alleles	at 0.95	at 0.99
Acph-1	0.995	0.005					✗	✗
Acph-2	0.882	0.066	0.024	0.014	0.009	0.005	✓	✓
Est-6	0.979	0.012	0.010				✗	✓
Fum	0.986	0.014					✗	✓
Lap-5	0.551	0.326	0.119	0.004			✓	✓

Test yourself

1 Use the example of the *Heliconius* butterflies in Figure 12.3 to explain the term *genetic diversity*.
2 Suggest why genetic diseases caused by recessive alleles are more common in inbreeding populations.
3 What increases the genetic diversity in a species and what reduces it?
4 In a study of marine worms, 11 out of 39 gene loci were found to be polymorphic at the 0.95 level. What is the value for *P* for this population?

Key term

Habitat A place where a species lives, expressed in one or a few words; for example, pond, lake, river, coral reef, forest, grassland, sand dune, heathland.

Niche The role of a species in an ecosystem; for example its position in the food web. A niche includes a description of all of a species' interactions with the different factors in its environment.

Tip

It is a good idea to practise using keys to identify different groups of organisms.

Another level can be assessed, because the genetic code is degenerate – there is more than one base triplet that codes for each amino acid (see page 264). Analysing genetic diversity is now almost exclusively at the level of base sequences. Investigations usually concentrate on specific sequences in the DNA in the nucleus and mitochondrial DNA (mt DNA); there is more about this in Chapter 14, page 268.

Habitat diversity

Three terms are used for the places where organisms live: biome, ecosystem and **habitat**.

Biome refers to a region of the world that has a dominant type of vegetation that is associated with the climate.

Ecosystems are areas that support communities of organisms that interact with each other and with their surroundings. Ecosystems can be large: the dominant ecosystem on Earth is open ocean. However, the term is also used for small areas such as rock pools. At low tide, rock pools contain water – unlike the surrounding shore, which dries out quickly as the tide goes out. The physical conditions within the rock pool allow some species to survive that cannot withstand long exposure to the air at low tide.

A habitat is a place where one species lives. Habitats provide the resources that organisms need. Plants require light, carbon dioxide, water and mineral ions; animals require food, water, shelter and somewhere to breed.

The terms *ecosystem* and *habitat* often refer to the same place. Ponds, grasslands, sand dunes, woodland, forest, meadows (fields of grass and other plants), streams and coral reefs are ecosystems, but these names are often used for the habitats of the species that live there.

Habitat diversity is harder to measure than species or genetic diversity, because the boundaries of a species' living space are often difficult to determine. Think of species that make use of two or more ecosystems. Many bird species feed in one ecosystem and roost in another; sea birds such as gannets, guillemots and puffins are examples. Some bird species spend part of the year in one ecosystem and migrate to breed in another. However, if species can be identified satisfactorily, it is possible to use the idea of niche to explore the different habitats that are available. A **niche** is an organism's role within an ecosystem. The more species that are present in an ecosystem, the more niches must be available and therefore the greater the habitat diversity.

A study in a forest in the Dominican Republic discovered several species of lizard that occupied different niches at different heights on the trees. This indicates that the trees provide several different habitats with slightly different conditions.

Local biodiversity

Imagine you are in a habitat like that in Figure 12.4 – an area of long grass with some trees and shrubs. The most obvious species are the large plants such as the trees and the buddleia bush that you can see at the left of the photo. Some of the larger animals, particularly birds and insects, are also easy to see. How do you investigate the biodiversity?

The first task is to identify and catalogue the types of organisms and build up a species list. Biologists use keys to identify the organisms that they find. There are different forms of key: some simply have drawings or photos

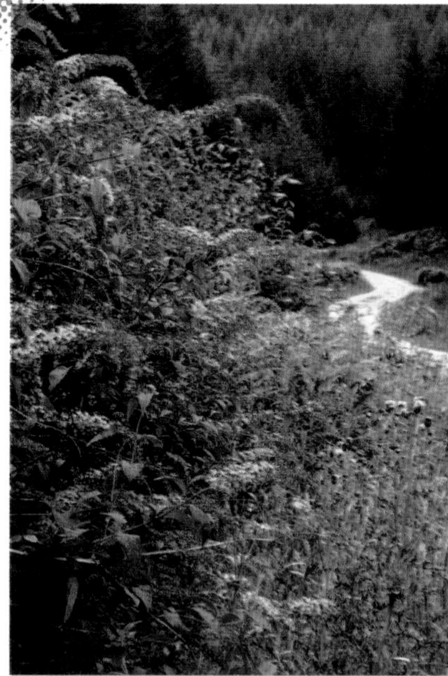

Figure 12.4 Some students investigated the vegetation of a piece of land dominated by the butterfly bush or buddleia, *Buddleja davidii*. How would you measure the biodiversity of this habitat?

Figure 12.5 Coral reefs are one of the most biodiverse ecosystems on the planet.

Figure 12.6 Biodiversity decreases with increasing distance from the equator and with increasing altitude. Few plant species are adapted to survive at high altitude, and this limits the diversity of habitats available for animals to colonise.

Key term

Species richness The number of species present in an area.

with identifications; others are written keys that may or may not have illustrations. The most common key is a dichotomous key. Some keys are available for use on mobile devices in the field.

One way to make a species list is to use a timed search. Each member of a group spreads out across the area and finds as many different species as possible. After a set time (e.g. 20 minutes), group members pool their results to give a species list. Species that cannot be identified can be photographed and the species later identified from the photos. Species that still cannot be identified can simply be called species A, species B, etc.

Table 12.2 lists some of the plants and animals found by students in a survey. A species list gives an indication of the **species richness** of an area – how many species are present. However, a species list made on one day in the summer does not give a full catalogue of the biodiversity of an area. For example, the only animals on the list are birds and butterflies. The students could extend their survey by searching for nocturnal animals. They could do this using a moth trap and some humane traps for small mammals. If they did this they would be likely to add the brimstone moth, *Opisthograptis luteolata*, the field mouse, *Apodemus sylvaticus*, and the pigmy shrew, *Sorex minutus*, to their list. Bacteria and many fungi can only be identified by placing soil samples onto agar in Petri dishes and allowing them to grow. This is time-consuming and specialised work not usually done by professional ecologists assessing biodiversity for governmental and local organisations. If it is impossible to sample every species that lives in a particular habitat, it is important to sample all the different types of organism, including the microscopic. It is also important to identify occasional visitors or migrants that may use the habitat for a short period each year.

Table 12.2 Species list for an area of waste ground.

daisy	ragwort	house sparrow	peacock butterfly
dandelion	spurge	blue tit	tortoiseshell
plantain	stinging nettle	hedge sparrow	butterfly
clover	willowherb	chaffinch	
buddleia	groundsel	blackbird	

Distribution and abundance of species

Species richness says something about the biodiversity of an area, but not everything. For example, one species may dominate the area almost to the exclusion of everything else. A species list is **qualitative**; it does not give the number of each species and so show which species are common and which are rare.

The second aspect of biodiversity is therefore recording how many of each species are present – their **abundance**.

The number of species *and* their abundance is another measure of biodiversity – **species evenness**. High species evenness indicates high biodiversity.

The standard piece of apparatus for assessing abundance is the quadrat. The different types of quadrat are:

- an open frame quadrat (Figure 12.8), which is a square of known area; a typical school or college quadrat is $0.25\,m^2$ or $1\,m^2$
- a gridded quadrat (Figure 12.7), which is divided by wires or string into smaller sections, perhaps 5×5 or 10×10

Key term

Species evenness A measure of the relative abundance of the different species in an area.

Figure 12.7 Using a 1 m^2 gridded quadrat to assess percentage cover in a meadow.

Figure 12.8 Open frame quadrats like the one top left of this photo are useful for assessing biodiversity in long grass. In many habitats, it is often much easier to use quadrats divided into smaller squares than it is to use an open frame quadrat, especially when estimating percentage cover.

- much larger quadrats that can be marked out with tape or by using GPS to pinpoint their location.

Some biodiversity surveys are done over much larger areas. For example, ecosystem diversity in the Amazon region was assessed with quadrats that were 1000 km × 1000 km.

Quadrats can be used to assess abundance in several different ways, by:

- species frequency
- species density
- percentage cover.

The simplest way to use a quadrat is to record the presence or absence of the species on a species list. **Species frequency** is then calculated as the percentage of quadrats that included each species. The number of individual organisms within a quadrat can be counted to give the **species density**, which is usually expressed as numbers per unit area, for example per m^2.

Species frequency and species density do not take into account the size of the organisms present. It is often difficult if not impossible to count the numbers of very abundant species such as grass. A quadrat divided into a grid makes it relatively easy to assess the area of the quadrat that is occupied by each species. This is **percentage cover** – the percentage of the area of the quadrat in which the plant or animal occurs.

Example

The table below shows the results of a quadrat survey to assess the abundance of dandelion, *Taraxacum officinale*, in a lawn. The area of the quadrat was 0.25 m^2.

Quadrat	1	2	3	4	5	6	7	8	9	10
Number of plants	2	6	0	3	7	3	0	0	8	4

1 Calculate the species frequency and the species density of *T. officinale* in the lawn. In both cases, show your working.
2 State another way in which the abundance of dandelion plants could be assessed using a quadrat.
3 Explain how species evenness differs from species richness.

Answers

1 *Species frequency*: dandelion plants were present in 7 out of 10 of the quadrats. The species frequency is 70%
Species density:

total number of dandelion plants = 33

total area of the 10 quadrats = $0.25 \times 10\,m^2 = 2.5\,m^2$

density of dandelion plants = $\frac{33}{2.5}$ = 13.2 plants per m^2

2 percentage cover
3 Species richness is the number of species found in an area; it does not take into account the abundance of any of the species. Species evenness is a measure of the relative abundance of each species in an area. If each species is present in roughly similar numbers, then the area has high species evenness; if there is one very abundant dominant species and all the other species are rare, then the area has low species evenness.

Activity

Estimating percentage cover

Figure 12.9 shows a quadrat placed on a lawn.

1 The quadrat is divided into smaller squares.
   a) How many small squares are there?
   b) What is the area of each small square?
   c) What percentage of the total area within the quadrat is covered by each smaller square?

2 Estimate the percentage cover of species A, species B and the grass.

3 Explain how this quadrat could be used to estimate the percentage cover for species A and species B across the whole lawn.

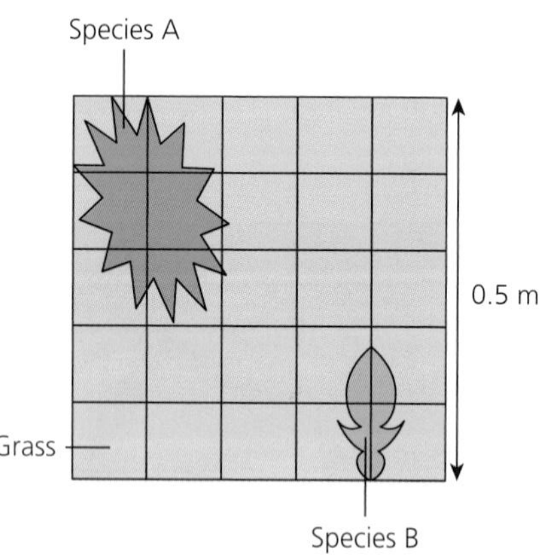

Figure 12.9 Estimating the percentage cover of two species, A and B, growing in a lawn using a $0.25\,m^2$ quadrat divided into a 5×5 grid.

Tip

If you use a quadrat divided into a 10 x 10 grid, then each square of the grid represents 1% of the total area.

Random sampling

Why is it important to take samples? Often the habitat is too large to take results from everywhere. It would take too long. If the habitat is uniform, then a percentage of the area (say 5 or 10%) should be representative of the whole area and the abundance of the species within the area.

Many of the plants that grow in waste ground are individual plants that can be counted. It would take too long to count all the plants in the whole area, so it is necessary to take representative samples. To do this, a quadrat is used to delimit an area of ground. If all sampling was done in the easiest places to sample, that would give a biased set of results. (The easiest places to sample may have the fewest plant species!) So **random samples** must be taken to avoid any bias on the part of the person doing the sampling. Random sampling can be done by placing tape measures at right angles to each other along two sides of the sample area. Random numbers are generated by using an app on a mobile phone to give coordinates where the quadrats can be placed.

Many quadrats have grids, like the one in Figure 12.9. The number of individual plants within the grid can be counted and the information recorded in the table. When sampling plant species, it is often necessary to record percentage cover, because it is impossible to see individual plants. In that case, you estimate how much of the quadrat encloses each plant and express the answer as a percentage of the area of the quadrat.

Activity

What size of quadrat to use?

Quadrats enable comparable samples to be obtained from areas of consistent size and shape. To be effective, each quadrat should include a reasonable number of the species present in an area. This activity shows how to choose an appropriate area of quadrat to use.

Quadrat number	Area of quadrat/m^2	Number of plant species found within each quadrat
1	0.25	10
2	0.50	14
3	1	16
4	2	19
5	4	21
6	8	23
7	16	24
8	64	26

Students used tent pegs and string to make eight quadrats of different sizes. They counted the number of different species in each quadrat.

1 Plot the data on a graph.

2 a) State how many additional species are present when the quadrat area is increased from $1\,m^2$ to $8\,m^2$.

b) The students decided that the best size for the quadrats would be $8\,m^2$, but instead they decided to use $1\,m^2$ quadrats. Comment on their decisions.

c) Suggest two ways in which the students could assess the abundance of the plant species.

3 When studies like this have been carried out in tropical rainforests, scientists have found that the number of species can continue to increase without ever becoming constant with increasing area.

a) Explain why this is so.

b) Suggest a way in which scientists studying a very large area of tropical rainforest could take random samples.

Figure 12.10 Shaking the tree dislodges many small invertebrates, particularly insects, which are easy to see on the white sheet and can be caught with a pooter.

Figure 12.11 Pond nets have to be very strong to withstand use like this.

Many animals in most ecosystems are much more difficult than plants to find and identify. Most animals are small and many only appear at night. It is not possible to find them by simply putting a quadrat on the ground. Often it is impossible to use quadrats to sample the animals. Some other pieces of apparatus are:

- Beating tray: A large white sheet is placed on the ground or supported by struts and held below a tree. The tree is shaken to dislodge the animals. Very small animals can be collected from the beating tray with a pooter.
- Pooter: Scientists can use this to suck small animals into the glass or plastic tube.
- Sweep net: Large nets are used to catch flying insects and insects that live in long grass.
- Pond net: These are stronger nets that can lift volumes of water from ponds or rivers; the water drains through the net to leave vegetation and animals behind.
- Pitfall traps: Cans or jars are buried in the ground, filled with paper or cardboard to provide shelter, and covered with a lid or stone to keep out the rain; these are useful for collecting ground-dwelling insects, which are often nocturnal (see Figure 12.12).

Some species are far too numerous to count. In the case of plants, it is difficult to isolate individual plants, so an abundance scale such as ACFOR (abundant, common, frequent, occasional, rare) is used. To use this scale you have to decide first how to apply each description – roughly how many plants or animals have to be present before you record their abundance as 'common' or 'frequent', for example. Once you have done that, you can make your recordings far more quickly than if making counts of individual species.

Figure 12.12 A pitfall trap. Rolled-up newspaper is put in the bottom to allow small insects to hide from their predators, which might also fall in.

Estimating population sizes of mobile animals

It can be difficult to assess the numbers of some animal species. One way to estimate the numbers of animal species is to catch a certain number and mark them. This is the first sample (S1). These animals are released back into the environment. After a suitable length of time to allow the released animals to mix thoroughly with the unmarked individuals in the population, the same sampling procedure is repeated. Some of the animals caught in the second sample (S2) will be marked individuals and some unmarked. The smaller the number of marked individuals, the larger the total population. The estimate of the population is the Lincoln index:

$$\text{population size} = \frac{\text{number in first sample (S1)} \times \text{number in second sample (S2)}}{\text{number marked in second sample (S2)}}$$

Example

A student estimated the number of water snails in a school pond by following this procedure:

- Pond nets were used to catch as many ramshorn and pond snails as possible using the timed search technique. The number of each species was recorded.
- Each snail was marked with a dot of non-toxic paint on its shell and left for the paint to dry. Then these marked snails were released back into the pond.
- After 24 hours, as many ramshorn and pond snails as possible were recaptured. The numbers of marked and unmarked snails were recorded and then the snails were returned to the pond.

The student's results are in shown in the table.

	Species	
	Pond snail	**Ramshorn snail**
Number marked and released	31	19
Total number recaptured after 24 hours	69	5
Number of recaptured snails that were marked	6	2

1 Use the data in the table to estimate the number of pond snails and ramshorn snails in the pond. Show your working.

2 The student was investigating the biodiversity of the pond. Two measurements of species diversity are species richness and species evenness.

a) Explain how the student would determine the species richness of the pond.

b) Explain how the student could use the mark–release–recapture technique to assess the species evenness of animal species in the pond.

Answers

1 number of pond snails $= \frac{31 \times 69}{6} = 357$ (to the nearest whole number)

number of ramshorn snails $= \frac{19 \times 5}{2} = 48$ (to the nearest whole number)

2 a) Use the pond net to take more samples and record the identities of the species found in the pond. In reality, the pond net will probably only collect the macroscopic organisms. Also, not all the macroscopic organisms will be identified to the level of the species; many will only be identified to higher taxonomic groups, for example family or order.

b) Repeat the mark, release, recapture method described for the snails, but with other animal species. The procedure could be repeated several times to take a mean of these estimates. The student could then compare the estimates for each of the species. The student can use the results to calculate Simpson's index of diversity (D). The diversity of the pond will be high if the numbers of the different species are very similar. This seems unlikely because there are many more pond snails than ramshorn snails, although the equivalent relative numbers for other pond species may be very different.

Test yourself

5 Distinguish between habitat diversity, species diversity and genetic diversity.
6 What are the differences between a biospecies and a morphological species?
7 Criticise the results obtained by the students who carried out the investigation into species diversity of a piece of waste ground (see Table 12.2 on page 216).
8 State five ways in which the abundance of animal and plant species can be measured.
9 Explain the advantages of random sampling.
10 Distinguish between species richness and species evenness.
11 Explain how a sample could be biased.
12 In a survey of trees in a dry tropical forest, some students identified five tree species (A–E). They counted the number in a large quadrat, with these results:

tree species	number
A	56
B	48
C	12
D	6
E	3

Calculate the Simpson's index of diversity for the trees in the forest.
13 Explain the advantage of using data on species richness and species abundance in calculating a diversity index.

Animals should be marked in such a way that they are not harmed (e.g. by toxic paint) or made more obvious to their predators. Snails can be marked with non-toxic paint just inside their shells and small rodents by clipping a small part of their fur.

Simpson's index of diversity

Species richness and species evenness generate quite a lot of data. An exhaustive study of an ecosystem might include the abundance of hundreds or even thousands of species. This is far too much data to absorb if you want to know the biodiversity of an area. These numbers can be reduced to one figure: an index of (bio)diversity. There are many of these indices; one of the simplest to use is Simpson's index of diversity.

Calculating Simpson's index of diversity

Some students counted the number of plants growing in ten randomly positioned quadrats on some grassland. Their results are shown in Table 12.3. The table also shows how to calculate Simpson's index of diversity.

The formula is $D = 1 - \Sigma\left(\frac{n}{N}\right)^2$

The symbol Σ means 'the sum of'.

The relative importance of each species is calculated by dividing the number by the total number of individuals. This figure is then squared and written into the fourth column of the table. Next, the total of all these figures is calculated and then subtracted from 1 to give the index.

Table 12.3

Species	Number of plants in 10 quadrats (n)	Number of individuals of each species (n) ÷ total number of individuals (N)	$(n/N)^2$
dandelion	40	0.154	0.0237
daisy	150	0.579	0.3352
buddleia seedlings	3	0.012	0.00014
ragwort	5	0.019	0.00036
plantain	18	0.069	0.0048
groundsel	15	0.058	0.0034
willowherb	11	0.043	0.0018
spurge	17	0.066	0.0044
Totals	259		0.3738

In this example of the plants in the waste ground, the index of diversity (D) is 1 – 0.3738 = 0.63 (to 2 significant figures).

What does this mean? When the index is small (near 0) there is low diversity. When the number is high (near 1) there is high diversity. The number (D) is the chance that any two individuals picked at random will be from different species. The probability in this case is 0.63, or 63%. Of course, you should realise that this calculation is only made for the plants on the waste ground. The students have not surveyed any other groups such as invertebrates above and below the ground or the birds and mammals. Nor have they sampled microscopic organisms, including soil bacteria and fungi.

You can see that the diversity depends on the number of different species and also the abundance of each of those species. A community with ten species, but where only one species is present in large numbers and the other nine are rare, is less diverse than one with the same number of species but where the species have a similar abundance (see the activity below).

Comparisons using this diversity index should be on a like-for-like basis – the communities should be similar, and the organisms chosen should also be similar. For example, the diversity index should not be used to compare the species diversity of seaweeds and invertebrates on a rocky shore with that of the ground flora in a woodland.

Communities with low diversity are less stable than communities with high diversity. These communities with high diversity tend to withstand any pressures on them, such as changes in environmental factors or pollution.

There can never be one measure of biodiversity. The appropriate measure of biodiversity depends what you are measuring, why you are measuring it, and to what use the measurement will be put (see page 240).

Non-random sampling

Tip

If you are studying geography or psychology, you may know about these methods of non-random sampling. The principles are the same.

Sometimes random sampling is not possible or takes too long. Under certain circumstances, it may be appropriate to carry out non-random sampling. This might be done in three ways, by:

- opportunistic sampling
- stratified sampling
- systematic sampling.

Opportunistic sampling

Imagine you are a TV producer making a programme about biodiversity in back gardens. It would be impossible to take a map of a local area and choose gardens to film by using the sort of process described for random sampling. Instead you might take a walk around the area until you find someone with an interesting garden who just happens to be at home. That is an example of opportunistic sampling.

You can probably appreciate the disadvantages of this approach. You may have chosen a garden that is representative of the area, but you cannot be sure, and you did not have the time to do anything else. Often students on field trips take samples from places that are easy to reach, are safe and that do not involve walking too far.

Stratified sampling

In some ecosystems, there may be different habitats occupying different proportions of the total area. For example, in an area of woodland there may be a dense area of coniferous trees, an area planted with native broadleaved plants with plenty of ground-level vegetation, a stream, and open tracks with wide verges. To obtain a fair representation of the biodiversity in this ecosystem, it is best to estimate the proportion of the total area occupied by these habitats and then sample accordingly. If 10% of the area is occupied by open verges, then 10% of the samples should be taken in this habitat.

Systematic sampling

Random sampling is not suitable for every ecosystem. It may be obvious that there is an abrupt change from one habitat to another or that there is a gradual change in conditions across the area you are studying and that random sampling will not reflect the way plants and animals are distributed. For example, a field may be bordered by an area of woodland, and or there may be a mountain and you may be interested in how species are distributed. The best way to sample areas like this is to use a transect, which is a line placed across the area sampled. A transect is used in two ways, as a:

- line transect
- belt transect.

Line transect

A line transect is simply a straight line marked out across a habitat. A long rope or measuring tape is laid across a habitat such as a rocky shore or a sand dune system. The choice of place to start the line should be decided randomly, but this is not always possible if, for example, obstacles are in the way. The species that touch the line at regular intervals (e.g. every 0.5 m) are identified and recorded. The results are converted into a drawing that shows the distribution of organisms.

Line transects are used to show how communities change along a gradient, which could be a slope or a change in an abiotic feature. They are a good way to show the changes qualitatively.

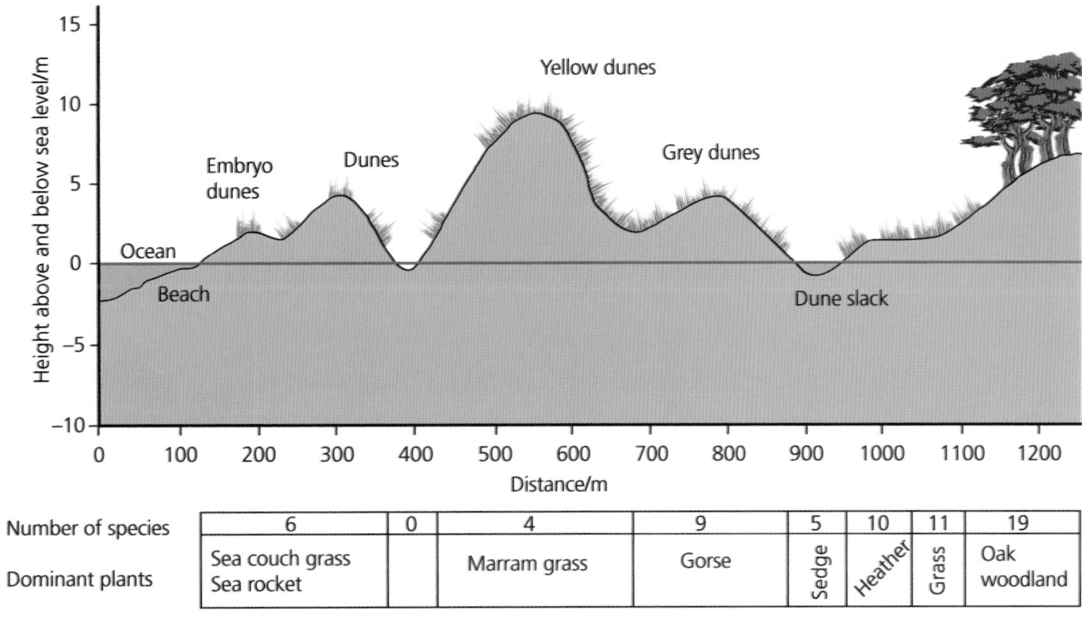

Figure 12.13 A line transect of a sand dune ecosystem.

Belt transect

A belt transect is a quantitative way to sample an ecosystem. It involves placing quadrats along a line taken through an ecosystem. If the distance is short, the quadrats may be placed continuously to give a **continuous belt transect**. If the distance is long, then placing quadrats continuously can be very time consuming; placing quadrats at intervals gives enough information on changes in populations. This is an **interrupted belt transect**.

Data from belt transects are presented as kite diagrams, which give a very good visual representation of changes along the transect.

Table 12.4 shows some data on four plant species collected by students from a belt transect across some heathland. They placed a $0.25\,m^2$ quadrat at 1 metre intervals.

Table 12.4 Percentage cover of plant species on heathland.

Distance from a path/m	Percentage cover in $0.25\,m^2$ quadrats			
	Holcus lanatus	*Agrostis capillaris*	*Viola riviniana*	*Calluna vulgaris*
1	22	54	0	0
2	14	42	7	0
3	21	71	14	0
4	0	58	20	0
5	21	13	9	0
6	36	24	2	19
7	6	14	0	2
8	3	5	0	0
9	13	25	0	24

Figure 12.14 shows you how to use data like that in Table 12.4 to draw a kite diagram. The data collected by the students is shown in the kite diagram in Figure 12.15.

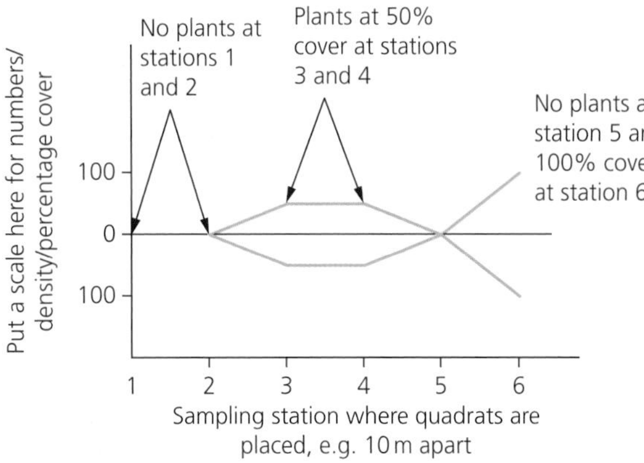

Figure 12.14 How to draw a kite diagram.

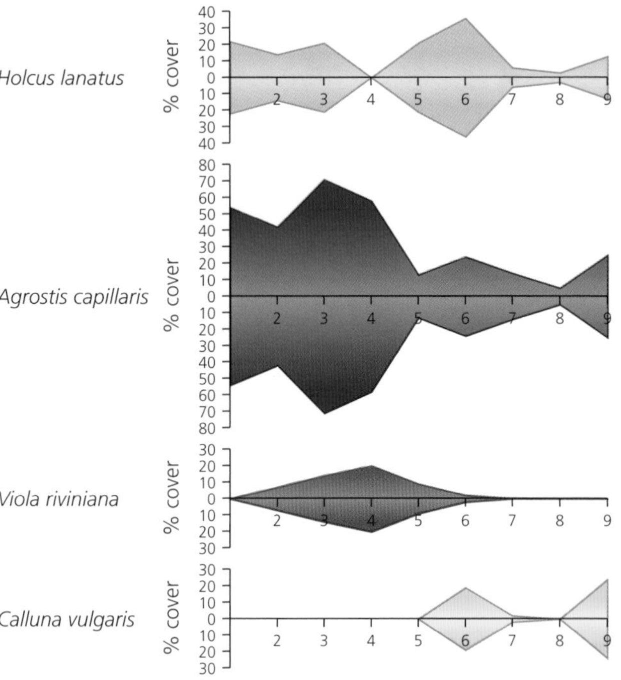

Figure 12.15 Distribution and abundance of four plant species collected by students on heathland using an interrupted belt transect.

Obviously, this method is biased, because the area you choose to put down the transect is not necessarily representative, and you may choose not to put it over a very boggy area or under low-growing trees, for example. However, this method is more straightforward than random sampling, because it is not necessary to measure coordinates for each sample.

Another way to do systematic sampling is to choose sampling sites at set intervals, for example to sample the lichen growing on trees in a wood by putting small quadrats, e.g. 0.01 m^2 (10 × 10 cm), at the same height on every 10th tree along transect lines across the wood.

Exam practice questions

1 The species richness of an area is defined as:

A abundance of different species
B distribution of different species across the area
C number of dominant species
D number of different species *(1)*

2 Which one of the following is **not** an advantage of random sampling?

A most of the species in the habitat are included in the results
B patterns of distribution across a habitat are revealed
C samples are representative of the whole area studied
D sampling bias is avoided *(1)*

3 In an assessment of genetic diversity, the proportion of polymorphic gene loci in an isolated population of plants was 0.30. What was the proportion of monomorphic loci?

A 0.15 **B** 0.30 **C** 0.60 **D** 0.70 *(1)*

4 The bluebell, *Hyacinthoides non-scripta*, grows across much of the British Isles. It is at risk of interbreeding with the invasive Spanish bluebell, *Hyacinthoides hispanica*. Students assessed a population of *H. non-scripta* in a nature reserve that showed no evidence of the invasive species. They saw that the bluebells grew more densely in an area of woodland than in more open grassland with isolated trees. They used a stratified sampling technique to assess the size of the population. The table shows the data that the students collected.

Area/m^2	**Habitat**	
	Woodland	**Open grassland**
	240	**60**
Number of random 1 m^2 quadrats	12	3
Mean number of bluebells per quadrat	105	11
Standard deviation	24	3

a) i) Explain why the students used stratified sampling and not random sampling across the whole area. *(2)*

ii) Explain why the students used 12 quadrats for the woodland and three for the open grassland. *(2)*

b) Calculate the density of bluebells across the area as a whole. Show your working. *(2)*

c) In another area the land was very steep, so the students decided to place their quadrats next to the path rather than using random sampling across the hillside. State the name given to this type of sampling. *(1)*

d) In another reserve, the difference between the densities of bluebells in different areas of woodland was not as large as in the table. Suggest how students should use their data to see whether there is a significant difference between the densities of bluebells in different areas. *(4)*

e) Suggest why it is important to conserve bluebells of the species *H. non-scripta*. *(3)*

5 The Amur leopard, *Panthera pardus orientalis*, is one of the world's rarest mammals.

There is a very small population of these animals in the wild in the Russian Far East. There are more Amur leopards in captivity in zoos in Europe, China and North America. A study compared the genetic diversity in wild and captive Amur leopards with that of two other subspecies: *P. pardus pardus* from Africa and *P. pardus kotiya* from Sri Lanka. The researchers studied the base sequences in the DNA of 25 different gene loci. The results are in the table below.

Use the data in the table to answer these questions.

a) i) State the proportion of monomorphic gene loci in the sample of *P. pardus kotiya*. *(1)*

ii) State the limitation of using values of *P* to assess genetic diversity. *(2)*

iii) Explain the advantage of using the other two wild leopard populations in this study. *(3)*

iv) Explain how the mean heterozygosity was determined. *(4)*

b) Discuss the limitations of the study as reported in the results table. *(5)*

Subspecies	Estimated population numbers (2014 unless otherwise stated)	Number in the study sample	Proportion of polymorphic loci (*P*)	Mean number of alleles per locus	Range of alleles per locus	Mean heterozygosity for the 25 gene loci
P.p. orientalis (wild)	less than 40	12	0.92	2.60	1 and 4	0.365
P.p. orientalis (captive)	200	21	1.00	3.12	2 and 6	0.490
P.p. kotiya (wild)	700–950 (2007)	11	0.96	3.52	5 and 15	0.500
P.p. pardus (wild)	500 000–700 000 (2007)	17	1.00	8.52	1 and 7	0.783

Stretch and challenge

6 a) What factors must be considered before carrying out a survey to assess the biodiversity of an area?

b) Species diversity is assessed by measuring species richness and species evenness. The diagram shows the animals found in two habitats after a quick survey.

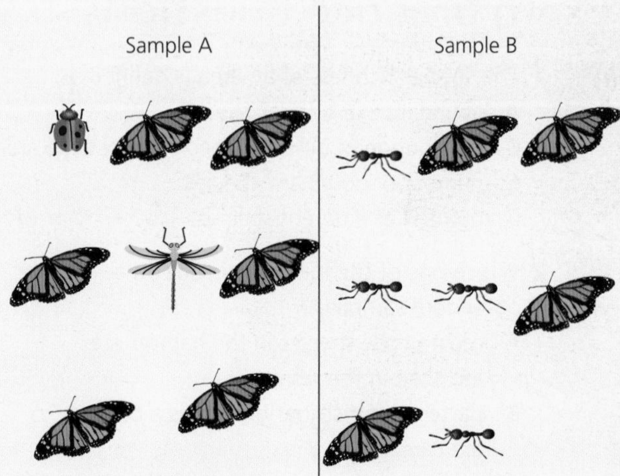

How many species are present in each habitat? What is the probability that any two organisms picked at random are the same species? Explain what your answers from this simple example tell you about two ways to assess biodiversity.

c) Explain why data on variation in DNA gives more information about gene diversity than differences in polypeptides.

Chapter 13

Maintaining biodiversity

Prior knowledge

Before you start, make sure that you are confident in your knowledge and understanding of the following points.

- Definitions of the following terms: ecosystem, habitat, biodiversity, species, population, gene and allele.
- Biodiversity exists on three different levels: ecosystem (habitat), species and genetic.
- Biodiversity is important because many of our resources are derived from biological sources; ecosystems provide services such as waste treatment and drinking water; organisms are sources of new drugs and foods; ecosystems with high biodiversity are better able to resist damage from storms and recover from such damage.
- Interactions between species in ecosystems are complex.
- Measuring changes in the distribution and abundance of organisms is a way of measuring and monitoring changes in ecosystems.
- Humans have many positive and negative interactions with ecosystems.
- Carbon dioxide and methane are greenhouse gases.
- Variation (both genetic and phenotypic) exists **between** species and **within** species.

Test yourself on prior knowledge

1. Outline what is meant by *ecosystem diversity.*
2. Explain the importance of maintaining genetic diversity in species of plants and animals.
3. Suggest why we try to eliminate some microorganisms but encourage the survival of others.
4. Why do animals become extinct?
5. Islands often have many endemic species that are found nowhere else. Why is this?

Elephants are too valuable to lose to poachers

Figure 13.1 Poaching of elephants for their valuable ivory, even in protected areas such as this national park in Central Africa, is a great threat to their survival.

The African elephant is the largest land animal on Earth. African elephants live in two distinct habitats: forest ecosystems and savannah ecosystems. It is estimated that there were between 3 and 5 million elephants across Africa in the 1930s, which decreased to 1.3 million in 1980 as a result of killings like that shown in Figure 13.1. Elephants are prized for the ivory of their tusks. Trade in ivory is now banned, but between 1973 and 1989, before the ban, Kenya lost 85% of its elephants. No-one knows exactly how many African elephants remain in the wild; estimates in 2012 varied between 470 000 and 690 000.

There have been five mass extinctions in the history of the Earth when a very large number of species became extinct in a relatively short period of geological time. The last mass extinction occurred 65 million years ago when the dinosaurs and other groups became extinct. Experts think that the current extinction rate is 1000 times higher than the rate was a century ago, and suggest that we are causing another mass extinction. Elephants and many other species need human intervention to ensure that they do not become extinct.

Factors affecting biodiversity

The biggest threat to biodiversity is the growth of the human population. In 2014, the human population numbered over 7 billion, having grown from 1 billion in 1800.

> **Tip**
>
> When writing about ways in which biodiversity is maintained, it is good to highlight the threats to biodiversity that make necessary the methods described later in this chapter.

Some of the threats to biodiversity from human population growth are:

- habitat destruction and the degradation of the environment
- the overexploitation and unsustainable use of resources
- modern agricultural practices, including monoculture, the use of chemical fertilisers and crop protection chemicals
- global climate change.

Maintaining biodiversity involves actions at different levels. To understand the types of action required, this chapter looks first at the different factors that threaten biodiversity and then at the reasons for maintaining biodiversity, before considering some of the ways in which biodiversity is conserved.

Figure 13.2 Deforestation in Malaysia. The natural forest is cleared and replaced by an oil palm plantation – a habitat with much less biodiversity than this forest.

Habitat destruction

The destruction of the natural environment leads to **habitat loss**. The clearing of land for agriculture, industry, housing, transport, leisure facilities, waste disposal and water storage removes the vegetation, and many species of plant and animal either lose their habitats completely or their habitats become divided into small areas, a process known as **habitat fragmentation**. Populations that get subdivided by this process are in danger of inbreeding and local extinction.

Much of the forest in the northern hemisphere was cut down by humans when people began to grow their own food – a process that started about 10000 years ago. Vast forests remained in the southern hemisphere until well into the twentieth century, but most of these in South-East Asia, Africa, Amazonia and Central America have been cut down, often to be replaced with cattle ranches and plantations of oil palm, which have much less biodiversity. **Deforestation** has had a devastating effect on the biodiversity of some countries (see Figure 13.2). Madagascar, famed for its unique plant and animal life, has lost almost all of its natural forest. Roots bind soil particles together and absorb much of the rainfall. When forests are taken away, water tends to run straight off the land leading to flooding. Deforestation can lead to severe land degradation as a result of soil erosion and loss of soil nutrients.

Figure 13.3 Without the habitat to which they are so well adapted, animals like orang-utans will only survive in a few forest reserves and in captivity.

Habitat loss is now occurring most rapidly in tropical rainforests, tropical dry forests and savannahs (see Table 13.1). Less visible to most of us are the marine ecosystems that have been destroyed by human activity. Some causes of the loss of marine ecosystems are:

- dynamiting coral reefs – an extreme way to catch fish
- fishing by using trawl nets that are pulled across the sea bed; very little if any of the natural ecosystem of the sea bed is left in the North Sea as a result of this activity

- dredging of coastal waters and development along coastlines for industry, housing and tourism
- removal of trees, ploughing and the run-off from roads and urban areas increasing the sediments in rivers that flow into coastal waters.

Activity

Satellite imaging is used to follow changes in land use.

Researchers collected data about forest cover across the world in four different categories as shown in Table 13.1. (Boreal forests are the natural coniferous forests in high latitudes across North America, Europe and Asia.)

1 Use the figures in Table 13.1 to calculate
   a) the area of each type of forest in 2000 as a percentage of the global total
   b) the percentage loss of cover from each type of forest over the period from 2000 to 2005
   c) the mean percentage loss over the same period.

   Present the results of your calculations in a table.
2 Explain why the actual area of forest cover lost between 2000 and 2005 might not be the same as you calculated from the figures in Table 13.1.
3 Suggest reasons for clearing forests.
4 Some forests or parts of forests are cleared completely of trees to provide timber. Suggest the problems that are encountered in replanting and restoring these forest ecosystems.

Table 13.1 Forest cover in 2000 and forest cover lost between 2000 and 2005 for four types of forest.

Type of forest	Forest cover in 2000/km^2 (thousands)	Gross forest cover lost between 2000 and 2005/km^2 (thousands)
Boreal	8723	351
Humid tropical	11564	272
Dry tropical	7135	204
Temperate	5265	184
Total	32687	1011

Overexploitation

Much of the timber that is taken from forests for use in industry is from managed forests of fast-growing coniferous trees. When these trees are removed, the areas are replanted. But not all timber is from conifers that are replanted; some is from slow-growing hardwood trees. Hardwood trees take many years to reach the required size and many, such as teak and mahogany, are felled at a faster rate than they can regenerate. This is an **overexploitation** of natural resources.

Fish stocks are another example of overexploitation. Several stocks of Atlantic cod, *Gadus morhua*, collapsed in the 1990s due to overfishing, declining by over 95% of their maximum historical biomass. These stocks of cod have failed to recover, even with no further fishing (see the activity below).

The response of the fishing industry to the steep decrease in large predatory species such as cod is for fishermen to target smaller fish further down the food chain (i.e. at lower trophic levels). The fishing of smaller fish, which also occurs in inland fisheries, is most pronounced in the northern hemisphere. Many fisheries, such as those for cod on the Grand Banks in the North Atlantic, herring in the North Sea and a variety of species in the East China Sea, have declined or collapsed. Fishing at lower trophic levels leads at first to an increase in catches and then to declining catches. This shows that the present exploitation of many fish stocks is unsustainable.

Tip

A sustainable resource is a resource that is renewed by the activity of organisms so that there is always sufficient for us to take from the environment so that it does not run out.

Another problem is that by removing large predatory fish such as cod, less food is available for larger fish, marine mammals and seabirds, with further loss of biodiversity.

It is thought that the oceans may now contain only 10% of the large animals that they contained before humans started exploiting the oceans. A solution to the problem of falling fish stocks is fish farming, but that too brings environmental problems such as habitat degradation as a result of increased salinity of the land, as has occurred in Bangladesh where shrimps are farmed intensively.

Activity

Loss of biodiversity in the oceans

Cod, *Gadus morhua*, is a popular fish. Figure 13.4 shows some of the feeding relationships in the North Atlantic Ocean.

1 a) Figure 13.4 shows some of the feeding relationships within an ocean ecosystem. State the name given to this type of diagram.
  b) Explain the change in position of cod as they get older as shown in the diagram.
2 The catches of cod between 1950 and 2012 are shown in Figure 13.5.
  Calculate the percentage decrease in catches of cod between 1968 and 2012. Show your working and express the answer to the nearest whole percentage.
3 Use Figure 13.5 to describe the changes in catches of cod between 1968 and 2012.
4 Use Figure 13.4 to explain the likely effect of a decrease in the population of cod on the biodiversity of the ocean.
5 Diagrams like that in Figure 13.4 oversimplify an extremely complex system. Explain why it is important to understand the complexity of interactions in any ecosystem that humans exploit or are likely to exploit in the future.

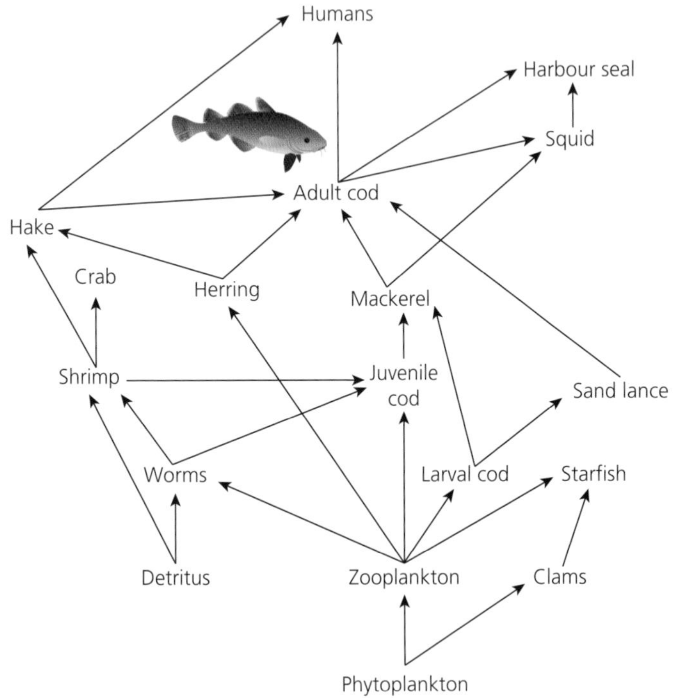

Figure 13.4 The cod feeds on larger and larger prey species as it grows older. Similarly, its predators increase in size.

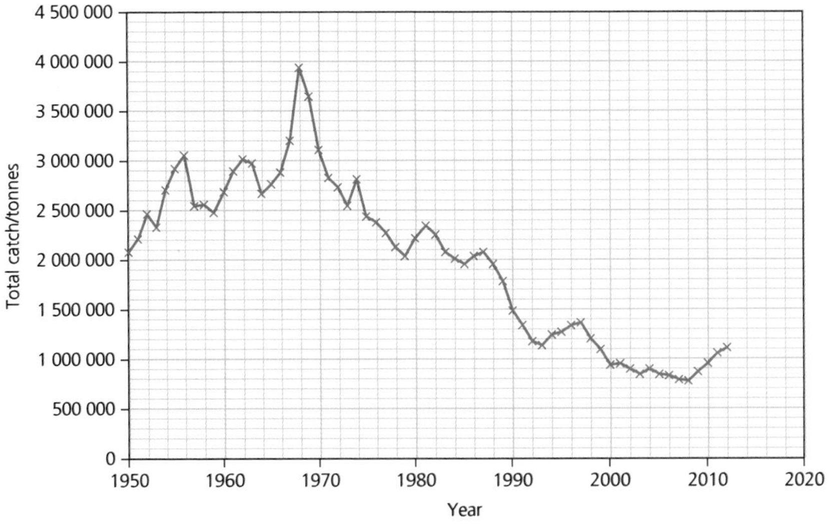

Figure 13.5 Catches of Atlantic cod, *Gadus morhua*, between 1950 and 2012.

Figure 13.6 Bush meat is a source of protein in many parts of the world, especially Africa and South America.

Hunting

Overexploitation also involves poor people in many parts of the world taking wild animals for 'bush meat'. The development of roads into forests for logging has made larger areas available for hunting bush meat. Particularly at risk are primate species such as monkeys and chimpanzees, but also many other species of mammal and reptile.

Plants are susceptible to the equivalent of hunting – removal from their habitats for sale or as food. Surveys of vegetation in East Africa have revealed that wild populations of the African violet, *Saintpaulia*, a very common cultivated plant, are dangerously low as a result of it being collected from the wild.

Agriculture

Key term

Monoculture Growing the same crop, such as wheat, maize or soya, over a large area of land, often for several years. The same term applies to intensive livestock rearing and to forests that consist of one type of tree – usually coniferous trees for timber and paper pulp production.

Arable farmers often plant the same crop year after year in large fields. This allows them to specialise in one or a few crops; hence they can use large machinery for cultivation and harvesting, employ few people and guarantee large yields per hectare. Intensive farming of this sort relies on high levels of inputs such as fertilisers, crop protection chemicals and the fossil fuels necessary to power machinery and manufacture agrochemicals. Intensive arable farming is a type of **monoculture**.

The term monoculture is also used for livestock farming where many animals of the same species are reared together, often in semi-industrial units. Rearing beef cattle over large tracts of land in the tropics following clearance of the natural vegetation is also a monoculture. Plantations of coniferous trees to provide softwood and paper pulp are monocultures, as are oil palm plantations in the tropics.

Monocultures allow farmers to cultivate land more efficiently, so less land is required than when farmers mixed arable crops and livestock and when wastage was much higher. However, monocultures often have much reduced biodiversity compared with the natural habitats that they replace.

Farming on an industrial scale is needed because of the huge numbers of people to feed.

Figure 13.7 Soya bean plants, as far as the eye can see. Monocultures like this cover many square kilometres of the Earth's most productive agricultural land.

Crop plants need to be supplied with resources. Farmers cannot do much about light intensity and carbon dioxide concentration, apart from choosing the most appropriate crop variety for their local conditions. However, growing the same crop year after year soon exhausts the supply of minerals in the soil; farmers replenish these with high inputs of chemical fertilisers. Crops also need irrigating, and this increases the demand for water, which may be a scarce resource. Farmers also have to contend with plants that compete with the crop for resources. Herbicides are sprayed to control weeds so that they do not reduce yields and interfere with mechanised harvesting of the crop. Plant pathogens such as species of fungi, bacteria and viruses are a continual threat to successful crop growth and final yields. Pesticides such as fungicides are used to control outbreaks of disease.

Monocultures present opportunities for pest species. Numbers of pests such as boll weevils and aphids increase exponentially when environmental conditions are favourable, because the crop provides an unlimited supply of food. Insects are one of the main groups of pests, and these are controlled by insecticides.

Agrochemicals have many negative impacts on the wild species that remain:

- Fertilisers make the soil nutrient-rich; it is thought that this encourages the growth of some plants, which shade out slower growing species, depriving them of light.
- Herbicides are used to kill weeds that compete with crop plants for resources and may be sources of pathogens and pests of crop plants. Herbicides may be sprayed on areas bordering fields and kill many non-target species, so reducing biodiversity in the surrounding areas. Broad-spectrum herbicides introduced since the 1970s kill a wide range of plants, species.
- Insecticides kill insect pests, but they also kill non-target species that may not have any negative effects on the crop, and may indeed be predators or parasites of pest species. Insects that are important pollinators may also be killed. Pesticides may persist in the soil and kill invertebrates that are detritivores (animals that shred dead material such as leaves into smaller particles).

Figure 13.8 Pushed into the highlands of Ethiopia by the pressure of population increase, this is farming far removed from the industrialised monocultures of more affluent countries, but is just as damaging to biodiversity.

In the UK, modern agriculture has resulted in a loss in the biodiversity of plants, arthropods (mainly insects and spiders), birds and mammals. The habitat that you can see in Figure 13.9 does not support a rich community of microorganisms, plants and animals, as used to be the case with traditional forms of agriculture.

Figure 13.9 A field of maize at ground level. Notice how there are lots of maize plants but not much else.

A loss of biodiversity has been caused by:

- habitat destruction, by removal of hedges, small areas of woodland and scrub to make larger fields, which removes nesting sites
- ploughing right up to the edges of fields, so removing the habitats of many plants and insects
- a reduction in habitat structure and diversity; for example, some bird species associated with farmland prefer tussocky grass species that are not favoured by livestock farmers
- improved methods for cleaning crop seeds, reducing the chances of weed seeds being resown
- loss of wild food plants for insects such as butterflies and hoverflies as a result of herbicide use
- conversion of pasture to arable land and the resultant decline in soil invertebrate numbers, meaning less food for farmland birds and mammals
- land drainage causing soil invertebrates to live deeper in the soil, so making them less accessible to birds and mammals
- farmers starting to sow cereal crops in the autumn instead of leaving stubble in the ground until spring (which started in the 1970s); this reduced the availability of food for bird and mammal species.

Changes in biodiversity have been most obvious in farmland birds. Records have been kept since 1970 of the numbers of 19 of these species. Figure 13.10 shows how these numbers have changed.

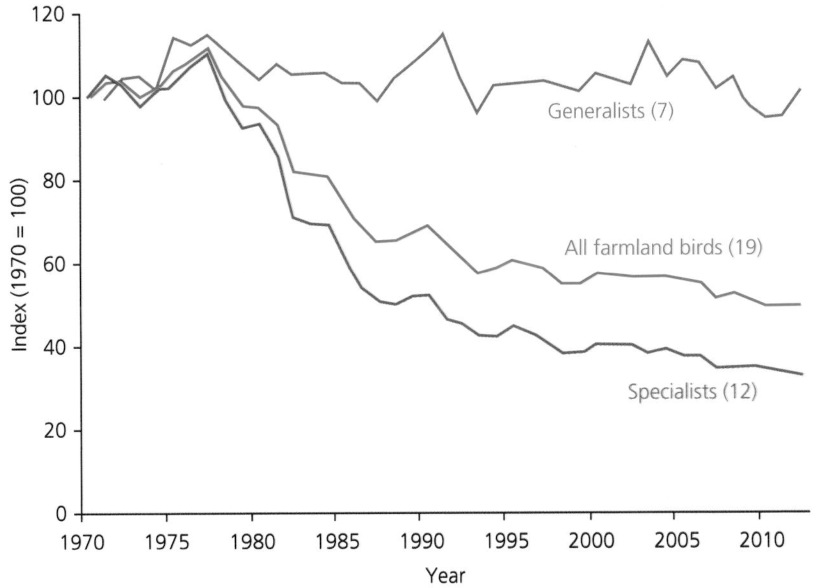

Figure 13.10 Changes in the index of farmland birds between 1970 and 2013. The arbitrary figure of 100 is given for 1970. Generalists are species that live in a variety of habitats including farmland. Specialists are species that live almost entirely in farmland habitats.

Figure 13.11 Population numbers of the corn bunting, *Miliaria calandra*, in the UK have declined by about 85% since the 1970s. This is partly because much less spring barley is grown. This crop makes a favourite nesting site for this species.

Pollution from agriculture

The three main sources of pollution from agriculture are

- fertilisers
- pesticides
- waste products from intensive livestock production.

Fertilisers are applied to crops to increase their yield. But farmers need to know how much fertiliser to add. Problems can occur if a farmer uses too much fertiliser or if the fertiliser is added before a period of heavy rain. The result of either of these is that fertiliser can drain from the land into rivers and lakes. Fertiliser can cause water pollution and **eutrophication** (which means that waters are enriched with plant nutrients, mainly nitrate ions and phosphate ions).

Nitrate is the main nutrient in run-off from arable land. This is not usually a limiting factor for growth of algae and plants in fresh waters, such as streams and rivers, but it is in marine waters. Nitrate causes algal blooms in the sea. Algal blooms are naturally occurring but happen more frequently as a result of eutrophication.

Fertiliser and sewage run-off also cause huge growth of plankton. These plankton quickly die and are consumed by bacteria that deplete the waters of oxygen. For example, the Gulf of Mexico has a 22 000 square kilometre 'dead zone' every spring due to run-off from the Mississippi River. It is estimated that over 400 areas in the oceans are regularly starved of oxygen because of decomposition of algal blooms fuelled by nitrogen in sewage and fertiliser run-off, and that the number of these dead zones is increasing.

Pesticides can cause the selection of resistant strains of weed and insect pests (see page 293). The widespread use of organochlorine pesticides such as DDT was responsible for eggshell thinning and poor reproductive success in peregrine falcons, *Falco peregrinus*, from the 1950s. In India, populations of vultures decreased by 95% between 1980 and 2010 due to poisoning by the veterinary drug diclofenac, used to treat cattle. The nine species of vulture in India are now at great risk of extinction.

Climate change

Throughout the Earth's history the climate has always changed, with ecosystems changing and species evolving and becoming extinct. However, these changes occurred over relatively long periods. Now, climate change is occurring much faster and is likely to be too fast for species to adapt, which may result in a great loss of biodiversity.

The two main effects of global climate change are a modification of weather patterns, and an increasing number of extreme weather events such as hurricanes, typhoons, floods and droughts.

Figure 13.12 Monteverde Cloud Forest in Costa Rica. Can the species in this ecosystem adapt to a warming world with competition from species migrating up from lower altitudes?

With global warming, the distribution of many species is changing – they are moving towards the poles and to higher altitudes. Species moving like this are entering new ecosystems where they may not be as well adapted and where they are in competition with much better adapted species. However, as conditions become warmer, species adapted to current conditions at high altitude and high latitudes may not survive the competition with species from warmer places (see Figure 13.12).

Global climate change poses specific threats to biodiversity in polar regions. The loss of Arctic ice may lead to the loss of an entire biome. The Arctic biome includes the algae that live on the underside of the ice; these are the producers for this very productive ecosystem, which includes many invertebrates, fish, birds and marine mammals.

Tip

Phenology is the study of timing when changes occur in nature. There is increasing evidence for changes in the timing of many natural events which are closely correlated with changing temperature.
Nature's Calendar is the website of the UK phenology network – **www.naturescalendar.org.uk.** You can read much more about the effect of climate change on British wildlife.

Another threat to marine habitats is the decrease in the pH of sea water (often called the acidification of the oceans). Carbon dioxide is very soluble in water. As the concentration of carbon dioxide in the atmosphere has increased, more has dissolved in sea water, decreasing its pH. The lower pH is making it more difficult for organisms to make shells of calcium carbonate; such organisms include tiny planktonic creatures that have shells, and coral polyps, some of which make coral reefs. The calcium carbonate of the organisms' skeletons acts as a 'sink' for carbon, which lasts millions of years. If this activity decreases, then less carbon dioxide is taken out of the atmosphere into these long-term stores in the oceans. Ocean acidification is one of many threats to the survival of coral reef ecosystems, which are predicted to disappear by 2050. Coral reefs hold about 25% of the biodiversity of the oceans.

Warming of ocean water is likely to cause stratification, so that surface waters will not mix with nutrient-rich water from deep in the ocean. If this happens, the phytoplankton – the main producers in the sea – will not grow to provide food for the great diversity of the oceans.

Tip

Other factors that have a negative impact on biodiversity, not covered here, are industrial and domestic forms of pollution, invasive or alien species, and the spread of plant and animal diseases.

Test yourself

1 Explain what is meant by *habitat fragmentation* and suggest the problems associated with it for maintaining biodiversity.
2 How can people in developed countries such as the UK justify campaigning to save rainforests when all woodlands in the UK have, at some time, had trees cut down by humans?
3 Suggest the steps that can be taken to halt deforestation.
4 Explain why monocultures have a negative impact on biodiversity.
5 Outline the effects of climate change on biodiversity.

Reasons for maintaining biodiversity

Why should we maintain biodiversity? Why is it necessary to maintain populations of species? Some of the reasons are:

- ecological
- economic
- agricultural
- aesthetic.

Ecological

The loss of species leads to an imbalance in natural communities. It is thought that ecosystems with high biodiversity have a higher stability and are better able to withstand environmental changes such as storms than those with less biodiversity. They are also thought to recover more rapidly and more thoroughly after environmental damage. Ecosystems with high biodiversity are said to have high resistance and high resilience.

Tip

Resistance is the ability of ecosystems to withstand environmental change; resilience is the ability to recover after damage.

Ecosystems become unstable when there are decreases in populations of organisms at different trophic levels. For example, the loss of top predators such as lions, tigers and leopards often leads to an increase in herbivores followed by overgrazing, land degradation, erosion and loss of biodiversity. These top predators fulfil roles in maintaining a balance in ecosystems far beyond what their low population numbers would suggest.

Important roles are played in ecosystems by **keystone species**. Some keystone species are top predators, but many are not. The Pacific sea otter, *Enhydra lutis*, was one of the first to be identified as a keystone species. The loss of sea otters from the Pacific coast of North America by overhunting removed an important predator of sea urchins, which feed on the stalks of kelp – a large seaweed. As the numbers of sea otters decreased, the population of sea urchins exploded, which led to devastation of the kelp forests and the loss of many species associated with kelp. When numbers of sea otters increased following legislation to protect them from hunting, the sea urchins were kept under control and the kelp forests recovered.

Key term

Keystone species A species whose presence contributes more to the function of an ecosystem than its size or number suggests. If a keystone species disappears from an ecosystem, species diversity decreases, interactions between species become far less complex, and the ecosystem becomes unstable.

In forest ecosystems of West and Central Africa, the forest elephant, *Loxodonta cyclotis*, is the only animal able to disperse the seeds of many trees – the trees depend on the elephants for their dispersal. Without elephants, it is reckoned that 30% of the tree species would disappear; hence the composition of the forests would change and so their biodiversity would decrease. By contrast, savannah elephants, *Loxodonta africana*, prevent the change in the habitat from grassland to forest by eating or uprooting acacia trees before they can become established. If the trees grew, a canopy would develop, grasses would be shaded out and die, and the large grazers of East Africa such as zebras and wildebeest would disappear. Elephants are keystone species in two quite different ecosystems.

Without the work of these keystone species, the habitat changes significantly. When a keystone species disappears from its habitat, there are many changes to the habitat's biodiversity and to the complex interactions between species at different trophic levels. The keystone species' disappearance triggers the loss of other species, and the intricate connections among the remaining residents begin to unravel. In a 'domino effect', species losses cascade through the habitat, as the loss of one species prompts the loss of others.

Some keystone species are top or **apex predators**. These are easy to distinguish, because they are usually large, fierce animals with small populations extending over wide areas. Their effects on an ecosystem can be profound, extending from the populations of their principal food sources to plants at the producer trophic level. This 'trophic cascade' can be seen when apex predators disappear from ecosystems, as has been the case with those hunted by man, such as tigers, and the reintroduction by humans of apex predators. The loss of apex predators can have serious effects on humans. A decrease in populations of lions and leopards in Ghana led to an increase in olive baboons, the numbers of which were previously kept in check by these predators. With fewer predators, the baboon population increased and baboons began attacking people's livestock, damaging crops and spreading intestinal parasites to the human population.

The effects of keystone species and apex predators can be seen when they are reintroduced to an ecosystem. The gray wolf, *Canis lupus*, was reintroduced to the Yellowstone National Park in the USA, with effects that spread right through the food web. The wolves have reduced the population of elk, *Cervus canadensis*, and this has led to decreased grazing of the vegetation along river banks, so increasing habitats for beavers. Carcasses abandoned by wolves provide food for bears and for many other species of bird and mammal.

Identification of keystone species is much more difficult than determining an apex predator. To identify a keystone species often involves long-term monitoring of an ecosystem following loss or removal of the organisms concerned and determining whether there are substantial changes in biodiversity, community structure and energy flow.

Example

Investigating a keystone species

The Chihauhuan Desert in Mexico has several species of seed-eating rodents. Chief among these is Merriam's kangaroo rat (see Figure 13.13), which has been identified as a keystone species because of its influence on plant diversity and energy flow in the desert ecosystem. This species has a very large skull for its overall body size.

Researchers investigating feeding relationships among the different rodent species put up fences around their experimental plots (50 m × 50 m) that small rodents could enter but kangaroo rats could not. In a long-term experiment lasting over 20 years, the species composition and energy use of the rodent populations was measured and compared with that of kangaroo rats in control plots identical in every way, except that kangaroo rats could enter through the fences.

Figure 13.13 An unlikely keystone species: Merriam's kangaroo rat, *Dipodomys merriami*.

Six seed-eating species colonised the plots, with population densities twice as large on the experimental plots as on the control plots. In 1996, the plots were invaded by another larger rodent, a species of hopping mouse, *Chaetodipus baileyi*, from a population about 5 km away. The population of these mice on the experimental plots was 20 times larger than the populations on the control plots. At no time during the experiment did uneaten seeds accumulate.

The graph in Figure 13.14a shows the energy consumed by the rodents in the experimental plots as a percentage of the energy consumed by kangaroo rats in the control plots. Figure 13.14b shows the energy use by the kangaroo rats in the control plots and by hopping mice in both experimental and control plots.

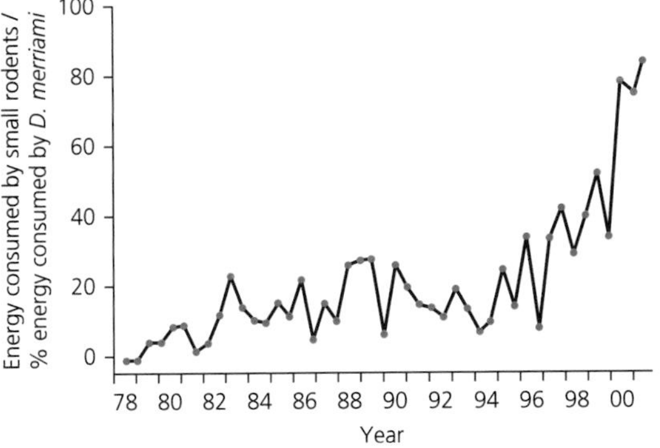

Figure 13.14 (a) Energy consumed by kangaroo rats

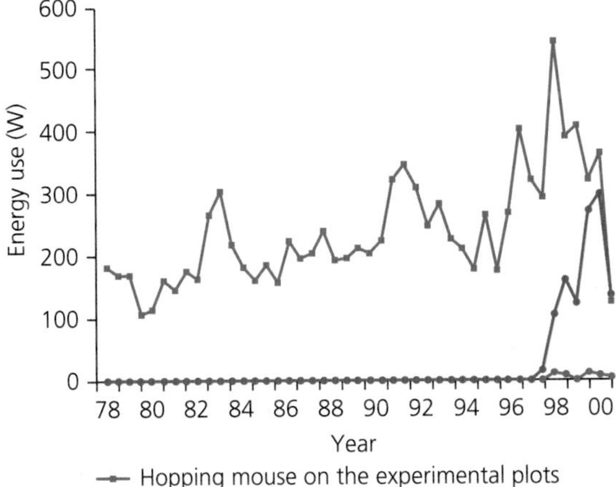

Figure 13.14 (b) Results of an exclusion experiment with desert rodents.

1 a) Describe the changes in energy consumption by the rodents in the experimental plots, as shown in Figure 13.14a.
   b) Suggest why the hopping mice consumed only a small percentage of the energy compared with the kangaroo rats in the control plots.

2 a) Explain why the hopping mice were able to colonise the experimental plots.
   b) Suggest the likely effects of excluding kangaroo rats on energy flow in the experimental plots compared with the control plots.
   c) Explain why the population of hopping mice was much higher in the experimental plots than in the control plots.

3 Explain why it is much easier to determine whether an animal is an apex predator rather than whether it is a keystone species.

Answers

1 a) Energy consumption increased from 0% of that taken by kangaroo rats to just over 20% in 1983. The energy consumed fluctuated between 5% and 30% until 1996 when it increased steeply towards 80% by 2000.
   b) The rodents did not invade and feed in the experimental plots until 1978 or 1979. Even without competition from kangaroo rats in the experimental plots, the hopping mice did not eat more than about 30% of the energy consumed by the kangaroo rats. This may be because their population was small or because they are much smaller animals that cannot crush or break open the larger seeds that kangaroo rats can eat because they have larger skulls. The energy consumed increased in 1996 because hopping mice invaded the control plots and took the larger or harder seeds that the smaller mice could not. In 2000, all the rodents in the experimental plots were taking nearly 80% of the energy taken by kangaroo rats in the control plots.

2 a) The hopping mouse has a smaller skull than the kangaroo rat, so could squeeze through the holes in the fence around the experimental plots.
   b) Less energy flows from producers (via their seeds) to rodents. Because seeds did not accumulate, other animals must have eaten the seeds. These animals might have been invertebrates or birds. With less energy flow to rodents, there would be less energy flow to their predators.
   c) These mice competed successfully with other animals for seeds. Plenty of seeds were also not taken by rodents, so hopping mice must have competed successfully against non-rodent seed eaters. On control plots hopping mice were outcompeted by the kangaroo rats.

3 An apex predator does not have any predators; it is usually a big animal; it usually has a variety of prey; it usually has a small or very small population; apex predators tend to be large and fierce.

Economic reasons

Materials from natural ecosystems

Although monocultures provide most of the human population with food, natural ecosystems still provide us with food and materials, for example fish and timber. Important drugs have been discovered in plant and animal species (see page 205 in Chapter 11). The natural world is a source of chemicals that are difficult for us to produce. The thermophilic bacterium *Thermus aquaticus* is the source of a heat-stable DNA polymerase used in the polymerase chain reaction to increase the quantity of DNA in many aspects of industry, medicine and forensics. Some species have numerous uses. Sisal, *Agave sisalana*, is the source of fibres used to make a range of products such as rope, paper, carpets and dart boards. If habitats disappear, then the wild populations of these commonly grown plants are reduced. This is happening to plants such as African violet, *Saintpaulia*, a common house plant. Without genetic diversity in the wild, plant breeders find it harder to breed new varieties, which reduces the income made by the horticultural industry.

Ecosystem services

It seems strange to say that ecosystems provide *services* for us as well as providing materials. Yet we all rely on natural ecosystems to provide us with the basic requirements of life: fresh water, oxygen and a suitable climate to live in. Here are some examples of these ecosystem services:

- Plants transpire water vapour, which contributes to the water cycle to provide us with drinking water.
- Water is filtered through soils and rock before it enters the water supply.
- Soil fertility is maintained by nutrient cycling, for example by decomposers and microbes that convert various forms of nitrogen, phosphorus and sulfur.
- Organic waste material added to waters is broken down.
- Reefs and mangrove forests protect coasts from erosion.
- Habitats moderate floods, droughts, and extremes of wind and temperature.
- Insect and other pollinators ensure that crop plants and orchard crops are fertilised. Examples are fruit trees, oilseed rape and sunflowers.
- Habitats support a wide variety of organisms that interact in ways we do not fully understand to continue life on this planet, for example by keeping pests and diseases in check.
- Forests and peat bogs absorb carbon dioxide and may help to reduce the effect of increases in carbon dioxide in the atmosphere. They are carbon sinks. The Flow Country in northern Scotland is an example.

It is easy to put values on the material products provided by animals, plants and microbes. It is not so easy to put values on these ecosystem services. This seems a cynical approach to conservation, but if many people can see the value of diverse ecosystems in monetary terms they are more likely to realise that conserving ecosystems and species is worthwhile.

Agricultural reasons

Biodiversity is important in maintaining soil structures. Continual monocultures reduce soil biodiversity. A survey of soil bacterial biodiversity in monocultures of soybeans in Argentina showed that there were fewer species than in both local natural ecosystems and in farmland managed with crop rotation. Maintaining a good soil structure with plenty of rotting material is good for binding soil particles together to prevent soil erosion and loss of nutrients.

Aesthetic reasons

Areas of natural wilderness and managed countryside are appreciated by many as beautiful places that should be conserved for future generations. Many people appreciate the aesthetic appeal of many species. There are many amateur ornithologists and botanists who enjoy wildlife. The natural world continues to provide inspiration for artists, photographers, poets, writers and other creative people.

E.O. Wilson, the world's greatest authority on biodiversity, coined the term *biophilia* to encapsulate the human love of nature. Someone else coined an uglier term – nature-deficit disorder – for those who have no or very little interaction with nature. Now that more than half the world's population live in urban environments rather than rural ones, it has become increasingly important to provide natural places for people to experience.

Ecotourism is booming. Countries such as Costa Rica and Belize in Central America, Dominica in the Caribbean, and the Maldives in the Indian Ocean all promote ecotourism and provide an important part of the income of these countries. Whale-watching alone was estimated to generate over £1 billion in 2008, with over 13 million people undertaking the activity in 119 countries.

Often people visit places to see one or more spectacular species – as they do when visiting reserves in East Africa, such as the Masai Mara. Charismatic species that become the subject of conservation campaigns, such as the mountain gorilla in Rwanda, the giant panda in China and the osprey in Scotland, are sometimes known as **flagship species**. Conservation of organisms, usually animals, in their natural habitat inevitably involves conserving their life-support systems. This means conserving the biodiversity of their ecosystems right down to the microscopic level. For example, the breeding populations of golden eagles, *Aquila chrysaetos*, require extensive open upland and a good supply of prey. Protecting this species indirectly protects many other species in its habitat. Golden eagles therefore serve as an 'umbrella' for both their prey and for other species that depend on large tracts of its mountain habitat. The term 'umbrella species' is often used for these animal and plant species.

Tip

In many parts of the world, traditional forms of medicine make use of the healing properties of many plants and some animals. Well-known drugs such as aspirin were originally discovered in plants. The diversity of plant species faces many threats. This is discussed in Chapter 11.

Test yourself

6 What is meant by *ecosystem services*?
7 Give three examples of such services.
8 List five practical reasons for conserving biodiversity.
9 Explain why it is important to conserve genetic diversity within species.
10 Monellin is a very sweet protein extracted from the serendipity berry.
  a) Suggest why it has limited use as a sweetener.
  b) It is difficult to grow the plant and to extract the protein. Suggest how else monellin can be produced.

Ways of maintaining biodiversity

Conservation is defined as protecting and maintaining ecosystems and species. The best way to conserve species is to maintain flourishing populations in their natural habitats. If numbers decrease so much that survival in the wild is precarious, then plants and animals may be removed from their natural habitat to a botanic garden or zoo.

In situ conservation

Key term

***In situ* conservation** The protection and maintenance of an area so that species can continue to live in their original habitat.

The conservation of a species in its natural habitat is ***in situ* conservation**. The Latin phrase *in situ* means 'in their original place'. Maintaining the natural habitat means that all life-support systems are provided. Conservation tends to concentrate on individual species or groups of species. High-profile programmes have centred on flagship species, usually mammals such as giant pandas and whales. As important, if not more important, are ecosystems threatened by development; the most talked about of these is the tropical rainforest, although many other less well-known ecosystems should be conserved.

In situ conservation must always be the preferred option for conserving individual species. This is because species have all the resources that they need in an environment to which they are adapted. Animal species kept in captivity and plants grown in botanic gardens have to adapt to an artificial environment; sometimes they are unable to. Also species kept *in situ* continue to evolve in their natural environment, whereas captive organisms are protected from environmental changes, because they are kept in relatively unchanging conditions. Larger species have to be kept in much smaller spaces than their natural habitat, and this may mean that the individuals that survive to breed are those most adapted to these conditions and therefore less likely to survive if returned to the wild. Animals are more likely to breed in their natural environment. On the other hand if populations are small they may be put at risk by poaching, making *in situ* conservation a more risky choice for long-term survival.

This section gives information about *in situ* conservation in the UK. Areas of land and water are protected in many ways for *in situ* conservation. Many of these areas are protected for a variety of reasons; species conservation is not necessarily the most important reason.

Internationally designated areas

Some areas have been designated internationally by organisations such as UNESCO (United Nations Educational, Scientific and Cultural Organization).

Biosphere reserves are areas recognised under UNESCO's Man and the Biosphere programme to promote sustainable development based on the work of local communities and sound science. Currently there are five biosphere reserves in the UK, including the Dyfi biosphere reserve in Wales and the north Norfolk coast. World Heritage Sites are identified by UNESCO as important physical or cultural sites. There are 28 World Heritage Sites in the UK, including St Kilda off the northwest coast of Scotland, where there are breeding sites for seabirds.

Nationally protected areas

There are three types of nationally protected area in the UK:

- National parks – 15 in England, Scotland and Wales
- Areas of Outstanding Natural Beauty – 46 in England, Northern Ireland and Wales
- National Scenic Areas – 40 in Scotland.

National parks are areas of countryside, which include villages and towns. A national park authority is in charge of each national park, with powers that include planning controls.

The Dark Peak moorland in the Peak District National Park is the nearest thing to wilderness in England.

National nature reserves

Nature reserves are areas that protect sensitive ecosystems and provide 'outdoor laboratories' for research. They are administered by a national body such as Natural England and their management is geared to maintaining conditions for some of the UK's rarest animals and plants. Wicken Fen National Nature Reserve in Cambridgeshire provides a habitat for one of Europe's rarest birds – the bittern, *Botaurus stellaris*.

Figure 13.15 Habitat creation. Planting reed beds to create a habitat for the bittern, one of the UK's rarest birds.

Local nature reserves

Local nature reserves are under the control of local authorities through ownership or by lease or agreement with the owner of the land. The main aim is to care for the natural features that make the site special. Many local nature reserves are managed by country wildlife trusts.

Pagham Harbour in West Sussex, is a local nature reserve that provides an important habitat for migrating birds.

Marine Conservation Zones

The Marine and Coastal Access Act 2009 set up Marine Conservation Zones to add to other protection schemes for the seas around the UK. Currently, there are 28 Marine Conservation Zones covering an area of about 10 000 km^2. These zones protect habitats and species that are representative of the biodiversity in our seas.

The seas around Lundy Island in the Bristol Channel, a former Marine Nature Reserve, became the first Marine Conservation Zone in January 2010. Similar schemes operating in Wales and Scotland contribute to a UK-wide network of marine protected areas; soon such areas will be designated in Northern Ireland, too. Management of these areas involves reducing the negative impact of fishing, pollution and other factors.

No-take zones in other areas of the world, notably in the Goat Island Marine Reserve in New Zealand, have shown that areas of sea bed almost completely devoid of life can be recolonised to support a diverse community within a short time of being declared a no-take zone. Many countries, realising the potential for ecotourism and making commercial fisheries sustainable, are setting up reserves. The president of Palau in the Pacific has proposed banning all commercial fishing to create one of the world's largest marine reserves covering an area roughly the size of France.

Sites of Special Scientific Interest (SSSI)

SSSIs give legal protection to the best sites for wildlife and geology. The protection is such that landowners cannot change the management of the area without permission. An example is the Gwithian sand dunes in West Cornwall. Sand dunes in many parts of Britain are ecosystems at risk of erosion and degradation as a result of overuse. The smallest SSSI is Sylvan House Barn in Gloucestershire, which is a space of just 4.5 square metres that houses 200 lesser horseshoe bats. Among the largest SSSIs, and possibly the busiest, is the Humber Estuary. It drains a fifth of England's fresh water and is the site of the largest breeding colony of grey seals in the UK.

Other methods of ecosystem protection

In addition to areas protected by these schemes, ecosystems are protected in other ways, for example by being owned by bodies such as the National Trust and English Heritage. Protected ecosystems include man-made ecosystems such as chalk downland – a common landscape in parts of southern England.

Many farmers and private landowners implement a wide range of conservation practices to encourage wildlife. One such scheme is the **Environmental Stewardship Scheme**, which provides funding to farmers and other land managers in England who deliver effective environmental management on their land. Introduced in 2005, this scheme replaced Environmentally Sensitive Areas and the Countryside Stewardship Scheme, which date back to the 1980s and 1990s respectively. Farmers may, for example, provide wildlife habitats on farmland, such as ponds, hedges and buffer zones around crops. In addition, farmers must:

- ensure that land is well managed and retains its traditional character
- protect historic features and natural resources
- conserve traditional livestock and crops
- provide opportunities for people to visit and learn about the countryside.

Tip

Traditional (rare) breeds of livestock and old cultivated varieties of crops are genetic resources that should be conserved for the future. Wild relatives of crop plants also represent an important genetic resource. Many are at risk of extinction.

Summary

In situ conservation does not just involve putting a line around a particular area of natural interest and controlling or preventing development, logging, poaching, fishing and hunting. It also involves a variety of other activities, such as:

- reclaiming ecosystems that have been damaged by human activities and in natural catastrophes such as volcanic eruption, hurricanes and flooding
- creating new habitats by allowing vegetation to take over land abandoned by people, digging ponds, and deliberately sinking ships to provide new surfaces for corals to colonise; for example, rock from digging the Channel Tunnel was used to extend the coastline near Folkestone and make the Samphire Hoe Country Park
- maintaining habitats by using fire, grazing or flooding to prevent the growth of plants that would change the structure of the community.

Activity

Saving the red squirrel

The red squirrel, *Sciurus vulgaris*, is the native squirrel in the British Isles. Numbers have decreased as a result of competition with the grey squirrel, which was deliberately introduced into Britain in the nineteenth century. Not only is the grey squirrel an effective competitor, but it also carries a disease that red squirrels have no protection against. Red squirrels are now locally extinct in most of England and large areas of Ireland, Wales and Scotland.

Captive breeding occurs at several places in the UK, including Belfast Zoo, Wildwood Trust in Kent and Paradise Park in Hayle, Cornwall. Some captive-bred squirrels were used to re-establish a population in Newborough Forest on Anglesey in Wales. This reintroduction scheme involved establishing a zone around the forest that was free of grey squirrels. Two squirrels from a wild population in Cumbria and six captive-bred individuals were genetically tested before assigning them to large breeding pens in the forest.

1 Explain the need to maintain a zone around the reserve in which there are no grey squirrels.
2 The squirrels used in the reintroduction programme were genetically screened before assigning them to breeding pairs. Explain why this was important.
3 Red squirrels are very difficult to see in their natural habitat. Suggest how modern technology could be used to make accurate estimates of populations of red squirrels.
4 Suggest why it is important to conserve red squirrels in the UK.

Ex situ conservation

Worldwide, the list of endangered protected areas is growing in number, and additional human-dominated activities such as water development, mining, road construction and resulting development, livestock grazing, poaching, logging, habitat loss and habitat degradation continue to be threats to biodiversity. Sometimes it is necessary to remove animals and plants from the wild and keep them in captivity or grow them in botanic gardens. An example is the Madagascan pochard, *Aythya innotata*, a duck that in 2014 had a population of just over 20. A plea was made for a suitable organisation to take the last remaining ducks to preserve them and start a captive breeding programme to increase their numbers.

Sometimes it is impossible to conserve a species in its natural habitat, because that habitat is shrinking, fragmented or there are so few specimens left in the wild that they must be removed to safeguard their future. This **ex situ conservation** is sometimes called 'storage of germplasm', even if it involves keeping live animals and plants.

Key term

***Ex situ* conservation** Removal of a species to a protected place that is not its normal habitat. Botanic gardens, seed banks, zoos, gene banks and 'frozen' zoos are examples of this form of conservation.

Tip

Germplasm is any form of genetic resource, such as seeds, sperm, embryos, tissue samples or live animals and plants.

Figure 13.16 One of the original breadfruit trees brought from Tahiti in 1793 by Captain William Bligh to the botanic garden in Kingstown, St Vincent, which was the first botanic garden to be established in the western hemisphere. His first attempt in HMS *Bounty* in 1788 had to be aborted after a mutiny. From here in St Vincent, breadfruit trees were planted all over the Caribbean as a source of food.

Figure 13.17 The botanic gardens in Singapore has an orchid breeding and conservation biotechnology laboratory where research is carried out on conserving these beautiful species.

Figure 13.18 This pitcher plant is in the collection of the Glasgow Botanic Garden. Notice the backward-pointing white hairs in the modified leaf: these ensure that any insect that crawls down into the pitcher cannot crawl back up again.

Botanic gardens

Some of the roles of botanic gardens in maintaining biodiversity are listed here:

- Keeping examples of wild plants either as living plants or as seeds: Some botanic gardens concentrate on plants from specific regions or have collections of specific taxonomic groups. The Karoo has a collection of South Africa's desert and semi-desert plants; the Bedgebury National Pinetum in Kent has the most complete collection of conifers in one botanic garden in the world.
- Growing plants that are extinct in the wild, in the hope that they can be reintroduced: Botanists at the Royal Botanic Gardens Edinburgh think there are only four plants of *Rhododendron kanehirae* from Taiwan. These four plants are in botanic gardens in Scotland.
- Protecting wild populations of plants collected from the wild: It is often easier to grow plants in places that are not their natural habitat than it is with animals. Growing such plants commercially provides sufficient for people to buy and reduces the need to take rare and endangered plant species from the wild. The Australian Botanic Garden in New South Wales in Australia is researching the horticultural development of the endangered Wollemi Pine, *Wollemia nobilis*.
- Protecting plants that are threatened by habitat loss: An example is a rare species of birch, *Betula chichibuensis*, from Japan. The Ness Botanic Garden in Liverpool acquired seeds from the few plants left in the wild and now has about 40 plants in the garden in large groups.
- Researching methods of reproduction and growth so that species cultivated in botanic gardens can be grown in appropriate conditions and can be propagated: The Royal Botanic Garden Edinburgh has collected Vireya rhododendrons from mountain habitats in South-East Asia and grown them in its research collections to understand their biology and evolution (see Figure 13.19).
- Researching conservation methods so plants can be introduced, perhaps to new habitats if their original habitat has been destroyed: This includes studying how natural communities change as new species become established, so that plants can be reintroduced in an appropriate sequence over time.
- Reintroducing species to habitats where they have become rare or extinct: The Huaruango woodlands in southern Peru were restored with the help of the Royal Botanic Gardens, Kew. Similarly, in parts of the North American prairies, grassland communities have been restored by the use of native species, and Atlantic rainforest has been restored in Brazil. The dragon tree, *Dracaena draco*, has been planted in areas of Gran Canaria from which it had been removed by the Botanic Garden in Las Palmas.
- Educating the public in the many roles of plants in ecosystems and their economic value to us.

Seed banks

Many botanic gardens have seed banks where seeds are stored. Seed banks are also associated with research institutes, such as those that study our major crop plants (see Table 13.2). Seeds are collected from the wild or from crops, sorted, dried and stored in very cold conditions. They are

Figure 13.19 *Rhododendron christi* – a Vireya rhododendron from New Guinea. Red-flowered species like this one are pollinated by birds.

Key term

Seed bank A long-term store of germplasm in the form of seeds. Collections of seeds are dried and then stored at temperatures near –20 °C.

Collection of seeds
Seeds packaged and labelled
↓ Transport of seeds to the seed bank
Removal of seeds from fruits
Seed samples examined under a microscope in the seed bank
↓ Removal of all debris from seed sample
Seed sample cleaned and checked for damage and infestation ⇄ Details of seeds logged into the Seed Information Database
↓ Removal of all damaged and infected seeds
Seeds placed into drying room at 15 °C and 15% RH
↓
Seeds checked to see if their moisture content is 5–7%
↓ X-ray analysis to check on seed health
Seeds packed in glass jars for long-term storage and labelled
↓
Seeds placed in store at –20 °C for 20 years or more
↓ Germination tests at:
- one month
- five years
- ten years

Seeds taken out of store, germinated and grown to produce more seeds or to be used in research or conservation projects

Figure 13.20 Collecting and preparing seeds for storage in a seed bank.

checked at intervals to see whether they are still viable. This ensures a supply of plants for the future and also is a store of genetic diversity – an important store of genes and alleles for future breeding programmes or to use for genetically modifying plants of economic importance. It is also a store of plant material that may be useful in providing chemicals such as medicines for the future.

Botanists carry out expeditions to particular areas of the world to collect seeds. There are also professional seed-collectors around the world. Seeds are collected from plants in the wild and also from the locally adapted crop plants. If possible, seeds of the same species are collected from different sites, so that the stored samples contain a good proportion of the total gene pool for that species. Figure 13.20 shows the stages involved in preparing the seeds for storage.

Removing water from seeds slows down their metabolism so that they remain viable for many years (although not forever). With this small water content there is little danger that cells in the seed will be damaged by ice crystals during freezing and thawing. Collections continue to be made, if possible, to top up the bank for each species.

Table 13.2 Some of the world's seed banks for crop plants.

Seed bank	Species	Location of seed bank	Organisation
Rice seed bank	rice, *Oryza sativa*	Philippines	International Rice Research Institute
Australian PlantBank	Australian species	New South Wales, Australia	Royal Botanic Gardens, Australia
Wellhausen-Anderson Plant Genetic Resources Center	wheat and maize (corn)	Mexico	International Maize and Wheat Improvement Center (CIMMYT)
ICRISAT genebank	sorghum, pearl millet, chickpea, pigeonpea and groundnut	India	ICRISAT
IITA genebank	African crops, e.g. maize (corn) and legumes	Nigeria	International Institute of Tropical Agriculture (IITA)
Millennium Seed Bank	many and various	UK	Kew

The only way to find out whether stored seeds are still viable is to try to germinate them. Seed banks carry out germination tests at five-year intervals. When fewer than 85% of the seeds germinate successfully, plants are grown from these seeds so that fresh seed can be collected and stored.

When such plants are grown from samples of stored seed, there is the possibility of altering the genetic diversity that was originally stored. Small samples of seeds from rare plants present a particular problem, because even smaller samples of the original are taken to test for viability or to grow into plants to increase the number of seeds in store. Such samples are unlikely to contain all the genetic diversity of the original sample. The only answer to this problem is to put as large and diverse a sample as possible into store in the first place.

Many species have seeds that do not survive freezing and/or drying so cannot be stored in seed banks as described above. Many tropical crops, such as cocoa trees, *Theobroma cacao*, produce seeds of this type. Instead of storing seeds, other ways have to be used to save their genes for the future. Some can be kept as tissue culture and others just have to be grown as mature plants in the ground in field gene banks. In the case of cocoa, there are gene banks in Trinidad and Costa Rica.

Zoos

Zoos have existed for thousands of years. Wealthy people used to collect animals and keep them in menageries for their private enjoyment. Other zoos are owned and run by zoological societies or other groups. These zoos have a variety of different functions in addition to providing enjoyment and interest for visitors, who can see and study animals they would not otherwise be able to see.

Animals from very small populations have a very small gene pool in which to mix their genes. Inbreeding is a serious problem. Zoos try to solve this by exchanging specimens or by artificial insemination where possible.

The roles of zoos in *ex situ* conservation are:

- protecting endangered and vulnerable species; the Jersey Zoo is involved with captive breeding of tamarins from Brazil (see Exam practice question 5 on page 250)
- taking part in breeding programmes for those species that will breed in captivity; a good example of such a programme is that for cheetahs, which are bred successfully at the Fota Wildlife Park in Cork in Ireland
- researching the biology of species to gain a better understanding of breeding habits, habitat requirements and genetic diversity; the Zoological Society of London, like many large zoos, has a scientific programme of research and is involved with many *in situ* conservation projects throughout the world, including the establishment of small marine protected areas in the double reef at Danajon Bank in the Philippines
- contributing to reintroduction schemes; the Emperor Valley Zoo in Trinidad has successfully reintroduced captive bred blue-and-gold macaws, *Ara ararauna*, to the Nariva Swamp in collaboration with Cincinnati Zoo in the USA
- Przewalski's horse, *Equus ferus przewalskii*, has been bred very successfully in many zoos and has been transferred to the Dzungarian Gobi Strictly Protected Area in Mongolia, where this wild horse became extinct 30 years ago
- educating the public about wildlife and conservation.

Tip

Any method of conservation that keeps whole organisms, gametes, embryos, seeds, tissues or any other part of an organism is known as a gene bank. It is not a store of bits of DNA but of whole genomes in one of the ways listed.

Zoos cooperate so that breeding programmes generate genetic diversity and species do not become inbred – a risk when maintaining small populations. As part of many breeding programmes, animals are transported from one zoo to another. This prevents inbreeding and promotes genetic diversity in the animals kept in captivity.

A much cheaper option is collecting sperm and keeping it frozen for many years in a sperm bank. When required, sperm samples are then thawed and used for artificial insemination. Young animals of over 30 different species, including rhinoceros, cheetah and a Chinese pheasant, have been produced using sperm from sperm banks.

Test yourself

11 State the difference between *in situ* and *ex situ* conservation.

12 Explain the advantages and disadvantages of *in situ* conservation.

13 Why is it important to keep seeds in seed banks?

14 Explain why it is not possible to keep the seeds of all species in seed banks such as those in Table 13.2.

15 Why is it important to clean and disinfect seeds before they are put into cold storage?

16 Four species of Vireya rhododendron are endemic to Mount Kinabalu in Sabah, Malaysia. Why is it important to study their biology?

17 Explain the problems with captive breeding of animal species such as the Amur leopard (see page 226).

Tip

You can find more information about the species described in this book by looking at their entries in the database compiled by the IUCN at **www.iucnredlist.org**.

Eggs (oocytes) and embryos can also be stored in much the same way as sperm. Eggs are more difficult to freeze because they are more likely to be damaged by the freezing or thawing processes. Eggs are large cells with lots of water that tends to form ice crystals, which damage internal membranes. Eggs are fertilised *in vitro* and then frozen until a surrogate mother becomes available. This technique of embryo transplantation has been used for many species, including wild ox and several species of African antelope.

'Frozen zoos' now hold genetic resources from many endangered and vulnerable species in case they are ever needed. San Diego Zoo has one of the largest facilities for storing sperm, embryos and cell cultures from endangered species. The zoo calls it its 'frozen zoo'. These facilities can hold much more genetic diversity than a normal zoo, and the material can be kept for a very long time.

International conservation

The International Union for the Conservation of Nature (IUCN) assesses the status of many of the world's species of animal and plant. The IUCN now classifies Przewalski's horse as endangered, when it used to be classified as extinct in the wild. That is quite an achievement for captive breeding and international cooperation. Figure 13.21 shows how the IUCN classification system works.

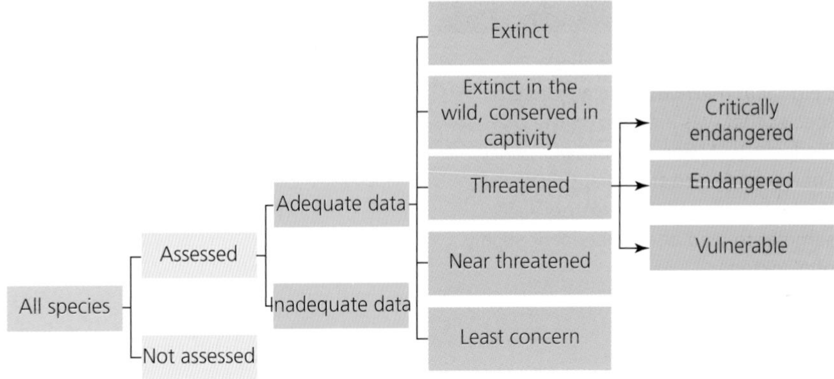

Figure 13.21 The IUCN classification of species.

The IUCN is just one of many international organisations working in the field of conservation. Other organisations include umbrella organisations that coordinate the activities of local, regional and national conservation bodies, some of which trace their origin back to the 1992 UN Conference on Environment and Development (UNCED) in Rio de Janeiro in Brazil (the so-called Earth Summit).

The **Convention on Biological Diversity** (**CBD**) was signed at the 1992 Earth Summit and ratified in 1993. The convention has three main strands, which are:

- conservation of biological diversity by use of *in situ* and *ex situ* conservation methods
- sustainable use of biological resources
- fair and equitable sharing of benefits arising from genetic resources.

The Secretariat of the Convention on Biological Diversity (SCBD), which is based in Montreal, Canada, was established to oversee implementation of the goals of the convention, and organise international cooperation and further meetings. Countries that signed the convention had to develop national strategies for the conservation and sustainable use of biological diversity by writing and implementing biodiversity action plans. The UK biodiversity action plan listed over a thousand species at risk, including the basking shark (see Figure 13.22).

Figure 13.22 The basking shark, *Cetorhinus maximus*, is an endangered species that was included on the UK Biodiversity Action Plan List of Priority Species and Habitats. A flagship species, it migrates into Scottish waters each summer to feed on the abundant supply of plankton.

At the 10th meeting of the SCBD in 2010, at Nagoya in Aichi Province, Japan, countries agreed to take action to halt the alarming global decline of biodiversity. The resulting Strategic Plan for Biodiversity 2011–2020 set a new direction for conservation; among its 20 targets (known as Aichi targets) for 2020 are some that relate to topics covered in this chapter:

- **Target 5** – the rate of loss of all natural habitats, including forests, is at least halved and where feasible brought close to zero, and degradation and fragmentation is significantly reduced.
- **Target 6** – all fish and invertebrate stocks and aquatic plants are managed and harvested sustainably, legally and applying ecosystem-based approaches, so that overfishing is avoided, recovery plans and measures are in place for all depleted species, fisheries have no significant adverse impacts on threatened species and vulnerable ecosystems, and the impacts of fisheries on stocks, species and ecosystems are within safe ecological limits.
- **Target 12** – the extinction of known threatened species has been prevented and their conservation status, particularly of those most in decline, has been improved and sustained.

As a consequence of this meeting in Japan, the UK biodiversity action plan was replaced by the UK Post-2010 Biodiversity Framework in 2012, and detailed plans were produced by the governments of each of the four countries of the UK.

Trade in wild animals and plants

The trade in animals for the pet trade and in animal materials such as ivory is huge. So is the trade in plants lifted straight from the wild rather than grown under cultivation in plant nurseries. Much of this trade is illegal. In 1973, 145 countries signed an agreement to control the trade in endangered species and any products from them such as furs, skins and ivory. More countries have joined since. This agreement is called the **Convention on International Trade in Endangered Species of Wild Fauna and Flora** (**CITES**). CITES came into force in 1975 to protect animals and plants from various forms of exploitation. Trade in endangered species, and the products made from them, is restricted in various ways. CITES considers the evidence presented to it about endangered and vulnerable species and assigns species to one of three appendices. Over 30 000 animal and plant species are protected by being placed on one of the appendices. Appendix I has the species most at risk of extinction. Appendices II and III list those species that are less threatened with extinction but may be so in the future if trade persists.

The species listed in the CITES appendices are reviewed regularly by committees of experts, and the list is growing.

Test yourself

18 What makes a species endangered?

19 There are people who call for trade in ivory from elephants to be legalised. Outline the arguments for and against this idea.

20 What protection is afforded by listing a species on Appendix I of CITES?

21 What advantages are gained by governments signing international agreements on biodiversity?

Concern has been expressed that a CITES listing does not always benefit a species. If trade in a species or its products becomes illegal, then the price that can be obtained for those products rises, and this is likely to make it worthwhile for people to break the law. Particular problems arise when it is announced in advance that a species will go on the list; in the months between the announcement and the introduction of the new law, trade in that species tends to increase.

You can learn a lot more about CITES and see photographs and information about the species that are listed on their website (**www.cites.org**).

In spite of the protection afforded by a CITES listing, many problems remain with the illegal trade in endangered animals, plants and their products. Poaching of elephants has increased steeply in recent years as the illegal trade in ivory has risen in response to increased demand from China. The elephant at the beginning of this chapter is just one of many slaughtered for this illegal trade: estimates run as high as 50 000 a year. Will the time come when all elephants are extinct as a result of human greed?

Exam practice questions

1 Some conservationists argue that ecosystems should be protected from development because they provide various services. Which of the following is an ecosystem service?

A providing a potential source of medicines
B recycling of plant nutrients
C releasing methane into the atmosphere
D supplying fibres for manufacturing *(1)*

2 The term *monoculture* is often applied to large areas of land planted with the same crop year after year.

a) Discuss the impact of monocultures on soils and biodiversity. *(6)*

b) Explain the role of botanic gardens in conserving rare and endangered species. *(5)*

c) Suggest the limitations of zoos in conserving endangered animals. *(3)*

d) Describe the roles of the Convention on International Trade in Endangered Species (CITES) and the Rio Convention on Biological Diversity. *(5)*

3 The following are examples of *in situ* conservation, *except*

A botanic garden
B local nature reserve
C marine conservation zone
D Site of Special Scientific Interest *(1)*

4 Climate change is having effects on the distribution of plants and animals.

a) Outline ways in which climate change is having a **negative impact** on plant and animal populations. *(6)*

b) The common eider duck, *Somateria mollissima*, used to be hunted for its feathers, which were used to make bedding. The gene for blood albumen was used to assess genetic diversity in a population of this species. The gene has two alleles, **A** and **a** that code for two slightly different albumen molecules. The albumen coded for by **A** travels faster when the molecules are separated by electrophoresis. Albumen was extracted from the blood of 67 eider ducks and separated by electrophoresis as shown in the diagram. The numbers of ducks in the population with each result are shown below the diagram.

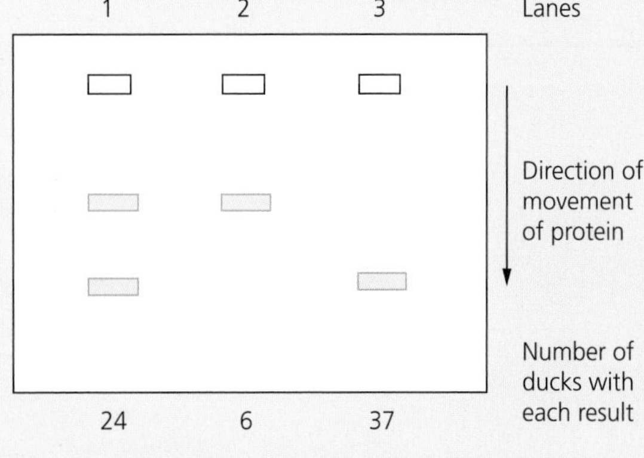

State the genotypes of the ducks that provided the blood in lanes 1, 2 and 3. *(1)*

c) Calculate the proportions of the population of eider ducks that have each of the genotypes. (3)

d) i) State a conclusion about genetic diversity in the population of eider ducks that can be made from these results. (1)

ii) Explain your answer. (2)

e) Explain why maintaining genetic diversity is important in conserving a species like the eider duck. (3)

5 Numbers of golden lion tamarins have declined considerably as their habitat in the Atlantic rainforest in Brazil has been cut down to 2% of its original area. The golden lion tamarin conservation programme is one of the largest global captive breeding and reintroduction schemes. The wild population consists of about 600 individuals dispersed among 11 isolated forest refuges, the largest being the Poço das Antas Biological Reserve, which contains about 350 golden lion tamarins. Movement of tamarins from the most severely threatened populations to a newly established reserve established a second large protected population.

Molecular analysis indicates that small populations of golden lion tamarins have little genetic diversity.

a) Explain the dangers of maintaining very small populations of golden lion tamarins in small forest reserves in Brazil. (3)

Golden lion tamarins are monogamous (have only one sexual partner). A genetic analysis of the 500 golden lion tamarins in captivity across 140 zoos showed that about two-thirds of the genes in this scattered population were derived from one breeding pair.

b) Suggest how the captive golden lion tamarins were managed as a result of this genetic analysis. (4)

c) Outline the possible problems of using captive animals for reintroduction programmes. (4)

d) Explain why animals such as golden lion tamarins, which serve no obvious benefit to humans, should not be allowed to become extinct. (4)

6 There are many seed banks around the world working to conserve seeds for the future. The vault at Svalbard in Norway is a seed bank that takes contributions from other seed banks. Critics of this 'seed bank of last resort' state that, even with the best storage conditions, most seeds lose their ability to germinate after about 20 years.

a) Explain why seed banks are described as an *ex situ* method of conservation. (2)

b) Describe the best conditions for the long-term storage of most seeds. (3)

c) Explain how the problem of the limited storage life of many seeds is overcome. (3)

d) Suggest why seeds of crop plants should be kept in long-term store in seed banks. (5)

e) Suggest reasons why species that are **not** crop plants should be stored. (4)

Stretch and challenge

7 Microorganisms fulfil many important roles for the continuation of life on Earth.

a) Outline some of these roles.

b) Suggest how to assess the biodiversity of microbes and find out whether any are endangered.

8 Ecological surveys are carried out as part of environmental action plans. Explain how you would survey the ecology of a greenfield site that is likely to be developed as a new housing estate.

Chapter 14

Classification

Prior knowledge

Before you start, make sure that you are confident in your knowledge and understanding of the following points.

- The binomial system is the international way to name species.
- Organisms can be classified into groups according to shared characteristics.
- The variety of life is a continuous spectrum, which makes it difficult to place organisms into distinct groups.
- Natural classification systems reflect the way in which organisms have evolved.
- Artificial classification systems are devised for simple purposes, such as for identifying organisms.
- Proteins have four levels of organisation: primary structure, secondary structure, tertiary structure and quaternary structure.
- The primary structure of proteins is the sequence of amino acids joined by peptide bonds.
- Base sequences in DNA code for the assembly of amino acids to make proteins.
- Evolutionary relationships between organisms can be displayed using evolutionary tree diagrams.

Test yourself on prior knowledge

1 Explain why biologists classify organisms.
2 Explain why it is important to give organisms scientific names.
3 What do all the animals that are classified as *Carabus problematicus* have in common?
4 What do molecules of protein and molecules of DNA have in common?
5 List five ways in which a polypeptide differs from a molecule of DNA.

Figure 14.1 The rock hyrax, *Procavia capensis*. Biologists classify this animal, which is about 55 cm long and 3 to 4 kg, in the same group with elephants.

Male hyraxes share a feature with elephants: they do not have a scrotum. Instead of being suspended outside the main body cavity, the testes of hyraxes and elephants are inside the abdominal cavity. This means that they are kept at core body temperature rather than at a slightly lower temperature as is the case with mammals that have scrota. The tusks of hyraxes develop from the incisor teeth, as do the tusks of elephants; in most other mammals, tusks develop from canine teeth. Hyraxes, like elephants, have flattened nails on the tips of their digits, rather than curved, elongated claws which are usually seen on mammals. As a result of these similarities, hyraxes and elephants are classified in the same group. Molecular evidence obtained by sequencing the amino acids in proteins and sequencing the nucleotide bases in DNA supports the close relationship between hyraxes and elephants.

Figure 14.2 *Arbutus unedo*, the strawberry tree, was known as *Arbutus caule erecto, foliis glabris serratis, baccis polyspermis* (Arbutus with upright stems, hairless, saw-toothed leaves and many-seeded berries).

Classifying the natural world is as natural to us as breathing. Early humans separated plants into those that were good to eat, those that were not good to eat, and those that were poisonous. Our system of scientific classification is useful, because it makes sense of the huge variety of life on Earth.

Making order out of chaos

In the seventeenth century, the English scientist John Ray (1627–1705) surveyed many plant species and gave them short descriptive Latin names. By the middle of the eighteenth century these names had grown longer, for example, the strawberry tree (Figure 14.2). Latin was used because it was the international language of science. These polynomials (many names) were long-winded and impossible to remember. Many species were known by many common names in different languages, so it was difficult for scientists to ensure that they were talking or writing about the same organism.

In the eighteenth century the Swedish biologist Carl Linnaeus (1707–78) standardised the **binomial system** for naming species, which we still use today. Linnaeus gave every known species two names – similar to a surname followed by a first name. In all, he named over 11 000 different species. The first name reflected the way in which he classified the species and the second often gave some information about it. Figure 14.3 shows the binomial for the common dandelion and how to write it.

Key term

Binomial system A system of naming species in which each species has two names: a generic name and a specific epithet.

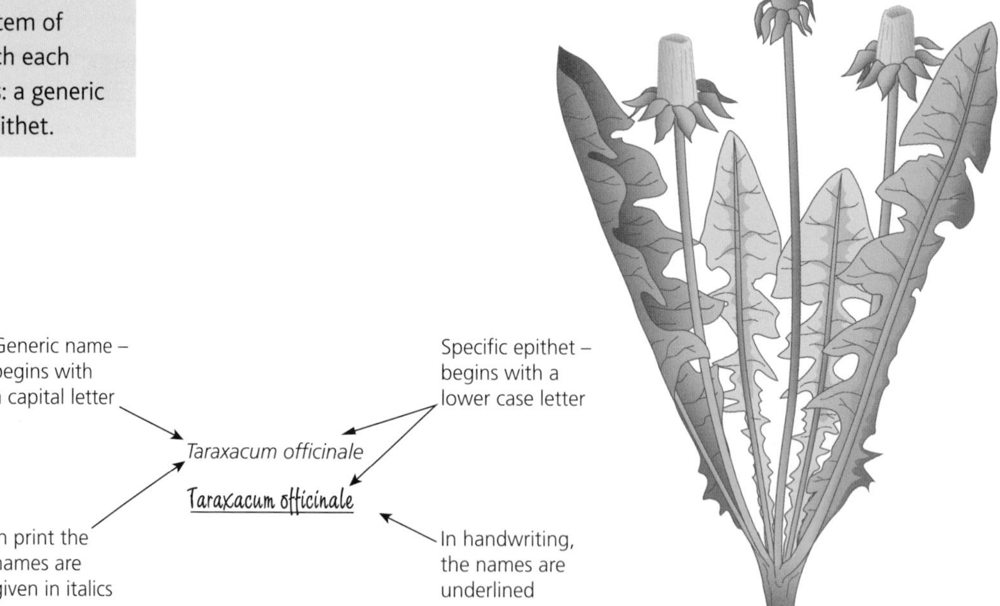

Generic name – begins with a capital letter
Specific epithet – begins with a lower case letter
Taraxacum officinale
Taraxacum officinale
In print the names are given in italics
In handwriting, the names are underlined

Figure 14.3 How to write the scientific name of the dandelion. Although all dandelions produce flowers and seeds, they do not reproduce sexually because the female gametes are diploid and develop into embryos without first fusing to a male gamete delivered by pollen grains. So, is the dandelion a true species?

Scientific names use Latin grammatical forms although they often originate from ancient Greek as well as other languages including English. The names usually tell us something about the species. The wood mouse is *Apodemus sylvaticus*, which translates as 'not house, wood', because it was always being confused (and still is) with the house mouse, *Mus musculus*. Some species are subdivided into subspecies; if this is the case, then a third name is given. You can find many examples of species names in this book.

The first name is the genus (or generic) name. This indicates a rank higher than species in the classification system. Species that show many similarities and are closely related are classified together in the same genus. The second name is the specific epithet. The generic name is one word, for example *Apodemus*; the species name is *both words*, for example *Apodemus sylvaticus*, which refers to one species within the genus *Apodemus*. If you have already used the name once, then it is permissible to abbreviate it when you use it again. The abbreviated form is the first letter of the generic name followed by the specific epithet, for example *A. sylvaticus*. If abbreviating the generic name may cause ambiguity – perhaps because you are writing about another species that has a generic name beginning with A – then you have to write out the name in full.

Generic names are not unique. The same name cannot be used for more than one animal genus or for more than one plant genus; however, it can be the name of an animal genus and a plant genus, as is the case with *Pieris*. The cabbage white butterfly is *Pieris brassicae*, and the plant known as Japanese andromeda is *Pieris japonica* (see Figure 14.12 on page 258).

Some specific epithets are used for many species. For example, 'vulgaris' is a Latin adjective that means 'common'; this epithet is applied to species such as the plant *Prunella vulgaris*, which is known in English as self-heal, and also to *Sciurus vulgaris*, the red squirrel.

The names of many species are now generally established. These codes devised by international organisations are used for regulating the naming of organisms:

- International Code of Zoological Nomenclature for animals
- International Code of Nomenclature for Algae, Fungi and Plants
- International Code for Nomenclature of Bacteria.

Taxonomists are biologists who specialise in describing, naming and classifying living and extinct organisms. **Taxonomy** is the study of classification and the way in which features are used to distinguish between different species and to group them together.

In biology, **classification** is the organisation of living and extinct organisms into groups that are arranged in a hierarchy. Some classifications are natural, in that they attempt to show relationships between species based on a study of many features. Others are artificial and are devised for some special purpose. Natural classifications use many features of organisms, whereas artificial classifications use only a small number.

Linnaeus devised a **hierarchical classification** system in which large groups were continually subdivided down to the level of the species (see Figure 14.4). His biggest group was the kingdom, and there were two: the plant and animal kingdoms. Within this hierarchical system, species are classified into a genus, genera are classified into a family, families are grouped into an order, orders are subdivisions of a class, classes are classified within a phylum, and phyla are classified into a kingdom (see Figure 14.4). There are far fewer species within any genus than there are species within a kingdom. Each of these groups is known as a **taxonomic rank**. The term **taxon** is used for examples of these levels; for example, the class Mammalia is a taxon, as is the order Rodentia and the genus *Apodemus*.

Tip

You will notice that throughout this book organisms are named using the binomial system. When writing, use scientific names when you can and present them as shown in Figure 14.3.

Key terms

Taxonomy The study and practice of naming and classifying species and groups of species within the hierarchical classification scheme.

Biological classification The organisation of living and extinct organisms into systematic groups based on similarities and differences between species.

Hierarchical classification The arrangement of organisms into groups of different rank. The lowest rank is the species; similar species are grouped together into the next rank, which is the genus. This continues to the highest rank, which is the domain, which groups many diverse species together.

Taxonomic rank Any level within the hierarchical classification, for example domain, kingdom, phylum, class, order, family, genus and species.

Taxon A group of organisms at any rank in the hierarchical classification scheme, for example any named species, or any group such as mammals and plants at higher taxonomic ranks.

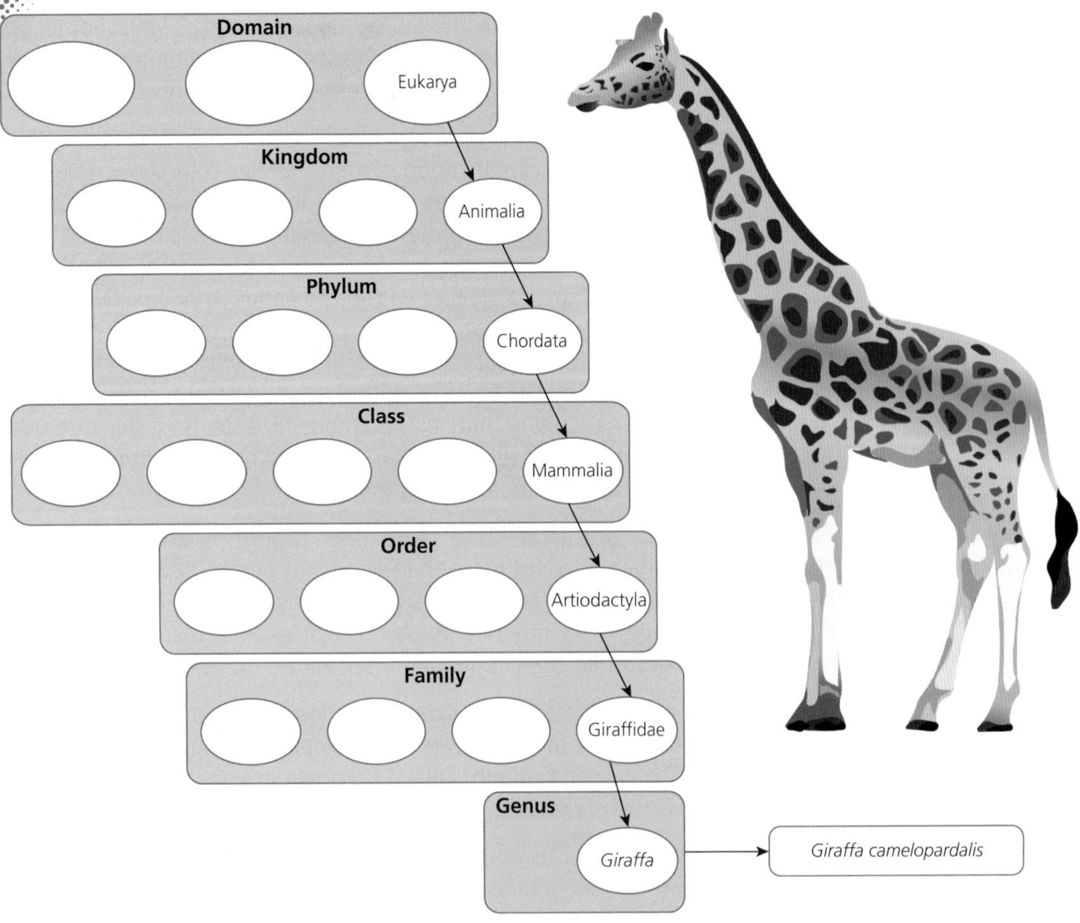

Figure 14.4 Hierarchical classification of the giraffe, *Giraffa camelopardalis*. Note that there are many more phyla in the animal kingdom than shown and there are ten families in the Artiodactyla.

Table 14.1 shows the classification of three species of mouse. The wood mouse and the house mouse are placental mammals. Macleay's marsupial mouse, *Antechinus stuartii*, also known as the brown antechinus, is, as its common name tells you, a marsupial.

Table 14.1 The hierarchical classification of three species of mouse. *Antechinus stuartii* is a marsupial mouse that lives in Australia. It is carnivorous, unlike the other two species. Superficially it looks like a mouse, but it is not closely related at all, as you can see below.

Taxonomic rank	Wood mouse	House mouse	Macleay's marsupial mouse
Domain	Eukarya	Eukarya	Eukarya
Kingdom	Animalia	Animalia	Animalia
Phylum	Chordata	Chordata	Chordata
Class	Mammalia	Mammalia	Mammalia
Order	Rodentia	Rodentia	Dasyuromorphia
Family	Muridae	Muridae	Dasyuridae
Genus	*Apodemus*	*Mus*	*Antechinus*
Species	*sylvaticus*	*musculus*	*stuartii*

Tip

You do not have to remember many of the names used in this chapter. You should learn the names of the five kingdoms and the three domains (see page 259) and the taxonomic ranks in Table 14.1. In an exam, you may have to write the names of taxa given in the questions, in which case always spell them correctly and use the abbreviated form, for example *T. officinale*, if appropriate.

The names for taxa higher than the genus can be written as Latin names starting with capital letters (as in Table 14.1) or as anglicised names, such as the animal kingdom, the chordate phylum and the rodent order.

Test yourself

1 The okapi, Okapia johnstoni, and the giraffe are both classified within the family Giraffidaea. Name the other taxa that they are classified within.
2 Giraffes and elephants are both mammals. Which rank in the hierarchical classification is this?
3 Suggest why elephants and hyraxes are not classified in the same order as giraffes.
4 What is meant by the term *hierarchical classification*?
5 Figure 14.4 shows that there are five classes in the phylum Chordata. Name the other four classes.
6 State the two important features of the system that Linnaeus developed for organising knowledge of living things.
7 List some features that the two mice in Figure 14.5 have in common.

Figure 14.5 *Mus musculus* (the house mouse) and *Antechinus stuartii* (the brown antechinus). These two species are an example of two different types of mammal that show many similar features.

The five kingdoms

With the improvement in microscopes, biologists began studying microorganisms. They discovered that organisms were built on two basic body plans: prokaryote and eukaryote. They also realised that Linnaeus's two kingdoms were not sufficient – there was so much diversity within these groups that in 1969 Robert Whittaker (1920–80) proposed in an article in the American journal *Science* the five-kingdom classification shown in Table 14.2.

Table 14.2 Some of the features used to categorise organisms into the five kingdoms. Note that viruses do not fit into this classification – they have their own system.

Features	Kingdoms				
	Prokaryota (Monera)	**Protoctista**	**Fungi**	**Plantae**	**Animalia**
Type of body	mostly unicellular	unicellular and multicellular	mycelium composed of hyphae; yeasts are unicellular	multicellular, branching body; not compact	multicellular, most have a compact body
Nuclear envelopes	✗	✓	✓	✓	✓
Cell walls	✓ (made of peptidoglycan)	present in some species	✓ (made of chitin)	✓ (made of cellulose)	✗
Cell vacuoles	present in a few species	algae have large permanent vacuoles; protozoans have small, temporary vacuoles	large permanent vacuoles	large permanent vacuoles	small temporary vacuoles, e.g. lysosomes and food vacuoles
Organelles and fibres, e.g. microtubules	✗	✓	✓	✓	✓
Type of nutrition	autotrophic and heterotrophic	autotrophic and heterotrophic	heterotrophic	autotrophic	heterotrophic
Motility (ability to move themselves)	some bacteria have flagella	some protoctists have flagella or cilia	✗	gametes of some plants have flagella	✓ muscular tissue
Nervous coordination	✗	✗	✗	✗	✓
Examples	bacteria and cyanobacteria (blue-green bacteria)	*Amoeba*, algae, slime moulds	mould fungi (e.g. *Aspergillus*), yeast	liverworts, mosses, ferns, conifers, flowering plants	jellyfish, coral, worms, insects, vertebrates

Tip

See Figure 1.24 and Table 1.3 on page 15 to recall how the cell structure of prokaryotes differs from the structure of eukaryotic cells.

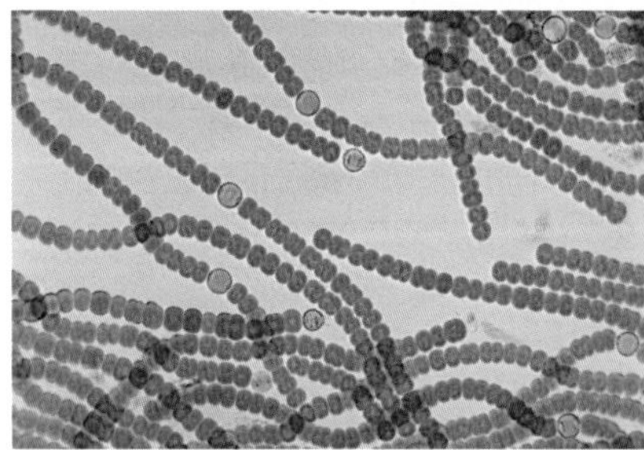

Figure 14.6 The filamentous blue-green bacterium, *Anabaena variabilis*. (×200) Blue-green bacteria fix carbon dioxide in photosynthesis, and also fix nitrogen by converting N_2 into organic forms of nitrogen. The wider, lighter green cells is where nitrogen fixation occurs.

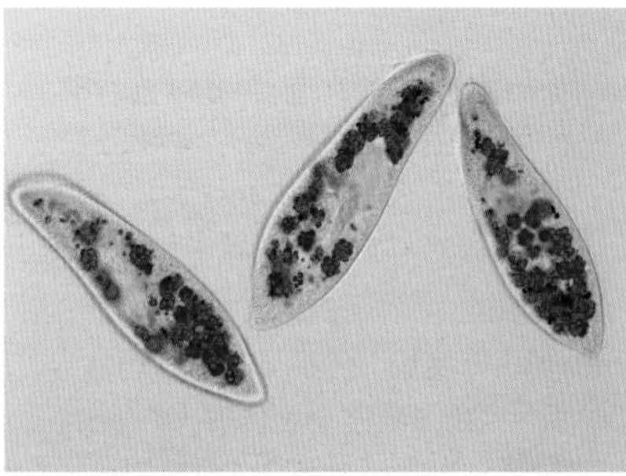

Figure 14.7 *Paramecium caudatum* is a single-celled protoctist that is covered in cilia which move it through the water and move food into an opening where food vacuoles are formed. These paramecia have been fed yeast cells stained red, which are now being digested inside food vacuoles (×85).

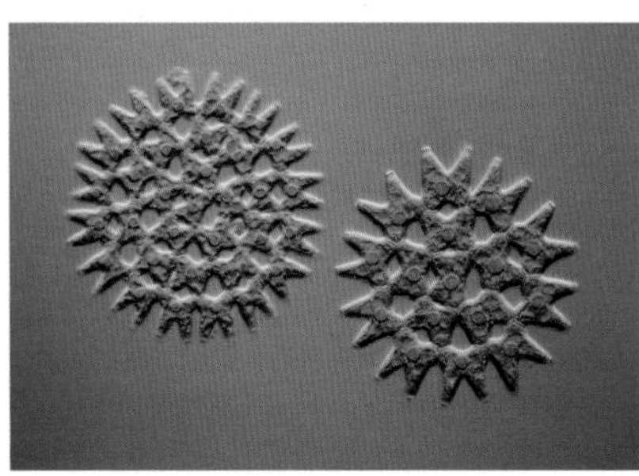

Figure 14.8 *Pediastrum duplex* is a colonial green protoctist. The body is composed of separate cells in a round, flat shape. These cells work together in a coordinated way, but there is little specialisation of cells. (×360)

Kingdom Prokaryota

The kingdom Prokaryota includes the bacteria and blue-green bacteria (see Figure 14.6). Most prokaryotes exist as single cells, filaments of cells or groupings of similar cells known as colonies. Prokaryotic cells are about 1 µm in diameter, some 10 or 100 times smaller than cells of eukaryotes. Although they are very small, prokaryotes make up about 90% of the total biomass of the oceans, and they live in a great variety of habitats, including many with extreme conditions of heat, pH and salinity that kill eukaryotes.

There are a great range of metabolic processes in this kingdom. Blue-green bacteria and some bacteria are autotrophic as they are photosynthetic and fix carbon dioxide and produce oxygen in the same way as green plants. Many are heterotrophic, feeding as decomposers on organic materials, both living and dead. Some are parasites, including those like *Neisseria meningitidis* and *Mycobacterium tuberculosis* that cause diseases in humans. Some prokaryotes can fix nitrogen gas (dinitrogen – N_2) to form ammonia from which they can synthesise many nitrogenous compounds. Others can make use of other inorganic substances in place of oxygen in respiration. Prokaryotes are important in recycling elements such as nitrogen, phosphorus and sulfur. Some live in anaerobic conditions and produce methane (CH_4) as a waste.

Prokaryotes do not have linear chromosomes like eukaryotes, and they do not divide by mitosis. Their DNA replicates in the same way as in eukaryotes (see page 52), but there is no nuclear envelope to break down and there is no separation of chromosomes as there is in anaphase (see page 99). Cell division in prokaryotes is by binary fission, which is a form of asexual reproduction. The closest that prokaryotes come to a form of sexual reproduction is the transfer of genetic material from one individual to another. This happens when bacteria join together and exchange DNA. Plasmids are also exchanged between bacteria, even between bacteria from different species. As a result, bacteria can gain new genes such as those for antibiotic resistance.

Mitochondria and chloroplasts are organelles in eukaryotes that have many similarities with bacteria. Indeed, it is now generally agreed that mitochondria and chloroplasts evolved from bacteria that invaded or were taken in by eukaryotic cells more than a billion years ago.

Kingdom Protoctista

This kingdom is composed of a diverse range of eukaryotic organisms, which include those that are often called protozoans ('simple animals') and algae such as seaweeds. To put it simply: any eukaryote that is not a fungus, plant or animal is classified as a protoctist. Many, such as *Paramecium* (see Figure 14.7), are single celled; some are filamentous and some, such as *Pediastrum duplex*, are groups of similar protoctists known as colonies (see Figure 14.8). Seaweeds are the most complex multicellular protoctists. Their bodies are not differentiated into organs such as roots, leaves and stems, but different areas of the body are specialised for attachment, photosynthesis and sexual reproduction.

Many organisms in this kingdom may actually be more closely related to organisms in other kingdoms than they are to each other. For example, there are strong arguments for classifying algae, such as seaweeds, in the plant kingdom.

Protoctists are found in many different natural and artificial environments. Algae are important photosynthetic organisms in aquatic ecosystems. Ciliates are important in sewage-treatment works, where they feed on bacteria, keeping their numbers in check. Some, such as *Plasmodium*, which causes malaria, are important human and animal pathogens.

Figure 14.9 This false-colour scanning electron micrograph shows spores being produced by *Aspergillus versicolor*, a fungus that can cause lung infections in humans. (×300)

Kingdom Fungi

All fungi are heterotrophic, obtaining energy and carbon from dead and decaying matter or by feeding as parasites. Fungi are important as decomposers aiding the recycling of carbon (as carbon dioxide) and mineral elements such as nitrogen. None of the fungi can photosynthesise. A few species are parasitic on animals, but many are parasites of plants and are economically damaging pathogens of crop plants.

Some fungi, such as yeasts, are single celled, but most are composed of microscopic threads or hyphae that grow over or through their food source. Each hypha has a cylindrical shape which in some species is subdivided into separate cells (see Figure 14.9). Hyphae secrete enzymes onto their food source. Complex compounds such as cellulose, starch and proteins are digested externally and the soluble products absorbed. The hyphae form a fungal body known as a **mycelium**. When grown on agar in Petri dishes, these mycelia are circular. In the soil or inside dead trees, the hyphae grow in all directions and it is very difficult to trace the extent of a mycelium (but see Figure 14.22 page 266).

Figure 14.10 A bracket fungus.

Fungi exhaust their food sources and need to find new ones. They do this by producing millions of spores, a few of which may land on a suitable substrate. The fruiting bodies of mushrooms, puffballs, toadstools and bracket fungi release huge numbers of spores. These fruiting bodies are formed from very compact hyphae.

Fungi have a vast range in size, from the microscopic yeasts to what may be the world's largest organism. Specimens of the honey fungi *Armillaria gallica* and *A. ostoyae* grow over huge areas in Oregon, Washington, Wisconsin and Michigan in the USA. The first of these 'humongous' fungi to be discovered grows in a forest in Wisconsin and spreads over 160 000 m^2. Not only are these honey fungi possibly the largest organisms on Earth, but they are also the oldest, at 1500 to 10 000 years old; the estimated mass of these fungi is 100 tonnes. Figure 14.22 shows some fruiting bodies of *A. gallica* from which many thousands of spores are dispersed into the air.

Fungi reproduce asexually and sexually. Yeasts reproduce asexually by budding (see page 102) and also reproduce sexually by producing haploid cells of different mating strains that fuse together.

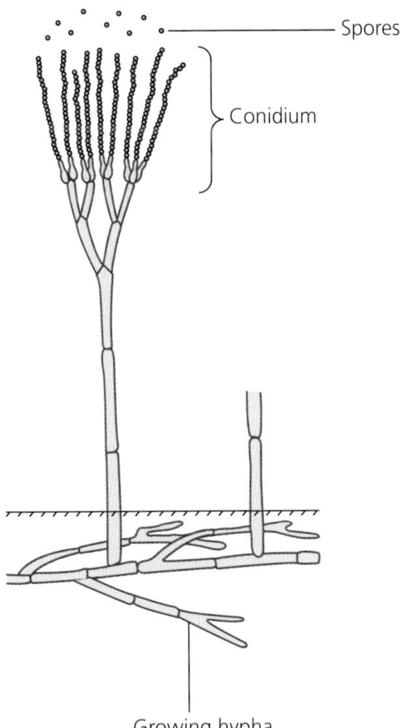

Figure 14.11 Structure of the mould fungus, *Penicillium chrysogenum*.

Kingdom Plantae

All plants are multicellular photosynthetic organisms. They have complex bodies that are often highly branched, both above and below ground. There are few types of specialised cells and fewer types of tissues compared with animals. However, biochemically, plants are diverse and are able to carry out a wider range of metabolic reactions than animals. For example, they carry out photosynthesis as well as respiration. In addition they can synthesise many substances from simple raw materials – carbon dioxide, water and ions such as nitrate, sulfate and phosphate.

Figure 14.12 Japanese andromeda bush, also known as Flame of the Forest because new leaves are red not green. The scientific name of this plant is *Pieris japonica*.

Almost all plants are immobile, because their bodies spread out to cover a wide area. To spread and to avoid competition they all have means of dispersal, usually linked to their form of sexual reproduction. All have a haploid stage that alternates with a diploid stage in their life cycle. In the seed plants, the haploid phase is reduced.

Plants are not motile, but some show limited movement. The Venus fly trap is a carnivorous plant that has modified leaves that can snap shut rapidly to trap insects. The leaves then secrete enzymes to digest the insects. All plants release spores that help to spread the species. The spores of seed plants such as conifers and flowering plants are pollen grains that carry male gametes in sexual reproduction.

Plants dominate most terrestrial ecosystems. In fact, biomes and ecosystems are often identified by the dominant vegetation; examples are oak woodlands and grasslands.

Kingdom Animalia

Animals are multicellular organisms that are all heterotrophic with many ways of obtaining their food. Bodies are usually compact, with a wide range of tissues that form complex organs. Organs work together in organ systems. The nervous system is unique to the animal kingdom. In the simplest animals, the nervous system consists of a net of nerve cells; in the most complex, it consists of a brain with huge numbers of nerve cells and even larger numbers of interconnections between them.

There is a great diversity of forms within the animal kingdom. The animal kingdom is usually divided into two groups: invertebrates and vertebrates. However, as you saw in Table 14.1 and Figure 14.4, invertebrates and vertebrates are not major taxonomic ranks. The vertebrate classes (fish, amphibian, birds, reptiles and mammals) are all classified into the phylum Chordata, which also includes some invertebrate species. The fundamental feature that animals in the phylum Chordata share is a *notochord* – a stiffening rod that supports the body during early development and is replaced by the backbone. Tunicates, better known as sea squirts, are one group of invertebrate chordates that do not develop bone, and therefore have no backbone. All other invertebrates are classified into about 30 phyla that represent the great diversity of form and function in this kingdom.

Figure 14.13 Swimming with the nautilus. The nautilus is related to squids, cuttlefish and octopods. They are fascinating animals to study in their natural environment.

All animals are heterotrophs, although some enter partnerships with autotrophic organisms that live within their cells. The best example of heterotrophs that partner with autotrophic organisms is corals, which are the basis of a very biodiverse ecosystem – the coral reef. The importance of coral reefs is discussed in Chapters 12 and 13.

Example

Table 14.3 shows the classification of *Nautilus pompilius* (see Figure 14.13). The taxonomic ranks are not in the correct sequence.

Table 14.3 Classification of *Nautilus pompilius*.

Number	Taxonomic rank	Taxon
1	species	*Nautilus pompilius*
2	kingdom	Animalia
3	family	Nautilidae
4	phylum	Mollusca
5	order	Nautilida
6	class	Cephalopoda
7	genus	*Nautilus*

1 Complete the table by entering the numbers indicating the taxonomic ranks in the correct sequence. The first one has been done for you.

2						

2 State three features of the kingdom Animalia that are shared with organisms classified in the kingdom Protoctista and kingdom Fungi.

3 Explain how the body of a fungus differs from the body of a plant.

4 Discuss whether the Protoctista should be considered a true kingdom.

Answers

1

2	4	6	5	3	7	1

The sequence must be completely correct to gain the mark in an exam.

2 Animals are eukaryotic, so each of their cells has a nucleus and membrane-bound organelles such as mitochondria. All animals are multicellular; there are multicellular protoctists and fungi.

3 The fungal plant body is a mycelium, made of hyphae. These are long thin threads that spread through and over the fungus's food supply. Plant cell bodies are made of cells arranged into tissues. Plant cell walls are made of cellulose; hyphae have walls made of chitin.

4 The organisms classified into this kingdom do not all share the same fundamental structures. For example, some protoctists are unicellular, e.g. *Paramecium* and *Amoeba*; others are multicellular, e.g. seaweeds. Many plant-like autotrophic protoctists have cell walls and chloroplasts; also, many animal-like protoctists are heterotrophic and motile.

Tip

The answer to Question 4 could include information about molecular taxonomy from further on in the chapter. See page 262 and Figure 14.19.

Three domains

During the latter part of the twentieth century, scientists discovered more about molecular biology, biochemistry and cell structure. This revealed that the prokaryotes are not one uniform group. In the 1970s, some prokaryotes were discovered in extreme environments, for example at high temperatures in hot springs, and in salt lakes, highly alkaline conditions and around 'black smokers' (see Figure 14.15). These *extremophiles* as they are called were found to share features with both bacteria and eukaryotes. New techniques of molecular biology allowed scientists to show that features such as ribosomal RNA (rRNA), aspects of protein synthesis and the structure of cell membranes and flagella indicated that these extremophiles had eukaryote and bacterial features.

In 1990, Carl Woese introduced the domain as a new taxonomic rank above the rank of kingdom, giving greater weight to molecular biology than to other features. Extremophiles were classified in a separate domain, the Archaea, which is at the same taxonomic rank as Bacteria and Eukarya. Since then, other archaeans have been discovered, not all of them living in extreme environments.

Table 14.4 shows some of the features of Bacteria, Archaea and Eukarya. Each domain is subdivided into other taxonomic ranks, although taxonomists debate whether kingdoms should be included. Figure 14.14 shows how the domains are related to one another.

Table 14.4 Features of Bacteria, Archaea and Eukarya.

Features	Domain		
	Bacteria	**Archaea**	**Eukarya**
Cell structure	prokaryotic – no nucleus, no membrane-bound organelles	prokaryotic	eukaryotic
Cell wall of peptidoglycan	✓	✗	✗
Cytoskeleton	✗	✗	✓
Membranous organelles (mitochondria, endoplasmic reticulum, etc.)	✗	✗	✓
DNA	circular	circular	linear (in the nucleus); circular in chloroplasts and mitochondria
Non-coding nucleotide sequences within genes (introns)	✗	in some genes	✓
Histone proteins in combination with DNA	✗	✓	✓
Types of RNA polymerase	one	several	several
Ability to grow at 100 °C	✗	some species	✗
Poisoned by diphtheria toxin (a bacterial toxin)	✗	✓	✓
Sensitivity to streptomycin (an antibiotic)	✓	✗	✗

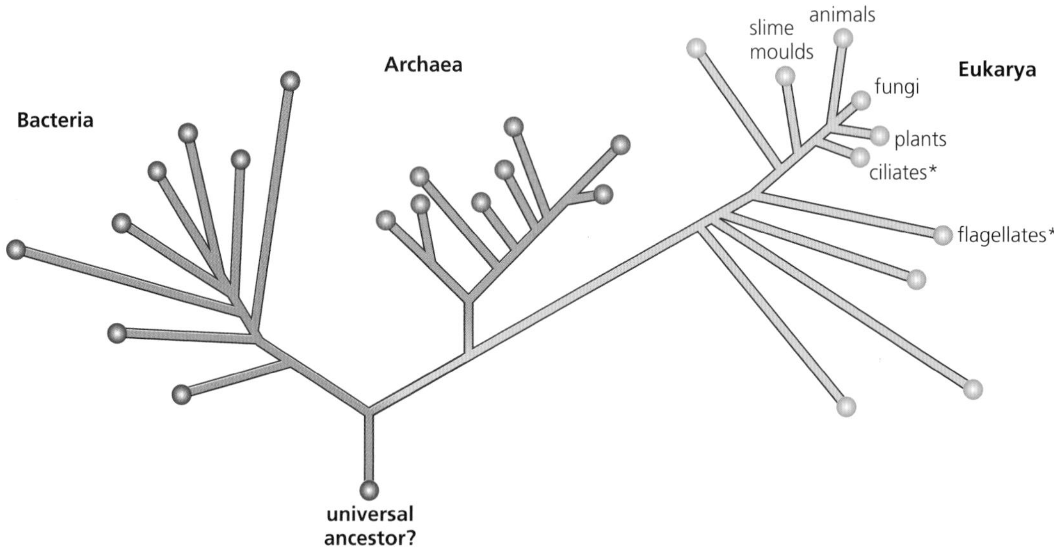

Figure 14.14 Bacteria, Archaea and Eukarya. The branches within each domain indicate different lines of descent. Taxonomists do not agree about what taxonomic rank these should be. * Ciliates and flagellates are included in the Protoctista in the five-kingdom classification.

Figure 14.15 A 'black smoker'. In 2010, a team from Southampton University discovered hydrothermal vents like these at 6000 metres on the Mid-Cayman Rise in the Caribbean Sea.

Figure 14.16 The giant tubeworm, *Riftia pachyptila*. These tubeworms are known only from hydrothermal vent communities. As adults they do not feed: they live entirely on the products from bacteria that live inside their cells, oxidising sulfur to gain energy to fix carbon dioxide and make organic compounds.

Key terms

Homology The existence of shared features that are inherited from a common ancestor. The shared features may not be used for the same function; for example, bat wings and human arms are homologous structures because they have the same pattern of bones, but they are used in different ways.

Phylogeny The evolutionary history of any particular taxon.

Test yourself

8 List the features that archaeans have in common with:
 a) bacteria
 b) eukaryotes.

9 Extremophiles have proteins with high numbers of amino acids with polar R groups. Suggest how this relates to their ability to survive at high temperatures.

10 Suggest how organisms such as bacteria and some yeasts can survive in lakes with high salt concentrations.

11 Explain the reasoning behind the introduction of the domain as a taxonomic rank above the rank of kingdom.

12 Which eukaryotic kingdoms contain the following organisms?
 a) Autotrophic organisms
 b) Heterotrophic organisms

Phylogeny

Natural classification systems aim to group organisms according to features that they share. In many cases, these groupings reflect the ways in which organisms have evolved. Closely related species are grouped together in the same genus because they share many features in common. However, some of the features they share may be adaptations to the same type of environment or way of life, as is the case with the three species of mice in Table 14.1. Although, superficially, marsupial mice look like placental mice, they are not closely related. Their 'mousiness' is a result of adaptation of animals with different ancestry to a similar way of life in different parts of the world. Mice such as *Apodemus sylvaticus* and *Mus musculus* are placentals related to rats and other rodents; the brown *Antechinus* has fundamental features in common with other marsupials.

Natural classification systems are based on **homology**. Features that are homologous are shared by organisms because they have been inherited from a common ancestor. The more recently two species have shared a common ancestor, the more homologies they share, and the more similar these homologies are. Until recent decades, the study of homologies was limited to anatomical structures and to patterns of embryonic development. In the past, the only features that biologists could study were those that they could see (including those that they could see with microscopes) and were therefore limited to the external appearance (morphology), the internal structure (anatomy), some details of cell and tissue structure, and patterns of development from zygote to adult. Classification systems were based on physical features, often of dead specimens collected by naturalists, explorers and scientific expeditions. In spite of this, classification systems reveal the **phylogeny** of taxa because they group together organisms with many shared features. Phylogeny is the evolutionary history of organisms.

In making classifications, it is important to distinguish between features that are similar because of common descent and features that are the result of convergent evolution. Figure 14.17 in the activity below is a tree diagram that shows the classification of the mice discussed earlier (see Table 14.1 and Figure 14.5).

Tip

Examples of two structures that are not homologous are bird wings and insect wings. These are both adaptations for flight, but they are completely different in structure. A bird wing is based on the same bones as in your arm. Insects do not have bones. These two wings are analogous structures; they are **not** homologous.

Tip

Do all classification systems reflect phylogeny? No! Artificial classifications are purely utilitarian – hunter-gatherer peoples classified plants and animals into the edible and the inedible. People who write keys for amateur naturalists often classify flowering plants according to the colour of the flowers, and use this feature to make their keys.

Activity

Phylogeny of rats and mice

Look closely at Figure 14.17.

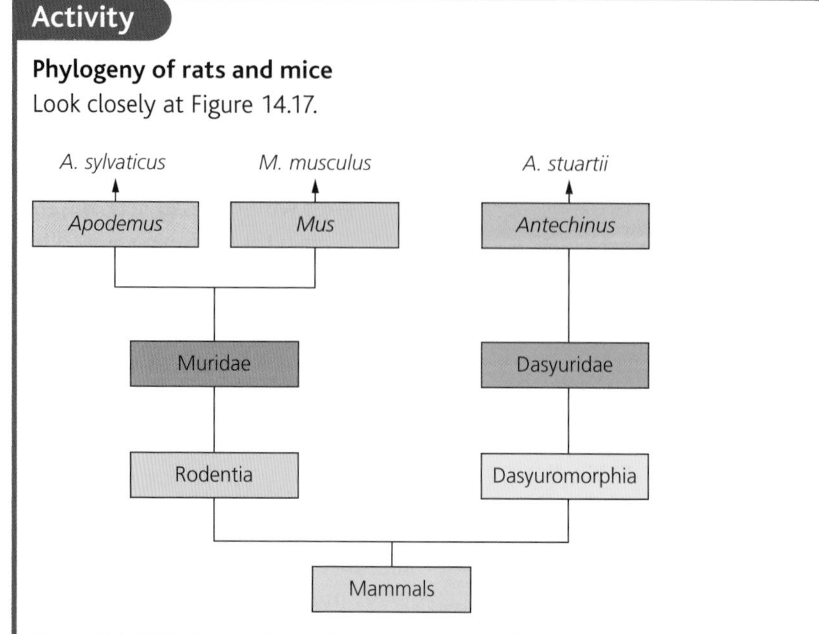

Figure 14.17 Phylogenetic tree for two species of placental mice and the marsupial mouse, *Antechinus stuartii*.

1 State the taxonomic ranks of Mammalia, Dasyuromorphia, Muridae and *Antechinus*.
2 With reference to Figure 14.17, explain how the phylogenetic tree indicates that *Antechinus* is not closely related to *Apodemus sylvaticus* and *Mus musculus*.
3 Rats are rodents in the family Muridae. There are 64 different species within the genus *Rattus*. All the *Rattus* species are larger than the species in the genera *Apodemus* and *Mus*. Suggest how you would adapt Figure 14.17 to include the genus *Rattus*.

Using molecular evidence in classification

Until the middle of the twentieth century, taxonomists relied on data from anatomy, morphology, behaviour, physiology and cell structure. With the development of new technologies, it became possible to study the structure of macromolecules – more especially the sequences of amino acids in proteins and the sequences of nucleotides in DNA. This led to the development of a new branch of biology: molecular phylogeny.

Using antibodies

Tip

To remind yourself about the differences between antigens and antibodies see Chapter 11.

Very early studies in molecular phylogeny used the ability of antibodies to detect the similarity between complex compounds such as proteins; research on this began as early as 1904 in Cambridge. As you may recall from Chapter 11, the variable regions of antibodies have a specific shape that is complementary to their respective antigens. Antibodies to a specific antigen can be produced by injecting an animal such as a rabbit or a mouse with that antigen. After a week or so, blood is taken from the animal and prepared as an anti-serum by removing all the cells and adding an agent to stop it from clotting. This anti-serum contains antibodies from different clones of plasma cells. This anti-serum can then be used to test the blood of other species to see whether the same antigen is present.

Figure 14.18 shows how this can be done using antibodies produced against the plasma proteins in human blood. Some blood is spun in a centrifuge to remove the red and white blood cells and it is treated to prevent it clotting. A small sample of the plasma is injected into a mouse or a rabbit. Several days later a sample of blood is taken from the rabbit and treated in a similar way. This anti-serum contains antibodies against all the proteins in human blood plasma. The anti-serum is mixed with human blood in a test tube. The rabbit antibodies attach to the human antigens and precipitate. This process is repeated with blood plasma samples taken from other species. The degree of precipitation is compared with that between rabbit antibodies and human blood plasma. The results are shown in Figure 14.19.

Tip

Try Question 5 in the Exam practice questions in this chapter for another example of the principle of using antibodies to detect similarities and differences between species.

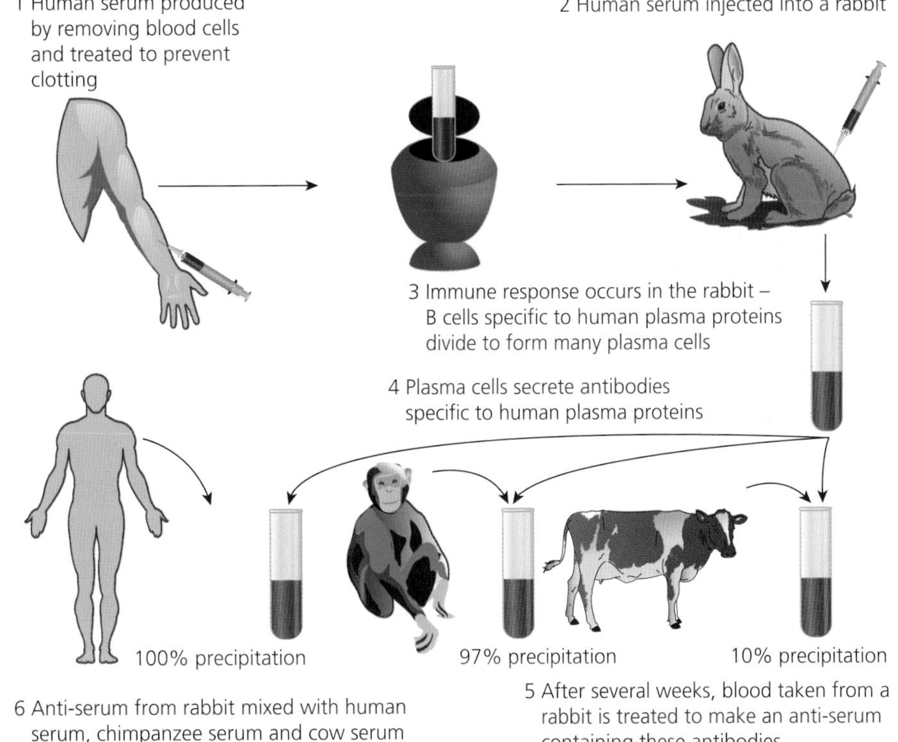
1 Human serum produced by removing blood cells and treated to prevent clotting
2 Human serum injected into a rabbit
3 Immune response occurs in the rabbit – B cells specific to human plasma proteins divide to form many plasma cells
4 Plasma cells secrete antibodies specific to human plasma proteins
100% precipitation
97% precipitation
10% precipitation
5 After several weeks, blood taken from a rabbit is treated to make an anti-serum containing these antibodies
6 Anti-serum from rabbit mixed with human serum, chimpanzee serum and cow serum

Figure 14.18 Using antibodies to detect similarities between the proteins in blood plasma of different species.

Protein sequencing

An alternative method to identify similarities and differences between proteins is to sequence the amino acids. Cytochrome c is found in many organisms, where it plays a key role in respiration. In eukaryotes, it is found in the inner membranes of mitochondria (see Figure 1.15 on page 9). This protein is a single polypeptide of about 100 amino acids wrapped around a haem group, in a similar way to the globins in haemoglobin (see Figure 3.9 on page 46). Relationships among proteins are discovered by aligning the primary sequences from a number of species; this involves comparing the sequences amino acid by amino acid and looking for similarities and differences.

To read these amino acid sequences you need to know the one-letter code for amino acids. Table 14.5 shows the two codes used for amino acids and the DNA codons. These are the codons that are identical to the mRNA that is produced when genes are transcribed, except that T replaces U (see page 50).

Table 14.5 The 20 different amino acids that are used to synthesise proteins, their three- and one-letter codes and their DNA codons.

Amino acid	Three-letter code	One-letter code	DNA codons (on the coding or non-template strand of DNA)
Alanine	Ala	A	GCA GCC GCG GCT
Arginine	Arg	R	AGA AGG CGA CGC CGG CGT
Asparagine	Asn	N	AAC AAT
Aspartic acid	Asp	D	GAC GAT
Cysteine	Cys	C	TGC TGT
Glutamic acid	Glu	E	GAA GAG
Glutamine	Gln	Q	CAA CAG
Glycine	Gly	G	GGA GGC GGG GGT
Histidine	His	H	CAC CAT
Isoleucine	Ile	I	ATA ATC ATT
Leucine	Leu	L	TTA TTG CTA CTC CTG CTT
Lysine	Lys	K	AAA AAG
Methionine	Met	M	ATG
Phenylalanine	Phe	F	TTC TTT
Proline	Pro	P	CCA CCC CCG CCT
Serine	Ser	S	AGC AGT TCA TCC TCG TCT
Threonine	Thr	T	ACA ACC ACG ACT
Tryptophan	Trp	W	TGG
Tyrosine	Tyr	Y	TAC TAT
Valine	Val	V	GTA GTC GTG GTT

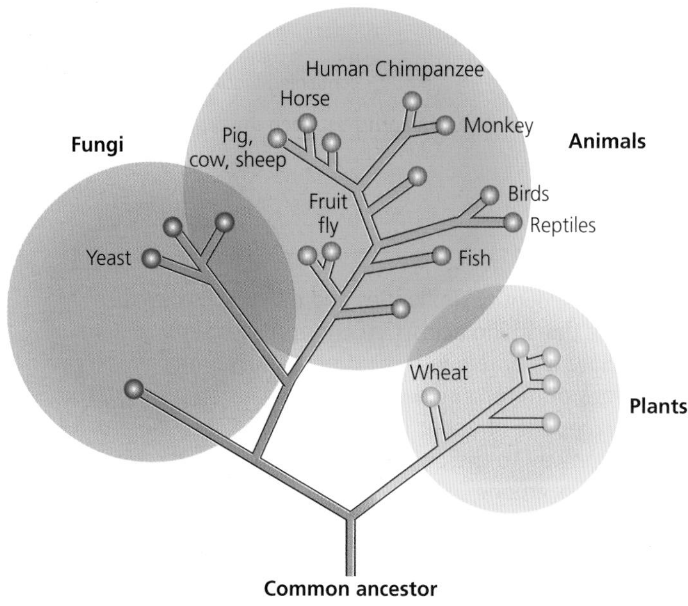

Figure 14.19 Phylogenetic tree based on the amino acid sequences of cytochrome c.

Data from investigations on the structures of proteins such as cytochrome c and haemoglobin are used to draw tree diagrams. The lengths of the branches in the tree in Figure 14.19 are drawn proportional to the number of differences in the primary sequence of cytochrome c. The tree clearly reveals the three main kingdoms of eukaryotes: fungi, animals and plants. Such trees tend to agree closely with those constructed by biologists using traditional methods of studying morphology and anatomy. They provide independent evidence of common descent.

DNA hybridisation

Each DNA molecule is made of two strands of nucleotides. Heating molecules of DNA causes the hydrogen bonds between the bases to break. When this happens the strands separate and can be isolated. If the molecules are allowed to cool, the attraction of the nucleotides will make the nucleotides reform hydrogen bonds.

> **Tip**
>
> This would be a good place to look at the structure of DNA to remind yourself about base pairing and the hydrogen bonds between bases that hold the two polynucleotide strands of the DNA molecule together.

To compare different species, scientists cut the DNA of two species into small segments, heat it to a temperature of about 90 °C, separate the strands, and mix the single-stranded DNA from the two species together. The two types of DNA bond together, but the match between the two strands is not perfect, because there are genetic differences between species. The more imperfect the match, the weaker the hydrogen bonding between the two strands. These bonds can be broken with just a little heat, while closer matches require more heat to separate the strands.

The technique of DNA hybridisation has been used to measure the similarities between the DNA of different species. This technique showed that there were greater similarities between the DNA of chimpanzees and humans than between either species and gorilla DNA or orang-utan DNA. This supported the evidence from earlier research using antibodies that showed that humans and chimpanzees have more similarities than either have with gorillas or orang-utans.

DNA sequencing

The sequence of amino acids in proteins such as cytochrome c and haemoglobin is determined by the sequence of bases in the genes that code for these proteins. It is now possible to sequence DNA more easily than it is to sequence proteins. Also, the data is much more detailed thanks to the degenerate nature of the genetic code (see Table 14.5). DNA sequence data has largely replaced protein sequence data in providing useful information for taxonomy. It also means that protein sequences have been derived from DNA data rather than the other way around. Comparing the DNA sequences of different species really gets to the fundamentals of classification and finding common ancestry.

> **Tip**
>
> The genetic code is a degenerate code because there are two or more codons for most of the amino acids. Only two amino acids are coded by a single codon. See Table 14.5 on page 264 to verify this.

One advantage of reading DNA sequences over reading protein sequences is that amino acid sequences can be exactly the same in different species, but because the genetic code is a degenerate code this may obscure the fact that the nucleotide sequences for these proteins are different. If you look at Table 14.5 you can see that different codons code for the same amino acid. Changes (mutations) to base pairs that result in the same amino acid as a result of degeneracy are known as **silent mutations**. Those that change the amino acid but not the type of R group and have no functional importance are known as **neutral mutations**. More significant are those changes that result in a functional change which spreads through the population – eventually becoming 'fixed' in the genome, in that almost all individuals possess it. Therefore, comparing gene sequences of different species gives much more information than comparing protein sequences.

The rate of change of DNA sequences over the very long periods for which life has existed on Earth varies from one part of the genome to another. For example, the genes that code for ribosomal RNA (rRNA) change slowly. These genes have therefore proved useful in sorting out relationships between the higher taxa such as domains, kingdoms and phyla. More rapidly changing genes help in showing relationships between recent speciation events and also between isolated populations of the same species, as has been done with groups of humans such as the indigenous human populations of the Americas.

Some of the most informative studies using DNA sequencing have been done with genes from the DNA in mitochondria, such as the DNA that codes for RNA that becomes part of the ribosomes in mitochondria. In eukaryotes, much of the DNA in the nucleus does not code for proteins. These regions of DNA have a variety of other functions or none that we yet know about. These regions of DNA are relatively free to vary without changing any protein

Figure 14.20 The thylacine. These marsupial wolves were driven to extinction. The thylacine thought to be the last one died in Hobart Zoo, Tasmania, in 1936. Remains of these animals were preserved, and the DNA of this species has been sequenced.

such as an enzyme or a receptor that must have a certain structure in order to function. These **non-coding sequences** provide much useful information for taxonomists. Mitochondrial DNA does not have these non-coding sequences, so making it easier to align sequences from different organisms.

DNA is more stable than protein. This greater stability means that taxonomists can sequence DNA from extinct organisms such as mammoths, which became extinct about 4500 years ago. Sequencing this DNA has helped to work out the phylogenetic relationships between mammoths and elephants. DNA has been extracted from the remains of Neanderthal people dated to 38000 years ago. These samples have been sequenced and compared with DNA from modern humans and from chimpanzees and gorillas.

The most useful genes for further analysis are the so-called 'housekeeping' genes that code for proteins that are required for fundamental features all organisms have.

Activity

Analysing DNA sequence data

In an investigation of gene sequences in marsupial and placental mammals, data was collected from databases on genes for a molecule of rRNA in the small subunit of the 70S ribosome in mitochondria. The eight sequences in Table 14.6 were aligned so that the bases between nucleotide position 29 and position 41 are shown. All the species in the table share about 50% of the nucleotides. This cannot come about purely by chance and shows that all these species share a common ancestor.

Look carefully at the table before answering the following questions.

1 List the positions that have the same base in all eight species.
2 Which **base pair** is the most common in this section of DNA? Give evidence for your answer.
3 Comment on the bases found at position 31 in these eight species.
4 What do you notice about position 37?
5 What conclusion would you make about the position of base A at position 34?
6 What do you conclude about position 32 among the placental species and position 38 among the marsupials?
7 Suggest why the information in Table 14.6 alone may be of limited value in helping to improve the classification of any one of the eight species.
8 It is thought that the thylacine is extinct. What are the advantages of sequencing the DNA from extinct organisms?

Table 14.6 DNA sequences in the genes that control the synthesis of rRNA in the mitochondria of eight mammalian species – four marsupials and four placentals. The thylacine was a wolf-like animal, now extinct (see Figure 14.20).

Group of mammals	Species	Position of base in rRNA												
		29	30	31	32	33	34	35	36	37	38	39	40	41
Marsupials	thylacine, *Thylacinus cynocephalus*	T	A	A	T	T	C	T	T	A	T	T	A	G
	quoll, *Dasyurus geoffroyi*	T	A	A	T	T	T	T	T	A	T	T	A	G
	dunnart, *Sminthopsis psammophila*	T	A	A	T	T	T	T	T	A	T	T	A	G
	marsupial mole, *Notoryctes typhlops*	T	A	A	T	T	A	T	T	G	C	T	A	G
Placentals	dog, *Canis familiaris*	T	A	G	T	T	T	T	T	A	G	T	A	G
	cat, *Felis sylvestris*	T	A	G	T	T	A	T	T	A	A	T	A	A
	shrew, *Sorex cinereus*	T	A	G	T	T	G	T	T	A	G	T	A	A
	European mole, *Talpa europaea*	T	A	G	C	T	G	T	C	A	G	T	A	A

Figure 14.21 Spore-producing fruiting bodies of *Armillaria gallica*. The use of barcodes such as the one shown in Figure 14.22 has shown that fruiting bodies of this species found many miles apart belong to the same mycelium.

Gene sequencing provides the means to identify species by isolating and testing their DNA. This has proved possible with the honey fungi *Armillaria gallica* and *A. ostoyae* (see page 257). Researchers sequence the bases of specific genes and publish them as 'barcodes'. These barcodes are also deposited in databases so that they are freely available to others. In the case of *A. gallica*, the barcode in Figure 14.22 comes from the Barcode of Life Data Systems (BOLD).

Figure 14.22 The first 193 bases in the gene for the mitochondrial protein cytochrome oxidase from the humungous fungus *Armillaria gallica*. Each base (A, T, C and G) is represented by a different colour.

Databases

Information about species is held in many databases such as:

- The Tree of Life at **http://tolweb.org**

  The Tree of Life project provides information about biodiversity, the characteristics of different groups of organisms and their evolutionary history.

- The taxonomy database at **www.ncbi.nlm.nih.gov/taxonomy**

  This holds information on about 10% of all described species. It is linked to other databases that hold data on DNA sequences and protein sequences – all of them can be found at the National Center for Biotechnology Information at **www.ncbi.nlm.nih.gov**

- Animal Diversity Web at **http://animaldiversity.ummz.umich.edu**

  This database is curated by the University of Michigan and holds data on animal natural history, distribution, classification and conservation biology.

- The National Herbarium of The Netherlands online at **http://vstbol.leidenuniv.nl**

  This is the largest single database of plants, with records of over 2 million specimens.

- The Encyclopedia of Life (EoL) at **http://eol.org**

  This database promises to have a web page for every species. The database brings together information from museums, scientific societies, experts and others, with a single, easy-to-use online portal.

Tip

Find out how successful the EoL has been in having a web page for every species. Try searching for some of the species described in this chapter. Look for the barcode for *Taraxacum officinale* when trying Exam practice question 8.

Researchers deposit their results in these databases, and the information is freely available to all. DNA sequence data allows researchers to compare their results with sequences already available in databases. Researchers can therefore compare closely related species that maybe do not have many structural differences. This molecular evidence also works at the other end of the taxonomic scale; for example, it shows that fungi are more closely related to animals than they are to plants. This is something that structural and growth habits of fungi do not suggest.

It is important to remember that biologists classify organisms based on what they believe to be their evolutionary relationships. Taxonomic information is continually being reinterpreted as new evidence is discovered. Not only will more species be discovered, but the way in which organisms are named and grouped within the system will change. Classification systems are in a constant state of flux because our understanding is continually being enhanced.

Example

Antipodean cicadas

Cicadas are insects that make a lot of noise. Biologists believe that cicadas migrated to New Zealand from Australia within the last 11.6 million years. To investigate the phylogeny of cicadas, researchers sequenced the mitochondrial DNA of 14 species.

Figure 14.23 shows the clapping cicada from New Zealand.

Small sections of the mitochondrial DNA in cicadas from Australia, New Caledonia and New Zealand were examined for similarities and differences. The results were used to construct the diagram in Figure 14.24a, which shows the relationships between these species. It is suggested that the eight cicada species in New Zealand originated from two migrations from Australia, as shown in Figure 14.24b.

1 a) Explain the advantages of studying gene sequences rather than protein sequences.
  b) Explain the advantage of using mitochondrial DNA rather than DNA from the nucleus in this phylogenetic analysis.
2 Suggest how phylogenetic trees such as the one in Figure 14.24a are constructed from DNA sequence data.
3 Explain how the results in Figure 14.24a support the idea that the cicada species in New Zealand originated from two migrations, as shown in Figure 14.24b.

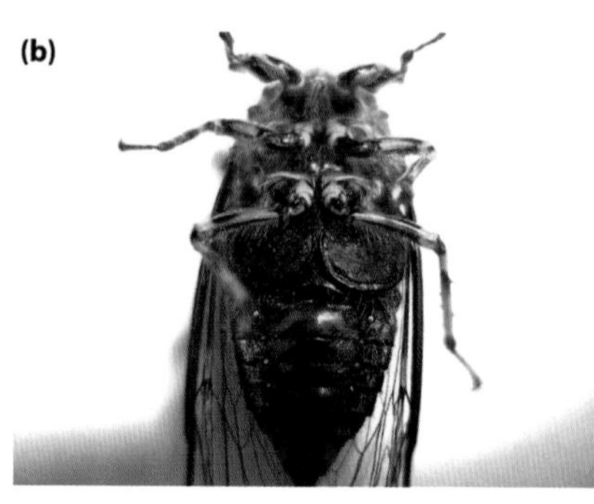

Figure 14.23 The clapping cicada, *Amphipsalta cingulata*, endemic to New Zealand: (a) dorsal view showing the animal's two pairs of wings; (b) ventral view showing the tymbals behind the last pair of legs, which produce all the noise.

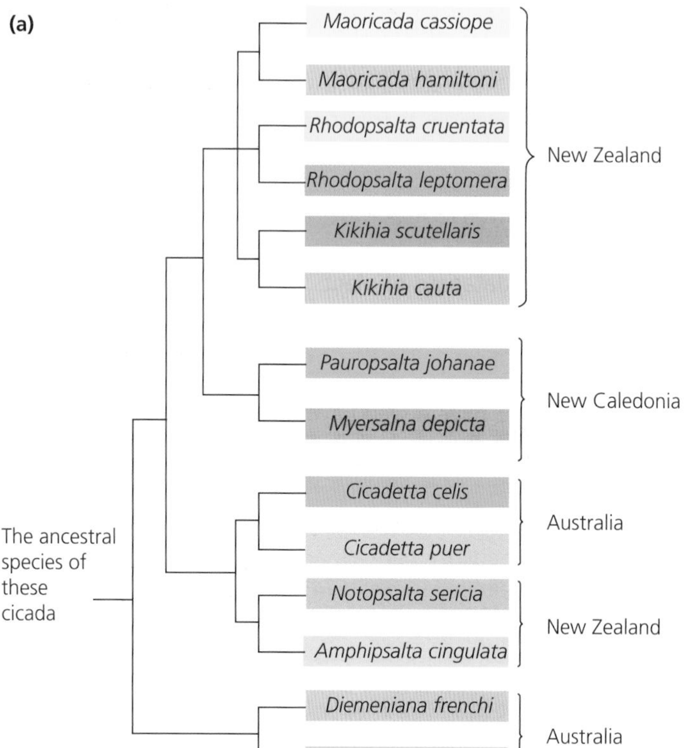

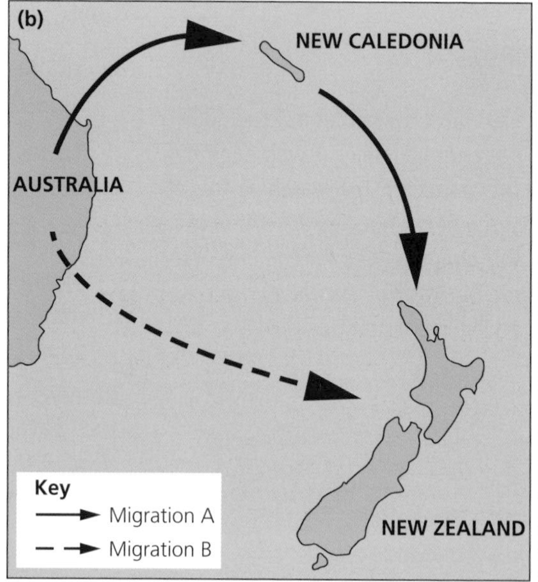

Figure 14.24 (a) Phylogenetic tree showing the relationships between 14 cicada species. (b) Map showing the possible migration routes of ancestral cicadas from Australia to New Zealand.

Answers

1 a) Gene sequences give more detailed information. Two species may have the same amino acid at a certain position in a polypeptide, but may have different codons for this amino acid. Some amino acids have as many as six codons. Differences in the sequences of nucleotides in DNA may therefore indicate relationships between species that are not detectable in amino acid sequences in proteins.

b) Mitochondrial DNA is not very extensive, unlike DNA from nuclei. The DNA is 'naked' – it is not combined with histones. There are no non-coding nucleotide sequences within the genes, so this makes aligning the sequences from different species much easier. There are multiple copies of DNA in mitochondria. Mitochondrial DNA from many species has been sequenced. This DNA is inherited relatively unchanged in the maternal line. However, mitochondrial DNA has a high mutation rate, so is only useful for comparing individuals within a population and species within the same genus or family. When comparing individuals from widely separated groups (e.g. animals and fungi) there are so many differences in gene sequences that it is impossible to draw any conclusions.

2 DNA sequences are aligned and compared. The number of differences is counted. Groups of organisms with the same sequences are grouped together. The more differences there are between sequences, the further apart the organisms, and branching points should be put at greater distances than between organisms that have few differences.

3 The diagram shows that species 13 and 14 are most distantly related to the other species. These species have been separated from the other species for a long time. Species 9 and 10 are most closely related to each other. The diagram shows that two groups (species 1–6 and 11 and 12) migrated to New Zealand or are descended from cicadas that migrated. Species 1–6 may have migrated indirectly from Australia via New Caledonia, because they are closely related to the two species (7 and 8) from New Caledonia. Species 11 and 12 migrated directly from Australia.

Test yourself

13 Explain the difference between DNA hybridisation and DNA sequencing.

14 DNA from two species was hybridised. The two strands of DNA separated at a very low temperature. Explain what this tells you about the two species.

15 Explain how taxonomists ensure that their classifications reflect the phylogeny of the species in the groups that they study.

16 Explain why 97% of plasma proteins in chimpanzees are the same as those in humans, whereas only 10% of the same proteins in cattle are identical.

Exam practice questions

1 The scientific name for any taxonomic rank higher than a genus is

A one name
B printed in italics
C two names
B underlined *(1)*

2 Which of the following sequences shows the correct hierarchy of classification, going from the group with the largest number of species to the group with the smallest number of species?

A Domain, phylum, order, class, genus, family
B Kingdom, phylum, class, family, order, genus
C Kingdom, phylum, class, order, family, genus
D Phylum, kingdom, order, family, class, genus *(1)*

3 Unicellular eukaryotic organisms are classified into the following kingdoms

A Animalia and Plantae
B Fungi and Plantae
C Protoctista and Fungi
D Protoctista and Plantae *(1)*

4 The organisms classified into the kingdom Protoctista include both unicellular and multicellular organisms. A handbook for biologists states that the kingdom is 'defined by exclusion'.

a) i) State one feature that is shared by all species classified into this kingdom. *(1)*

ii) Explain what is meant by 'defined by exclusion' with respect to the kingdom Protoctista. *(2)*

The following is taken from a website about poisoning events.

Blue-green bacteria grow very fast in waters that are polluted by fertilisers. The blue-green bacteria use up all the oxygen in the water and release toxins that kill animals, such as fish, that live in water and any animals that drink the water. In one serious incident some 20 sheep and 15 dogs died of poisoning by toxins released by blue-green bacteria. The cause was confirmed with the isolation of the toxin from the bacteria and from the dead animals. It was shown to be produced by microcystis aeroginosa, which increases explosively when conditions are suitable.

b) Identify the mistakes that the author of this passage has made in writing the name of the blue-green bacterium responsible for the toxin. *(2)*

c) Explain why this organism is not an alga and why it classified in a different domain from the true algae. *(3)*

d) Explain how DNA data is used to identify this species. *(3)*

5 Ruminants, such as cattle, sheep, goats and giraffes, are classified in the order Artiodactyla. Albumen is a plasma protein that is synthesised by liver cells. Albumen was isolated from blood samples from a cow and a human. These albumens were used to immunise two different rabbits. The rabbit anti-sera were collected several days later. Agar plates were prepared and the albumen solutions and anti-sera placed into wells cut into the agar, as shown below. A third rabbit was immunised with another plasma protein, transferrin, from a cow.

The lines on the figure indicate where antigens and antibodies formed complexes together.

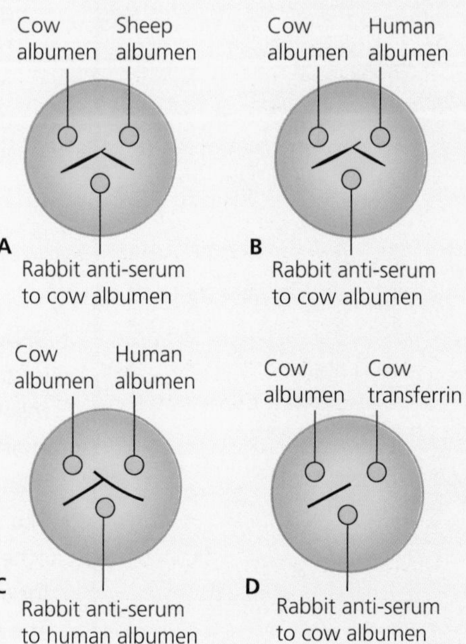

a) Describe and explain the results in each of the agar plates, **A**, **B** and **C**. *(6)*

b) What do these results suggest about how cattle, sheep and humans should be classified? Explain your answer. *(2)*

c) Explain why there is no precipitation between the rabbit anti-serum and bovine transferrin. *(2)*

d) Predict the results that you would expect when rabbit anti-serum to cow albumen is used with the following:

i) rabbit albumen
ii) rabbit transferrin
iii) giraffe albumen
iv) rock hyrax albumen. *(4)*

6 Chymotrypsin C is a protease enzyme synthesised by many animal species. The protein sequences for equivalent regions of this enzyme from four species were aligned. The nucleotide (DNA base) sequences corresponding to these regions were found in databases and also aligned. The tables at the bottom of the page show the results of this investigation.

Use the information in the tables to help you find the answers to the following questions:

a) State how many amino acid differences there are between the following pairs
- **i)** *H. sapiens* and *M. musculus* *(1)*
- **ii)** *H. sapiens* and *D. melanogaster* *(1)*
- **iii)** *D. melanogaster* and *A. aegypti* *(1)*

b) Part of the sequence for chymotrypsin C is identical in each of the species.
- **i)** Identify the amino acids in this sequence. *(1)*
- **ii)** Suggest why part of the sequence is exactly the same in the different species. *(3)*

c) What can you conclude from your answers to (a)? *(2)*

d) Suggest a possible DNA base sequence for the missing codon in the space for *D. melanogaster*. *(4)*

e) Explain the advantage of using DNA sequence data rather than protein sequence data in studying phylogenetic relationships between organisms. *(3)*

Stretch and challenge

7 Explain why it is important to be able to assign individual organisms to a particular species.

8 a) Dandelions reproduce by forming diploid female gametes which develop into embryos without being fertilised. This form of asexual reproduction is known as apomixis. Is the dandelion a true species or many thousands of different species? How could modern methods of taxonomy be used to find out?

b) The domain was introduced in the 1990s as an additional taxonomic rank. Why was this considered necessary?

Amino acid and nucleotide sequence for human, *Homo sapiens*.

213														225
N	A	D	S	G	G	P	L	N	C	Q	L	E	N	G
atg	gct	gac	tcc	ggt	ggc	cca	aac	gtt	tgc	cag	ctg	gag	aac	ggt

Amino acid and nucleotide sequence for house mouse, *Mus musculus*.

213														225
N	G	D	S	G	G	P	L	N	C	P	V	E	D	G
atg	gct	gac	tct	ggt	ggc	cca	ctg	aac	tgc	cca	gtg	gaa	gac	ggc

Amino acid and nucleotide sequence for mosquito, *Aedes aegypti*.

189														203
S	A	D	S	G	G	P	L	V	K	Q	S	G	E	E
tcg	gct	gat	tca	ggt	ggc	ccg	ttg	gtg	aaa	cag	tcc	ggt	gaa	gaa

Amino acid and nucleotide sequence for fruit fly, *Drosophila melanogaster*.

210														224
T	G	D	S	G	G	P	L	V	L	K	D	T	Q	I
act	ggt	gac	tct	ggc	ggt	cca	ctc	gtt	ctc	aag	gac	act	caa	

Chapter 15

Evolution

Prior knowledge

Before you start, make sure that you are confident in your knowledge and understanding of the following points.

- Life has evolved over a very long period of time on Earth.
- The evolution of new species occurs over time through natural selection.
- An organism's genotype is the genetic make-up of that organism; its phenotype is all its physical and biochemical features.
- A gene codes for the sequence of amino acids in a polypeptide. Different alleles of a particular gene code for the same polypeptide but with slightly different sequences of amino acids.
- All individuals of the same species have the same genes, but they do not necessarily have the same alleles of those genes.
- Some aspects of phenotype are determined by genotype alone; others are determined by environment alone. Still others are determined by the interaction between genotype and environment.
- Variation (both genetic and phenotypic) exists both between species and within species.
- Mutant alleles that are present in gametes may result in new features in the phenotype of any offspring that inherit them.
- Bacteria have evolved resistance to antibiotics and some pest species have evolved resistance to pesticides.
- Adaptation is a modification to an organism that allows it to 'fit into' its particular habitat.
- Examples of the types of evidence that support the idea that life has evolved are fossils, comparative anatomy and molecular biology.
- Humans use selective breeding (artificial selection) to improve the plants and animals used in agriculture and horticulture.

Test yourself on prior knowledge

1 Approximately how long has life existed on planet Earth?
2 Outline how natural selection acts as a mechanism for evolution.
3 Explain the difference between *gene* and *allele*.
4 List the causes of genetic variation in organisms that reproduce sexually.
5 Explain why there is less variation in a population of eukaryotic organisms that reproduces asexually compared with one that reproduces sexually.
6 What is the difference between natural selection and selective breeding (artificial selection)?
7 Give two adaptations of plants for survival in dry habitats, such as sand dunes.

Imagine that you have had a bright idea, an idea that explains everything. Now imagine that you have spent over 20 years finding evidence for your bright idea, enough evidence to convince your colleagues and the general public. You also know that many people, including some in your own family, would not like your bright idea one bit because of all the implications that it has for society. Now imagine that one morning the postman delivers a letter from across the other side of the world. You

Figure 15.1 Charles Darwin in 1840, a few years after he returned to England following his voyage on HMS *Beagle*.

open and read the letter to find that someone else has had exactly the same bright idea. After all the work you have done on your bright idea, how would you feel? This is exactly what happened when the postman delivered a letter to Down House in Kent one day in early summer 1858.

Down House was the home of Charles Darwin (1809–1882), a geologist, naturalist and explorer. As the scientist on the survey ship HMS *Beagle*, Darwin had the good fortune to make many observations of the natural world and to collect specimens of plants, animals and fossils in South America, and crucially, as it turned out, the Galápagos Islands in the Pacific, Australasia and Mauritius. His observations and tentative conclusions convinced him that species were not immutable – they changed over time. He had found enough evidence to convince himself that life evolved.

On his return from the expedition, Darwin spent many years corresponding with fellow scientists and others across the world, collecting more and more evidence, drafting his thesis and putting off the day when he would have to publish his great idea. For one thing, he was fully aware of the upheaval that publication would cause.

What is evolution?

When people talk about evolution they are usually referring to two distinct ideas:

1. The general theory of evolution. This states that organisms have changed over time and continue to change. As scientists in the eighteenth and nineteenth centuries learned more about the Earth, there came a general realisation that the Earth was very old and that life had changed over time. For example, discoveries made by Mary Anning in the Jurassic rocks at Lyme Regis in Dorset showed them to be full of fossils of animals that were no longer alive; in fact, they had been extinct for millions of years.
2. The special theory that evolution occurs by the process of natural selection.

Figure 15.2 A statue of Alfred Russel Wallace, one of the greatest naturalists of the nineteenth century. This records the moment he first saw a male golden birdwing butterfly, *Ornithoptera croesus*, in the rainforest of Bacan Island, Indonesia. The statue stands just outside the Darwin Building at the Natural History Museum in London; if you could follow the statue's gaze you would see a model of the butterfly on the wall.

Charles Darwin collected information about many groups of organisms, most famously during his voyage around the world on HMS *Beagle* from 1831 to 1836. Another scientist, Alfred Russel Wallace (1823–1913), also spent many years travelling and collecting specimens in South America and in South-East Asia. It was Wallace who had independently proposed natural selection as a way in which evolution could occur and had written that fateful letter to Darwin in 1858. A joint scientific paper proposing their theory of natural selection was presented in 1858 at the Linnaean Society in London. In November 1859, Darwin published *On the Origin of Species by Means of Natural Selection, or the Preservation of Favoured Races in the Struggle for Life*, to give the first edition of the book its full title. This book presented the special theory of natural selection, with much supporting evidence painstakingly amassed by Darwin over many years.

Although our ideas about natural selection have changed over the past 150 years and been much criticised or even denigrated by those who remain unconvinced, it remains the best explanation of the numerous observations made by scientists about species and how they change over time. Interestingly, in the first edition of *On the Origin of Species*

Darwin did not use the word 'evolution' until the last paragraph. Instead he referred to 'descent with modification', which is a good way to think of evolution as you read this chapter.

Key term

Variation The differences between species and the differences within species.

A central pillar of Darwin's argument in *On the Origin of Species* is the **variation** that exists within species. He spent much of his life collecting and recording examples of this variation. We will first consider several aspects of variation, before dealing with the idea of natural selection.

Variation

No group of organisms is completely identical. Some organisms are clones – they all have the same genotype. But even though their genotypes are identical, there will still be slight differences in phenotype as a result of the different environments that they live in or the different interactions with other organisms that they have had.

Tip

See Chapter 12 for a definition of *species*. Remember that there are many different definitions for this term. Perhaps by the end of this chapter you may realise why it is a concept that is so hard to define.

Genetic variation is the variation in genotypes that exists within a species (see Chapter 12). Phenotypic variation is the variation that is visible, or detectable, within a species. We usually assume that an organism's phenotype is its outward appearance; such things as height, mass, shape and colour. However, as we will see, an organism's phenotype includes all aspects of its biology except its genes. Your blood group is as much a part of your phenotype as the shape of your face.

Key terms

Interspecific variation The variation *between* species.
Intraspecific variation The variation *within* species.

There is variation both *within* and *between* species. **Interspecific variation** is variation *between* species. This variation is used to identify different species and to classify them, as discussed in Chapter 14. Much more important for the discussion of natural selection and evolution is **intraspecific variation**, which is variation *within* species. This variation is due to genetic differences between individuals and also the effect of the environment, as well as interactions between genes and the environment. This variation is the raw material for natural selection.

Interspecific variation

The great cormorant and the shag are birds from the same genus that live near the sea and dive for their food (Figure 15.3). Shags are the smaller and slimmer of the two species; they also have a thinner bill. In the breeding season both species have a dark, glossy green plumage, but shags develop a single crest on the tops of their heads. The shag's tail has 12 feathers; the cormorant's tail has 14 feathers. Out of the breeding season, it is quite difficult for the non-specialist to tell the two birds apart. While we may have problems, the birds don't. There is no interbreeding between the two species.

Tip

Exam practice question 6 on page 295 has information about how *Phalacrocorax carbo* and another related species, *P. auritus*, co-exist in the same areas of Nova Scotia in Canada.

Figure 15.3 Interspecific variation. Two closely related species of bird: (a) the shag, *Phalacrocorax aristotelis*; and (b) the great cormorant, *Phalacrocorax carbo*.

Figure 15.4 (a) Goose barnacles; and (b) an acorn barnacle – two species with plenty of interspecific variation that makes it easy to tell them apart.

Not only are these two species closely related, but they tend to live in the same habitat in the same coastal areas. No two species occupy the same niche; however, they may occupy the same physical area but feed in different ways and roost in different places so that they avoid direct competition.

Interspecific variation can be much more obvious than this. Figure 15.4 shows two species of barnacle that not only look quite different, but are also found in different habitats.

Visible differences are one way that biologists can measure variation and this includes microorganisms. Although many species of bacteria look very much alike, there are morphological differences between them. For example, *Vibrio cholerae*, the bacterium that causes cholera, has a flagellum whereas *Mycobacterium tuberculosis*, which causes tuberculosis, does not. There are differences in the structure and chemical composition of their cell walls and also in their metabolism. *M. tuberculosis* produces a compound that prevents lysosomes fusing with phagosomes and *V. cholerae* releases the toxin choleragen, which is taken up by intestinal cells and causes them to secrete ions and water into the gut.

There is more about interspecific variation in Chapter 14, as the differences between species are used in their classification and identification.

Intraspecific variation

Phenotypic variation between individuals of the same species is often more difficult for us to detect than variation between different species. Darwin spent a lot of time studying variation in domesticated animals, as he realised that selective breeding results in much greater variation within a species than occurs in the wild. He acquired some different breeds of pigeon and made a detailed study of their biology, and also carried out breeding experiments to investigate the inheritance of characteristics. He even joined societies of pigeon fanciers in London and gained much information by talking to them. The first chapter of *On the Origin of Species* is almost exclusively devoted to pigeons.

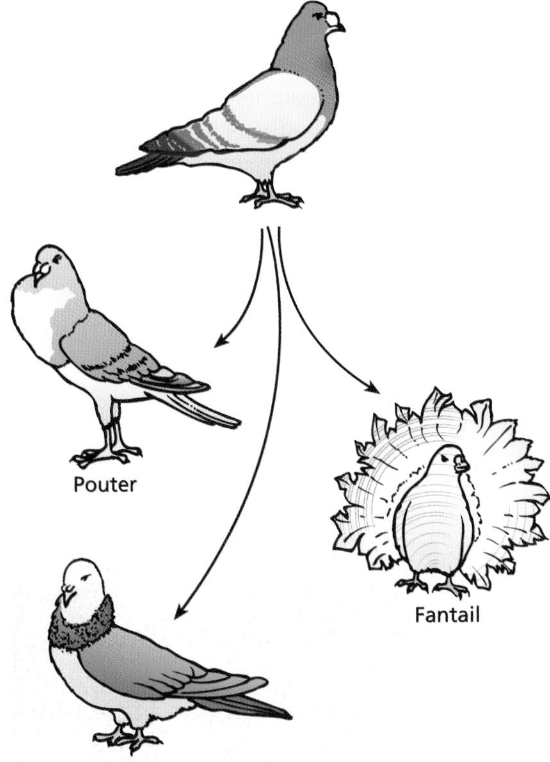

Figure 15.5 Darwin bred pigeons at Down House in order to study variation in domesticated animals.

Whether we look at horses, cattle, sheep, pigs, dogs, cats, mice or rats, we see the same principle at work. We can select and breed different varieties, often 'releasing' variation that is rarely visible in wild populations. An example is the alleles of various genes that give rise to albinism – the lack of any skin pigment. Albinism is rarely a feature that helps an organism to survive in its natural environment (although there are some examples where it is not an impediment to survival). However, albinism has been artificially selected and is common in varieties of animals such as mice, rats and rabbits that are kept as pets.

Morning glory, *Ipomoea purpurea*, is a flowering plant that produces flowers that open in the morning (see Figure 15.6). Figure 15.7 shows the variation in flower colour in this species.

Other types of intraspecific variation are less obvious. The most obvious features that we can see in the yeast cell in Figure 15.8 are the developing bud and the bud scars left when previous buds broke away from this parent cell. The number of these scars is a phenotypic feature partially controlled by the environment. Yeast cells only reproduce by budding when conditions are favourable.

Figure 15.6 Morning glory, *Ipomoea purpurea*, is a climbing plant that originates from Mexico. This is a cultivated variety.

Figure 15.7 This variation in flower colour and appearance in morning glory is genetic – the interactions between three genes are responsible for these different colours and patterns.

Microorganisms such as yeast show much variation in their metabolism. Within the yeast species, *Saccharomyces cerevisiae*, there is variation in the types of enzymes and other proteins involved in respiration. This variation in yeast is of great economic importance as we rely on this species for many industrial processes, such as baking, brewing, wine-making and biofuel production. Some strains of *S. cerevisiae* are better at making ethanol for biofuel; other strains are more suited to making alcoholic drinks such as beer and wine. In a study of over 100 closely related strains of *S. cerevisiae*, researchers identified more than 50 different genes that control the levels of specific metabolites, including genes that encode enzymes and those that control the rate of metabolism by determining the flow of compounds through membranes.

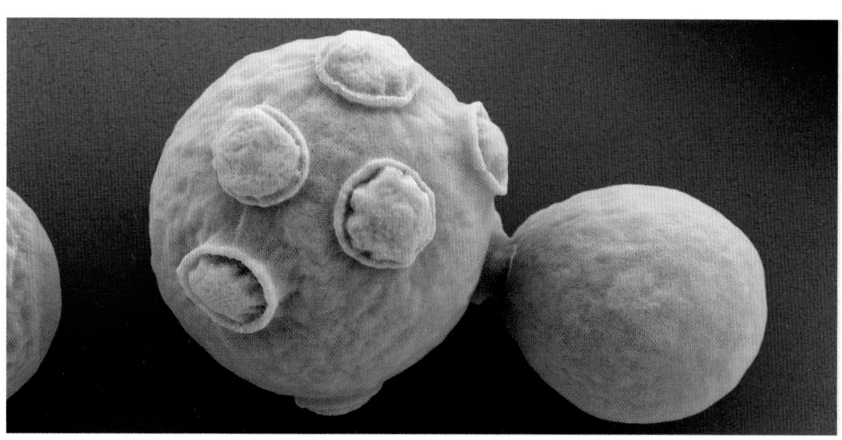

Figure 15.8 A scanning electron micrograph of a yeast cell that is reproducing asexually by budding. There are many bud scars over its surface where earlier buds broke away from their parent. (x 11 860)

Test yourself

1 Describe the features that show variation in the different breeds of pigeon shown in Figure 15.5.
2 Explain why many breeds of domesticated animal would be unlikely to survive if they were all suddenly released into the wild.
3 Flower colour in *Ipomoea purpurea* is controlled by three genes. Use Figure 15.7 to suggest the features of the flowers that these genes control.
4 Some yeast cells have a higher tolerance of alcohol than others. Suggest how you would develop a strain of yeast that is tolerant of high concentrations of alcohol.

Key term

Discontinuous variation Variation in a feature that has discrete categories without any intermediates, e.g. human blood groups (A, B, AB and O).

Continuous variation Variation in a feature that shows a range of phenotypes between two extremes with many intermediates, e.g. leaf length, body mass, height.

Types of variation

If you look at different features of plants, animals and microorganisms, you can distinguish between two forms of variation:

- **discontinuous variation**
- **continuous variation**.

Discontinuous variation

This type of variation is sometimes known as qualitative or discrete variation. There are distinct categories without any intermediates. Examples are:

- the human blood groups (ABO and Rhesus)
- red, pink and white flowers in the snapdragon, *Antirrhinum majus*
- drug-resistant and drug-susceptible forms of *Mycobacterium tuberculosis*.

Most examples of discontinuous variation are controlled solely by genes; the environment usually has no effect on the expression of the genes in the phenotype. Each of the examples given above is controlled by a single gene with only two or a few different alleles.

Tip

Do not assume that because a feature shows discontinuous variation, it must be controlled simply by one gene with two (or more) alleles. The inheritance of many human features that show discontinuous variation, such as tongue rolling, 'hitchhiker's thumb' and attached and unattached earlobes, is more complicated than this.

Human blood groups

Do you know your blood type? It is a useful thing to know, especially if you decide to travel abroad. Blood typing is done easily by placing a drop of blood on each of four sections on an EldonCard™, as you can see in Figure 15.9.

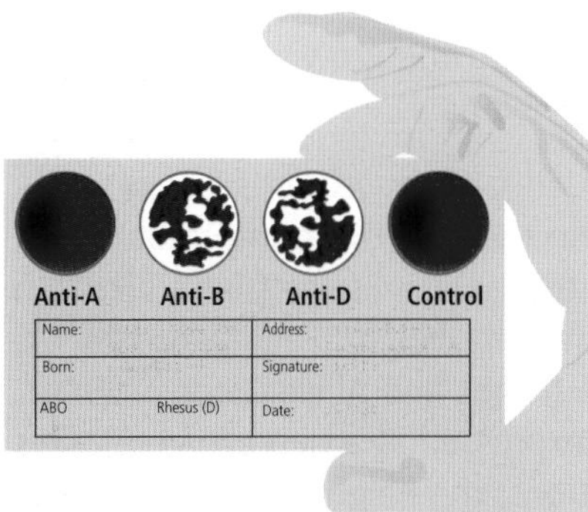

Figure 15.9 An EldonCard™ is used to test people's blood to find out their blood group. Three of the circles contain the antibodies anti-A, anti-B and anti-D that between them can identify the ABO blood group and the Rhesus blood group. The fourth circle is a control to check the test is working properly.

There are two genes involved in controlling human blood groups. A gene on chromosome 9 controls the ABO system, which has the phenotypes A, B, AB and O. The gene codes for an enzyme that adds a sugar molecule to a glycoprotein situated in the cell surface membrane of red blood cells. There are three alleles of this gene: $\mathbf{I^A}$, $\mathbf{I^B}$ and $\mathbf{I^O}$. $\mathbf{I^A}$ and $\mathbf{I^B}$ are co-dominant, and both are dominant over $\mathbf{I^O}$. This means that a person born with the genotype $\mathbf{I^AI^B}$ has the phenotype AB, and someone with the genotype $\mathbf{I^AI^O}$ is blood group A.

The other blood group system is the Rhesus system, which is controlled by a gene on chromosome 1 that has two alleles, **R** and **r**. This gene codes for a red blood cell transmembrane protein. The two phenotypes are Rhesus positive and Rhesus negative. The allele **r** does not code for anything – no protein is produced at all in someone who is homozygous recessive, **rr**.

Activity

Intraspecific variation in blood groups

People routinely have their blood typed; the information is used by doctors, clinics, hospitals and blood banks. Table 15.1 shows the frequency of eight different blood groups in the population of the UK.

Data on examples of discontinuous variation such as this can be presented in different ways – for example, as pie charts or as bar charts.

1 What is the blood group of the person being tested in Figure 15.9? Suggest the person's genotype.
2 Plot a bar chart of the data in Table 15.1.
3 Suggest why a bar chart is the appropriate method of data presentation for this example.
4 Explain how this form of variation is controlled.
5 a) How many chromosomes are there in the nucleus of a human cell?
   b) Human red blood cells do not have nuclei. Explain how a gene can determine a feature that is expressed in these cells.
   c) Explain how it is possible to have every combination of ABO blood group and Rhesus blood group.
   d) Make a table to show the genotypes of each blood group shown in Table 15.1.

Table 15.1 The frequency of ABO and Rhesus blood groups in the UK.

Blood group	Percentage of the population of the UK
A +	35
A –	7
B +	8
B –	2
AB +	3
AB –	1
O +	37
O –	7
Total	100

Continuous variation

This type of variation is sometimes known as quantitative variation. Any feature that can be measured, for example with a ruler or a balance, shows this type of variation. Examples are mass and linear measurements of organisms, such as height of plants, width of leaves and length of tails. For each feature there is a range of measurements between two extremes, with no easily identifiable intermediate groups as in discontinuous variation. Continuous variation is controlled both by genes and by the environment.

The effects of the environment are relatively easy to appreciate. The mass and height of an animal is dependent on the quantity and quality of food that it eats. An animal with many genes for 'large size' will not reach its genetic potential if it is starved during the time when it should be growing. Similarly, a plant that has the genotype for dwarfism is not going to grow into a tall plant, however good the growing conditions and the supply of light, water and mineral ions.

There are usually many genes that influence these features and often these genes have two or more alleles, so the genetic contribution to the variation is often complex. In a simple example, imagine that a plant has four genes, **A**, **B**, **C** and **D**, that control its height. Each gene has two alleles: Allele 1 adds 10 mm to the height and Allele 2 adds 20 mm. The alleles are co-dominant (both are expressed in a heterozygote). This means that, given good growing conditions, the heights of plants with some selected genotypes are as shown in Table 15.2.

Table 15.2 The heights of three plants, each with a different genotype (see text for details).

Genotype	Phenotype: height/mm
$A_1A_1B_1B_1C_1C_1D_1D_1$	80
$A_1A_1B_1B_2C_1C_2D_1D_1$	100
$A_1A_2B_1B_2C_1C_2D_1D_2$	120

Genes that control features in the way shown in Table 15.2 are known as **polygenes**. The feature, in this case height, is described as **polygenic**.

Data on features that show continuous variation is more complex than that for discontinuous variation. Table 15.3 shows the lengths of 50 leaves taken from a plant. The leaves were measured to the nearest millimetre.

Table 15.3 The lengths of 50 leaves from the same species.

Length of leaves/mm				
31	39	43	12	30
26	21	37	29	32
13	22	12	28	35
26	23	27	8	34
33	21	27	31	18
19	36	25	28	24
19	24	27	30	22
17	24	25	31	23
28	25	26	31	40
22	24	25	18	24

This data does not mean much as it is. Data like this can be entered into a spreadsheet and ranked, for example in ascending order. The data needs to be organised and simplified even more to help us see any pattern. Continuous data does not fall into discrete groups as data for features that show discontinuous variation does. To organise the data, it is necessary to make some categories, known as classes.

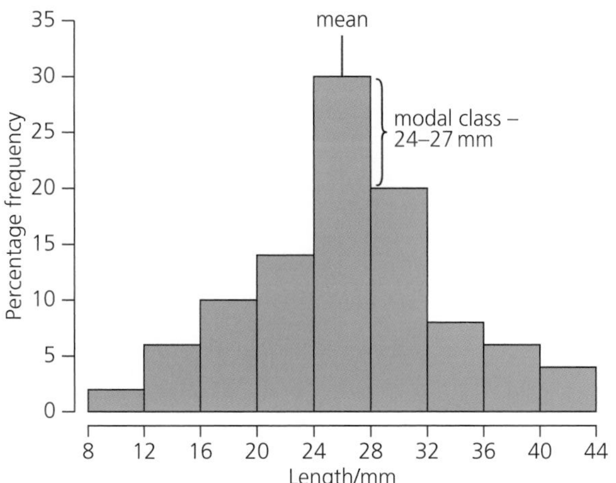

Figure 15.10 Histogram to show the variation in length of leaves.

The data can now be arranged into classes as shown in the tally table, Table 15.4.

Table 15.4 Tally table for the data in Table 15.3.

Class/mm	Tally	Frequency	Percentage frequency
8–11	|	1	2
12–15	|||	3	6
16–19	𝍸	5	10
20–23	𝍸 ||	7	14
24–27	𝍸 𝍸 𝍸	15	30
28–31	𝍸 𝍸	10	20
32–35	||||	4	8
36–39	|||	3	6
40–43	||	2	4
Total	50	50	100

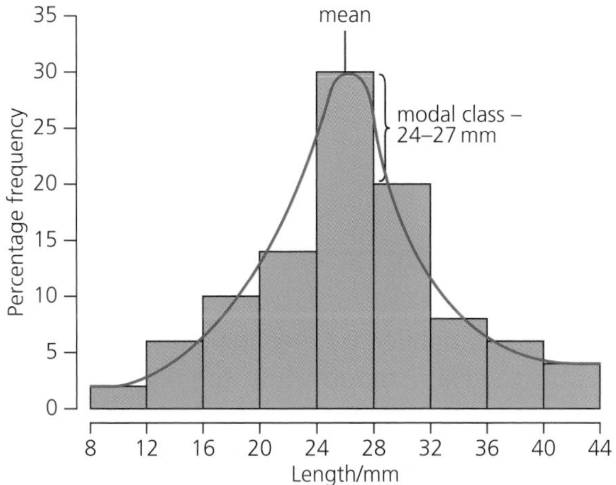

Figure 15.11 Smoothing out the histogram in Figure 15.10.

The information about variation in leaf length is now in a form that can be presented as a frequency histogram. Figure 15.10 shows a frequency histogram of the data in Table 15.4.

The data can be simplified in other ways, for example by finding different measures of 'averageness'. These are: mean, median and mode. The most appropriate 'averages' for the data in Tables 15.3 and 15.4 and Figure 15.10 are mean and modal class, and these are shown on Figure 15.11.

Activity

Measuring blue-rayed limpets

The blue-rayed limpet, *Patina pellucida*, is a marine mollusc (see Figure 15.12). It is found living on kelp, *Laminaria* spp., at low tide on rocky shores. A student investigated variation in shell length of this species and measured to the nearest 0.01 mm using a pair of calipers. The results are recorded in Table 15.5.

Table 15.5 Shell length of 110 blue-rayed limpets found at low tide on a rocky shore.

Lengths of blue-rayed limpets/mm							
5.54	6.59	7.24	9.91	8.81	10.10	5.70	4.01
7.23	9.07	11.20	5.81	6.75	7.00	11.61	7.68
9.11	6.98	4.38	7.82	6.92	6.73	11.00	3.90
7.26	7.11	7.07	9.84	6.64	6.62	4.29	6.76
11.99	9.25	9.23	7.32	7.32	6.29	7.01	8.72
9.62	11.18	5.73	9.12	7.57	4.64	11.04	6.52
6.97	5.24	11.89	6.64	8.04	7.39	5.41	5.72
6.66	9.87	14.81	5.30	12.12	7.16	10.95	8.72
9.66	4.95	5.84	5.40	7.85	9.77	11.67	8.79
5.27	6.60	7.67	10.33	8.45	4.45	5.82	6.03
8.45	9.57	12.04	5.02	8.12	6.65	5.36	8.49
10.12	8.08	4.75	9.70	6.55	12.39	6.21	7.98
5.81	10.42	4.39	10.25	6.15	13.20	6.10	6.65
9.16	10.71	5.43	9.53	6.31	13.80		

Figure 15.12 Blue-rayed limpets, *Patella pellucida*.

1 Organise the data shown in Table 15.5 so that it can be plotted as a histogram.
2 Draw a histogram to show the variation in shell length in *P. pellucida*.
3 **a)** Use your histogram to describe the variation in shell length in *P. pellucida*.
  b) Suggest an explanation for the variation in this population of *P. pellucida*.
4 The standard deviation for the data in Table 15.5 is 2.313 mm. What is the advantage of calculating the standard deviation?

Frequency histograms are often simplified by drawing a line through the centre of each of the blocks. This 'smooths out' the data. Smoothing out the data for the leaves shows that the variation is a **normal distribution** – the line is symmetrical with most of the leaves in the middle of the range and the mean within the modal class. This shows that 50% of the leaves are smaller than the mean and 50% are larger than the mean. Distributions do not have to be like this. They can be skewed, with more individuals at one extreme of the range or the other. The distribution can also be bimodal, with most individuals towards either end of the range and very few in the middle of the range.

The standard deviation (SD) is calculated to show how widely the data is dispersed. No matter what the data, 68% of results are within 1 SD of the mean and 95% are within 2 SD of the mean. In our example, the SD is 4.3 mm.

Test yourself

5 Explain how continuous variation is different from discontinuous variation.
6 Suggest how one gene with two alleles controls flower colour in snapdragons (*Antirrhinum*).
7 Use the information in Table 15.2 to state the height of a plant with the genotype $A_1A_2B_1B_1C_2C_2D_1D_1$.
8 Why do the plants have two alleles of each gene, A, B, C and D?
9 When plants with identical genotypes are grown in gardens they do not all have the same height. Explain why this is so.
10 What can be learnt by growing plants with the same genotype in different environments?
11 What can be learnt by growing plants with different genotypes under exactly the same environmental conditions?

Activity

Geographical variation in *Achillea lanulosa*

A long-term project entitled *Experimental Studies on the Nature of Species* (ESNOS) studied variation in plant populations from the 1930s for about 20 years. One of the best-known studies was on variation in yarrow, *Achillea lanulosa* (Figure 15.13). Researchers, led by Jens Clausen, examined this species along a transect that ran from the Pacific coast of California, where the climate is very mild, into the Sierra Nevada mountain range to altitudes of over 3000 metres. Climatic factors such as rainfall and temperature vary greatly over short distances across these mountains. The researchers measured the heights of plants in the populations of *A. lanulosa* at sampling stations along the transect.

Figure 15.13 Yarrow, *Achillea lanulosa*, which has successfully colonised nearly every climatic region in the northern hemisphere.

Figure 15.14 shows that these plants exhibited a characteristic pattern of variation in the different environments, with taller plants in the San Joaquin Valley and shorter plants at high altitude.

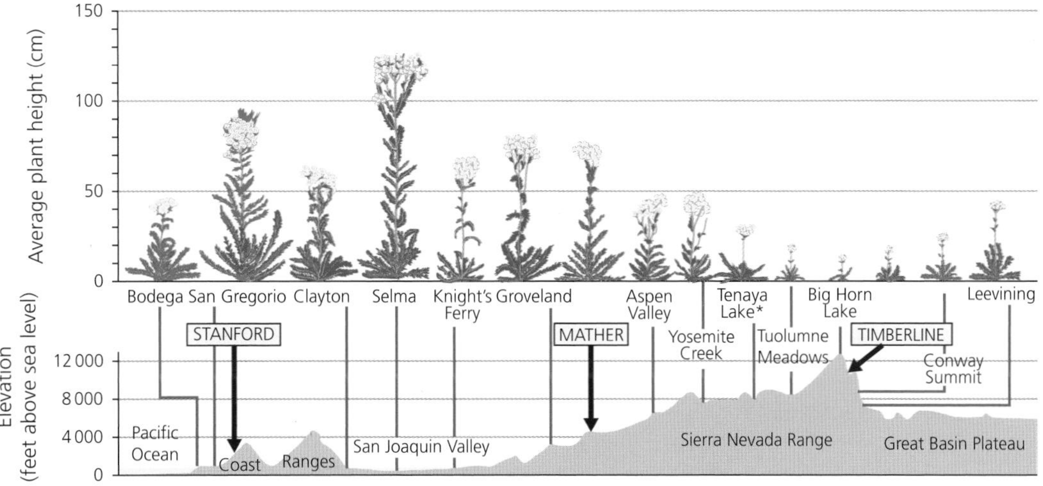

Figure 15.14 Intraspecific variation in populations of *Achillea lanulosa* at different altitudes across California. The mean height of the plants at each sampling station is shown.

The researchers took cuttings from the plants from six of the populations shown in Figure 15.14 and grew them to maturity under the *same* experimental conditions in garden plots at Stanford in California, which is just above sea level.

They made sure that the plants all grew in identical environmental conditions. The researchers measured the heights of the plants that grew from the cuttings. Figure 15.15 shows the variation in the heights of the plants in each group as histograms; the drawings of the plants represent the mean height of the plants in these transplant experiments.

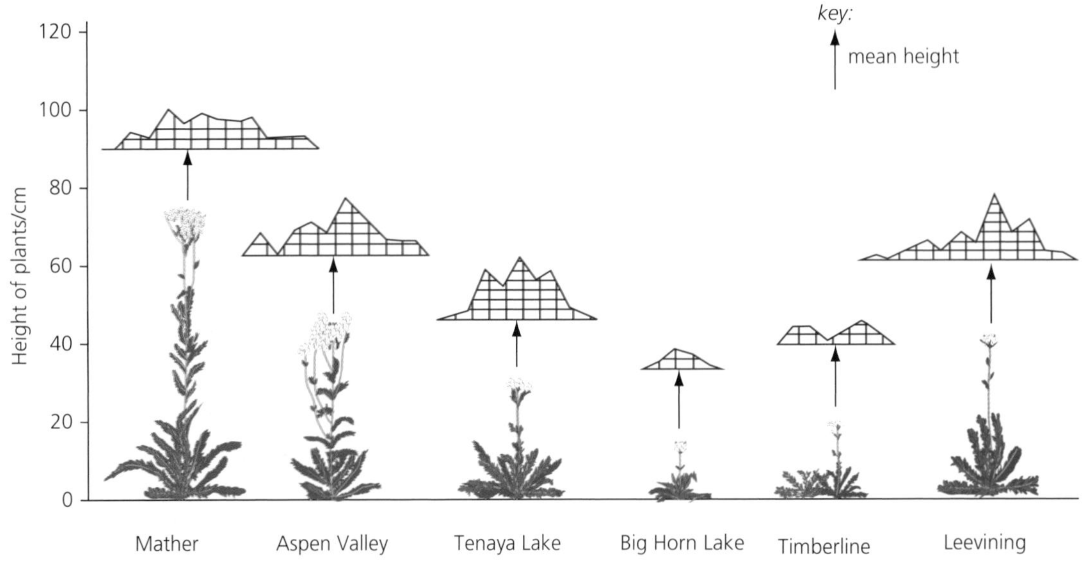

Figure 15.15 The results of growing plants collected from six populations of *Achillea lanulosa* from stations across the Sierra Nevada Mountains in garden plots at Stanford.

Plants from five of the populations were also transplanted and grown in garden plots at Timberline and Mather. The results are shown in Figure 15.16.

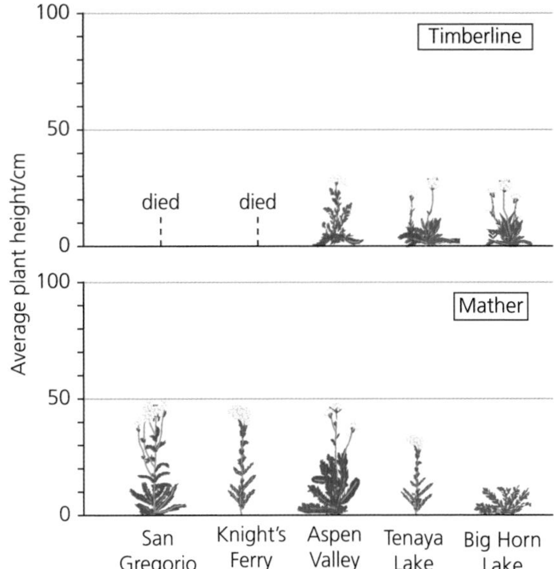

Figure 15.16 The mean heights of plants of *A. lanulosa* grown in gardens at Mather and Timberline (see Figure 15.14 for their original locations).

1 These populations of *A. lanulosa* in California all belong to the same species. Explain why.
2 Suggest environmental factors that influence the height of plants. Explain how each of the factors you suggest influences the height of plants.
3 Explain why the researchers grew plants from the different populations in the same garden plots at Stanford.
4 Genes also influence plant height. Suggest how this may happen.
5 Describe the results shown in Figures 15.15 and 15.16.
6 What can you conclude from these results?
7 What does variation in *A. lanulosa* tell us about how organisms might evolve?
8 *A. lanulosa* has successfully invaded almost all climatic regions in the northern hemisphere. Explain what these experiments reveal about this species that has allowed it to do this.

Causes of variation

Features that show discontinuous variation are controlled mainly or entirely by genes; the environment has little or no effect. Your blood group, for example, is not something that can change according to the environment in which you grow up or the food you eat. It is determined by the genes you inherit at conception and there is nothing that can be done to change it.

Tip

The results from the long-term studies of *A.lanulosa* of Clausen, Keck and Hiesey (see Activity on pages 281–2) show that there are both genetic and environmental factors involved in intraspecific variation.

Features that show continuous variation are influenced both by genes and the environment. As a further example, consider the factors that determine the body mass of mice: availability of food, environmental temperature and quantity of stored fat. There are also genes that influence body mass; an allele of one of these genes gives rise to obese mice (Figure 15.17). The gene concerned codes for a hormone, leptin, that is produced by fat tissue as it is stored. It helps to regulate food intake by controlling appetite. The brain responds to an increase in the concentration of leptin in the blood by reducing the appetite. The obesity allele is the result of a mutation and is recessive – it does not produce any leptin. A mouse that does not have at least one copy of the dominant allele never stops eating and puts on excessive quantities of fat.

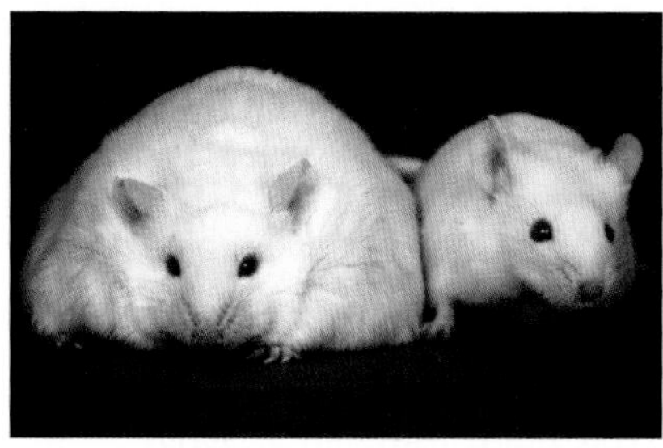

Figure 15.17 An obese mouse and a normal mouse. The obese mouse has a mutation on chromosome 7, which has increased fat deposition in its body. Mice with the mutation grow 35% to 50% fatter by middle age than normal mice, even when they are given a low-fat diet.

There are some aspects of variation that are caused only by the environment. Damage that is inflicted on an organism might remain as a scar, for example where a branch has broken from a tree or an animal has been wounded in a fight. Sperm whales often carry scars inflicted by giant squid.

In the latter part of the twentieth century, a variety of methods used to study species became available. For example, it is now possible to isolate substances from blood and tissue samples and separate them using the technique of electrophoresis, as explained in Chapter 12. This revealed even more examples of variation between individuals and extended what we understand by the term 'phenotype' to include variation at the biochemical level.

Many genes have two or more alleles. Allozymes are enzyme variants that are coded by different alleles at the same gene (see Chapter 12). Cells may contain two slightly different versions of the same enzyme because they are heterozygous for the gene concerned. In addition to this form of genetic variation, there may be several very similar genes that code for the same type of enzyme. These isozymes all catalyse the same reaction but are coded for by different genes. They may be able to function at different temperatures, so giving an organism the ability to acclimatise when it migrates from one area to another or when the seasons change. This could well be an example of biochemical adaptation to the environment or it may be just an example of the variation that exists within species – the variation that is the raw material for selection.

Adaptation

Key term

Adaptation The way a structure, a function or an aspect of behaviour helps an organism to survive in its environment or perform an activity, such as feeding or evading predators.

For an organism to exist successfully in an environment, it must possess features that help it to survive. Any modification of a structure, function or aspect of behaviour that helps an organism survive in its particular habitat is an **adaptation**. This involves all aspects of an organism – its external appearance (morphology), internal structure (anatomy), the function of its body systems (physiology), the chemistry of its cells (biochemistry) and its behaviour, reproduction and life cycle.

Figure 15.18 Southern blue gum, *Eucalyptus globulus*, is endemic to Australia and grows in hot, dry environments.

Figure 15.19 Fennec foxes live in the deserts of North Africa.

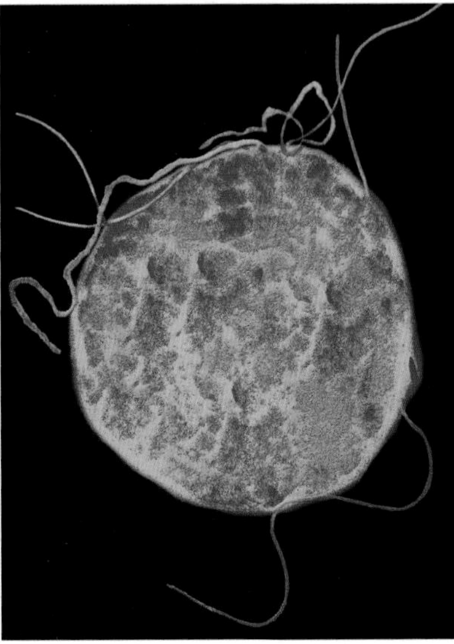

Figure 15.20 A scanning electron micrograph of *Sulfolobus acidocaldarius*. This thermophilic archaean lives in volcanic springs with temperatures of about 80 °C, a pH of about 3 and a high concentration of sulfur. These are conditions that would instantly kill almost all other organisms except other thermophilic prokaryotes. (x85 000)

Some of the best places to find structural, behavioural and physiological adaptations are in extreme environments. Eucalyptus trees, fennec foxes and the thermophilic archaean *Sulfolobus acidocaldarius* all survive in habitats with high temperatures.

Southern blue gum, *Eucalyptus globulus*, adaptations:

- **Structural:** The leaves hang vertically – this reduces the amount of light that the tree is exposed to, thus reducing transpiration. Thick bark protects the tree from fire.
- **Behavioural:** The eucalyptus tree releases seeds after there has been a fire. This gives the tree seedlings less competition from other plants that cannot survive after fire. The young trees therefore have a better chance of survival.
- **Physiological:** The leaves produce toxic compounds to deter grazing animals. However, not all animals are deterred. The koala, *Phascolarctos cinereus*, and the greater glider, *Petauroides volans*, have adaptations that protect them from being poisoned by these compounds.

Fennec fox, *Vulpes zerda*, adaptations:

- **Structural:** The fox has large ears and eyes to give it good hearing and vision to help catch prey. Large ears also help to lose heat during the day. Its thick fur retains heat when hunting during the cold nights.
- **Behavioural:** The fox is nocturnal, remaining underground in a burrow during the heat of the day. This also helps it avoid being caught by the eagles that are its main predator.
- **Physiological:** The fox's kidneys reabsorb much of the water in its urine so it is very concentrated. This helps the animal avoid dehydration. Fennec foxes can survive without drinking water.

***Sulfolobus acidocaldarius*, a thermophilic archaean, adaptations:**

- **Structural:** It has cell membrane lipids that have ether linkages rather than ester linkages; ether linkages are stronger than ester linkages. It also has DNA with a high proportion of C–G base pairs and an enzyme that supercoils DNA to make it very compact to withstand the effect of heat. (At temperatures near boiling, DNA is denatured as the hydrogen bonds between the two polynucleotides break apart.) Its proteins have a large number of polar amino acids to form hydrogen bonds and ionic bonds to stabilise the tertiary structure and reduce chances of denaturation.
- **Behavioural:** The formation of a matrix that sticks the bacteria into biofilms helps them to withstand high temperatures.
- **Physiological:** When the temperature rises the bacteria produce heat shock proteins to protect the cell against damage. A heat-resistant DNA polymerase enzyme allows replication at temperatures near 100 °C.

Convergent evolution

Different parts of the world provide very similar niches. Sometimes the animals and plants that fill those niches are closely related species and share a common ancestor which had much the same adaptations.

The mammals that migrated to Australasia about 50 million years ago were marsupials and these evolved to fill many niches including the subterranean. The two species of marsupial mole live in the deserts of western and northern Australia. Just like the European mole, which is a placental mammal, they burrow through soil and very rarely emerge into daylight.

Some of the adaptations that they share are:

- short and powerful limbs with huge front claws for digging
- no external ears: the openings into the ear canals are just under the fur
- limited eyesight: the marsupial mole has tiny, non-functioning eyes and the placental mole has small eyes with vision limited to the detection of light
- similar fur: the marsupial mole's fur is described as 'silky' and that of the placental mole as 'velvety'.

The similarities between these species, and the golden moles from southern Africa, might make one think they are all descended from a common mole-like ancestor. This is not the case; the common ancestor of these animals was nothing like a mole. These features have evolved independently in different taxonomic groups as adaptations to similar ways of life exploiting the opportunities offered by living underground.

These species are not entirely alike; the marsupial mole has a hard area on the front of its head, while the European mole has a thin snout. The marsupial mole has a feature unique to mammals: its neck vertebrae are fused together which is an adaptation for digging that has not evolved in the other species of mole.

Test yourself

12 Explain the difference between genetic and environmental variation.
13 Suggest how fennec foxes can survive in a desert habitat without drinking water.
14 Explain the advantage of having DNA with a high proportion of C:G base pairs rather than A:T base pairs.
15 Explain what is meant by the term *adaptive feature*.
16 Suggest the advantages that are gained by moles of living underground.

Now that we have finished exploring variation, it is time to return to Darwin and natural selection.

Darwin's observations

In developing his theory of natural selection, Darwin made four observations:

1. All organisms reproduce to give far more offspring than are ever going to survive. He calculated that a pair of elephants could leave 19 million descendants after about 740–750 years.
2. Populations of organisms fluctuate, but they do not tend to increase and decrease significantly over time – their numbers remain fairly constant.
3. There is variation among individuals in many of their characteristics. This is what we have called intraspecific variation. This variation arises in each generation when organisms reproduce sexually to produce offspring.
4. Offspring resemble their parents. Features are transmitted from one generation to the next.

Tip

Meiosis is one of the causes of variation. Look back to Chapter 6 for details of crossing over and independent assortment.

Overproduction:
left unchecked the numbers of organisms in a population would increase exponentially
However, the environment provides limited resources, e.g. food, water and space, and has predators and diseases

Intraspecific competition:
competition in a population is at its most intense between individuals of the same species, as they occupy the same niche

Individuals that are successful in competition for resources **survive and reproduce** to leave more offspring than those that are less successful
Population now contains a higher proportion of individuals that are better adapted to the environmental conditions than the others

Figure 15.21 The theory of natural selection.

Darwin explored the consequences of these observations and made the following inferences:

1 In the words of the full title of *On the Origin of Species*, there is a 'struggle for life'. In other words, there is competition between organisms in the same population for limited resources.

2 The organisms that are successful in competing for resources have features that mean they are adapted to their environment. They have a higher chance of surviving than individuals that are not so well adapted. Those that are successful in competition live long enough to reproduce.

3 Organisms that reproduce pass on inherited characteristics to the next generation. This next generation contains a higher proportion of offspring from the better adapted parents than from those that are less well adapted.

4 This unequal ability of individuals to survive and reproduce leads to a gradual change in a population, with certain adaptations increasing in frequency over the generations.

Tip

As we will see, selection does not always lead to changes in frequency. Most of the time the environment does not change and selection acts to stabilise the population from generation to generation.

Competition

There are finite quantities of resources for organisms. Individuals of the same species need the same resources and have similar ways of gaining them. If there are more organisms than the environment can support, many will die, so that populations remain fairly constant from generation to generation. Some of the environmental factors that control the sizes of populations of heterotrophic organisms are food, water, disease and predation. Populations of autotrophs are controlled by disease and grazing, but also by access to light, carbon dioxide, water and mineral ions. Poorly adapted organisms have a lower chance of surviving and a higher risk of dying before they have a chance to reproduce.

Darwin's great idea cannot easily be summarised in a snappy phrase. It is often reduced to 'survival of the fittest', which in some people's minds reads as 'survival of the strongest'. This is not so; think of wily predators that not only outsmart their prey, but are better at it than others of their species. In biology, fitness is a term with a specific meaning: it means that organisms survive to adulthood and reproduce. By this definition the fittest *must* be the survivors, so the phrase is meaningless. However, the big idea behind it is not.

Key term

Natural selection The effect of environmental factors, such as competition, predation and disease, on a population such that those individuals that are best adapted to the conditions at the time have a higher chance of surviving, reproducing and passing on their advantageous alleles. Natural selection is the main mechanism that brings about evolutionary change.

Tip

Remember that 'survival of the fittest' is like a catchphrase. It is far more important to explain the principles of natural selection as outlined here than to use this phrase on its own in an answer.

Natural selection

The organisms that are better adapted to gain resources, avoid catching lethal diseases and escape being eaten are likely to survive long enough to find a mate and pass on their alleles to future generations. These organisms are better adapted than others to the conditions prevailing at the time. They are selected by the environment.

When the environment is stable, **natural selection** acts to maintain the features of a species, but if the environment changes then selection pressures in the environment also change. This may mean that organisms with features that were previously disadvantageous are now the ones that compete well, survive and breed. The generation of variation within each generation is necessary for this to happen and for species to adapt to changing conditions. This shows that variation, adaptation and selection are important components of evolution.

In a changing environment the individuals within a population that are best adapted to the new conditions have a higher chance of survival. Often they are at one extreme of the natural range of variation. For example, in cold conditions it is larger mice with a smaller surface area to volume ratio that survive, although they are not as large as the obese mouse in Figure 15.17!

Example

The oldfield mouse, *Peromyscus polionotus*, lives in old, overgrown agricultural fields in the USA. These mice are dark brown in colour. Along the coasts of Florida and Alabama there are populations of this species that have colonised the light-coloured sand dunes and barrier islands. These mice are known as beach mice and they all have much lighter-coloured fur than the populations of oldfield mice that live inland.

Beach mice feed mainly at night on seeds. They also dig extensive burrows. Their main predators are owls, foxes and domestic cats. Predation experiments showed that the most conspicuously coloured mice were captured most often by owls, which are night-hunting predators.

1 Explain how beach mice are most likely to have evolved from oldfield mice.
2 The different beach mice along the coasts of Florida and Alabama are classified as subspecies of *P. polionotus*.
   a) Explain why they are classified as part of the same species as oldfield mice.
   b) Suggest how beach mice could evolve into a different species.

Tip

There are many different ways to ask questions about natural selection. First, recognise that you are being asked about natural selection and then apply the points about overproduction, variation, competition, predation (as in the example), disease, survival of those best adapted at the time, reproduction and the inheritance of alleles by the next generation.

Answers

1 Among the oldfield mice there will be variation in coat colour. Mice that colonise the sand dunes and beaches are susceptible to predation. Night-hunting owls are more likely to catch the darker-coloured mice, as they will show up against the light-coloured sand. Mice with lighter-coloured coats are better adapted to the conditions on the beaches, as they have better camouflage than mice with darker coats. Lighter-coloured mice are therefore more likely to survive, reproduce and pass on their alleles to the next generation. This means that, over time, the alleles for white, yellow or light-brown coat colour become more frequent in the population. With time the population of mice living on the sand dunes will be a different colour from the inland mice.

2 a) Beach mice breed with oldfield mice to give fertile offspring. They share many features in common. The DNA of both types is very similar.

b) Beach mice could evolve into a different species if they become reproductively isolated from the oldfield mice. This could happen if there is a geographical barrier between the areas that the beach mice have colonised and the inland habitats of the oldfield mice. Within the beach mice population mutations would then occur. Some of these would be beneficial and improve the way in which the mice are adapted to their habitat. These mutations would become 'fixed' as a result of natural selection. Over time, more differences would accumulate so that eventually the beach mice and the oldfield mice could not interbreed.

Test yourself

17 Sketch a graph to show how a population increases if there are no limiting factors.

18 What factors help a young mammal to survive?

19 What factors help a young tree seedling to survive?

20 A pair of small birds may have 15 young each year. Explain why so few of them survive.

21 Explain what *selective advantage* means.

22 What is wrong with 'survival of the fittest' as a strapline or slogan for the theory of evolution by natural selection?

Evidence for evolution

Tip

The general theory of evolution refers to the idea that life has evolved over time. Natural selection is the mechanism by which evolution has occurred. See page 291.

There are many lines of evidence that support the general theory that evolution has occurred. Some of these are:

- comparative morphology – the similarities in outward appearance of organisms; for example, all birds have beaks
- comparative anatomy – for example, the same basic pattern of bones in the limbs of all tetrapods (amphibians, reptiles, birds and mammals)

Figure 15.22 Fossilised fern, *Sphenopteris laurenti*. This specimen dates from the Middle Coal Measures of the Carboniferous period about 300 million years ago. The frond is 90 millimetres in length.

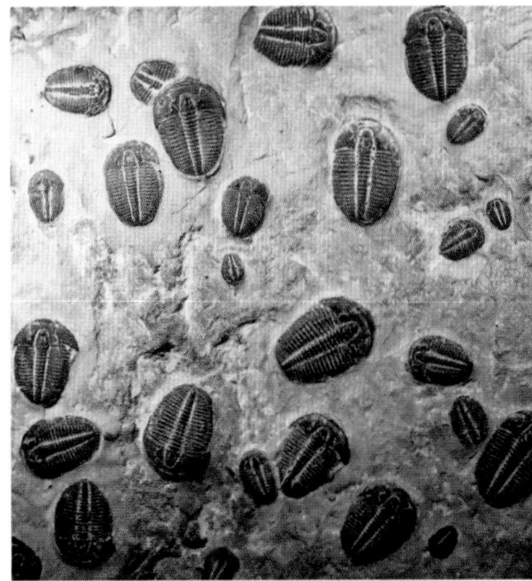

Figure 15.23 Trilobites – animals that are only known about from fossils. They thrived for 300 million years but became extinct about 240 million years ago.

- fossilised remains of organisms and other fossilised traces (such as footprints, burrows and organic chemicals made by organisms) that are discovered in rocks
- comparative biochemistry – many biochemicals are found in all organisms, which suggests that they have a common origin; similarities and differences between proteins and DNA provide evidence for the evolutionary relationships between organisms
- classification reflects the phylogeny of groups of organisms, as discussed in Chapter 14.

There are also examples of selection in action that provide evidence for this as the mechanism for evolution (see pages 291–4).

Evidence from fossils

A fossil is the mineralised or otherwise preserved remains of an animal, plant or microorganism. Fossils may also consist of other traces of organisms, such as footprints, burrows and faeces (coprolites). Fossils are found in sedimentary rocks and chemical traces of fossils have also been detected in metamorphic rocks. The oldest fossils are those of prokaryotes, which have been found in rocks that are 3.5 billion years old. Chemical traces of prokaryotes have been found in rocks even older than this, indicating that life began as long ago as 3.9 billion years.

In the Grand Canyon in Arizona, the Colorado River has cut a deep gorge through layers of rock. At the base there are fossils of prokaryotes that are 1250 million years old. Near the top there are fossils of more recent origin, including coral and molluscs, that are 250 million years old. In the middle there are fossils of reptiles, amphibians and terrestrial plants. Fossils tell us that environments and organisms have changed over millions of years. Climbing up the Grand Canyon is like walking a timeline of life on Earth, from its simplest beginnings to the diversity of more complex forms that we have today. The rocks contain a record of the changes that have occurred over the long expanse of geological time.

There are a variety of ways in which rocks can be dated, such as by using different methods of chemical analysis and also by using fossils. By looking at the alignment of layers, or strata, of rocks, we can put fossil organisms into a sequence from oldest (lowest stratum) to youngest. This allows scientists to follow the evolution of different groups, by arranging the fossils in a sequence that reveals the changes that have happened in different lines of descent. For example, it may be possible to see that a species known only from fossils that are millions of years old has similarities with many species in existence today. It is hard to prove that this fossil species was the ancestor of present-day species, but it provides evidence for what that ancestral species would have looked like. The fossil evidence is not complete and never will be, because it is unlikely that every species has been fossilised, especially those without hard parts.

Throughout Earth's history there have been extinction 'events' that have nearly wiped out most organisms – there have been seven great extinctions in the history of the Earth. But life has always recovered and many new species have evolved each time. It is thought that humans are responsible for the eighth great extinction event that is happening now. Scientists have estimated that over 99% of the species that have lived on Earth are now extinct.

Studying fossils provides evidence for the gradual change from simple life forms (Archaea and Bacteria) to more complex life forms (Eukarya) and the changes that have occurred within different groups. However, not all change is 'simple' to 'complex', as some species have lost structures over time. You can see an example of this if you compare parasitic worms with their free-living relatives. Structures such as sensory organs are redundant when living inside a host animal, so worms with reduced sense organs were the ones to survive as successful parasites.

Figure 15.24 Paleontologists from the Smithsonian and the Universidad de Chile study one of the most complete fossil baleen whale skeletons from Cerro Ballena, a rich site of fossil marine mammals in the Atacama region of Chile.

Biologists in the nineteenth and early twentieth centuries compared the morphology and anatomy of species in order to show their evolutionary relationships. For example, the bones in the front limbs of all tetrapods (amphibians, reptiles, birds and mammals) have basically the same pattern even though there are differences between them. This indicates that these animals had a common origin. However, over the past 60 years or so a much wider range of techniques has become available to take this study of similarities and differences among organisms much further.

Evidence from biochemistry

The biochemistry that you have studied (see Chapters 2 and 3) provides some evidence that all life has a common origin. Here are some pieces of evidence:

- The atoms in amino acids can be arranged to give two molecules that are mirror images of each other – so-called left-handed and right-handed molecules. All amino acids are left-handed; no right-handed amino acids exist in nature.
- There are many possible different types of amino acids, but only 20 are used to make proteins – they are the same 20 in all organisms.
- The molecule of inheritance in all cellular organisms (prokaryotes and eukaryotes) is DNA.
- The genetic code that specifies an organism's amino acids is basically the same in all organisms.
- ATP is the universal currency for energy in the cells of all organisms.

Analysis of the amino acid sequences of proteins reveals that proteins from closely related organisms are very similar. The active site of an enzyme like catalase tends to be identical whatever organism it comes from, as no other arrangement of amino acids gives the right 3D shape to fit the specific substrate. But there are unlikely to be such constraints on other parts of the molecule, and looking at enzymes from different organisms reveals differences that become greater the less related they are (see Chapter 14).

Some enzymes perform very basic functions found throughout all life forms, such as DNA polymerase, the enzyme that repairs and copies DNA. As expected, DNA polymerase has only relatively small differences in its amino acid sequence between phyla and even between kingdoms. Significant changes in this enzyme reflect significant events in the evolutionary history of organisms.

The primary structure of proteins is determined by the sequence of bases in DNA. Data obtained from sequencing the nucleotides in DNA shows the relationship between different species. Nucleotide sequences in the genes of closely related species are found to be very similar. Similarities and differences can be used to group species, and then deciding the extent of the differences gives an idea of when speciation occurred, as described in more detail in Chapter 14.

Natural selection in action

Evolution is sometimes assumed to happen very slowly, partly because of the huge periods of time over which it has happened throughout history. However, it can actually happen relatively quickly. We have already seen one example of the evolution of a new species, *Spartina anglica*. The development of drug resistance in bacteria and insecticide resistance in insect pests are examples of rapid change in populations brought about by natural selection.

Drug resistance

As we have seen in Chapter 11, antibiotics are the main group of drugs used by doctors and vets to treat bacterial diseases. Antibiotics became widespread in the late 1940s. They proved hugely successful in treating diseases such as TB. However, soon after their introduction, some antibiotics became less effective as bacteria developed antibiotic resistance. This happened because a few bacteria naturally possessed genes that helped prevent the effect of the antibiotic. For example, penicillin is effective because it prevents the growth of cell walls in some bacteria. Resistant bacteria have enzymes that can break down penicillin. When antibiotics are used, any resistant bacteria are clearly at an advantage, as they are adapted to the new conditions. The bacteria that are susceptible to the antibiotic die and the resistant bacteria survive and reproduce to pass on their genes to future generations.

Tip

In any discussion of evolution, whether it is the effect of natural selection on a species as described in this section, or whether it is bigger changes such as speciation, remember that variation, competition, adaptation and selection are the major themes to write about.

There are two ways in which a bacterium gains resistance to an antibiotic:

- a gene on the chromosome may spontaneously mutate to give a form that codes for a polypeptide that is not affected by the antibiotic
- the bacterium may gain a plasmid with a gene for resistance from another bacterium, even from an individual of a different species.

Antibiotic resistance works in a number of ways. Genes may code for enzymes that break down the antibiotic, as mentioned earlier – penicillinase is an enzyme that breaks down the antibiotic penicillin. Other resistance genes code for membrane proteins that pump out any antibiotic molecules that enter the cell. Many antibiotics work by inhibiting the action of enzymes and other proteins involved in replication, transcription and translation. Mutations may change the structure of these proteins so that there is nowhere for antibiotics to bind, protecting the bacteria from the effect of the antibiotic.

Bacteria have only one copy of each gene since they only have a single loop of double-stranded DNA. They are essentially haploid. This means that a mutant gene for antibiotic resistance will have an immediate effect on any bacterium possessing it. These individuals have a tremendous selective advantage in an environment that contains an antibiotic. Bacteria without this mutant gene will be killed, while those resistant to the antibiotic survive and reproduce. Bacteria reproduce asexually by binary fission; the DNA in the bacterial chromosome is replicated and the cell divides into two, with each daughter cell receiving a copy of the chromosome. This happens very rapidly in ideal conditions, and even if there was initially only one resistant bacterium, it might produce 10 000 million descendants within 24 hours.

When someone takes an antibiotic to treat a bacterial infection, bacteria that are susceptible to that antibiotic will die and the few that are not susceptible will be eliminated by the immune system. In most cases, if the dose is followed correctly, eventually the entire population of the disease-causing bacteria will be killed. However, if the dose is not followed, perhaps because people stop taking the antibiotic when they start to feel better, then some less susceptible bacteria survive. The next time there is an infection of this strain of bacteria, the antibiotic may not be effective.

The consequences of antibiotic resistance are serious and of great concern to medical authorities. It appears that we are running out of options as far as using antibiotics to treat disease is concerned. Antibiotics must not be overused if we are to avoid resistance. Some are kept to be used as a 'last resort' when all others have failed. But despite this, there are now strains of bacteria that are untreatable using antibiotics.

There are various steps that can be taken to reduce cases of antibiotic resistance:

- only prescribe antibiotics when absolutely necessary and do not prescribe them for non-bacterial diseases or 'just in case'
- ensure people finish their course of antibiotics
- rotate antibiotics so that one type is not used continuously in the treatment of specific diseases
- keep some antibiotics to use as a 'last resort'
- invest in research to find new antibiotics.

The future for antibiotics looks grim. No new class of antibiotic has been discovered since the 1980s. Bacteria such as *Mycobacterium* are becoming resistant to many antibiotics and other drugs. Medical experts have warned that there may soon be no antibiotics available to treat some diseases.

Tip

Researchers announced in January 2015 that they may have discovered a new class of antibiotic. *Teixobactin* was isolated from soil bacteria that previously could not be cultured in the lab. It acts against *Mycobacterium* and *Staphylococcus aureus* without any 'detectable resistance'.

Tip

There are various types of pesticide classified according to the type of organism that they are used against – insecticides for insect pests, fungicides for fungal pathogens, molluscicides for slugs and snails, rodenticides (such as warfarin) for rodents such as rats, and herbicides for weeds.

Key term

Selective agent Any factor in the environment of an organism that influences the survival of that organism and so brings about natural selection.

Pesticide resistance in insects

Pesticides are chemicals that kill any sort of pest, pathogenic organism or weed. We use them to control insects – both pests of crops and vectors of disease (for example, *Anopheles,* which transmits malaria and tsetse flies, which transmit sleeping sickness).

When crops are grown over large areas of land as a monoculture, food is no longer a limiting factor for insect popluation growth. For example, a field of potatoes provides huge quantities of food for Colorado potato beetles. If other conditions, such as temperature, are favourable, the number of these pests increases exponentially and they cause massive damage to crop plants, reducing yields significantly. Farmers spray insecticides to control the populations of insect pests. Any insect that is resistant to the pesticide will survive and reproduce, as there will be less competition (Figure 15.25).

In these examples, antibiotics and insecticides are acting as **selective agents**.

Pesticide resistance poses the same problems as antibiotic resistance. If we cannot control insects and other pests then food supplies are not secure for the future. The evolution of resistant strains means that we cannot carry on using the same chemical controls indefinitely. We have to use them sparingly or rotate them so they are not used all the time.

Figure 15.25 The selection of insecticide resistance in Colorado potato beetles.

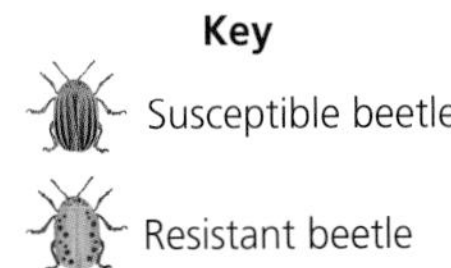

In this population of Colorado beetles there are a few that are homozygous for an allele that gives them resistance to an insecticide.

The population has increased so much that farmers need to spray insecticide to control the beetles.

The insecticide kills most of the susceptible beetles, but resistant beetles are unaffected.

The resistant beetles have little competition so survive to reproduce and pass on the allele for resistance to their offspring. Spraying the insecticide again increases the chances that resistant beetles will survive and the allele for resistance increases in the population. Eventually the population may be composed entirely of resistant beetles.

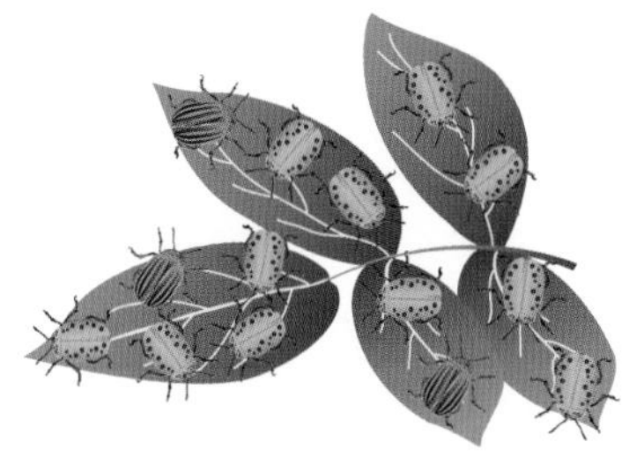

Pesticide manufacturers may advise that no more than a specified number of applications of a pesticide are used. After that the farmer or grower should switch to another type of pesticide for a while. This is intended to extend the useful life of pesticides. Two or more pesticides that kill insects in different ways may be mixed together before spraying. Using a combination of pesticides delays the time it takes for resistance to emerge and spread. Farmers are also encouraged not to apply insecticides as an insurance policy to prevent any likely invasion by pests. Instead they are advised to apply insecticides only when they know their crops are at risk.

Test yourself

23 Criticise this statement: 'Using pesticides causes pesticide resistance in insects'.

24 Explain why antibiotic resistance and pesticide resistance are not examples of artificial selection.

25 Why is it difficult to give a satisfactory definition of the term *species*?

Farmers can use other forms of pest control, such as biological control. This involves releasing or encouraging natural parasites and/or predators of the pest species. Plant breeders also continue to use selective breeding to develop new varieties of crop plants that are resistant to pests. Genetically modifying crops to make toxins that kill the pests that eat them is another strategy that is increasingly being used. Integrated pest management uses these strategies together with the limited use of pesticides as and when absolutely needed.

Example

Insecticide resistance

1 Define the term *variation*.

2 The Colorado potato beetle, *Leptinotarsa decemlineata*, is a serious insect pest of potato plants. This pest species was controlled by the insecticide DDT until the 1950s, when it gained resistance that became widespread. Since then, many types of insecticide have been used to control it but the beetle is now resistant to all of them. Some populations of beetle may not be resistant to all types of insecticide, but within the species as a whole there is resistance to all insecticides.

a) Spraying with an insecticide rarely kills all the pests in a field of potatoes. Apart from resistance, suggest why this is the case.

b) Explain how resistance to an insecticide arises and spreads in Colorado potato beetles.

Answers

1 Variation is the differences between species (interspecific variation) and the differences between individuals within a species (intraspecific variation).

2 a) Insecticides are sprayed from above, so any beetles on the underside of the leaf may not be affected. Insecticide may land on some beetles, but maybe not enough to kill them.

b) A gene mutates spontaneously to give an allele that enables a beetle to resist the insecticide. Most mutant alleles are recessive. This mutant allele may pass on for several generations before it is expressed in a beetle. If this beetle is part of a population that is sprayed with insecticide, it will not be affected and will survive. All beetles that are not homozygous recessive are susceptible and die. The resistant beetles now have much less competition for resources, especially food. They are more likely to survive to reproduce. If a resistant beetle mates with another resistant beetle (as is likely after the insecticide has been sprayed) then the next generation will all be resistant. This increases the proportion of resistant beetles in the population, making it harder to control them with the same insecticide. Natural selection has occurred. The insecticide is the selective agent.

Exam practice questions

1 Which of the following statements is part of Darwin's theory of natural selection?

A Asexual reproduction generates very little variation.
B Interspecific variation is used to classify organisms.
C Poorly adapted individual organisms never leave offspring.
D There is competition between individuals of the same species. *(1)*

2 There are many breeding pairs of small birds in an area of woodland. Which of the following is not a selective agent for this population?

A the availability of nesting sites
B the number of eggs laid each breeding season
C the population of sparrowhawks that are predators of these small birds
D the quantity of food available in the wood *(1)*

3 The smallest biological unit that evolves over time is

A a genus
B an individual organism
C a population
D a species *(1)*

4 Individual organisms within a population that reproduces sexually show genetic variation. The main cause of this type of variation is

A environmental change
B meiosis
C mutation
D natural selection *(1)*

5 Natural selection acts on

A a species
B genes and their alleles
C the genotype of an individual organism
D the phenotype of an individual organism *(1)*

6 Cormorants are sea birds that dive for food (see Figure 15.3). A scientist in Canada investigated the diets of two closely related species, the great cormorant, *Phalacrocorax carbo*, and the double-crested cormorant, *P. auritus*, in waters around Nova Scotia where the two birds live in the same area. The birds catch fish and regurgitate pellets containing any hard pieces of their prey such as bone that they cannot digest. The scientist collected pellets from places where the birds roosted over several months in the summer of one year. Table 1 shows the different species of fish present in the pellets. The scientist also made other observations of the birds. The observations are summarised in Table 2.

a) i) Explain why the percentages of fish present in the pellets were calculated. *(2)*

ii) Suggest why the researcher collected data from regurgitated pellets. *(3)*

b) Suggest two limitations of the study on the composition of pellets. *(2)*

Table 1

Habitat of fish	Prey species	Numbers and percentages of fish found in pellets			
		Phalacrocorax carbo		*Phalacrocorax auritus*	
		Numbers	Percentage	Numbers	Percentage
Open water fish	pollock	155	24.0	228	30.7
Bottom-dwelling fish	cunner	269	41.7	109	14.7
	cod	36	5.6	30	4.0
	long-horned sculpin	40	6.2	9	1.2
	short-horned sculpin	66	10.2	131	17.6
Flat fish	flounder	73	11.3	47	6.3
	plaice	6	0.9	0	0
Eel-like fish	wrymouth	0	0	176	23.7
	rock gunnel	0	0	13	1.7
Total		645	100	743	100

Table 2

Observations	*Phalacrocorax carbo*	*Phalacrocorax auritus*
Feeding areas	Sea water	Sea water and freshwater
Total number of fish species caught	11	16
Mean diving time/s	51.0	25.1
Mean resting time/s	13.9	10.3
Mean foraging depth/m	10.7	4.7
Range of foraging depths/m	4.6–19.8	1.5–7.9

c) Suggest three adaptations that cormorants have for catching fish. *(3)*

d) Use the information in Tables 1 and 2 to explain how these two very similar species co-exist in the same ecosystem. *(5)*

7 Charles Darwin proposed the theory of natural selection as the way in which organisms evolved. He made four observations:

A Organisms have the ability to produce large numbers of offspring.

B Populations of organisms fluctuate but tend to remain fairly stable over time.

C There is variation within species.

D Offspring generally appear to be similar to their parents.

Explain the consequences of these four observations made by Darwin when proposing his theory of natural selection. You may identify the observations by the letters A to D above. *(8)*

8 a) Define the term *intraspecific variation*. *(1)*

Lower Liassic rocks are about 180 million years old. They contain large numbers of different species of the genus *Gryphaea*. Its slightly curved shape has earned *Gryphaea* the name 'the Devil's toenail'.

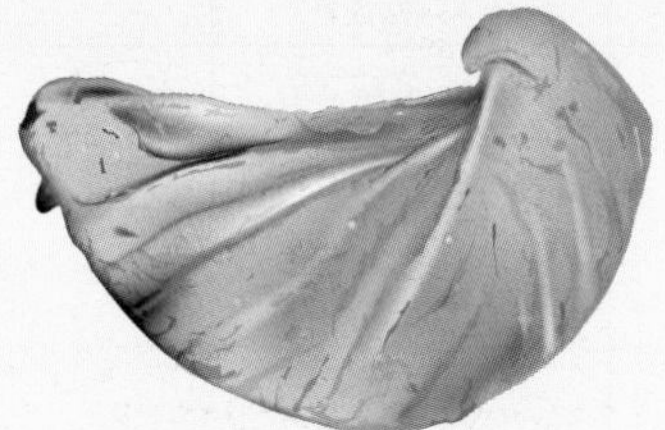

The uppermost layers of rock contain species of *Gryphaea* that look like the one in the figure. Lower layers contain specimens that have much straighter shells. So many fossils of *Gryphaea* have been collected that it is possible to study interspecific variation in the organism.

b) i) Explain the term *interspecific variation*. *(2)*

ii) Suggest why it is difficult to distinguish between different species of animals, such as *Gryphaea*, that have been extinct for millions of years. *(3)*

c) Explain how the study of fossils provides evidence for evolution. *(4)*

d) Outline two different pieces of molecular evidence for evolution. *(4)*

9 The table shows information about lengths of shells in three populations of the same mollusc species.

Population	Numbers	Range/ mm	Mean/ mm	Modal class/ mm	Standard deviation/ mm
A	65	60–84	71.2	72–74	5.2
B	57	61–80	67.2	63–65	5.3
C	54	60–86	74.4	78–80	6.1

a) i) Which population has the largest range in shell length? *(1)*

ii) What percentage of population A is within the range 71.2 ± 5.2 mm? *(1)*

iii) What are the smallest and largest shell lengths that include 95% of population B? *(2)*

b) Sketch smoothed histograms to show the variation of shell length in each of these populations. In each case indicate the position of the mean length. *(3)*

c) Use the Student's *t*-test to compare the difference between the means of population A and population B. Search online for the formula for the Student's *t*-test and the table of probabilities. Show your working. *(3)*

d) A student thought that populations A and B were different species. Suggest why shell length may not be a good feature to use to decide whether two populations are from the same species or from different species. *(3)*

e) A population of this mollusc on another beach had a bimodal distribution. Make a sketch of a bimodal distribution. *(1)*

10 Fleas are insects that are external parasites of humans. They feed on the blood of their hosts using mouthparts that are shaped like a hypodermic needle.

a) Suggest three other adaptations of these external parasites. *(3)*

b) Many domesticated animals are prone to several types of external parasite, such as ticks, mites and fleas. These can be controlled by pesticides that are either ingested by the animal or applied to its skin.

Describe how an investigation to compare the effectiveness of two different pesticides could be carried out. In your answer state the steps that should be taken to ensure that valid comparisons can be made. *(6)*

c) Outline the implications for humans of pesticide resistance in insects. *(5)*

Stretch and challenge

11 A study was carried out in New Zealand to investigate the effects of genes and the environment on the milk yield and milk fat content of dairy cattle. Forty milking herds of cattle were divided into two groups:

- group A contained all the herds that had high yields of milk
- group B contained all the herds that had low yields of milk.

There were two parts to the study.

- In Part 1, 240 young female calves were taken from the herds in groups A and B and reared together at the Ruakura Animal Research Centre until they were mated and joined the centre's milking herd. Careful records were kept of the quantity and quality of milk from each animal.
- In Part 2, 110 sets of identical twin calves were found from among the 40 herds. One calf of each twin was placed in a herd in group A (high yielding) and its twin was placed in a herd in group B (low yielding). Careful records were again taken of the milk produced by each cow.

The tables below show the results.

Part 1

Features of milk yield	Herd performance			Calves kept at Ruakura		
	A – high yielding herds	B – low yielding herds	Difference	From high yielding herds (A)	From low yielding herds (B)	Difference
Mean milk yield per cow/kg	2814	1960	854	2728	2736	–8
Mean fat yield per cow/kg	155	100	55	150	145	5

Part 2

Features of milk yield	Herd performance			Twin female calves		
	A – high yielding herds	B – low yielding herds	Difference	Twins sent to high yielding herds	Twins sent to low yielding herds	Difference
Mean milk yield per cow/kg	2789	1985	804	2640	1857	783
Mean fat yield per cow/kg	153	101	52	139	95	44

a) Explain the reasons for the design of the investigation, including the use of identical twins.

b) Summarise the data and explain what it shows about the variation in milk production in dairy cattle in New Zealand when the study was carried out.

12 Plants of *Achillea lanulosa* that grow at low altitude have finely divided leaves like those in the figure. At high altitude the leaves of this species are different in shape and size. They are more compact and are not finely divided. Explain how you would find out whether the features of the leaves of *Achillea* populations are determined genetically or by the environment in which they grow.

Chapter 16

Exam preparation

Overview

Your exam preparation begins from the moment you enter your first A/AS Level biology lesson. The most successful candidates drive their own learning and do not rely entirely on their teacher or the work they do in lessons. The exams include an element of recall, but simply knowing the facts will **not** get you a good grade. Many of the marks require a demonstration of understanding and the ability to apply your knowledge in a new context. It is important that you are thoroughly familiar with the topics covered, but also that you have good general biological knowledge. If something does not make sense or some information seems to be missing, you should investigate on your own using the internet and any available books. Scientific magazines and TV programmes should also be used wherever possible to broaden your knowledge base. This will bring you into contact with information, examples and new contexts that will reinforce your knowledge and understanding of the work on the A Level specification.

A thorough familiarity with the OCR specification is vital. The specification tells you precisely what you need to know (and therefore what you do not need to know) and gives details of the assessment process. You need to be familiar with the format of the exam papers and practical assessment.

If you constantly 'read around' the subject as you go through the course, you will need to spend less time on revision, because many of the basic facts will be hard-wired into your brain. Nevertheless, thorough and effective revision will be necessary. If you have a detailed strategy or plan for your revision, you can improve your grades.

Once you get into the exam, the mark you achieve will depend not only on how much you know and understand about the subject. Exam technique plays a big part in the final outcome, because it is essential that you answer precisely what the questions ask and explain yourself clearly. You cannot get marks for material you do not know or understand, but it is essential that you gain every possible mark from the content that you do know.

This chapter outlines some links and tips that will guide you in the right direction. However, different people learn in different ways, and you will need to find out the working practices and revision techniques that work best for you.

Figure 16.1 Watching documentaries about biology is a good way of broadening your knowledge and practising explaining what is said from your own knowledge. Reading scientific magazines and books also improves your knowledge base and helps to 'join up' your knowledge, improving your understanding.

Examinations

The AS Level biology course consists of four modules, as follows:

- Module 1 – development of practical skills in biology
- Module 2 – foundations in biology
- Module 3 – exchange and transport
- Module 4 – biodiversity, evolution and disease

At the end of the AS course you will sit two exams, each of which tests material from all four modules. Each paper lasts one hour and 30 minutes, and all papers have equal weighting. The papers have a different structure, however. Paper 1 is called 'Breadth in biology' and consists of two sections, one of multiple-choice questions and one of structured questions. Paper 2 is entitled 'Depth in biology' and consists entirely of structured questions and extended-response questions. There is no coursework at AS Level, but there are questions on enquiry and practical skills in both papers.

Synoptic assessment

Both papers contain an element of synoptic assessment, and it is important that you understand what this term means. Synoptic assessment tests the learners' understanding of the connections between different elements of the subject. You will be expected to see and explain links between different parts of the specification and apply the principles and concepts of experimental and investigative work to different areas. This will include the analysis and evaluation of data.

Candidates who have a good biological general knowledge and have read and researched around the basic content will be best prepared for synoptic questions. Just learning information will not be enough, and it is important that you try to extend your knowledge and keep up with any advances in biology during your course.

The examinations test three assessment objectives, which are described below.

Assessment objective 1

Demonstrate knowledge and understanding of scientific ideas, processes, techniques and procedures.

Assessment objective 2

Apply knowledge and understanding of scientific ideas, processes, techniques and procedures:

- in a theoretical context
- in a practical context
- when handling qualitative data
- when handling quantitative data.

Assessment objective 3

Analyse, interpret and evaluate scientific information, ideas and evidence, including in relation to issues, to:

- make judgements and reach conclusions
- develop and refine practical design and procedures.

The allocation of marks for different skill areas is shown in Table 16.1 below.

Table 16.1 Allocation of marks for different skill areas.

Component	% of marks		
	AO1	AO2	AO3
Breadth in biology (Paper 1)	22–24	19–20	7–9
Depth in biology (Paper 2)	13–16	21–24	13–14
Total	35–40	40–44	20–23

AO, assessment objective.

What this allocation of marks means is that there will be significantly more questions that require recall of facts in Paper 1, whereas Paper 2 will have a greater emphasis on the application of knowledge, analysis, interpretation and evaluation.

Your communication skills will be tested throughout both papers (except in the multiple-choice section of Paper 1). This will not be by specific questions, but will be integrated into the mark scheme of the paper as a whole. In other words, at all points you need to communicate ideas in detail, clearly and with precision. Examiners do not interpret what you say or give you the benefit of the doubt if you have not made yourself absolutely clear.

You will also be assessed on a range of mathematical skills (once again, these marks will be integrated throughout the whole paper). The full range of skills that you will be expected to have can be found in the specification on the OCR website.

Using the OCR website

The OCR website is a source of many useful documents and information. Here you can access the specification and download past papers and mark schemes. You can also find out exam dates.

The use of the specification has been mentioned above. It details exactly what you need to know. Only biological terms mentioned in the specification can be directly tested in the exams. The depth of understanding required is not always clear, and here the use of this textbook and revision guides can help.

Answering past-paper questions is a very important part of exam preparation. It gives you a good idea of the types of questions asked and, if you have the mark scheme, the depth of the answers required. There is little value in attempting past-paper questions if you do not have the relevant mark scheme. In the early years of a new specification, very few or no past papers will be available. Papers from the previous specification can still be useful, because much of the content will be the same. The awarding bodies usually publish specimen papers for new specifications to give teachers and students an idea of what the new exams will look like.

Workflow

The first stage of exam preparation is to become thoroughly familiar with the work when you first encounter it. A suggested workflow that will enable you to do this is shown in Figure 16.2. The coloured boxes give just one example, the study of prokaryotic and eukaryotic cells. Different suggestions are given in the sections on interacting with the material, asking questions, and researching answers and looking for new examples. These are ideas that you could use, although it is not intended to indicate that you should use all of them. Different students have different methods of working, and you should select working practices that suit your own interests and style.

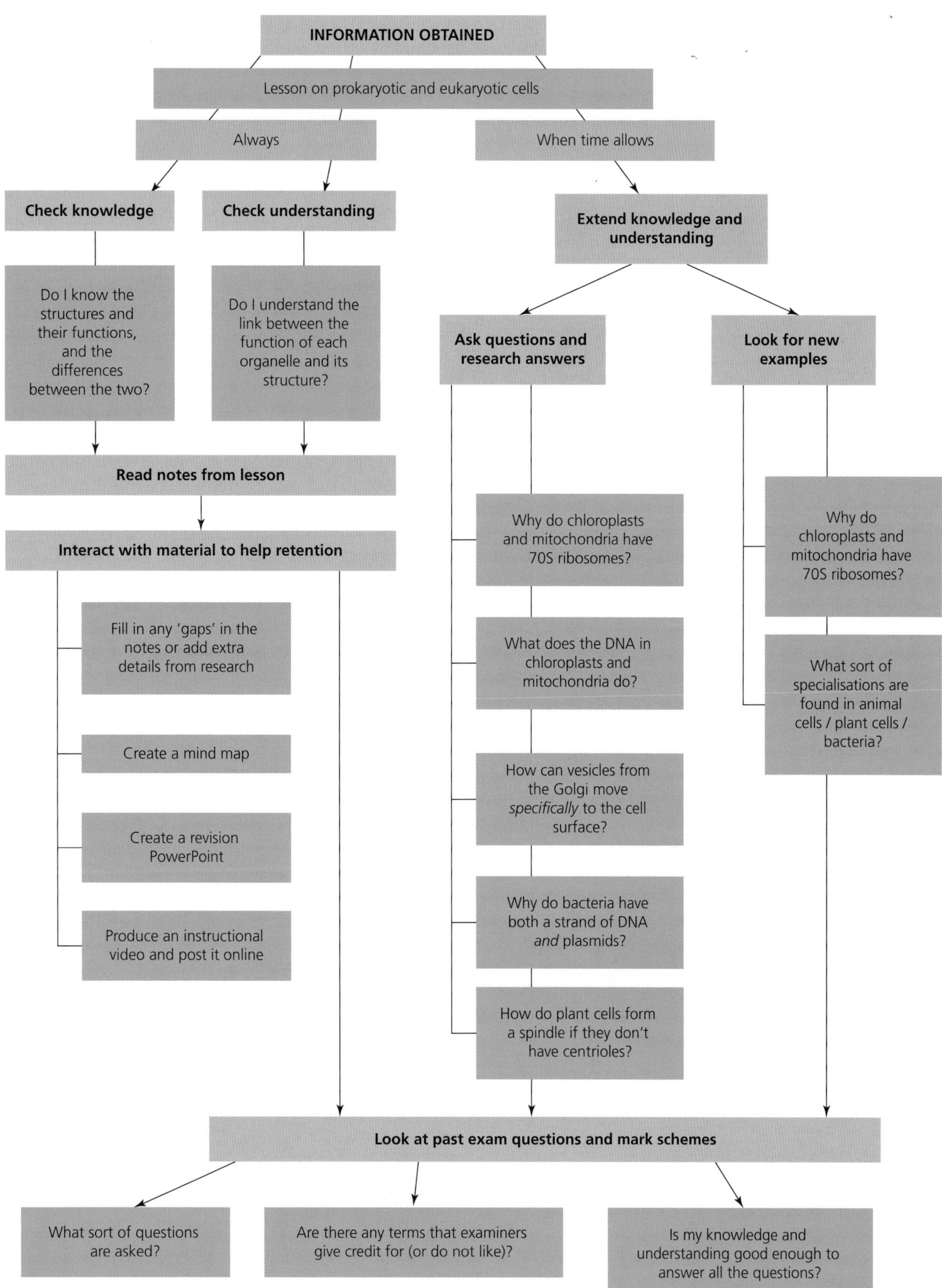

Figure 16.2 Suggested workflow for AS Level students.

As you go through the course, this type of workflow will ensure that you have a thorough understanding of the work and are familiar with a variety of biological terms and ways of explaining ideas. This will reduce the time you need to spend on revision before the exams, and help you develop the communication skills that are vital at AS Level.

Revision

To revise effectively, it is useful to have a basic understanding of how memory works. When you receive information, either for the first time or when you are revising something that you have completely forgotten, it enters your working (or short-term) memory. From the point of view of passing exams, working memory is useless. It fades very quickly so, if you are going to have any hope of remembering what you have learned, you have got to get it into your long-term memory. Fortunately, that is not difficult. All you really have to do is pay close attention to the information as it comes in and try to make some sense of it.

Information that is meaningless to you will not make it into your long-term memory. It is therefore important that you ask your teacher (or a computer search engine) questions if you do not understand the information. However, getting the information into your long-term memory is only half the battle. It is no good having it stored somewhere in your brain if you cannot retrieve it again. In other words, you have to try not to forget it. Evidence suggests that the information will stay in your brain for quite a long time, possibly forever, but your ability to access it may fade. Effective revision methods optimise the retrieval of information, but you will always forget a good proportion of what you have heard or read (see Figure 16.3).

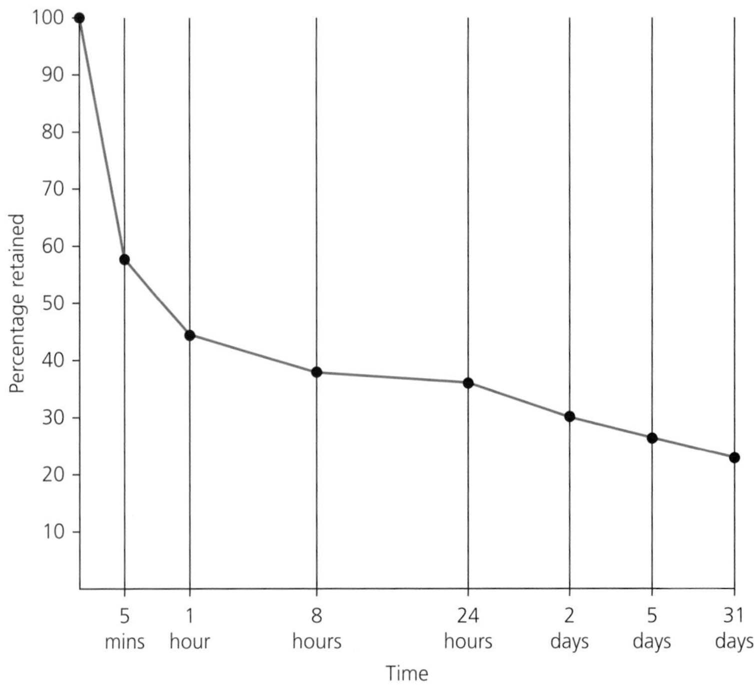

Figure 16.3 Percentage of information retained after different time intervals.

This is not quite as bad as it seems. A lot of what you hear or read is not important; for example, you need to retain the basic ideas but not the actual sentences used, word for word. A number of established learning techniques will greatly increase how much you remember:

- **Repetition**. Review new or revised material a week or so after encountering it. This need only be a brief refresh of the memory – you do not have to learn the material all over again. Try to do regular reviews as you go through the course, because this will mean that you already remember much of the material when you start intensive revision.
- **Form memory cues**. Link the information with something you already know. This may be other parts of the course, or personal experiences. The human aspects of biology are particularly suited to this – you have experience of things like breathing, sweating and reflexes; you may know someone with diabetes, etc.
- **Mnemonics**. A mnemonic is a formula or rhyme that assists the memory. A common form is to use the initial letters of a list of words to form a memorable sentence; for example, when memorising the stages of mitosis (prophase, metaphase, anaphase, telophase) you could use the sentence 'penguins march around trees'. When doing this, it is important that the sentence is easily memorable – it is better to choose things that are visual and a bit ridiculous! The mnemonic forms a memory cue, but this type of cue is mainly restricted to lists.
- **Translate the information into a different form**. Use the information to make a mind map, a picture or a presentation, or explain what you have learned to someone who knows nothing about A Level Biology. If you cannot make someone else understand the information, you may not fully understand it yourself.
- **Do not do long revision sessions without a break**. Most people's brains cannot take in information continuously for longer than about half an hour without a break. A short break of a few minutes will refresh your brain and allow you to revise more efficiently.
- **Structure your revision**. It has been established that people learn best at the beginning and end of a session. Start your revision with material that is quite difficult to understand or remember, finish it with similar material (or a review of what you learned at the start), and do the easier content in the middle. This is also a reason for taking breaks in your session: with breaks, you get more 'beginnings' and 'endings'.

Exam technique

Your grade will not only depend on how much you know and understand, it will also depend on good exam technique and the ability to communicate information clearly and concisely. The best way to acquire good exam technique is to try out past-paper questions, paying particular attention to the mark schemes, familiarising yourself with the things that are or are not credited. Practise extended writing, both in answering exam questions and in making notes, and get someone to read your efforts to see whether you have made things clear. Some basic points of exam technique are listed below:

- Read the question carefully and ensure your answer is precisely targeted to what it asks. If you look at mark schemes, you will see that the items required are very specific and correct, but irrelevant points will not be credited.

- Always read the whole question from the beginning – do not jump to the actual questions and then go back to the initial text. If you do, you are more likely to miss important information.
- Question writers do not include information that has no use. If you get to the end of the question and there is some information you have not used, alarm bells should ring in your head. If you get stuck on a question, look carefully through the question for clues to the answer.
- Communication must always be clear and detailed. Avoid vague and general comments that do not include factual information.
- The number of marks for a question indicates the number of points that need to be made.
    - In questions worth several marks, it is always best to give one more piece of information than the number of marks, because you cannot know exactly what is in the mark scheme.
    - If you are specifically asked for (for example) two reasons for something, there is no point giving three. Generally, only the first two you give will be marked.
- In your more extended answers, use scientific terms – but only if you are confident they are correct. If in doubt, it is safer to describe the structure or process.
- Although you should try to use scientific terms, do not try to write in 'scientific' language. This does not really exist; plain English will usually make your answer easier to understand.
- Do not spend a long time struggling with a question or part question you find difficult. Move on and come back to these items when you get to the end of the paper.
- Never leave a question worth three marks or more blank. Sometimes you may hit a marking point even if you think you do not know the answer. No answer at all absolutely guarantees no marks.

Activity

Communication skills

Improve these answers, all of which are factually correct but lack clarity or detail:

1 The folded membrane of a mitochondrion is an adaptation for more efficient respiration.
2 The reason for the higher blood pressure in the left ventricle is because it needs to pump blood all around the body.
3 Natural selection involves 'survival of the fittest'. Animal populations show variation and some variations make them better suited to the environment. The fittest animals will survive and pass on their features to the next generation.

Chapter 17

Preparing for practical assessment

Experimental science is the bedrock of modern Biology, and throughout history Biologists have increased our understanding by conducting research. Our understanding of genetics has been improved by many different researchers including: Hershey and Chase, who worked on the function of DNA using bacteriophages (viruses invading bacteria); Meselson and Stahl, who worked on DNA replication; and Beadle and Tatum, who worked on the relationship between the gene and proteins. In order to be successful they had to:

- investigate the scientific papers of the time
- develope a hypothesis
- test established hypotheses experimentally.

It is very difficult to actually prove a hypothesis: it can be tested and disproved or tested and modified, but actual proof is very hard to find. In many cases, a hypothesis remains a hypothesis until it is superseded by another or is modified in the light of further scientific evidence.

In this way, practical studies are an essential and fundamental aspect of biology at any level, and must therefore be an integral part of any biology course. The skills you learn at A Level will be vital preparation for studying biology at university and on other higher level courses.

In the new specifications, practical skills are integrated into the course, allowing a complete understanding of the skills that are essential to any biologist. A wide range of practical skills will be examined at a number of levels in the theory papers, as well as being assessed in practical experiences set out by the examining body. In the first year of your A Level course, or at AS Level, the practical skills you gain and the practical work you carry out will support your learning and will provide opportunities to develop the skills needed for the practical endorsement at A Level.

You must complete and record in a log book a minimum of 12 assessed practicals over the two years of an A Level course. The aim of the assessed practical activities is to allow learners to demonstrate an understanding of a wide range of skills while at the same time allowing the skills to be used in a practical situation.

Figure 17.1 Martha Chase, who worked on the function of DNA using bacteriophages with Alfred Hershey.

Essential practical skills

Planning an experiment

Planning is an important aspect of experimental biology. You should begin with some research to help identify other studies that have been carried out on the problem; this will allow you to select the techniques and apparatus that will be most suitable for the design of your experiment. Your selection must be based on sound scientific knowledge with a full understanding of the problem under investigation. For example, when designing an enzyme experiment, it is vital to have a good understanding of the fundamental principles of how enzymes work and the conditions necessary for enzymes to function, and an idea of how you might test them.

Figure 17.2 Part of planning your experiment should include a risk assessment.

Part of your experimental design must involve preliminary research and a preliminary study, because this informs you of many aspects of the actual procedure: what apparatus to use, the most suitable technique, the variables to be controlled, as well as both the independent variable and the dependent variables. You need to take into account any data processing you may need to carry out, because this will determine how much data you should collect. Finally, you must consider how you will record your data. Drawing a table for recording the data collected is another aspect to consider when planning your procedure; this must be decided during your planning and preliminary studies.

Preliminary studies

Having chosen the best technique for your experiment, based on your preliminary research, and some suitable apparatus for carrying out the experiment and testing the results, the next step is to carry out some trials or **preliminary studies** to help identify the variables that could affect the experiment, how you could control these variables, and other important aspects such as the best quantities and volumes needed to complete the experiment. It would be a waste of your time to carry out the experiment without already having investigated these fundamental aspects, since your findings may be completely invalid or you may find yourself nearing the end of the experiment and running out of the reagents or reactants before the study is complete.

Identifying variables

In order to consider that the outcome of your experiment is valid, it is vital to **control any variables** that may affect the results and therefore the outcome in any way.

In any experiment, there should only be one variable that is changed, and that is the **independent variable** – the one you change to determine its effect on the results. Any other variables must be kept constant, or at least known about to take account of the impact they may have. These are called the **controlled** or **confounding variables**. Therefore, you must have enough detailed information on how to control all these variables consistently and how to change the independent variable, so that it is possible to repeat the experiment in exactly the same way. Another scientist should be able to follow the instructions given exactly, without any need for further information, and still repeat the experiment in an identical way.

The **dependent variable** is what you measure to determine the outcome. How you measure this variable is important, because this will tell you the effect that the independent variable has on the data and provide you with the results.

A **control** is also a useful part of the experiment, because it will be used to show that the independent variable – not any other aspect of the experiment – has caused the changes. It may be, for example, that the changes would happen by themselves. This possibility needs to be eliminated, which is the main role of the control.

For example, in an enzyme experiment, the independent variable could be a change in the enzyme concentration. All other aspects of the procedure must be kept constant (all the confounding variables) to avoid any influence on the results (the dependent variable). The control would then involve repeating the experiment with distilled water (or boiled enzyme to denature it) in place of the enzyme. This would eliminate the possibility that the results would occur without the enzyme and without changing its concentration.

As part of the experimental design, you must be able to evaluate your experimental procedure to make sure it does test the problem, is appropriate and will provide results or outcomes that may be expected.

Evaluation of experimental method

Any **limitations** of the experiment must be identified and, where possible, removed or corrected. In this context, a limitation is a fundamental design fault. For example, when using any gas collection apparatus, some gas leakage is inevitable and could be said to be a limitation, because it prevents an accurate measure of the gas volume collected. The level of **accuracy** required must be evaluated and improved where possible; this is especially true of any experiment where the outcome is likely to have a significant impact, such as delivering the correct volume of gas as an anaesthetic to fruit flies that you want to use for genetic breeding. Failure to get this right would result in the death of the flies or in under-anaesthetising them so that they fly away during the procedure.

The **precision** of the apparatus used in the experiment as well as of any measuring equipment used to measure the outcome – the **dependent variable** – must also be considered as part of the evaluation of the procedure. Finally, if you carry out the experiment only once you cannot assess whether the results are reliable. If you carry it out twice and the results are different, you will not know which result is correct. The only way to determine the **reliability** of the data is to repeat the experiment several times and remove any anomalies. This will be looked at again later.

The procedure must be evaluated and changes made to improve it, where necessary. It is sometimes possible to evaluate a procedure before undertaking the main experiment, and so make the necessary improvements at the start; at other times, it is only possible to evaluate the procedure at the end of the experiment.

It is essential to record *all the steps* of your procedure to allow other researchers to carry it out exactly, without any further input required. An example of this is given later.

Implementing an experiment

Using apparatus and techniques

It is important that you understand how to employ a range of experimental techniques; you may need to practise some techniques before you begin the experiment. Reading instructions and carrying them out correctly and in full may seem simple, but in practice it is easy to misread or misinterpret instructions and follow them incorrectly, so you also need to practise following instructions.

There are two types of experiment: qualitative experiments, in which you collect and record observations without collecting numerical data, and quantitative experiments, where you collect and record numerical data.

Units

You need to know the correct and appropriate SI units of measurement when measuring substances during the procedure and when measuring the outcome. You also need to know the correct **symbols** to use for each SI unit. See the section on analysis and Chapter 18 for more on units and the correct symbols.

Presenting data

Qualitative data and quantitative data require different presentation formats, as do results taken from an ecology study or indeed drawings from microscope slides or of live specimens. The different forms of presentation must be learned and applied correctly in each situation.

It is important to enter any experimental outcomes in a table, which is in the correct format and is appropriate for your particular procedure. The table should be drawn with ruled lines to separate the cells. Use the correct headings and the correct units and symbols in the headings (not in the actual cells) where appropriate. The independent variable should be in the first column, with collected observations (dependent variable readings) in subsequent columns.

Identify at the planning and implementing stage what you will do with the data you collect and the format in which it will be presented.

Recording

Quantitative data must all be recorded to the correct number of **significant figures**. When using the data to calculate a mean, for example, the number of significant figures for the total of all the data determines the number of significant figures that should be used to express the mean. For example, the sum of 14, 14, 13, 12, 14, 13, 12, 11, 13 and 13 is 129. The mean to the same number of significant figures as the total is therefore 12.9.

Key terms

Accuracy How close a reading is to the true value.

Precision How similar repeated measurements are to each other (in other words how tightly clustered repeat readings are).

Resolution The smallest change in the quantity being measured that can be detected by the measuring instrument.

All data must be recorded to the same number of decimal places, but processed data may be recorded either to the same number of decimal places or to one more decimal place than the raw data. The reason for this is the **resolution** of the measuring instruments: if the measuring apparatus records to the nearest whole number, then it is not possible to be more precise than one decimal place when calculating the mean.

Factors such as the precision and resolution of the measuring apparatus and how competent the scientist is when handling the apparatus and reading the results determine the accuracy of the data.

Making and recording the measurements for a quantitative activity is an important skill. The measurements you are most likely to make include temperature, pH, time, volume, length and mass. However, you may need to make other measurements when you use apparatus for measuring light (a light meter records the unit of lux) or optical density (a colorimeter records either percentage transmission or absorbance in arbitrary units, or au).

Tip

When measuring time, you are restricted by your own reaction time. As a result it is usual to only record time to the nearest 0.5 of a second or at the very most to 0.1 of a second. You must therefore round up from the many decimal places that are given on a digital stop clock to one decimal place.

If your experiment demands that you make more **accurate** measurements, you may need to change the type of apparatus you use. A 10 cm^3 syringe will not give the necessary resolution to accurately measure 0.1 cm^3, and so a 1 cm^3 syringe with smaller graduations will be needed. In the same way, a watch allows you to measure time in seconds, but you need a timer with greater precision to measure in milliseconds.

Tip

A common error in converting units is not converting centimetres to micrometres at the end, so always make sure you are working in the correct units.

When measuring length, the usual unit is a millimetre, which you can measure using a ruler. To measure the length of a chloroplast under a microscope, a **graticule** is needed to measure the length in micrometres (μm). There is more on measuring microscopic structures in Chapter 18.

Centimetres should not be used to measure length; because the centimetre is not an SI unit, it is not suitable or precise enough to use in biology. Always measure using millimetres (1 millimetre is one-thousandth of a metre) and micrometres (1 micrometre is one-thousandth of a millimetre). Take care with the decimal points when converting centimetres to micrometres.

Analysis

Once an experimental procedure has been carried out, the results collected must be processed and interpreted in order to reach any valid conclusions from the outcomes.

Exactly how the **processing of data** is carried out depends on the type of experiment.

Drawings, whether from life or from microscope work, do not need to be processed; you may need to make measurements and work out scales, but no other processing is needed. Qualitative data cannot, by its nature, be processed mathematically, because there will not be any numerical data to process, but analysis of the observations will be needed. This analysis will generally be subjective but may be made more objective by setting standards to compare against or by further experimental work using a more quantitative method.

Tip

Record the numerical data to the same number of decimal places, taking note of the precision of the measuring instruments and of any instructions in the activity or question. Frequently there is an instruction such as to record to one decimal place. Read Chapter 18 for more detail on processing and recording for the assessment.

Quantitative experimental results must be processed in order to allow valid conclusions to be drawn.

The **mathematical skills** required at the simplest level will involve calculations of rates or means. Other mathematical skills needed may include calculating the standard deviation and standard error, or using statistical tests such as chi squared, Student's *t*-test or paired tests such as Spearman's rank correlation coefficient. The most important aspects when using these tests are to follow the steps in a logical sequence using the formula given, and then to understand the meaning of the results of each test and the impact the result has on any conclusion you have already drawn from your experiment.

As part of the analysis, presentation of the outcomes in the form of **graphs, bar charts or histograms**, whichever is the most appropriate, is essential. You must select the correct graphical format for your data. All graphs must be correctly scaled with ascending equidistant intervals, and axes correctly orientated and labelled with correct units and unit symbols, in order that data can be correctly read off the graph. A line graph may be drawn by using a smooth line or curve of best fit. This must be correctly drawn when there is confidence in the intermediate values. Approximately 50% of the plots should fall each side of the line. When there is no confidence in the intermediate values a ruled line from one plot to another – a plot-to-plot line – should be drawn.

Tip

The general rule is that the parameter (or the measurable factor that sets the value being used) is written in full followed by the unit symbol. You need to learn the correct symbols for each unit that you will use. Chapter 18 has more on this.

The size of the graph is also important. As a rough guide, you should make good use of the paper provided. The plotted points should occupy half of the graph grid line in both *x* and *y* directions. How the graph is then used depends on the type of experiment. For example, an experiment on enzymes may involve reading the **gradient** of the graph line in order to determine rate, while another experiment may need you to determine the water potential by reading the **intercept** of the graph.

Evaluation

Evaluation of the results is a different skill from evaluation of the procedure. You need this essential skill in order to draw valid conclusions from your results. As part of evaluating the results, you may rely on some of the mathematical processing you have already completed, especially where it involves certain statistical tests.

During the planning, you looked at identification of **limitations** in the procedure. Now you need to evaluate these limitations in terms of the impact they have on the data collected. Any experimental **errors**, otherwise known as operator errors or 'one off' errors, affect the results and may produce anomalies. You need to know how to identify these anomalies – see the example later in this chapter.

> **Tip**
>
> Anomalies are best identified by looking for any results that are out of line with the trend or with other replicates carried out at the time. These results will show a larger difference from the mean than the other results; this is often taken as a difference of greater than 10% from the mean value.

Precision and **accuracy** of the apparatus used and the measurements made will have an important impact on any certainty of the conclusions. The margins of error of the apparatus are usually given on any glassware used and can be used as part of a percentage error calculation. This gives you an idea of the magnitude of any error and therefore how much impact it may have on your results. A percentage error of less than 5% may be considered statistically not significant. Therefore, any conclusions drawn could be said to be valid. However, a percentage error of 25% would be considered too large to be ignored, and the impact on the results too important for the conclusions to be completely valid. Therefore, the conclusions may be rejected or they may need to be tested again by making **improvements** to the apparatus used or to the procedure carried out, to reduce the percentage error.

> **Tip**
>
> Percentage error is discussed in Chapter 18.

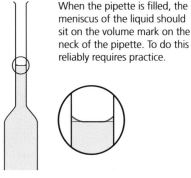

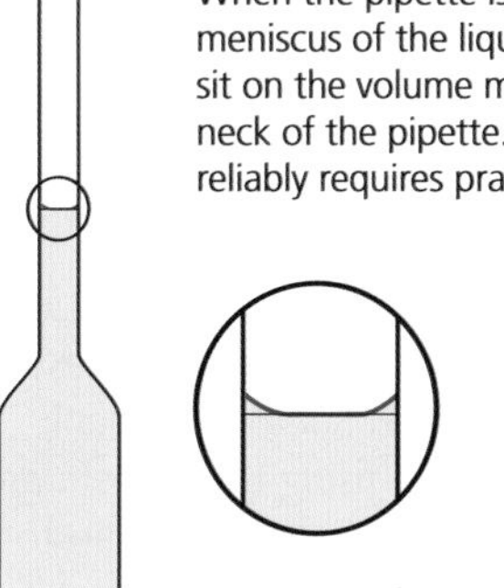

When the pipette is filled, the meniscus of the liquid should sit on the volume mark on the neck of the pipette. To do this reliably requires practice.

Figure 17.3 Reading the volume correctly with a meniscus.

Practising practical techniques

Using laboratory glassware

Laboratory glassware is used for a wide range of experimental techniques, and so it is important to practise using as many different types as possible. Using glassware such as measuring cylinders is a basic practical skill for accurate measuring of volumes. The appropriate size of measuring cylinder you need depends on the volume you need to measure. A small measuring cylinder often gives better resolution but may not be large enough for the volume you need to use.

In order to measure volumes accurately, it is important to know how to read the volume correctly, taking into account the **meniscus**. Volumes must be read at eye level, and the bottom of the meniscus should be lined up against the graduation line on the measuring cylinder.

Using a microscope

The ability to use a light microscope is an important skill for biologists because so many organisms and biological structures are extremely small and can only been seen if magnified by the light microscope. This applies to tissues, cells and cell organelles as well.

A light microscope operates by light being directed through a thin biological specimen or material that has been placed on a glass slide. The light then passes through several different lenses to allow an image to be seen through the eyepiece lens. There are usually several different objective lenses, which can be selected by turning a rotating disc holding them in position. This allows you to view a specimen under low-power lenses as well as under high power.

> **Tip**
>
> Always start viewing your specimen by focusing on low power and carefully positioning the specimen in the centre of the field of view to avoid it being lost from view as the higher power lenses are used. Once focused, it should be easy to turn to a higher power and find that the specimen is still in view and roughly focused – although it may need to be more finely focused under the high power to improve the clarity.

Microscope drawings

Drawing a specimen seen under the microscope is a vital and useful way to document and record the observations you make. Biological drawings made in this way should be scientific rather than artistic, since they represent a record of what you have seen. You should follow some strict guidelines when making these diagrams:

- Draw the structures seen under the microscope exactly as you see them, so that it is possible for another person to identify the same structures if shown the same slide.

- Clearly represent the exact shape, proportion and scale of the features.
- Draw the lines in pencil using a sharp point so that the lines are clear and continuous.
- Do not use shading or colour in any part of the drawing.
- Low-power drawings should be **tissue plans** and should include no cell detail at all.

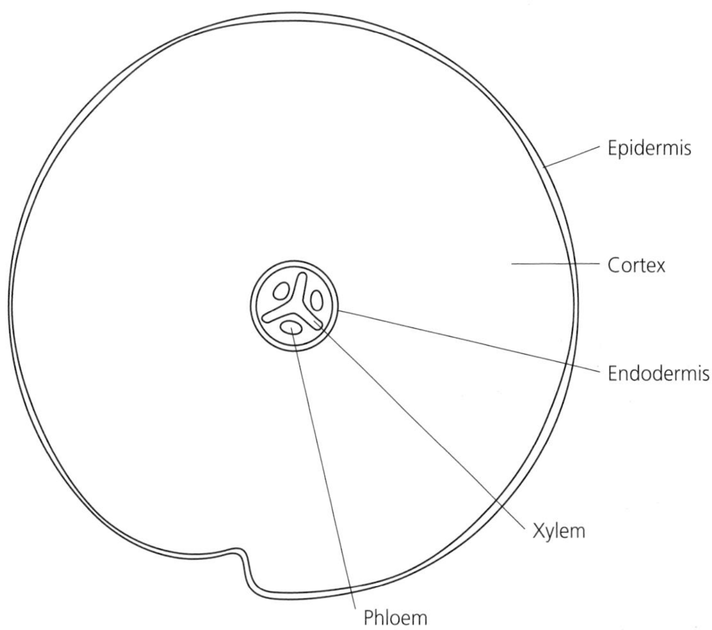

Figure 17.4 Low-power plan of a transverse section through a root.

Tip

High-power drawings must be labelled with the names of the cells drawn, not the tissues. Low-power plans should be labelled only with the names of the tissues, not the names of the cells. For example, in a plant stem, the low-power plan should include a label for the xylem, because this is the name for the tissue, while in the high-power drawing the xylem cells should be labelled as xylem vessels.

- High-power drawings should be detailed cellular drawings. It is usual to draw a specific number of cells. The instructions may include the number; however, if not, then the general rule is to draw no more than 15 cells, which should be connected and should also be representative of all the cell types visible on the slide.

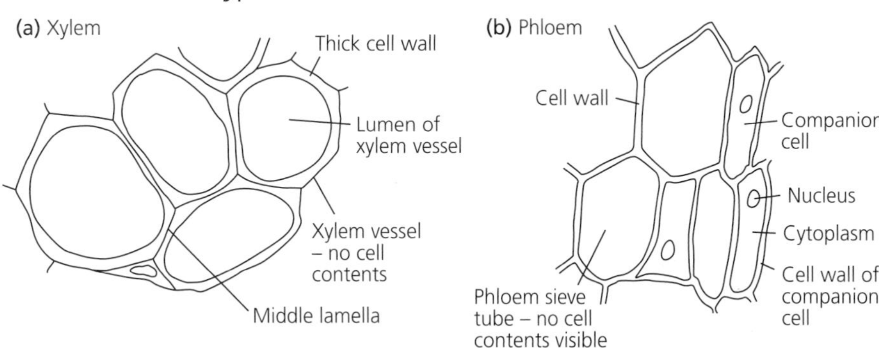

Figure 17.5 High-power detail of cells of xylem vessels and phloem sieve tubes.

- Include the magnification or an indication of scale, such as a scale bar.
- Make sure that any label lines are straight lines drawn with a ruler. Write the label at the end of the line, not along it.
- Use annotations on a drawing to give extra information about the features drawn, such as about the functions, properties or even features you have observed.
- When drawing, make good use of the available paper.

Other drawings

Some of the guidelines made for microscope drawings also apply to other biological drawings. The lines must be in pencil and be clear, sharp and continuous, with no shading or colouring; a scale must be included. The drawings should represent the structures exactly with the same shape, proportion and scale. Labels and annotations, where included, should have straight label lines drawn with a ruler, and again making good use of the paper.

Qualitative tests

Qualitative tests involving reagents allow you to identify biological molecules; they are simple and easy-to-carry-out chemical tests with resulting colour changes. The starch test using iodine solution and the reducing FI sugar test using Benedict's solution are two examples (see Chapter 2).

Some other qualitative tests involve an indicator that changes colour when a certain reaction takes place. For example, in the Hill reaction the electron acceptor changes from a coloured solution to a clear solution when it accepts electrons from the chloroplasts' electron transport chains.

An enzyme reaction may be used to change a cloudy solution such as milk to a clear solution when the milk protein casein is hydrolysed.

When a few drops of blood are added to water or to dilute solutions of sodium chloride, red blood cells burst releasing haemoglobin into solution (Tube A in Figure 17.6). When dropped into more concentrated solutions red blood cells remain intact (Tube B). Recording the effect of different concentrations of sodium chloride on red blood cells can be done by finding out whether the solutions are transparent or opaque using newsprint as in Figure 17.6.

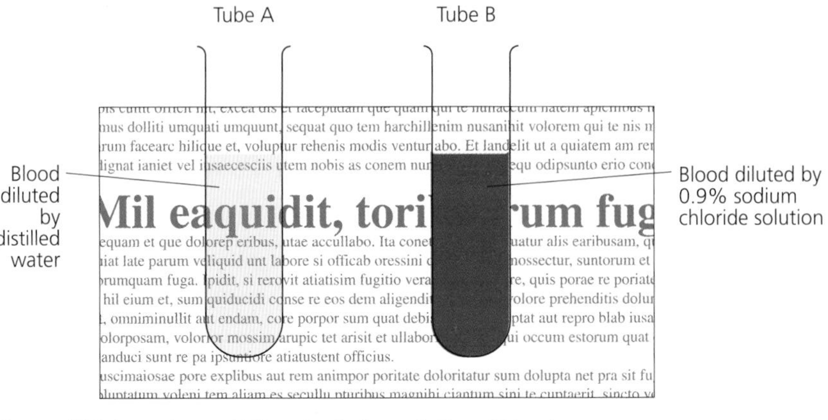

Figure 17.6 Assessing turbidity of solutions of diluted blood.

Tip

See page 91 for an explanation of the effect shown in Figure 17.6. Also see Question 5 on page 94 for some results collected from an investigation of the effect of solutions of different concentration on red blood cells.

Quantitative tests

Quantitative tests involve numerical values for the data collected and account for many of the biological experiments you will carry out.

Some experiments may be semi-quantitative, for example using colour standards to generate ranked data, as in the semi-quantitative Benedict's test.

In order to generate numerical data, a quantitative test must use apparatus that measures or collects this type of data. The simplest of these tests is to time a colour change or a change in appearance. Time then becomes the numerical data collected. The problem with using time is the uncertainty about the actual end point, which is subjective; you cannot be certain you have used the exact end point even if a comparator is used as a reference point.

Plotting a graph from a set of standards provides another opportunity to gather numerical data, but again may rely on the judgement of the end point.

Table 17.1 Results of timing the disappearance of pink colour from potassium manganite(VII) solution after adding different concentrations of glucose.

Glucose concentration/g 100 cm^{-3}	Time for pink colour to disappear/s
6	443
8	289
10	187
15	114
20	57

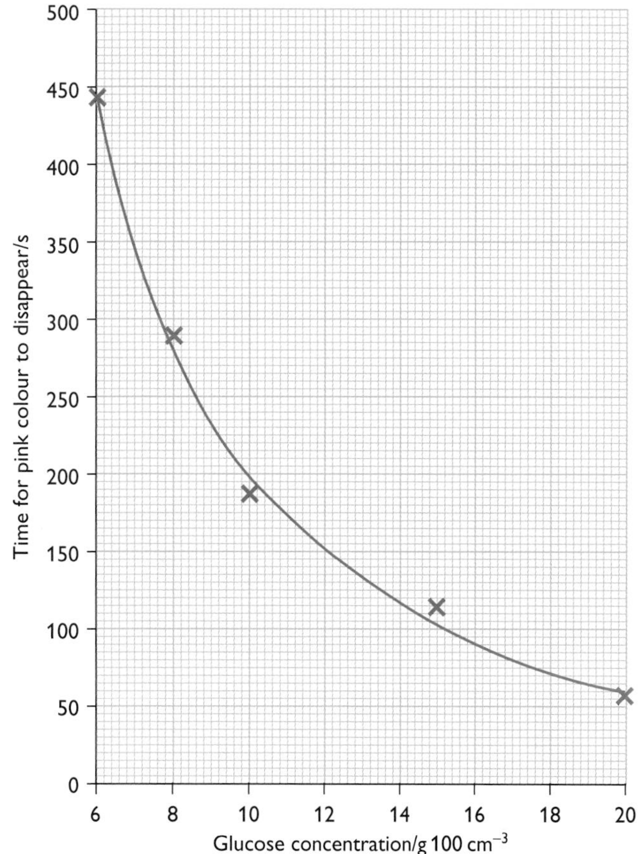

Figure 17.7 A calibration graph for the decolourisation of potassium manganate by different glucose concentrations, with the data.

Tip

You need to know how to use a colorimeter and to be aware of the limitations and possible errors that may result from using such apparatus, and how to reduce their impact.

Using a **colorimeter** is another suitable method of collecting numerical data. A colorimeter measures the amount of light passing through a coloured solution to determine how much of the colour is present. A range of different samples can be tested; for example, for solutions of betalain pigment obtained from a cylinder of beetroot tissue left in water of different temperatures for a set time. A colour filter is used in the light path to ensure that the correct wavelength of light is used to measure the optical density of that specific pigment. In the case of betalain (a purple pigment), the correct wavelength of light is green light.

The colorimeter must be zeroed before each test tube is inserted. This may be done using distilled water in a colorimeter tube, often called a cuvette. However, in some cases distilled water is not appropriate and the cuvette should be filled instead with the liquid that is the baseline for the reaction. For example, when using a colorimeter to determine the loss of starch from a mixture of starch and enzyme, the reagent used to determine the presence of starch is iodine. In this case, the blank (the tube used to zero the colorimeter) should be filled with iodine solution instead of distilled water.

Separating techniques

One separating technique is **chromatography**. This technique involves separating unknown mixtures using a solvent that dissolves the molecules, and a medium such as paper. The solvent moves through the medium and carries the molecules with it. The molecules spread out and separate, because they all move at a different rate. The mixture is carefully and slowly spotted one drop at a time on to a small area about 1 cm from the bottom of the paper (or other medium being used, such as thin layer plates). This area is called the origin.

Tip

Dry your spot between each drop application to create a small tight but intense spot. A hairdryer is useful.

Once the medium has been loaded with the mixture, the whole chromatogram is lowered into a container, with the solvent just below the level of the spotted origin. It is usual to add the solvent and cover the container some time before it is needed, to make sure the atmosphere is saturated. This ensures the solvent does not travel too fast through the chromatography medium, and the separation will be improved.

Once the solvent has moved through the medium, the chromatogram is removed from the container and a pencil is used to mark the line of the origin, the solvent front and the leading edge of the spots before they dry, in order to allow identification of the molecules.

Identification of the molecules in the unknown mixture is achieved by calculating the Rf value of each solute in the mixture and comparing the Rf values against known values.

Calculating the Rf value of each solute is carried out using the following formula:

$$\mathrm{Rf} = \frac{\text{distance moved by the solute}}{\text{distance moved by the solvent}}$$

Because the Rf value is a ratio, the value is always lower than 1 and has no units.

Electrophoresis is another separation technique. It uses the principle that molecules in a solute separate by size as a result of variation in the rate at which they move. In this case, the molecules are moved through a fluid-covered gel using an electric current. The smallest molecules travel the furthest, and so the solutes are separated along the line of the current.

Safe and ethical use of organisms to measure plant or animal responses or physiological functions

When using live animals you must take care to avoid any harm to them. Water-based organisms should only be observed for a short time and then returned to their main container as quickly as possible. Microscopic pond life should be observed using **cavity slides** to ensure enough water is available to protect organisms from dehydration or heat damage from the microscope light.

When carrying out investigations on other people, such as pulse rate experiments and studies of the effect of exercise, it is important to carefully plan the type of exercise and how it will be carried out in order to take the necessary care to prevent injury. The use of chemicals, such as caffeine and alcohol, is prohibited as these are toxins and should not be administered in a school situation.

Using aseptic techniques to investigate microorganisms

Sterile techniques are used to prevent contamination of the culture of microbe you are studying by any other microbes. This is especially important to prevent the accidental culture of human pathogens.

Aseptic techniques:

- The area surrounding the experiment must be carefully sterilised and the biologist must have clean hands and laboratory clothes.
- All apparatus, glassware and collecting loops must be fully sterilised before use.

- The culture medium – either a culture broth or an agar plate – must be sterilised and poured under sterile conditions.
- When sampling and transferring the microbes, any collecting loop should be flamed and cooled in sterile water to avoid collecting microbes in the atmosphere.
- After use, all areas and non-disposable apparatus must be cleaned and sterilised again, and any disposable apparatus should be sealed and autoclaved before disposal.

Tip

Always use sharp instruments to give a good clean dissection line with as little damage as possible. To avoid harming yourself, always cut away from you.

Dissection

Sometimes it is necessary to cut open an organism to see the internal structures. Opening up an organism, whether it is a plant or an animal, to see the internal structures is called dissection. You need to use sharp scissors and scalpels, which you must handle safely but also with confidence so that you do not damage the internal tissues when they are laid open to allow all the structures to be seen clearly. You will draw these structures as a record of your dissection.

Using sampling techniques

In the field, investigations on populations must involve measuring a **sample** of the population, because it would be impossible to measure the whole population. This sample is intended to be representative of the whole population; therefore it is important that the sample is random and avoids any bias such as the selection of certain individuals for their specific characteristics or because they are easier to sample, for example.

To make sure a random sample is selected and used for measuring, these steps should be followed:

- Use a quadrat (a square of known dimensions) to select and collect the sample.
- Select the correct quadrat size and number of quadrats to sample using specific quadrat selection techniques.
- Use random numbers to generate coordinates for placing the quadrats in the chosen area.
- Lay down a grid or double metre tapes to create a grid across the area to be sampled.

When sampling on a seashore, there may be certain requirements such as sampling from different zones of the beach that must also be factored into the collection.

Sampling plant leaves or insects to measure variation in size may involve taking account of other factors such as age. It would invalidate the study if the size variation was determined by the age of the organism and not the environmental condition under study, and so in this case only those specimens of certain ages may be selected. A good example here is a study on the ratio of height to width of the shell of dog whelk, where only the adults were selected with the juveniles ignored.

Sometimes a sampling technique can be used on a microscopic level, for example to study the density of stomata on a leaf surface. In this case, an engraved slide with a grid is used to count the number of stomata within certain squares, which are determined using random numbers for the coordinates. A specialised slide, similar to this and engraved with a detailed grid, is called a haemocytometer, because it was originally used in clinical laboratories to determine the blood cell numbers in a blood sample. Now blood counts are made using a specific laser and computer to count the blood cell number.

Using technology such as data loggers or computer modelling

Many experiments lend themselves to data collection using a data logger. The range of uses is enormous, from simple data collection such as on pulse rate or on the loss of water from a potometer to more complex data collection involving the use of a variety of probes linked to the main computer. The computer then tabulates the data, processes a mean and performs statistical tests, and plots graphs.

If you enter data on variation into a spreadsheet, the mean, standard deviation and some statistical tests will be carried out for you, so you can compare the outcomes quickly and efficiently.

There are many computer-based modelling exercises that you can use in the laboratory or at home to supplement your learning. These include studies on ecological sampling, as well as succession and zonation, which will allow you to compare and contrast the two ecological situations. Studies on the impact of diseases, predation or predator–prey relationships allow you to study theoretical effects on a population.

All the skills discussed here are an important part of the practical aspects of a scientific course such as biology. Practical skills will be tested in a general way throughout your course, as described here, and in a specific way using particular experiments as examples in the theory papers and also in the assessed practicals required for the practical endorsement. Hence, your knowledge and abilities will be tested both practically and theoretically. Any practical work carried out in the first year of study is vital to allow you to build up these necessary skills and to develop a level of experience ready to be tested in your second year.

Chapter 18

Maths skills

This chapter deals with some of the skills that you will use during the course. An equivalent chapter in OCR A Level Biology 2 deals with statistical tests.

Numbers

Biologists study the very large and the very small. Here are some examples from either end of the range of size that biologists study:

- mass of an adult blue whale = 190 000 kilograms
- length of the longest bony fish = 11 metres
- mass of a bacterium = 1×10^{-15} kilograms
- length of an influenza virus = 100 nanometres
- mass of DNA in a human cell = 6 picograms

Do not write 'size' without qualifying this by stating whether you are referring to a linear measurement or to area, volume or mass.

Numbers and powers

The letters *x* and *y* signify any number. You may know that x^1 means '*x* to the power of 1'; more likely you will know that x^2 means '*x* to the power of 2', which is *x* multiplied by *x*, or '*x* squared'. The number written as the superscript is the power, or index, or exponent. Indices are used in biology because the range of sizes is so great. It is not always easy to handle numbers that are much larger than 1, for example 100 000 000 000 or with numbers much less than 1, for example 0.000 001. Changing these from ordinary form into standard form using indices helps a great deal. In standard form, 100 000 000 000 is 1×10^{11} and 0.000 001 is 1×10^{-6}. In standard form, there is only one digit before the decimal place.

Negative indices are used both for writing very small numbers, for example 5.6×10^{-3}, and for writing units instead of 'per', for example the rate of reaction as a volume produced per second would be expressed as $cm^3\,s^{-1}$.

Here is an example from Chapter 3. There are four bases in DNA: A, T, C and G. These bases code for 20 different amino acids. The genetic code could consist of pairs of bases, for example AG, CG, TA, and so on. In this case, there would be 4^2 different ways in which the four bases could be organised:

4^2 (4 to the power of 2) is $4 \times 4 = 16$, so the genetic code cannot be like this.

4^3 (4 to the power of 3) is $4 \times 4 \times 4 = 64$. This is more than enough to code for the 20 different amino acids.

If the bases were in groups of four, there would be $4^4 = 256$, which is far too many.

You can see the complete genetic code in Table 14.5 and confirm that the number of different codons is 64. Three of them are stop codons that do not code for any amino acid, and the rest are divided between the 20 amino acids. The code is degenerate, because in most cases there is more than one codon for each amino acid. The maths tells us that there is no other way in which three bases can code for 20 amino acids.

When multiplying numbers in standard form, add the indices together:

$2.0 \times 10^4 \times 3.3 \times 10^5 = 6.6 \times 10^9$

When dividing numbers in standard form, subtract the indices:

$$\frac{4.6 \times 10^6}{2.0 \times 10^3} = 2.3 \times 10^3$$

Fractions and reciprocals

Some numbers are written as fractions, such as $\frac{1}{2}$, $\frac{1}{4}$ or $\frac{3}{4}$. In a fraction, the upper number is the **numerator** and the lower number is the **denominator**. These words are useful to use when describing some mathematical concepts or the relationship between two values, for example surface areas and volume. Fractions should always be simplified; for example, never write $\frac{2}{8}$ but divide both denominator and numerator by the common factor, which is 2, to give $\frac{1}{4}$. The fraction $\frac{15}{8}$ cannot be simplified, so remains as it is. To calculate a fraction of a number, multiply the number by the numerator and divide by the denominator.

Tip

Remember that the numerator is the upper number (both have a 'u'). Here is an example of calculating a fraction of a number: $\frac{1}{2}$ of 8 = s $\frac{1}{2}$ x 8 = (8 x 1) ÷ 2 = 4

Reciprocals are calculated by dividing 1 by a number. If the answer is a very small number then it may be multiplied by 100, 1000 or 10000 to give a whole number that is easier to use. For example, the reciprocal of 25 is 0.04, but the reciprocal of 2500 is 0.0004, which can be multiplied by 10000 and expressed as 4×10^{-4}. This makes it easier to plot reciprocals like this on a graph, for example. Reciprocals are used as rates of reaction (see page 66).

Decimals

The decimal point in scientific writing in English is given as a full stop. In other languages it is often a comma. You will write up your practical work and give your examination answers in English, so always use the full stop whatever you may use in your own language.

To multiply a decimal number, move the decimal point to the right for every 10. To divide a decimal number, move the decimal point to the left for every 10. For example:

$76.80 \times 10 = 768.0$

$$\frac{1652}{100} = 16.52$$

$0.3002 \times 1000 = 300.2$

How many decimal places?

The rule here is just to use common sense. If you are timing something with a stopwatch or the timing device on your phone, you may obtain answers to several decimal places, for example 6.534 seconds. It is highly unlikely that you can react fast enough to capture an event to the nearest 0.001 of a second. In this case, give your answer to the nearest second or maybe even the nearest 10 seconds. If you are carrying out a calculation on some data you have collected or been given, remember not to make the result more accurate (with more decimal places) than the data you are using. There is more about this below under the heading Significant figures.

Calculations – words of advice

Write out all the steps in your calculations. Do this even if you use a spreadsheet, calculator or an app on your phone. If you write out the steps in your calculations, you will gain some advantages:

- You take time to think about what you are calculating.
- You have a record of what you have done if you need to backtrack and look for possible errors.
- Anyone reading your answers (such as an examiner) can follow what you have done and maybe award some credit, even if your final answer is incorrect.

Estimating answers

Before you punch the numbers into your calculator or calculator app, make sure that you have a general idea of what the answer should be. This is making an **estimate**. You can do this in your head or by jotting down numbers on paper and doing the calculation with numbers that are rounded up.

Example

An electrocardiograph (ECG) shows that each heartbeat takes 0.8 seconds. How many heart beats are there in one minute?

First estimate the answer by rounding up 0.8 s to 1 s.

If each heartbeat takes 1 s, then there will be 60 in a minute.

Each heartbeat takes less than a second, therefore there must be *more* beats per minute than 60, but not very many more. The answer is:

$$\frac{60}{0.8} = 75 \text{ beats min}^{-1}$$

Rounding up and down

When you do calculations in a spreadsheet or with a calculator or an app, the result often appears with many decimal places. This does not indicate that you have achieved a greater degree of accuracy than in your original measurements, so you need to round the numbers up or down. This is easiest to explain with some examples.

Example

Rounding to one decimal place

A set of calculations give answers to two decimal places, but the data collected was given to one decimal place. If the second decimal place is between 0 and 4, round down. If the second decimal place is greater than or equal to 5, then round up. For example:

- 7.56 rounded to one decimal place is 7.6 (rounded up).
- 10.54 rounded to one decimal place is 10.5 (rounded down).

If there are three decimal places, then the same principle applies:

- 9.546 rounded to one decimal place is 9.6 (rounded up).

If the numbers after the decimal point are 44X, then a slightly different rule applies:

- 6.443 rounded to one decimal place is 6.4 because '3' in the third decimal place is less than 5 and 6.443 rounded to two decimal places is 6.44
- 6.447 rounded to one decimal place is also 6.4 because although 6.447 rounds to 6.45 and that should round up to 6.5, it does not because 6.447 is less than 6.5.

Answers are usually followed by the decimal places that have been chosen, for example 7.56 is rounded up and written as 7.6 (1dp).

Units of measurement

The units for scientific measurements are those of the SI system, which stands for *Système Internationale.*

Table 18.1 shows you the units that you are most likely to use in your biological studies. If you are studying chemistry and physics there are others from this system that you will know and use.

Table 18.1 Units of measurement: the SI system.

Measurement	Base unit	Symbol	Derived units used in biology (symbols)
Length	metre	m	10^3 m = kilometre (km) 10^{-3} m = millimetre (mm) 10^{-6} m = micrometre (μm)
Note that centimetre is missing from this list because it is not an SI unit. However, it is often used, but make sure you realise that 1 cm = 10^{-2} m and 100 cm = 1 m. Take care when using cm, because it is easy to forget that there are only 100 cm in a metre and not 1000.			
Volume	cubic metre	m^3	10^9 m^3 = cubic kilometre 10^{-9} m = cubic millimetre (mm^3) 10^{-18} m = cubic micrometre ($μm^3$)
The most common units to measure volumes are: • the cubic centimetre (cm^3), which is exactly the same as the millilitre (ml) • the cubic decimetre (dm^3), which is exactly the same as the litre (written as l or L). You will see ml and l or L written on glassware. Most scientists refer to the litre and the millilitre (pronounced 'mill') rather than the decimetre cubed or centimetre cubed. So the next row shows the most common way in which volumes are measured and expressed.			
Volume	cubic decimetre	dm^3	10^{-3} dm^3 = centimetre cubed (cm^3) 10^{-6} dm^3 = millimetre cubed (mm^3)
Area	square metre	m^2	10^4 m^2 = hectare (ha) 10^{-4} m^2 = square centimetre (cm^2)
Mass	kilogram	kg	10^3 kg = tonne (no symbol – sometimes called the metric ton) 10^{-3} kg = gram (g) 10^{-6} kg = milligram (mg) 10^{-9} kg = microgram (μg)
Time	second	s	60 s = minute (min) 60 min = hour (h)
Pressure	pascal	Pa	10^3 Pa = kilopascal (kPa)
Energy	joule	J	10^3 J = kilojoule (kJ) 10^6 J = megajoule (MJ)
Temperature	degrees Celsius	°C	
Amount of substance	mole	mol	10^{-3} mol = millimole
The word 'amount' has the scientific meaning of 'mole', so if you write 'amount of salt solution' you are referring to moles. Avoid using the word 'amount' in your writing – you should be precise and use volume, mass or concentration instead. Most uses of 'amount' in student answers are therefore ambiguous.			

Example

5.6×10^{-6} m is easier to see and read than 0.000 005 6 m, but 5.6 μm is much easier to read, write and understand than either of them. This is particularly the case when you are using numbers and units to plot graphs. It is difficult to scale an axis when the range is 0–0.000 005 m. It is much easier if this is converted into a range of 0–5 μm.

Measurements from microscopy

Magnification is the ratio between the actual size of an object and the size of an image such as a photograph or a drawing.

You may have to calculate magnifications or actual sizes. You should use these formulae:

$$\text{magnification} = \frac{\text{size of image}}{\text{actual size}}$$

$$\text{actual size} = \frac{\text{size of image}}{\text{magnification}}$$

$$\text{size of image} = \text{actual size} \times \text{magnification}$$

The easiest way to remember what to do is to use the magnification triangle.

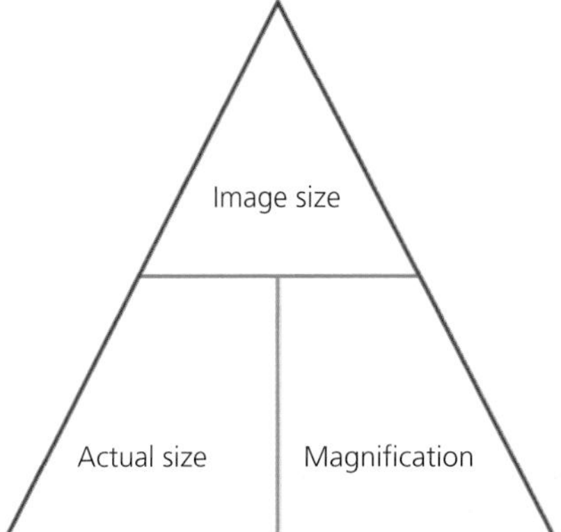

Figure 18.1 The magnification triangle.

If you are drawing an image of a macroscopic object, and the image will be smaller than the object, then magnifications are less than 1. For example, a drawing of a dissected heart is likely to be half life size $= \times \frac{1}{2}$.

You may have to find the actual size of a cell or organelle in an electron micrograph or a drawing made from an electron micrograph. Measure in millimetres and then divide by the magnification that you are given. Multiply your answer by 1000 to give an answer in micrometres (μm).

You may also be asked to calculate the magnification of a cell or an organelle in an electron micrograph. You will be told the actual size in micrometres. Here you need to measure the size of the cell, convert into micrometres and then divide by the actual size.

Drawings or photographs may have scale bars, as in Figure 18.2.

Figure 18.2 Use the scale bar to measure the actual size of these bacteria.

Proceed as follows:

1 Measure the length of the scale bar as printed in millimetres and record this.
2 Measure the length of the object in millimetres and divide by the length of the scale bar in millimetres.
3 Multiply the answer by the length represented by the scale bar.

Using the example in the diagram, the actual length of the bacteria is 2 μm.

Significant figures

You should not express the results of any calculations to more significant figures than the least accurate figure used, which will have the lowest number of sig figs. All digits are significant except the zeros that are used to place the decimal point:

- Zeros at the beginning of numbers are not significant; for example 0.032 has two sig figs.
- Zeros at the end of a whole number are significant; for example 7090 has four sig figs.

- Zeros between sig figs are significant; for example 601.4 has four sig figs.
- Zeros after a decimal point are significant; for example 0.780 has three sig figs.

If your input data has three sig figs, then your calculated answer should not have more than three sig figs. If the numerator has four sig figs but the denominator has two sig figs, then the answer should only have two sig figs. Do not round up the number until the very last stage of your calculation.

Descriptive statistics

Tip

If you are collecting data and then analysing it mathematically, it is much quicker to type it into a spreadsheet and then save it. You then have your data in a form that can be processed easily and presented as graphs and charts at the press of a few keys.

Table 18.2

Leaf	Length /mm	Width /mm
1	110	52
2	112	50
3	79	43
4	113	51
5	96	47
6	95	45
7	75	46
8	94	47
9	79	39
10	96	43
11	101	45
12	98	50
13	84	38
14	104	50
15	103	53
16	104	46
17	95	40
18	92	49
19	71	35
20	84	40

Rules for making tables

Before you start to draw a table, decide what you want to record. Decide on how many columns and how many rows you need. Make a rough table in pencil. Make sure you have read all the instructions before you draw the table outline. Follow these rules:

- Use the space provided, do not make the table too small.
- Leave some space to the right of the table in case you decide that you need to add more columns.
- During a practical, make the table ready to take observations or readings, so that you can write them directly into the table.
- Draw the table outlines in pencil; use a ruler for lines between the columns and rows, and around the whole table.
- Write brief but informative headings for each column.
- Columns headed with physical quantities should have appropriate SI units.
- When two or more columns are used to present data, the first column should be the independent variable; the second and subsequent columns should contain the dependent variables.
- Entries in the body of the table should be brief – they should be single words, short descriptive phrases, numbers, ticks or crosses.
- Data should be ordered so that patterns can be seen – it is best to arrange the values of the independent variable in ascending order down the column.

The solidus or slash (/) should **not** be used to mean 'per' in units. If you include concentrations in a table, **do not** write g/100 cm^3. This should always be written out in full as g per cm^3 or, better still, as g 100 cm^{-3}.

Correct	Incorrect
Concentration/g 100 cm^{-3}	Concentration g/dm^3
Rate of activity/cm^3 s^{-1}	Rate cm^3/s

Note that the solidus is used to separate what is measured from the unit in which it is measured. Brackets are also used to indicate the units. In OCR examination papers you will see graphs labelled in this style: Concentration (g 100 cm^{-3}).

Calculating averages

The term *average* means the centre of a range. There are three types of 'average' that you can determine from data like that in Table 18.2. These are:

- mean
- median
- mode.

Mean

The mean is calculated arithmetically by using this formula:

$$\bar{x} = \frac{\sum x}{n}$$

where $\bar{x}$ = mean, $\sum$ = the sum of, x = any value for the data collected, and n = the total number of values.

Never write 'average' in a table or in an answer to a question, because that is ambiguous. Always use the term 'mean'.

The mean is the best form of average to calculate if the data shows a normal distribution (a bell-shaped curve).

Tip

You can see an example of a histogram showing a normal distribution on page 329.

Median

If the data is not distributed normally and is therefore positively or negatively skewed, the median is the best average to use. The median is the middle number. To find the median, write out the values in order. If there is an odd number, the median is the middle value; if there are an even number of values then the median is the mean of the two central values.

Mode

The mode is the value that occurs most frequently in a set of data. This is quite useful for values that can only be whole numbers, such as the number of leaves on a plant or the number of a particular species of shrub in quadrats in a woodland.

The mode also applies to grouped data: see Histograms below. Mode is also useful if two values occur most frequently; such a distribution is called bimodal.

Example

Finding the centre

The lengths of 11 leaves from a bay tree, *Laurus nobilis*, were measured to the nearest millimetre. The results were:

110, 112, 79, 113, 96, 95, 75, 94, 79, 97, 84

The sum total for these leaves = 1034 mm

The **mean** = $\frac{1034}{11}$ = 94 mm

To find the median, reorganise the data:

75, 79, 79, 84, 94, 95, 97, 96, 110, 112, 113

The middle number = 95, so the **median** = 95 mm

The most frequent length is 79, so the **mode** = 79 mm

Dispersal of results about the mean

The **range** is the difference between the largest result and the smallest. You can express this in two ways: for example, for the length of leaves in the example above: 75 mm to 113 mm, or as the difference between the two values, which is 38 mm. The range is one way to show the spread of results about the mean.

Variables such as length and mass are likely to be distributed normally so that the mean is in the middle of the range. If you draw graphs of continuous data such as leaf length, they should show a bell-shaped curve, which indicates a normal distribution. There is more about this on pages 278–80.

If the data has a normal distribution, you can calculate how much spread there is about the mean. One way to do this is to calculate **standard deviation**. Standard deviation is derived from the variance of the data. The variance is determined by taking the difference between each result and the mean. All of these differences are squared to remove the negative signs. The sum of these differences is then divided by the total number of values, or by $n - 1$ if you have a sample from a much larger population:

$$\text{variance} = \frac{\Sigma(x-\bar{x})^2}{n-1}$$

The **standard deviation** is the square root of the variance:

$$\text{standard deviation (SD)} = \sqrt{\frac{\Sigma(x-\bar{x})^2}{n-1}}$$

You can use a calculator or a spreadsheet to calculate standard deviation for you. If the standard deviation is required in an examination, you will be given the equation. The standard deviation is used to show how widely the data is dispersed. Whatever the standard deviation that you calculate, 68% of the values are within 1 standard deviation of the mean and 95% of the values are within 2 standard deviations of the mean.

One important use of standard deviation is assessing the spread of replicate results. If replicates are widely dispersed, then the standard deviation is large compared with the mean, and this indicates that you cannot be very sure of the true value of the mean and therefore you cannot have much confidence in your results.

You can show this information on graphs by adding **error bars** to the points on a line graph or the bars on a bar chart. Error bars may show the:

- total range of results (maximum to minimum)
- standard deviation.

There are error bars shown on Figure 5.5 on page 85.

Percentages and percentage change

'Per cent' is short for *per centum*, which in Latin means 'by the hundred'. Calculating percentages is a way of making valid comparisons between items when the totals are different, for example 'about 10% of cells near the root tip of garlic are in stages of mitosis, but further up the root only 2% are dividing'.

Percentages are calculated as:

$$\% = \frac{\text{number with feature under study}}{\text{total number}} \times 100$$

Epidemiologists rarely use percentages; instead, they express data for morbidity (sickness) and mortality (death) *per thousand* or *per hundred thousand* of a population. This ensures that they have whole numbers to deal with, because the number of cases or number of deaths is very small when expressed as a percentage.

A way of making valid comparisons for changes is to calculate a percentage change, comparing the change to the original starting number. The percentage change is calculated as:

$$\%\ \text{change} = \frac{\text{difference between final and original numbers}}{\text{original number}} \times 100$$

Percentage changes can be positive or negative as you can see in the example below.

Example

Percentage increase and percentage decrease

A starter culture of cells was placed into a flask. At the start there were 600 cells; after a day the number of cells had increased to 950. The percentage increase in the number of cells:

Increase in number of cells = 950 – 600 = 350 cells

$$\text{percentage increase} = \frac{350}{600} \times 100 = 58.3\%$$

The population of capercaillie in Scotland in 2003/04 was estimated as 1980 birds; in 2009/2010 the population estimate was 1285. The percentage change in the population is:

$$\text{change in numbers} = 1285 - 1980 = -695$$

$$\text{percentage change} = \frac{-695}{1980} = -35.1\%$$

(Note that the percentage *change* must be a negative number to emphasise that the numbers have *decreased*.)

Calculating areas and volumes

You should know how to calculate the area and volume of a prism. A prism is a solid object with identical-shaped ends, flat sides and the same cross-sectional area along its whole length. Prisms can have all sorts of shapes but must have these three features. Cuboids and cylinders are examples of prisms. The formulae are given in Table 18.3.

You should also know how to calculate the circumference and surface area of a circle. It is very difficult to measure the height of trees as they grow. Instead growth is determined by measuring the increase in the circumference of the trunk. The formula is:

$$\text{circumference of a circle} = 2\pi r$$

The circular area that you see through a microscope is the field of view. You may have to count the number of cells in a field of view and then use the area to calculate the density of cells in a tissue. The formula for the surface area of a circle is:

$$\text{surface area of a circle} = \pi r^2$$

Ratios

Ratios are useful when actual numbers are not that important. The ratios of the base pairs in DNA provided one strand of evidence for determining its structure. In DNA, the ratio of A:T is always 1:1 and the ratio of C:G is also 1:1 (see Table 3.5 on page 56).

Surface area to volume ratios

The ratios between surface area and volume are of relevance to diffusion, gas exchange and transport (see Chapters 8 and 9). You have to know how to calculate the surface areas and volumes of cuboids and cylinders and calculate the surface area : volume ratio. If you need to calculate the surface area and volume of a sphere, the formulae will be given to you; however, they are given in Table 18.3.

Example

Surface areas and volumes

Table 18.3 shows three shapes, their surface areas and volumes, and the surface area : volume ratio for specific sizes. Work through the calculation to confirm the surface area : volume ratios given in the last column.

Table 18.3 How to calculate the surface area : volume ratios for three shapes that represent animals.

Shape		Surface area	Volume	Size	SA : V ratio
Cuboid		$2 \times lw + 2 \times wh + 2 \times lh$	lwh	l = 10 mm h = 10 mm w = 10 mm	0.6 : 1
Cylinder		$2\pi rh$	$\pi r^2 h$	d = 20 mm r = 10 mm l = 30 mm	0.2 : 1
Sphere		$4\pi r^2$	$\frac{4}{3}\pi r^3$	d = 20 mm r = 10 mm	0.3 : 1

Key:

- l = length
- w = width
- h = height
- d = diameter
- r = radius $\left(\frac{1}{2}d\right)$
- π = 3.14

Tip

These methods do not work for fungi (except yeasts which are near spherical) and plants that are highly branched. These methods work for some organisms, for example a rhino is fairly cuboidal, worms of many sorts are near cylindrical, and yeasts, some bacteria and many protoctists are spherical. Mould fungi and many plants have extensive branching bodies which are much more difficult to calculate.

Line graphs

In many practicals you investigate the effect of factor x on factor y (see Chapters 3, 4, 5, 8 and 9). For example, in Chapter 4 investigations of four factors on enzyme activity are described. These factors are temperature, pH, substrate concentration and enzyme concentration. In each case, the factor is the independent variable; enzyme activity, usually measured as the rate of reaction, is the dependent variable.

Line graphs are used to show relationships in data that are not immediately apparent from tables. The term 'graph' applies to the whole representation.

Following these guidelines:

- The plots must occupy at least half the grid provided in both x and y axes; do not make the graph too small.
- Draw the graph in pencil.
- Plot the independent variable on the x axis.
- Plot the dependent variable on the y axis.

Tip

The table headings should provide you with the correct label, unit and unit symbol to use.

- Mark each axis with an appropriate scale. Critically examine the data to establish whether it is necessary to start the scale(s) at zero. If necessary, you can displace the origin for one or both axes, indicated with a jagged line from the origin.
- Scale each axis using multiples of 1, 2, 5 or 10 for each 20 mm square on the grid. This makes it easy for you to plot and extract data. Never use multiples of 3.
- Label each axis clearly with the quantity and SI unit(s) or derived units as appropriate, e.g. time/s and concentration/g dm^{-3}. Derived units are those that you have calculated from the data that you have collected. For example, you may calculate rates from volumes and times that you have measured. The rate is the derived variable.
- Plot points clearly and make sure they are easily distinguishable from the grid lines on the graph. Use encircled dots (ʘ) or saltire crosses (x); do not use dots on their own. If you need to plot three lines, you can also use vertical crosses (+).

After plotting the points, you need to decide whether any of them are anomalous. Ask yourself 'do they fit the trend?' But what is the trend? You should know something about the theory behind the investigation, so you should be aware of the likely trend. If you think one or more result is anomalous, then it is a good idea to ring the points. Put a circle on the graph away from the line and put a key to state that the circled point(s) represent anomalous result(s).

The next thing to decide is how to present the line:

- It may be obvious that the points lie on a straight line. You must first decide whether to include the origin (0,0) if it is not a datum point in your results. If you decide that it is a point, then place a clear plastic ruler on the grid and draw a straight line from the origin, making sure that an even number of points are on either side of the line. If the origin is not a point or you are not sure, then start the line at the first plotted point. Do not continue the line past the last plotted point.
- You should only draw a smooth curve if you know that the intermediate values fall on the curve. You may be expecting the relationship to be a curve, and if the points seem to fit on a curve then draw one. Again decide whether the origin is a point and, if not, start at the first plotted point. The curve should go through as many points as possible, but try to make sure that there are an even number of points on either side of the line. Do not continue past the last plotted point.
- If you are not sure whether the relationship is a straight line or a curve, draw ruled straight lines between the points. This indicates uncertainty about the results for values of the independent variable between those plotted.
- If a graph shows more than one line, then each should be labelled to show what it represents.

Analysing and interpreting graphs

In an examination, you will almost certainly have to analyse and interpret graphs (see Chapters 4, 8 and 9). The analysis of a graph involves:

- giving a general description of any trend or pattern
- giving details of the trend or pattern and illustrating this with data taken from the graph
- processing data from the graph, for example calculating rates of increase or decrease or percentage changes.

Start any description with a qualitative statement that gives the overall pattern or trend, for example 'the volume of product increases with time and then remains constant'. It is a good idea to divide graphs into regions and rule lines on the graph to indicate these. This makes it easier to find exactly when there is a significant change in a curve, for example when a rate first reaches a constant value. Sometimes the dependent variable fluctuates rather than showing a consistent trend, for example transpiration rates increase and decrease over a 24-hour period.

Analysing graphs

In an investigation that gives a linear relationship, the line is expressed by the equation:

$$y = mx + c$$

where m = the gradient of the line and c = the intercept on the y axis.

If c = 0, the line goes through the origin and the relationship is one of direct proportion.

You can calculate values of y if you know the values of x, m and c.

Figure 18.3 shows you how to calculate the gradient of a straight line. However, most graphs that you will draw and use do not have straight-line relationships. The time–course graph of an enzyme-catalysed reaction is not a straight line. The rate of reaction decreases as the reaction proceeds because the substrate concentration decreases. In this case you can calculate the rates at specific times on the graph by drawing tangents as shown in Figure 18.4.

Example

Calculating rates

The rate is calculated as the gradient:

in this example,

$$\text{rate} = \frac{\text{change in mass}}{\text{change in time}}$$

$$= \frac{8\,\text{g}}{10\,\text{min}}$$

$$= 0.8\,\text{g min}^{-1}$$

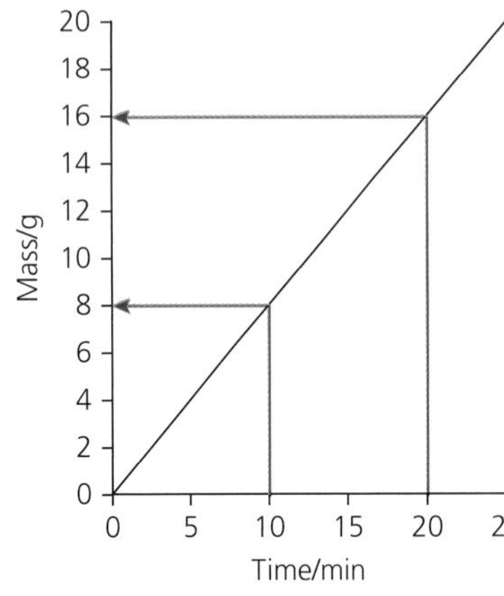

Figure 18.3 How to calculate the rate from a linear relationship.

The initial rate and the rate at 38 minutes are calculated by taking tangents to the line:

$$\text{initial rate} = \frac{\text{change in mass}}{\text{change in time}}$$

$$= \frac{24\,\text{g}}{14\,\text{min}}$$

$$= 1.7\,\text{g min}^{-1}$$

$$\text{rate at 38 minutes} = \frac{17\,\text{g}}{26\,\text{min}}$$

$$= 0.65\,\text{g min}^{-1}$$

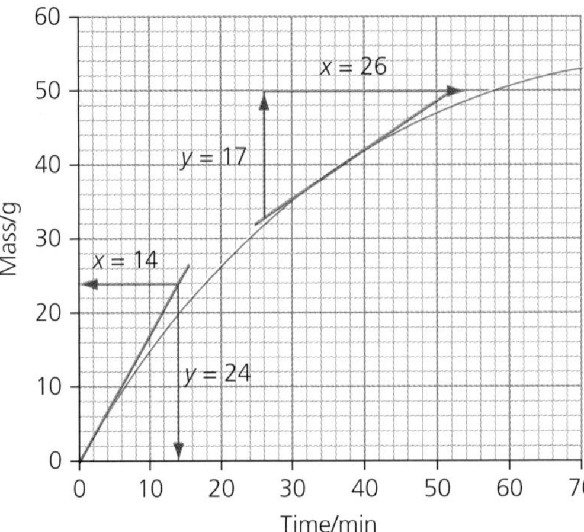

Figure 18.4 How to calculate the initial rate and a rate from the midpoint of a reaction using tangents.

Interpreting graphs

Look carefully at the two (or more) variables that have been plotted on the graph. You need to apply your knowledge of the topic to explain the trend you have described and analysed mathematically. Before you start interpreting a graph in an examination question, go back and read all the information you have been given to look for clues about the reasons for the relationship shown in the graph. When analysing all graphs, use a ruler to follow and mark the changes that occur.

Spirometer traces and ECGs can be analysed in a way similar to graphs. In both cases, the *x* axis is time; the *y* axis on a spirometer trace is the volume of air or volume of oxygen. The vertical axis on an ECG is the voltage, although this is rarely included. Questions on ECGs are likely to include a normal ECG and one or more abnormal ECGs.

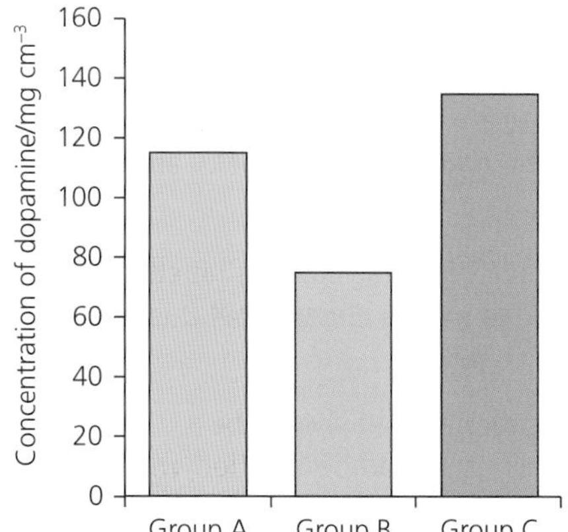

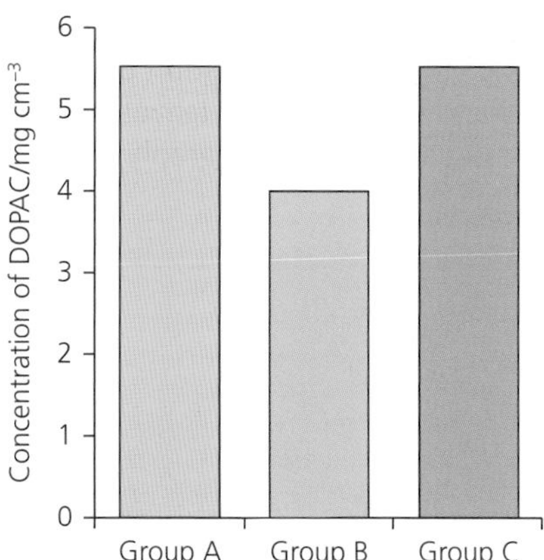

Figure 18.5 Two bar charts.

Bar charts

Rules for drawing bar charts (see Figure 18.5):

- Use at least half the grid provided; do not make the chart too small.
- Draw the chart in pencil.
- Bar charts can be made of blocks of equal width, which do not touch.
- The intervals between the blocks on the *x* axis should be equidistant.
- The *y* axis should be properly scaled with equidistant intervals.
- The *y* axis should be labelled with units.
- The lines or blocks can be arranged in any order, but it can help to make comparisons if they are arranged in descending order of size.
- Each block should be identified. There is no need to shade the blocks or colour-code them.

Blocks can be drawn horizontally rather than vertically. This is useful if the labels for each block are quite lengthy or if there are many blocks – writing the labels within the blocks makes it easier for others to read the information. In this case, the *x* axis goes on the left-hand side of the grid.

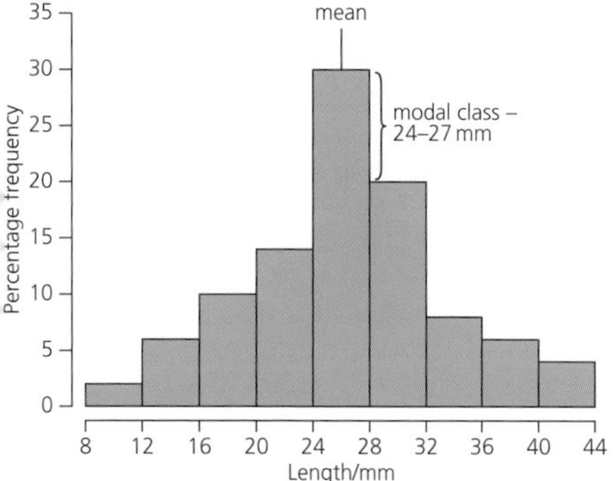

Figure 18.6 A histogram.

Histograms

Histograms are used when the independent variable is numerical and continuous data is collected (see Chapter 15). They are sometimes referred to as frequency diagrams (see Figure 18.6).

First the raw data needs to be organised into classes:

- The number of classes needs to be established. This will largely depend on the type and nature of the data.
- The rule for determining the number of classes is $5 \times \log_{10}$ total number of readings.
- The range within each class needs to be determined; this is usually the total range divided by one less than the number of classes.
- There should be no overlap in the classes, for example

  3.0–3.9

  4.0–4.9

The data should be organised using a tally chart and by drawing 'five-bar gates' as shown in Table 15.4 for the data in Table 15.3 on page 278.

Follow these rules when drawing a histogram:

- Use at least half the grid provided; do not make the histogram too small.
- Draw the histogram in pencil.
- The *x* axis represents the independent variable and is continuous. It should be labelled clearly with an appropriate scale.
- The blocks drawn should touch each other.
- The area of each block is proportional to the size of the class. It is usual to have similar-sized classes so the widths of the blocks are all the same.
- The blocks should be labelled; for example, the first block is labelled '3.0–3.9', which means that 3.0 is included in this class, but 4.0 is not. 4.0 will be included in the next class: 4.0–4.9.
- The *y* axis represents the number or frequency and should be properly scaled with equidistant intervals. It should be labelled with appropriate units.

Scattergraphs

Scattergraphs are useful for deciding whether there is a correlation between two variables, for example between widths and lengths of the leaves in Table 18.2. A correlation can be positive or negative. A scattergraph may not show any correlation.

Example

Scattergraphs

These scattergraphs show the sort of results you can expect if there is:

a) a positive correlation or
b) a negative correlation or
c) no correlation between two variables.

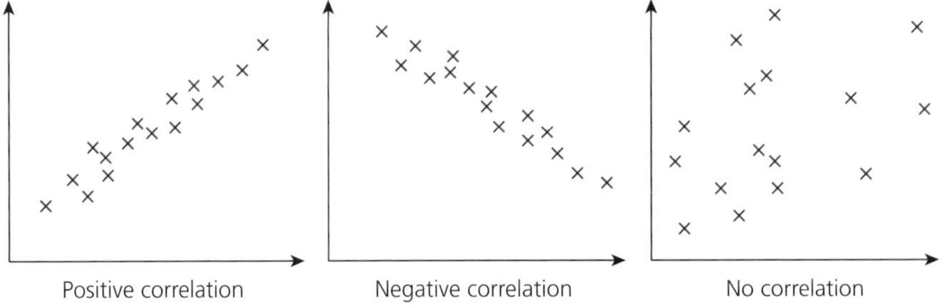

Figure 18.7 Three scattergraphs: (a) a positive correlation; (b) a negative correlation; (c) no correlation between two variables.

Correlation does not imply that one variable causes a change in another. However, a positive or negative correlation does identify a trend worth investigating further to find out whether there is a causal link between the variables. For example, there may be a correlation between the distribution of plants and soil moisture or soil pH. There may be many reasons for such a correlation, and these ideas should provide hypotheses that can be tested by experimental work.

Uncertainty in measuring

Uncertainty is expressed as half the smallest graduation on the apparatus; for example, if the smallest division on a syringe is 1.0 cm^3 then the uncertainty is +/− 0.5 cm^3. So if you start measuring at zero, the uncertainty applies where you take the measurement – say at 6.3 cm^3. The result is expressed as 6.3 +/− 0.5 cm^3. But if you have to start at a measurement other than zero (for example, when taking readings from a burette), the uncertainty applies at both ends, so it is multiplied by two because there is an error at each end; for example 7.5 +/− 1.0 cm^3.

It is possible to calculate the percentage error for the apparatus you used for measuring your results. Imagine that you have collected a gas and measured the volume with a syringe that has graduations every 1 cm^3. If you have measured 5 cm^3 of gas with your syringe, you can be certain that you have more than 4.5 cm^3 but less than 5.5 cm^3. Your error is +/− 0.5 cm^3 in 5 cm^3. This makes the percentage error:

$$\frac{0.5\ cm^3}{5\ cm^3} \times 100 = 10\%$$

If the volume of gas collected was 10 cm^3, then the percentage error would be 5%.

Index

Free online material

Answers for the following features found in this book are available online:

- Prior knowledge questions
- Test yourself questions
- Activities

You'll also find an Extended glossary to help you learn the key terms and formulae you'll need in your exam.

Scan the QR codes below for each chapter.

Alternatively, you can browse through all chapters at www.hoddereducation.co.uk/OCRABiology1

How to use the QR codes

To use the QR codes you will need a QR code reader for your smartphone/tablet. There are many free readers available, depending on the smartphone/tablet you are using. We have supplied some suggestions below, but this is not an exhaustive list and you should only download software compatible with your device and operating system. We do not endorse any of the third-party products listed below and downloading them is at your own risk.

- for iPhone/iPad, search the App store for Qrafter
- for Android, search the Play store for QR Droid
- for Blackberry, search Blackberry World for QR Scanner Pro
- for Windows/Symbian, search the store for Upcode

Once you have downloaded a QR code reader, simply open the reader app and use it to take a photo of the code. You will then see a menu of the free resources available for that topic.

1 Cell structure

3 Proteins and nucleic acids

2 Water, carbohydrates and lipids

4 Enzymes

5 Biological membranes

12 Biodiversity

6 Cell division, diversity and organisation

13 Maintaining biodiversity

7 Exchange surfaces and ventilation

14 Classification

8 Transport in animals

15 Evolution

9 Transport in plants

16 Exam preparation

10 Diseases of animals and plants

17 Preparing for practical assessment

11 Disease prevention

18 Maths skills

Acknowledgements

The Publisher would like to thank the following for permission to reproduce copyright material.

Photo credits:

p.1 *t* © Dr Keith Wheeler/Science Photo Library, *b* © Wim Van Egmond/ Visuals Unlimited, Inc./Science Photo Library; **p.2** *tl* © NIBSC/Science Photo Library, *tm* © Martin Shields/Science Photo Library, *tr* © Peter Bond, EM Centre, University Of Plymouth/Science Photo Library, *bl* © Dr Olivier Schwartz, Institute Pasteur/ Science Photo Library; **p.3** © Heiti Paves/Science Photo Library; **p.4** Public domain/http://commons.wikimedia.org/wiki/File:Bacillus_subtilis.jpg; **p.5** © Dr Keith Wheeler/ Science Photo Library; **p.6** © Biophoto Associates/Science Photo Library; **p.7** *t* © MEDIMAGE/Science Photo Library, *b* © MICROSCAPE/Science Photo Library; **p.8** *t* © Biophoto Associates/Science Photo Library, *b* © Dr Gopal Murti/Science Photo Library; **p.9** © Keith R. Porter/Science Photo Library; **p.10** *t* © Dr Jeremy Burgess/Science Photo Library, *b* © Steve Gschmeissner/Science Photo Library; **p.11** *tr* © Don W. Fawcett/ Science Photo Library, *tl* © Jean-Claude Rävy, ISM/Science Photo Library, *bl* © Steve Gschmeissner/Science Photo Library, *br* © Dr. Jeremy Burgess/ Science Photo Library; **p.15** © Scientifica, RMF, Visuals Unlimited/Science Photo Library; **p.16** © Dr Gopal Murti/Science Photo Library; **p.17** © Biophoto Associates/Science Photo Library; **p.18** © Richard Fosbery; **p.20** © Andrey Kuzmin – Fotolia; **p.26** © Richard Fosbery; **p.29** *both* © Andrew Lambert Photography/Science Photo Library; **p.30** © Richard Fosbery; **p.31** © Biophoto Associates/Science Photo Library; **p.34** © Richard Fosbery; **p.39** © Andrzej Wojcicki/Science Photo Library; **p.40** © Science Photo Library; **p.46** *t* © Laboratory Of Molecular Biology, MRC/ Science Photo Library, *m* © molekuu.be – Fotolia.com, *b* © Eye Of Science/Science Photo Library; **p.47** © Richard Fosbery; **p.52** © Andrzej Wojcicki/Science Photo Library; **p.60** *t* © molekuu.be – Fotolia.com, *b* © Getty Images/Hemera/Thinkstock; **p.62** *t* © David Munns/Science Photo Library, *b* © Noble Proctor/Science Photo Library; **p.66** Laurence Wesson and John Luttick, James Allen's Girls' School; **p.81** *t* © Claude Nuridsany & Marie Perennou/Science Photo Library, *b* © Sinclair Stammers/Science Photo Library; **p.91** © Claude Nuridsany & Marie Perennou/Science Photo Library; **p.92** *both* © Mike Samworth; **p.96** *t* © Eye Of Science/Science Photo Library, *m* © Science Pictures Limited/Science Photo Library, *b* © Steve Gschmeissner/Science Photo Library; **p.99** © Rieder & Khodjakov, Visuals Unlimited/Science Photo Library; **p.100** all © Michael Abbey/ Science Photo Library; **p.101** © Steve Gschmeissner/Science Photo Library; **p.102** © Eye Of Science/Science Photo Library; **p.103** © Dr. Robert Calentine, Visuals Unlimited/Science Photo Library; **p.104** © Pr Philippe Vago, ISM/Science Photo Library; **p.110** © Biophoto Associates/ Science Photo Library; **p.115** *t* © Dr Gopal Murti/Science Photo Library, *b* © Dr. Gladden Willis, Visuals Unlimited /Science Photo Library; **p.116** © Innerspace Imaging/Science Photo Library; **p.117** © Dr Keith Wheeler/ Science Photo Library; **p.118** © Dr Keith Wheeler/Science Photo Library; **p.121** *t* © Dr Keith Wheeler/Science Photo Library, *b* © Anton Denisov/ Ria Novosti/Science Photo Library; **p.122** © Claude Nuridsany & Marie Perennou/Science Photo Library; **p.126** *t* © Dr. Fred Hossler/Visuals Unlimited, Inc, *b* © Steve Gschmeissner/Science Photo Library; **p.130** © AJ Photo/Science Photo Library; **p.135** *l* © Biophoto Associates/Science Photo Library, *m* © Dr Keith Wheeler/Science Photo Library, *r* © Getty Images/Visuals Unlimited; **p.138** *t* © Biophoto Associates/Science Photo Library, *b* © PHT/Science Photo Library; **p.141** © Z220/Custom Medical Stock Photo/Science Photo Library; **p.143** *all* © Dr Vim Jesudason; **p.145** © AJ Photo/Science Photo Library; **p.147** © Chuck Brown/Science Photo Library; **p.148** © Visuals Unlimited/Science Photo Library; **p.149** © Biophoto Associates/Science Photo Library; **p.158** *t* © Dr Keith Wheeler/ Science Photo Library, *b* © petrenkoua - Fotolia; **p.159** *all* © Dr Keith Wheeler/Science Photo Library; **p.168** *t* © Ints Vikmanis – Fotolia, *b* © Gene Cox; **p.169** *t* © katchepix – Fotolia, *b* © Jon Bertsch/Visuals Unlimited/Science Photo Library; **p.171** *l* © slydgo1111 – Fotolia, *r* © Eric Isselée – Fotolia; **p.173** *t* © Jose Antonio Peñas/Science Photo Library, *b* © Kletr – Fotolia; **p.174** © Louise Gubb/Corbis; **p.175** *from t to b* © Kwangslin Kim/Science Photo Library © Science Photo Library, © Robert & Jean Pollock/Visuals Unlimited/Corbis, © Thomas Splettstoesser, Visuals Unlimited/Science Photo Library; **p.176** *t* © Jose Antonio Peñas/Science Photo Library, *b* © Dr. Hans Gelderblom, Visuals Unlimited/Science Photo Library; **p.177** *t* © Dennis Kunkel Microscopy, Inc/Visuals Unlimited, Inc., *m* © Dr Ken Greer, Visuals Unlimited/Science Photo Library, *b* © Nigel Cattlin/Alamy; **p.178** © Nigel Cattlin/Visuals Unlimited/Corbis; **p.181** © Nigel Cattlin/Alamy; **p.182** © GeoM/istockphoto; **p.188** *t* © Sean Sprague/Alamy, *b* © National Library Of Medicine/Science Photo Library; **p.189** © NIBSC/Science Photo Library; **p.190** © Eye Of Science/ Science Photo Library; **p.192** © Dr. Gopal Murti/Visuals Unlimited, Inc.; **p.194** © Dr. Gopal Murti/Visuals Unlimited, Inc.; **p.199** © Getty Images/ Science Faction; **p.200** *t* © Dan Race – Fotolia, *b* © ISM/Science Photo Library; **p.201** © Karen Kasmauski/Corbis; **p.205** © Cultura Creative (RF)/ Alamy; **p.206** © Sean Sprague/Alamy; **p.207** © ZUMA Press, Inc./Alamy; **p.208** © Biology Media/Science Photo Library; **p.210** *t* © Mathieu Joron, *b* © Getty Images/National Geographic Creative; **p.212** *l* © Stockbyte/ Getty Images, *m* © Mathieu Joron, *r* © Mathieu Joron; **p.216** *t* © Pam Collins/Science Photo Library, *m* © vlad61_61 – Fotolia, *b* © wildnerdpix – Fotolia; **p.217** © ephotocorp/Alamy; **p.219** *t* © Philippe Psaila/Science Photo Library, *b* © Sinclair Stammers/Science Photo Library; **p.226** © Stephen Meese – Fotolia; **p.227** *t* © Dan Burton/Alamy, *b* © Martin Harvey/CORBIS; **p.228** *t* © Richard Carey – Fotolia, *b* © tiero – Fotolia; **p.231** *t* © Patrick Landmann/Science Photo Library, *b* © Željko Radojko – Fotolia; **p.232** *t* © George Steinmetz/Corbis, *b* © Polarpx – Fotolia; **p.233** © pepjp – Fotolia; **p.234** © jptenor – Fotolia; **p.236** © Vibe Images – Fotolia; **p.241** © Getty Images/Nature Picture Library; **p.244** *t* © Richard Fosbery, *m* © Bruce Watson, *b* © Margot Fosbery; **p.245** © Richard Fosbery; **p.248** © Dan Burton/Alamy; **p.251** *t* © Richard Fosbery, *b* © SLDigi – Fotolia; **p.252** © Florapix/Alamy; **p.255** *l* © Iosif Szasz-Fabian – Fotolia, *r* © NHPA/Photoshot; **p.256** *t* © Carolina Biological Supply Co, Visuals Unlimited/Science Photo Library, *m* © micro_photo – Fotolia, *b* © Wim van Egmond, Visuals Unlimited/Science Photo Library; **p.257** *t* © Dennis Kunkel Microscopy, Inc./Visuals Unlimited, Inc., *b* © Bruce Watson; **p.258** *t* © John Keates/Alamy, *b* © David Fleetham/Alamy; **p.261** *t* © B. Murton/Southampton Oceanography Centre/ Science Photo Library, *b* © Woods Hole Oceanographic Institution, Visuals Unlimited/Science Photo Library; **p.266** © Natural History Museum, London/Science Photo Library; **p.267** © Andrew Darrington/Alamy; **p.268** *both* Phil Bendle http://tiny.cc/nzwide; **p.272** © gekaskr – Fotolia; **p.273** *t* © Lebrecht Music and Arts Photo Library/Alamy, *b* © Natural History Museum, London/Science Photo Library; **p.274** *l* © Erni – Fotolia, *r* © Colette – Fotolia; **p.275** *t* © Nancy Sefton/Science Photo Library, *b* © Alexander Semenov/Science Photo Library; **p.276** *t* © Donna Smith – Fotolia, *b* © Eye Of Science/ Science Photo Library; **p.280** © Paul Kay/Getty Images; **p.281** © mica – Fotolia; **p.283** © Oak Ridge National Laboratory/US Department Of Energy/Science Photo Library; **p.284** *t* © Tamara Kulikova – Fotolia, *m* © Hagit Berkovich – Fotolia, *b* © Dr. Terry Beveridge/Visuals Unlimited, Inc./Science Photo Library; **p.289** *t* © Natural History Museum, London/ Science Photo Library, *b* © gekaskr – Fotolia; **p.290** © James F. Parham. Courtesy Smithsonian Institute; **p.298** *t* © AntonioDiaz – Fotolia, *b* © Galina Barskaya – Fotolia; **p.305** *t* © Peter Muller/Cultura/Science Photo Library, *b* © Science Photo Library; **p.306** © Peter Muller/Cultura/Science Photo Library; **p.317** © Sergey Nivens – Fotolia.

t = top, *b* = bottom, *l* = left, *r* = right, *m* = middle

Every effort has been made to trace all copyright holders, but if any have been inadvertently overlooked, the Publisher will be pleased to make the necessary arrangements at the first opportunity.